Gas Chromatography-Mass Spectrometry for Omics Applications

Gas Chromatography-Mass Spectrometry for Omics Applications

Edited by

Flavio A. Franchina
Department of Chemical, Pharmaceutical, and Agricultural Sciences
University of Ferrara, Italy

WILEY

Published by John Wiley & Sons, Inc., Hoboken, New Jersey.

Library of Congress Cataloging-in-Publication Data:

Names: Franchina, Flavio A. (Flavio Antonio) editor
Title: Gas chromatography-mass spectrometry for omics applications / edited by Flavio A. Franchina.
Description: Hoboken, New Jersey : Wiley, [2026] | Includes bibliographical references and index.
Identifiers: LCCN 2025046972 (print) | LCCN 2025046973 (ebook) | ISBN 9781394209279 cloth | ISBN 9781394209286 adobe pdf | ISBN 9781394209293 epub
Subjects: LCSH: Molecular biology–Methodology | Gas chromatography | Mass spectrometry
Classification: LCC QH506 .G33 2026 (print) | LCC QH506 (ebook)
LC record available at https://lccn.loc.gov/2025046972
LC ebook record available at https://lccn.loc.gov/2025046973

Cover Image: © Flavio A. Franchina
Cover Design: Wiley

Set in 9.5/12.5pt STIXTwoText by Straive, Chennai, India

Contents

List of Contributors

Eloisa Bagnulo
Department of Drug Science and Technology
University of Turin
Turin
Italy

Djulia Bensaada
Organic and Biological Analytical Chemistry Group, MolSys
University of Liège
Liège
Belgium

Carlo Bicchi
Department of Drug Science and Technology
University of Turin
Turin
Italy

Cecilia Cagliero
Department of Drug Science and Technology
University of Turin
Turin
Italy

Andrea Caratti
Department of Drug Science and Technology
University of Turin
Turin
Italy

Chiara Cordero
Department of Drug Science and Technology
University of Turin
Turin
Italy

Marco De Poli
Department of Chemical Pharmaceutical, and Agricultural Sciences
University of Ferrara
Ferrara
Italy

Glaucimar A. P. de Resende
Instituto de Química, Universidade Estadual de Campinas
Campinas, SP
Brazil

and

Instituto Nacional de Ciência e Tecnologia em Bioanalítica
Universidade Estadual de Campinas
Campinas, SP
Brazil

Riccardo Di Stefano
Department of Chemical
Pharmaceutical, and Agricultural Sciences
University of Ferrara
Ferrara
Italy

Allyson L. R. dos Santos
Instituto de Química, Universidade Estadual de Campinas
Campinas, SP
Brazil

and

Instituto Nacional de Ciência e Tecnologia em Bioanalítica
Estadual de Campinas
Campinas, SP
Brazil

Allan Polidoro
Department of Environmental and Prevention Sciences
University of Ferrara
Ferrara
Italy

Angelica Fina
Department of Drug Science and Technology
University of Turin
Turin
Italy

Jean-François Focant
Organic and Biological Analytical Chemistry Group, MolSys
University of Liège
Liège
Belgium

Flavio A. Franchina
Department of Chemical
Pharmaceutical, and Agricultural Sciences
University of Ferrara
Ferrara
Italy

Micaela Galletta
Messina Institute of Technology, c/o
Department of Chemical, Biological
Pharmaceutical and Environmental Sciences
University of Messina
Messina
Italy

Virginia Galliani
Department of Chemical
Pharmaceutical, and Agricultural Sciences
University of Ferrara
Ferrara
Italy

Nadine Gawlitta
Department of Plant and Environmental Sciences
University of Copenhagen
Frederiksberg
Denmark

Barbara Giocastro
Chemical Kinetics and Analytics
Institute of Combustion Technology
DLR – German Aerospace Center
Stuttgart
Germany

Thomas Gröger
Chemical Kinetics and Analytics
Institute of Combustion Technology
DLR – German Aerospace Center
Stuttgart
Germany

Beate Gruber
Analytical and Material Science
BASF SE
Ludwigshafen
Germany

Leandro W. Hantao
Instituto de Química, Universidade
Estadual de Campinas
Campina, SP
Brazil

and

Instituto Nacional de Ciência e
Tecnologia em Bioanalítica
Universidade Estadual de Campinas
Campinas, SP
Brazil

Varaha Jayarama Krishna Jonnalagedda
Department of Energy and Mineral
Engineering
Pennsylvania State University
University Park, PA
USA

Do Hyun Lee
Department of Energy and Mineral
Engineering
Pennsylvania State University
University Park, PA
USA

Erica Liberto
Department of Drug Science and
Technology
University of Turin
Turin
Italy

Martina Lombardi
Department of Medicine
Surgery and Dentistry
"Scuola Medica Salernitana"
University of Salerno
Baronissi
Italy

and

Theoreo srl
Montecorvino Pugliano
Italy

and

European Institute of Metabolomics
(EIM) Foundation
Baronissi
Italy

Luigi Mondello
Messina Institute of Technology, c/o Department of Chemical, Biological Pharmaceutical and Environmental Sciences
University of Messina
Messina
Italy

and

Chromaleont s.r.l., c/o Department of Chemical, Biological, Pharmaceutical and Environmental Sciences
University of Messina
Messina
Italy

Damien Pierret
Analytical Chemistry Lab
Gembloux Agro-Bio Tech
University of Liège
Gembloux
Belgium

Giorgia Purcaro
Analytical Chemistry Lab
Gembloux Agro-Bio Tech
University of Liège
Gembloux
Belgium

Lourdes Ramos
Department of Instrumental Analysis and Environmental Chemistry
Institute of Organic Chemistry, CSIC
Madrid
Spain

Anaïs Rodrigues
Organic and Biological Analytical Chemistry Group, MolSys
University of Liège
Liège
Belgium

Christopher Rüger
Institute of Chemistry (Analytical Chemistry)
University of Rostock
Rostock
Germany

Pierre-Hugues Stefanuto
Organic and Biological Analytical Chemistry Group, MolSys
University of Liège
Liège
Belgium

Hilal Ezgi Toraman
Department of Energy and Mineral Engineering
Pennsylvania State University
University Park, PA
USA

and

Institute of Energy and the Environment
Pennsylvania State University
University Park, PA
USA

and

Department of Chemical Engineering
Pennsylvania State University
University Park, PA
USA

Peter Q. Tranchida
Messina Institute of Technology, c/o
Department of Chemical, Biological
Pharmaceutical and Environmental
Sciences
University of Messina
Messina
Italy

Fulvia Trapani
Department of Drug Science and
Technology
University of Turin
Turin
Italy

Jacopo Troisi
Department of Medicine
Surgery and Dentistry
"Scuola Medica Salernitana"
University of Salerno
Baronissi
Italy

and

Theoreo srl
Montecorvino Pugliano
Italy

and

European Institute of Metabolomics
(EIM) Foundation
Baronissi
Italy

Benedikt Weggler
Analytical and Material Science
BASF SE
Ludwigshafen
Germany

Ralf Zimmermann
Institute of Chemistry
Analytical Chemistry at University of
Rostock
University of Rostock
Rostock
Germany

Preface

The evolution of omics sciences represents a fundamental shift in analytical chemistry, challenging the chemical domain from the study of a defined set of isolated components to the comprehensive profiling of complex systems. This transition marks the dawn of an era of high-dimensional research, where the analytical goal is not only to find a needle in a haystack but to characterize the entire haystack itself. In this new landscape, the haystack represents a vast, interconnected web of molecular data that requires a unique combination of high-resolution techniques and harmonized analytical strategies, including well-thought experimental design and data elaboration, to unravel.

While the omics disciplines originated within the life sciences, they have rapidly expanded far beyond their biological roots. This evolution from living to inanimate omics is not a fragmentation, but a unification of methodologies that share common principles. For instance, whether we are analyzing a metabolic snapshot in biofluids, the components profile of a processed foodstuff, or the hydrocarbon fingerprint of a petroleum derivative, we are exploiting the same high-dimensional research principles to map a chemical space.

This potential becomes even more significant when we consider wide selectivity, high-resolution, and information-rich measurements across the entire analytical workflow, from sample preparation to chromatographic separation and detection. This can be seen today in industrial applications where, to refine processes and obtain specific characteristics in a novel product, the totality of its chemical constituents must be characterized and monitored with the same rigor reserved for clinical diagnostics. The suffix "-omics" has now confidently taken root in a variety of fields such as *Petroleomics*, where the goal is a complete chemical description within the extensive catalog of crude oil components (several dozens of thousands), and *Foodomics*, which deconstructs the intricate molecular landscapes of the products we consume. These are not isolated examples, they are part of an emerging scientific movement that treats all complex matter as an *-ome*, rich in information.

In this broad landscape, mass spectrometry stands as the unifying analytical foundation. The combination of its immense chemical information power with an upfront separation technique creates a hyphenated platform that is analytically unbeatable for detailed analysis at the molecular level.

With the awareness that no single analytical platform can capture the entirety of a sample's complexity, and that only a portion of the total chemical space is GC-amenable, GC-MS and its multidimensional evolution reveal a fundamental piece of information that is often inaccessible to other platforms.

While GC-MS inherently focuses on a specific portion of the available chemical space, that of volatile and semivolatile molecules (typically <700 Da), it remains the gold standard for robust, high-resolution separation and identification. Whether utilized for the discovery of novel biomarkers or the routine monitoring of contaminants, the binomial employment of gas chromatography and mass spectrometry is a pillar of modern analytical science.

By gathering the foremost experts in the omics sciences who also master GC-MS's theory and practice, this book provides the pathway and tools to unravel the challenges of modern omics research for biological, environmental, and inanimate samples. Their collective insights will help the reader to navigate the volatile and semivolatile chemical space with greater awareness, whether you are a researcher looking to integrate GC-MS into a multi-omics workflow or a specialist refining a targeted assay for clinical or industrial use.

Gas Chromatography-Mass Spectrometry for Omics Applications discusses the basis for understanding the meaning and importance of high-resolution separations and identifications in chemical analysis for omics sciences. In particular, it focuses on the chemical domain of small molecules and the GC-MS combo considering both conventional and multidimensional GC, and both low- and high-resolution mass spectrometry.

The sheer volume of information generated by these techniques supports the omics mission across both life sciences and the study of inanimate nature. Through recent discoveries, the intimate relationship between a measured metabolite, when considering a living system, or a constituent, when considering a processed or inanimate sample, and the biological, environmental, or industrial event is illustrated.

The book is a reference for researchers who need to integrate modern strategies of chemical analysis for molecular sample characterization, with both targeted and non-targeted approaches. However, more focus is devoted to the latter because the intrinsic nature of high-resolution techniques like GC(×GC)-MS makes them ideal for discovery studies and non-targeted investigations.

To guide the reader effectively, the book is structured into two distinct sections. The first section, ***Basics and Fundamentals***, provides the essential definitions: the philosophy of non-targeted and targeted analysis (Chapter 1), the gateway of sample preparation in the analytical process (Chapter 2), the principles of high-resolution GC and MS (Chapters 3 and 4), the related structure and interpretation of the data (Chapter 5), and the strategies to ensure high data quality (Chapter 6).

The second section, ***Omics Applications***, shifts from the theory to the deployment of these strategies in solving modern challenges. These explore the study of biological systems through metabolomics, whether involving plants (Chapter 7) or animal models (Chapters 8.1 and 8.2). It then moves into the surroundings, discussing the discovery of emerging contaminants in the environment and their exposure on individuals (Chapter 9). Finally, the book covers the characterization of refined samples to determine their value or to improve industrial processes, in the context of the food industry (Chapter 10), and the oil refinery and plastics (Chapter 11).

April 2026

Flavio A. Franchina
Ferrara, Italy

Acknowledgments

A project of this magnitude is never a solitary endeavor. It is the result of a collective intellectual effort, and I am grateful to the exceptional team of scientists who invested their time and expertise in these chapters. Their commitment to scientific rigor and their ability to translate complex methodologies into accessible insights have elevated the quality of this book.

I also want to extend my sincere thanks to the reviewers, whose critical eyes and demanding standards refined the technical accuracy of this book, and to my close coworkers, whose daily collaboration and shared passion for the nuances of gas chromatography and mass spectrometry provided the ideal environment for these ideas to flourish.

Beyond the laboratory and the desk, I must acknowledge the broader spectrum of support that fueled this journey. I am profoundly grateful to those who believed in my vision for this book from its inception: your encouragement was the catalyst for every page that follows. Equally, I owe a debt of gratitude to those who did not believe, the skeptics whose doubt provided a different, yet vital, form of motivation. Thank you for giving me a reason to sharpen my focus and for providing the challenge that allowed me to demonstrate the true value of this work and my own professional determination.

Finally, and most importantly, I owe my greatest debt to the partner in life, soulmate, my wife Delphine. Your unwavering support and motivation were the constants that allowed me to navigate the long hours and the intense focus required to bring this book to fruition. You are the foundation upon which my professional life is built, and this book is as much a testament to your patience and belief in me as it is to my efforts.

I am proud to present this book, and I trust it will serve as a valuable tool in your pursuit of novel discoveries.

Section I

Basics and Fundamentals

1

Definitions and Concepts for Targeted and Non-targeted Chemical Analysis

Marco De Poli[1], *Allan Polidoro*[2], *and Flavio A. Franchina*[1]

[1] *Department of Chemical, Pharmaceutical, and Agricultural Sciences, University of Ferrara, Ferrara, Italy*
[2] *Department of Environmental and Prevention Sciences, University of Ferrara, Ferrara, Italy*

1.1 Introduction

Over the past decades, chemical analysis has undergone extensive transformations. Advances in instrumentation and computational capacity have made possible the development of highly sensitive, robust, and high-resolution analytical platforms. These technological evolutions not only enhance the number and diversity of compounds that can be separated and detected but also reshape the strategies for resolving complex chemical mixtures.

In parallel, the record of chemical substances continues to grow larger and more diverse. The Chemical Abstracts Service (CAS) Registry currently lists over 219 million compounds, and more than 350,000 substances have been listed for global production [1]. Online resources, such as ChemSpider and PubChem, as well as mass spectral libraries like the National Institute of Standards and Technology (NIST), provide access to millions of chemical structures and their corresponding fragmentation patterns. This vast and growing chemical space places increasing demands on analytical workflows, particularly in fields such as metabolomics, environmental science, food chemistry, and the petroleum industry, where sample matrices often contain hundreds to thousands of components, many of which are unknown or unexpected [2]. Chapters 7–11 are dedicated to these fields of application.

Analytical chemists can approach chemical analysis with two complementary strategies: targeted analysis (TA), which focuses on detecting known predefined compounds with high specificity, and untargeted or non-targeted analysis (NTA), which aims to capture a broader chemical profile without prior selection of analytes [3]. Both untargeted and non-targeted terms can be used interchangeably and refer to the same concept.

NTA is, nowadays, the driving approach in modern omics sciences for the discovery and understanding of complex systems, encompassing not only living organisms (*e.g.*, metabolomics) and their intimate relationship with the environment (*e.g.*, exposomics) but also processing events and inanimate samples (*e.g.*, foodomics and petroleomics).

While this distinction is operationally well established and widely adopted, it does not, on its own, account for the full range of factors that influence which chemical information is ultimately retained, excluded, or transformed throughout the analytical process. A more integrated perspective should consider how sample preparation, chromatographic and spectrometric resolution, and data processing strategies interact to determine which components of a sample become chemically accessible and which are excluded. These methodological choices are not merely technical; they directly influence the scope, quality, and structure of the information obtained from an analytical system. In this context, it is important to recognize that what is detected, interpreted, and reported is shaped by the tools, procedures, and experimental approaches employed to investigate the sample [4].

In the case of NTA, however, it is important to be aware that GC-amenable analytes inherently introduce a bias in the realm of analyte chemistries that can be introduced in the analytical system, which can be separated and detected/identified. Even though this might seem a limitation, GC-based NTA significantly contributes to expanding and completing the chemical space, especially for volatile and semivolatile, as well as many nonpolar compounds that are poorly represented or entirely missed by methods relying on other techniques.

To articulate this view, in this chapter, we will discuss the concepts of TA and NTA, focusing on their practical application through GC- or GC×GC-MS, and introduce the concept of analytical information continuity (AIC) to describe the degree to which the analytical workflow and strategies retain the chemical complexity of a sample from its collection to data interpretation. From a future perspective, high-resolution techniques, which are well suited for NTA, allow a scale-down to targeted and simpler techniques for routine analysis.

At the same time, the chapter sets the basis for the upcoming chapters, in which the non-targeted approach is applied to various fields of application.

1.2 Conceptual Foundations

1.2.1 Analytical Strategies

The landscape of chemical analysis is defined by how information is selected, isolated, and interpreted across the various experimental conditions or workflows available. *TA* aims to detect and quantify a predefined set of compounds, often referred to as "known knowns" [5], using reference standards and optimized analytical conditions to achieve maximum sensitivity and specificity [3]. This approach is especially prevalent in fields such as environmental monitoring and food safety, where regulatory compliance and the unambiguous quantification of specific and highly characterized substances are required due to hazardous properties and safety issues [6].

The use of reference standards in TA enables the unequivocal identification of compounds, typically through matching retention times and the characteristic mass spectra or ion ratios, thereby ensuring analytical reliability and reproducibility. The entire analytical workflow is tailored for maximum selectivity and the required sensitivity: sample preparation is deliberately designed to isolate target analytes at regulatory limits while removing matrix components considered as interferences [7].

In contrast, *NTA* offers a different paradigm: instead of limiting the scope to predefined analytes, NTA includes an umbrella of strategies/schemes, which aims to detect, characterize, and, when possible, identify as many chemical constituents as possible in a given sample, the so-called "unknown unknowns", regardless of prior expectations or database availability [8]. This holistic approach is particularly valuable in contexts where unawareness or growing knowledge may be present, such as environmental monitoring for emerging pollutants or the discovery of novel biomarkers in biological matrices. To achieve a comprehensive chemical understanding of a given sample, multiple methodologies should ideally be combined, since each has its own selectivity, and no single technique is universally applicable.

Unlike TA, where selectivity is deliberately maximized to reduce complexity, NTA workflows emphasize chemical breadth. Sample preparation is usually designed to capture the widest possible spectrum of compounds, employing broad extraction methods. This necessity for inclusiveness, however, introduces a fundamental trade-off: as the range of recovered analytes expands, so does the potential for matrix effects and analytical challenges related to sample complexity [9, 10].

Moreover, the massive volume and complexity of generated data often demand time-consuming pre- and post-processing, as well as advanced chemometric and statistical approaches for data mining and interpretation. The lack of standardized protocols, incomplete reference databases, and the computational burden and uncertainty associated with handling thousands of features per sample contribute to the challenges of NTA [11].

An intermediate strategy between TA and NTA is *suspect screening*, which focuses on detecting a predefined list of compounds or chemical classes whose presence in the sample is uncertain. This terminology is mostly used for environmental applications, and unlike TA, suspect screening does not rely on the availability of analytical reference standards. Instead, it utilizes prior information, such as molecular formulas, exact masses, predicted fragments, or retention indices, often derived from databases or *in silico* predictions. These compounds are often referred to as "known unknowns," since their structures are known, but their occurrence in the sample is not. By narrowing the scope of inquiry while maintaining a non-targeted acquisition approach, suspect screening reduces the computational and interpretive challenge typical of non-targeted workflows, while still allowing for exploratory flexibility aligned with a specific research objective [10, 12]. A straightforward application of this strategy is demonstrated in the study by Casey *et al.*, who employed a suspect screening strategy using GC-HRMS to analyze incineration stack emissions that potentially contained PFAS. Even in the absence of analytical standards, several fluorinated compounds were tentatively identified by combining library matching, observation of diagnostic fragments, and confirmation of molecular ions across multiple ionization modes (EI, PCI, and NCI) [13].

Within the broad domain of NTA, the literature also distinguishes between *profiling* and *fingerprinting* strategies. Although the terms are sometimes used interchangeably, they represent distinct approaches [14]. Profiling is designed to extract comprehensive information about the constituents of a sample, including retention time, spectral data, and intensity profiles. When profiling is conducted without prior selection of analytes, it is typically described as non-targeted profiling [15–17]. For instance, Di Giovanni *et al.* developed a GC×GC-MS method for metabolite profiling in serum, identifying 33 unique candidate biomarkers to distinguish Crohn's disease patients and healthy controls [16].

In contrast, fingerprinting prioritizes pattern recognition rather than compound identification. The goal is not to identify or quantify individual analytes, but to capture global chemical signatures that differentiate samples based on similarities and differences in their overall signal profiles. This approach is particularly useful for classification problems, where analytical discrimination is achieved through the multivariate comparison of features, and the number of detected peaks or features directly contributes to the robustness of the fingerprint [15, 17, 18]. An example of this strategy is the study developed by Caratti *et al.*, who applied non-targeted GC×GC-MS/FID analysis, combined with multivariate fingerprinting, to differentiate the volatile profiles of balsamic vinegar. This approach distinguished samples based on geographical origin and fermentation practices through both unsupervised and supervised models [19].

Figure 1.1 offers a conceptual visualization of the analytical strategies discussed, mapped within a three-dimensional space defined by three typical descriptors: the number of analyzed samples (X-axis), the number of investigated analytes (Y-axis), and the overall volume of chemical information obtained from the sample (Z-axis). Each strategy is represented as a "cloud" because it is often difficult to

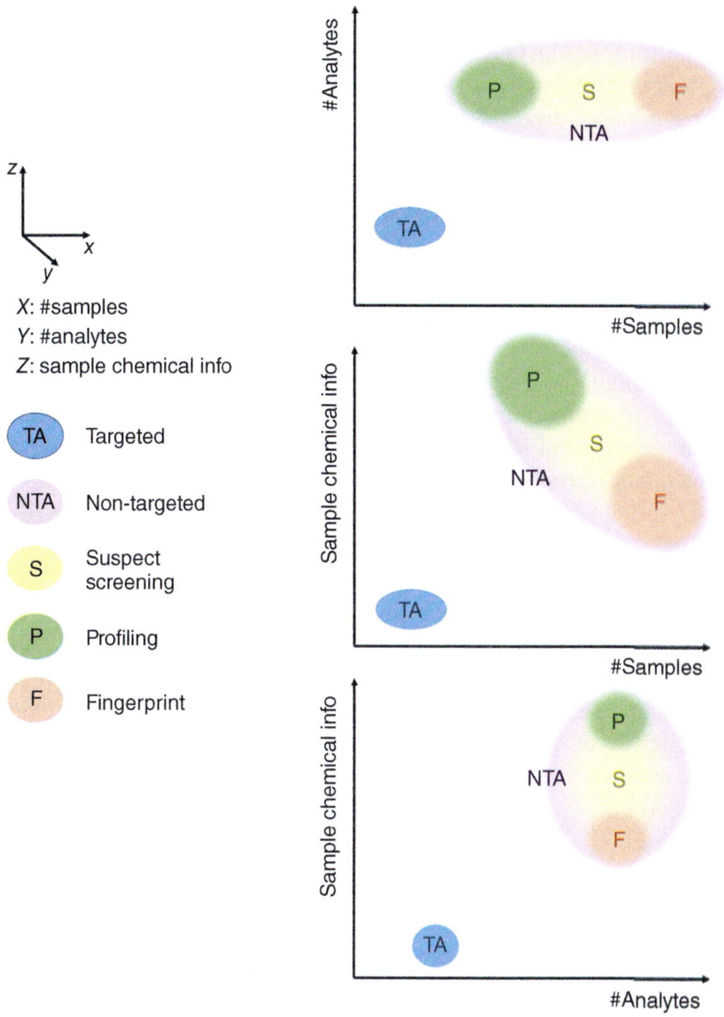

Figure 1.1 Analytical strategies positioned in a 3D space defined by three descriptors: number of samples (*X*), number of analytes (*Y*), and chemical information (*Z*), illustrating their overlap and gradation across methodological dimensions.

draw clear borders and extensions, emphasizing the absence of strict boundaries between such categories and groups.

In this representation, targeted analysis (TA cloud) occupies a relatively narrow region, defined by a limited number of analytes and a focused informational output. Even when large numbers of samples are processed, each one is typically assessed independently by comparing it to a calibration curve. Direct comparisons between samples are not inherent to the strategy, although they can be added later depending on the study design (*e.g.*, seasonal variation of contaminants in air).

In contrast, non-targeted analysis (NTA cloud) encompasses a much broader region, covering a wider array of analytes and samples, while generating a significantly larger volume of chemical data. Within this non-targeted domain, the three substrategies are illustrated. Suspect screening (S in Figure 1.1), located near the center of the NTA space, focuses on compounds with known structures but unknown presence (so-called "known unknowns") and typically relies on information such as accurate mass, predicted fragmentation patterns, or retention index to support identification. Profiling (P in Figure 1.1) is positioned higher along the Z-axis, as it emphasizes the detailed chemical characterization of each sample. Fingerprinting (F in Figure 1.1), in contrast, extends along the X-axis, reflecting its focus on comparing global signal patterns across multiple samples without requiring compound identification. This multidimensional perspective highlights that analytical strategies are not isolated categories but rather interconnected approaches that vary along continuous gradients of selectivity, coverage, and informational output. There are no fixed borders, only regions shaped by methodological choices and analytical goals.

If high-resolution data are recorded (*e.g.*, GC×GC-HRMS), it is up to the analyst to decide how to use them. Even owing all the chemical information, the acquired data can be used as fingerprinting (just patterns, no weight on chemical identification which might be accessible afterwards), targeted (by focusing only to the predetermined analytes of interest), or post-targeted (*e.g.*, looking retrospectively for the presence of newly regulated contaminants that were not regulated at the time of the acquisition).

1.2.2 Resolution

Resolution in analytical chemistry is more than a numerical parameter; it defines the level of detail with which chemical information can be accessed and understood. In this context, the chromatographic and mass spectrometric resolution defines how finely the chemical space can be resolved into distinguishable components.

1.2.2.1 Chromatographic Resolution

Chromatographic resolution quantifies how well a separation system can distinguish adjacent analytes based on retention time. It is formally expressed as follows:

$$R = \frac{t_{R2} - t_{R1}}{\frac{W_1 + W_2}{2}}, \tag{1.1}$$

where t_{R2} and t_{R2} are retention times, and W_1 and W_2 are peak widths. This parameter is influenced not only by retention but also by selectivity and efficiency, which play a significant role in overall chromatographic resolution [20]. Please refer to *Chapter 3* for a detailed discussion on the chromatographic process and fundamentals.

While this formula offers a standard definition, chromatographic resolution should be understood as a strategic decision rather than a fixed metric. In TA, chromatographic resolution is best seen as a fit-for-purpose parameter, as it only needs to be high enough to separate the analytes of interest from nearby interferences. Beyond that, increasing resolution may add time and complexity without improving results [21].

In contrast, non-targeted strategies require broader and more efficient separation capabilities. When the analytical aim is to explore or monitor a wide array of unknowns, unresolved coelutions can result in spectral entanglement, misannotation, or complete loss of signal. In such cases, resolution plays a defining role in determining how much of the sample's chemical complexity can be separated, detected, and interpreted. Chromatographic efficiency, a parameter that reflects how effectively a column can produce narrow, symmetrical peaks with minimal broadening, is also critical as peak sharpness and symmetry directly influence the separation, the quality of mass spectra, and the success of deconvolution algorithms [20, 22].

Multidimensional chromatographic systems, such as GC×GC or LC×LC, are especially valuable in this context. By combining multiple separation mechanisms and distributing analytes across two retention axes, they expand the peak capacity while maintaining or even increasing efficiency. Additionally, the use and combination of such multidimensional approaches/techniques contribute to the reduction of sample dimensionality [23].

Such multidimensional, high-resolution, and high-efficiency chromatograms not only enhance the number of components detected but also exhibit structured elution patterns that reveal meaningful chemical class information, enabling semi-supervised exploration even in the absence of analytical standards and facilitating downstream interpretation and multivariate analysis [22, 24].

Therefore, chromatographic resolution and efficiency together are not simply about how many peaks can be separated, but about how well the chemical information is organized for interpretation.

1.2.2.2 Mass Spectrometric Resolution

Mass spectrometric resolution is defined as follows:

$$R = \frac{m}{\Delta m} = \frac{m/z}{\Delta m/z}, \tag{1.2}$$

where m is the m/z of a selected ion and Δm is the smallest mass difference that can be resolved. Please refer to Chapter 4 for a detailed discussion on mass spectrometry fundamentals.

Low-resolution instruments, such as quadrupoles operating at unit resolution, are often well-suited for targeted workflows. This approach is useful for monitoring predefined m/z with high precision and reproducibility. However, their capacity to resolve complex mixtures or identify unexpected species is limited [25].

High-resolution mass analyzers, including time-of-flight (TOF) and Orbitrap systems, can resolve closely spaced ions and deliver mass accuracies often within a few parts per million. This enables the discrimination of isobaric compounds, supports the assignment of elemental formulas, and facilitates the annotation of unknown features in complex matrices. In contexts where spectral libraries are incomplete or unavailable, high mass resolution serves as a surrogate for chemical knowledge, enabling inference where direct identification is not possible [25].

Notably, mass resolution also improves signal-to-noise ratios and reduces the ambiguity of peak assignment, which are crucial for both qualitative and quantitative analyses. Furthermore, when combined with accurate mass and isotope pattern recognition, high-resolution data enhances confidence in structural hypotheses [26, 27].

In the same way as in chromatographic resolution, non-targeted strategies greatly benefit from high-mass spectrometric resolution, both in terms of qualitative and separation capabilities. Indeed, when mass spectral high resolution is combined with soft ionization techniques (*e.g.*, PCI and NCI), it supports more reliable chemical identification also for unknowns. On the other hand, the ability to accurately distinguish closely spaced ions becomes critical for spectral deconvolution [26, 28, 29].

Both chromatographic and mass spectrometric resolutions define the degree to which a given chemical domain can be detailed into its single components. Analytical techniques that operate with low-resolution settings may offer higher throughput at the cost of losing information about individual components. Conversely, maximizing resolution increases detailed analytical information, although it presents challenges in terms of data volume, complexity, and processing demands [26, 28, 30].

Understanding resolution, not only as a stand-alone traditional metric (chromatographic and mass spectrometric) but as a unified concept of an overall analytical resolution, is necessary for successful non-targeted strategies.

1.3 Analytical Information Continuity

The analytical workflow defines experimentally what can and cannot be accessed in the sample. In targeted and non-targeted strategies, methodological choices on the sample preparation, chromatographic separation, and mass spectrometric detection jointly determine which portions of the chemical space become analytically visible/accessible. These three dimensions govern the informational output of any analysis, not only shaping the detectability of compounds but also confining the knowledge that can be drawn from samples. They form a continuum of methodological selectivity, from the narrow focus of targeted protocols to the inclusiveness of non-targeted exploration. However, existing terminology often fails to capture how these three dimensions interact across the analytical workflow.

To address this conceptual gap, we refer to *AIC* as a concept to evaluate the analytical workflows based on their capacity to preserve and transmit the chemical complexity of a sample throughout all methodological stages: from sample preparation to data utilization, AIC refers to the degree to which each analytical decision retains the informational level from the original sample.

AIC can be defined as the extent to which the chemical information present in a sample is retained across the analytical workflow and made available for interpretation. It reflects a continuum, not a binary division. At one end, workflows are highly selective and narrow, optimized for specific analytes under controlled conditions, typically seen in classical TA. At the other end, methods attempt to capture the broadest possible chemical profile, as in NTA, using minimal sample manipulation and high-resolution techniques.

This concept shifts the evaluative focus from analytical intent (*what are we looking for?*) to analytical capacity (*how much of the sample's chemical content can be accessed?*). In doing so, AIC becomes not only a practical consideration but also a measure of informational depth; it quantifies what becomes analytically accessible.

This concept is illustrated in Figure 1.2, where the analytical workflow acts as a selective and transformative interface that governs how chemical reality is converted into chemical knowledge.

During this process, a loss of information is unavoidable due to the intrinsic selectivities of the analytical procedures and strategies, whatever their nature.

Although the concept of AIC can be applied across instrumental platforms, the following discussions are primarily framed within the context of GC- and GC×GC-MS, focusing on the particularities of workflows designed for volatile and semivolatile compounds. Many of the considerations discussed, however, can be adapted to other analytical domains.

1.3.1 Sample Preparation as an Information Filter

The sample preparation is the first and one of the most critical steps in defining the AIC of the analysis. Every extraction, purification, or derivatization step

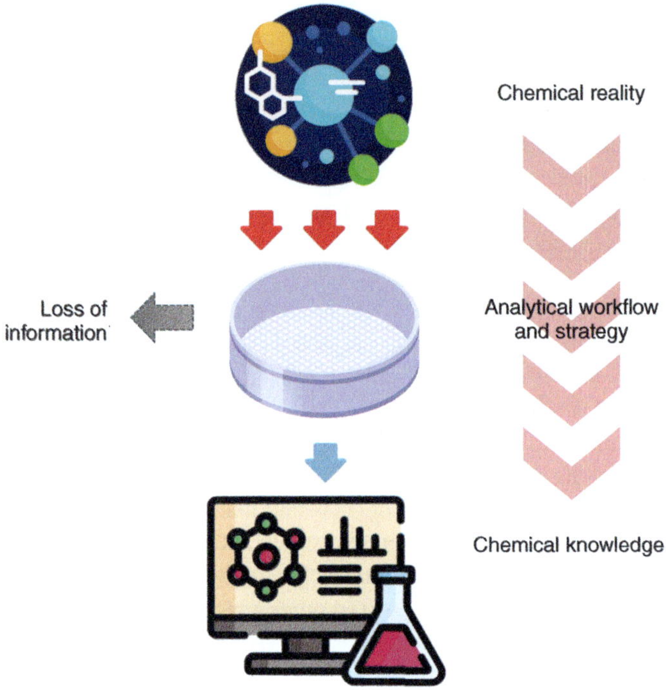

Figure 1.2 Schematic representation of how the analytical workflow mediates the transformation of chemical reality into chemical knowledge. Analytical information continuity reflects the inclusiveness and fidelity of this transformation.

acts as a filter, selectively enriching certain analytes while excluding others. Highly specific preparation procedures (*e.g.*, solid-phase extraction tailored for particular analyte classes) enhance sensitivity and reproducibility but reduce the scope of detectable compounds [31]. This results in low AIC, since much of the sample's chemical content is discarded.

Conversely, nonselective sample preparation strategies such as dilute-and-shoot, thermal desorption, or direct injections are designed to preserve the broadest possible chemical profile. While these approaches may introduce matrix effects and complicate downstream analysis, they maximize the initial AIC by allowing more of the sample's chemical reality to enter the analytical pipeline. In this context, the selectivity–universality axis of sample preparation becomes determinant of AIC.

However, it is important to emphasize that less selective or minimal sample preparation does not necessarily translate into greater AIC. While broader extraction strategies may allow more compounds to enter the analytical system, they also increase the risk of matrix effects, signal suppression, and co-elutions, all of which can compromise detection and interpretation [32]. The effectiveness of sample preparation must therefore be evaluated in conjunction with the resolution and sensitivity of the downstream analytical techniques. In this sense, sample preparation is not an isolated step, but rather a process that must be tuned with the entire analytical workflow and research objective. Please refer to Chapter 2 for a detailed discussion of sample preparation procedures used in GC-MS.

1.3.2 Sample Introduction as a Selective Interface

Even before chromatographic preparation, the act of introducing a sample into the analytical system functions as an inherently selective event. This has profound implications for NTA. Although NTA is often framed as an inclusive strategy aimed at capturing the broadest possible array of chemical constituents, the operational reality is that inclusion is conditional. For GC-based NTA, analytes must be volatile, thermally stable, and compatible with the carrier gas and liner materials. These requirements introduce implicit selectivity, constraining the accessible chemical domain even when no prior target list is defined [33]. Thus, while NTA is conceptually open-ended, its AIC is always instrumentally mediated.

In addition to acting as a selective gatekeeper, the injection step also plays a decisive role in chromatographic performance, since high injection efficiency ensures narrow, well-defined bands entering the column, which is fundamental for achieving high-resolution GC separations. From the perspective of AIC, the injection system is therefore not a passive interface but both a filter and a performance factor. This physical and chemical threshold regulates not only what becomes accessible but also how effectively it can be resolved downstream in the analysis process (i.e., in the column, in the mass spectrometer, and in the data elaboration and interpretation).

1.3.3 Chromatographic Separation

Chromatographic efficiency and resolution directly affect the number of distinguishable chemical signals that can be extracted from a sample, impacting the degree of AIC. High-efficiency systems generate narrow, symmetrical peaks and minimize band broadening, enabling a larger number of compounds to be separated, detected, and eventually identified.

Capillary GC inherently offers very high separation efficiency due to the narrow internal diameter, uniform stationary phase coating, and optimal flow dynamics. When coupled with the structured modulation of GC×GC, this efficiency is further amplified, producing exceptionally sharp and well-resolved peaks. Such performance not only maximizes peak capacity but also sets a strong foundation for accurate mass spectrometric detection, feature deconvolution, and reliable compound identification, all of which are indispensable for achieving high AIC.

In GC×GC, the first dimension acts as a preparative fractionation for the second one like a hierarchical sampling system. This recurring separative logic enhances detectability by distributing compounds more evenly across a 2D chromatographic space, resolving co-elutions that would remain hidden in a single column. Structured chromatograms, band compression, and various separation mechanisms all contribute to preserving the sample's complexity throughout the separation process. In contrast, workflows relying on conventional one-dimensional chromatography might sacrifice a portion of the chemical landscape due to peak overlap and limited resolving power. Although often sufficient for targeted applications, their AIC is constrained.

1.3.4 Mass Spectrometry and Informational Depth

Mass spectrometry contributes to the information continuity as well. For instance, excluding a range of *m/z* values from the acquisition in EI-MS could limit or erroneously affect the recorded information. MS also defines the level of structural information available for each detected signal. High-resolution instruments, such as TOF and Orbitrap analyzers, provide precise mass measurements and high resolving power, which are essential for distinguishing isobaric compounds, assigning elemental formulas, and deconvoluting co-eluting peaks, resulting in high AIC [25]. These capabilities are especially critical in non-targeted strategies, where analytes may not be present in spectral libraries and accurate mass becomes a necessary feature for tentative identification. In such contexts, high mass resolution extends the interpretive depth of the analysis, allowing greater confidence in structural assignments and facilitating more reliable annotation of unknown features [27, 34].

In contrast, unit-resolution instruments, such as quadrupoles, while effective for monitoring specific transitions in targeted analyses, they impose a limitation on what can be learned from the data. They lack the granularity to resolve fine mass differences and are generally unsuited for exploratory or high-complexity workflows [25]. Thus, mass resolution plays a decisive role not just in detection but in shaping the scope of chemical knowledge that can be extracted. When considering MS/MS systems, the choice mainly depends on the nature of the analyzers and the acquisition mode. For example, triple quadrupoles in tandem MS mode exclude most of the information by filtering only the selected parent/daughter ions and are used exclusively for target analytes, which, however, results in limited AIC.

In GC-based workflows, mass spectrometry also benefits from the extensive EI spectral libraries, which greatly enhance the identification power and are highly reliable and standardized across different instrument suppliers [35].

Beyond standard EI, the use of complementary ionization modes such as soft EI, negative chemical ionization (NCI), and positive chemical ionization (PCI) can selectively emphasize different structural features or compound classes [36]. This multimodal approach enhances AIC by expanding the range of detectable analytes, increasing confidence in identification, and facilitating the characterization of compounds that may be underrepresented or poorly ionized under a single ionization regime.

1.3.5 Data Governance

The continuity of analytical information is not solely dictated by instrumental selectivity or resolution but also by how the recorded data is processed, managed, and exploited. In workflows with high data density, such as those involving HRMS and/or multidimensional chromatography, the data analysis stage becomes an additional bottleneck, particularly in terms of informational depth. Without proper post-acquisition treatment, much of the chemical information embedded in raw data may remain hidden, misinterpreted, or entirely unused, thus undermining the information continuity. Data-processing strategies influence AIC through multiple stages, since deconvolution, noise reduction, baseline correction, peak alignment, and feature annotation algorithms not only determine the detectability and quantifiability of features but also influence how peaks are interpreted and assigned, as well as the reliability of identification [37–39].

In addition, the use of multivariate tools to highlight chemical patterns, especially unsupervised tools such as Principal Component Analysis (PCA) and Hierarchical Cluster Analysis (HCA), or machine learning algorithms, extends the interpretative power of datasets. These techniques can reveal relationships between features, sample origins, or biological outcomes that are otherwise inaccessible via direct compound annotation. Similarly, retention time indexing, predictive retention modeling, and mass defect scripting can enhance confidence in tentative annotations, particularly when analytical standards are unavailable [39, 40]. In this sense, AIC is not only enhanced by the quality of the raw data but also by the efficiency of the entire data treatment pipeline.

Moreover, as data volumes grow, data curation, metadata integration, and result reporting become essential to preserve analytical meaning. Workflows must strike a balance between information retention and interpretative clarity, especially in regulatory or clinical settings. Ultimately, the analytical power of a method is determined not only by what it detects but also by how the analyst structures and presents that information in a coherent and chemically meaningful form. For a more detailed discussion on data processing and quality of data for GC-MS data, the reader is directed to Chapters 5 and 6.

1.3.6 Cumulative Effects Across the Analytical Workflow

The continuity of analytical information is defined not by any isolated single step but by the cumulative effect of all stages. A method with high-resolution MS but selective sample preparation remains limited in what it can reveal. Likewise, universal sample extraction paired with low-resolution separation and detection yields an incomplete picture. It is the preservation of information across the entire analytical chain, from sampling to preparation, separation, detection, and interpretation, that defines a method's AIC. Because each of these steps introduces its own selectivity, achieving a truly comprehensive chemical picture often requires combining multiple analytical methodologies. No single technique can capture all compounds present in a complex sample due to the diverse physicochemical properties and concentrations. By integrating complementary approaches, such as GC and LC, or multiple ionization modes, analysts can mitigate limitations and expand the accessible chemical space.

To fully maximize the rich recorded chemical data, efforts must also be made during data analysis by selecting suitable tools and tailoring them to the analytical strategy and objective [39]. Otherwise, most of the data would remain unused, offering little benefit and not balancing the efforts of experimental conditions development [37].

This holistic view allows AIC to be conceptualized as a gradient shaped by trade-offs. Increased chemical breadth often comes with greater analytical noise, higher computational burden, and challenges in validation. Nevertheless, maximizing AIC is often essential when the analytical goal is exploratory, as in NTA strategies.

Finally, AIC provides a framework for examining how methodological design influences the scope and depth of chemical information that can be obtained. Every methodological choice, what to extract, what to separate, what to detect, and how to process, affects not only the data obtained but also the knowledge that can be produced.

Workflows with high AIC are better suited for exploratory studies, which aim to unravel unknowns and identify unexpected compounds. In contrast, low-AIC workflows are optimized for certainty, quantification, and regulatory compliance, but at the cost of discarding potentially relevant information. In this sense, AIC is a measure of how much of the chemical information a given method is able to unravel.

By integrating this concept into method development and evaluation, analysts can make more informed choices on how to balance sensitivity, selectivity, and informational richness, depending on whether the goal is to confirm the known, explore the unknown, or do both at once.

References

1 Ciccarelli, D., Samanipour, S., Rapp-Wright, H. et al. (2025). Bridging knowledge gaps in human chemical exposure via drinking water with non-target screening. *Crit. Rev. Environ. Sci. Technol.* 55 (3): 190–214.

2 Milman, B.L. and Zhurkovich, I.K. (2017). The chemical space for non-target analysis. *TrAC Trends Anal. Chem.* 97: 179–187.

3 Ballin, N.Z. and Laursen, K.H. (2019). To target or not to target? Definitions and nomenclature for targeted versus non-targeted analytical food authentication. *Trends Food Sci. Technol.* 86: 537–543.

4 Hacking, I. (1983). *Representing and Intervening: Introductory Topics in the Philosophy of Natural Science.* Cambridge University Press.

5 Little, J.L., Cleven, C.D., and Brown, S.D. (2011). Identification of "known unknowns" utilizing accurate mass data and chemical abstracts service databases. *J. Am. Soc. Mass Spectrom.* 22 (2): 348–359.

6 Devers, J., Pattison, D.I., Hansen, A.B., and Christensen, J.H. (2025). Comprehensive two-dimensional gas chromatography as a tool for targeted and non-targeted analysis of contaminants of emerging concern in wastewater. *Talanta* 282: 127032.

7 Hernández, F., Portolés, T., Pitarch, E., and López, F.J. (2011). Gas chromatography coupled to high-resolution time-of-flight mass spectrometry to analyze trace-level organic compounds in the environment, food safety and toxicology. *TrAC, Trends Anal. Chem.* 30 (2): 388–400.

8 Gruber, B., David, F., and Sandra, P. (2020). Capillary gas chromatography–mass spectrometry: current trends and perspectives. *TrAC – Trends in Analytical Chemistry* 124: 115475.

9 Hajeb, P., Zhu, L., Bossi, R., and Vorkamp, K. (2022). Sample preparation techniques for suspect and non-target screening of emerging contaminants. *Chemosphere* 287: 132306.

10 Pourchet, M., Debrauwer, L., Klanova, J. et al. (2020). Suspect and non-targeted screening of chemicals of emerging concern for human biomonitoring, environmental health studies and support to risk assessment: from promises to challenges and harmonisation issues. *Environ. Int.* 139: 105545.

11 Pollo, B.J., Teixeira, C.A., Belinato, J.R. et al. (2021). Chemometrics, comprehensive two-dimensional gas chromatography and "omics" sciences: basic tools and recent applications. *TrAC – Trends in Analytical Chemistry* 134: 116111.

12 Hollender, J., Schymanski, E.L., Ahrens, L. et al. (2023). NORMAN guidance on suspect and non-target screening in environmental monitoring. *Environ. Sci. Eur.* 35 (1): 1–61.

13 Casey, J.S., Jackson, S.R., Ryan, J., and Newton, S.R. (2023). The use of gas chromatography – high resolution mass spectrometry for suspect screening and non-targeted analysis of per- and polyfluoroalkyl substances. *J. Chromatogr. A* 1693: 463884.

14 Tranchida, P.Q., Franchina, F.A., and Mondello, L. (2017). Analysis of essential oils through comprehensive two-dimensional gas chromatography: general utility. *Flavour Fragr. J.* 32 (4): 218–227.

15 Murray, J.A. (2012). Qualitative and quantitative approaches in comprehensive two-dimensional gas chromatography. *J. Chromatogr. A* 1261: 58–68.

16 Di Giovanni, N., Meuwis, M.A., Louis, E., and Focant, J.F. (2020). Untargeted serum metabolic profiling by comprehensive two-dimensional gas chromatography-high-resolution time-of-flight mass spectrometry. *J. Proteome Res.* 19 (3): 1013–1028.

17 Stilo, F., Bicchi, C., Jimenez-Carvelo, A.M. et al. (2021). Chromatographic fingerprinting by comprehensive two-dimensional chromatography: fundamentals and tools. *TrAC Trends Anal. Chem.* 134: 116133.

18 Pierce, K.M., Kehimkar, B., Marney, L.C. et al. (2012). Review of chemometric analysis techniques for comprehensive two dimensional separations data. *J. Chromatogr. A* 1255: 3–11.

19 Caratti, A., Ferrero, F., Tabacco, E. et al. (2024). Translating 2D-chromatographic fingerprinting to quantitative volatilomics: unrevealing compositional changes in maize silage volatilome for robust marker discovery. *J. Agric. Food Chem.* 72 (42): 23616–23630.

20 Swinley, J. and De Coning, P. (2019). *A Practical Guide to Gas Analysis by Gas Chromatography*. Elsevier.

21 Dettmer-Wilde, K. and Engewald, W. (ed.) (2014). *Practical Gas Chromatography*. Berlin, Heidelberg: Springer Berlin Heidelberg.

22 Mondello, L., Lewis, A.C., and Bartle, K.D. (2001). *Multidimensional Chromatography*. Wiley.

23 Giddings, J.C. (1995). Sample dimensionality: a predictor of order–disorder in component peak distribution in multidimensional separation. *J. Chromatogr. A* 703 (1–2): 3–15.

24 Zanella, D., Focant, J.F., and Franchina, F.A. (2021). 30th Anniversary of comprehensive two-dimensional gas chromatography: latest advances. *Anal. Sci. Adv.* 2 (3–4): 213–224.

25 Tranchida, P.Q. and Mondello, L. (2020). *Hyphenations of Capillary Chromatography with Mass Spectrometry*. Elsevier.

26 Kaufmann, A. (2012). The current role of high-resolution mass spectrometry in food analysis. *Anal. Bioanal. Chem.* 403 (5): 1233–1249.

27 Schymanski, E.L., Jeon, J., Gulde, R. et al. (2014). Identifying small molecules via high resolution mass spectrometry: communicating confidence. *Environ. Sci. Technol.* 48 (4): 2097–2098.

28 Schymanski, E.L., Singer, H.P., Slobodnik, J. et al. (2015). Non-target screening with high-resolution mass spectrometry: critical review using a collaborative trial on water analysis. *Anal. Bioanal. Chem.* 407 (21): 6237–6255.

29 Kind, T. and Fiehn, O. (2010). Advances in structure elucidation of small molecules using mass spectrometry. *Bioanal. Rev.* 2 (1): 23–60.

30 Pani, O. and Górecki, T. (2006). Comprehensive two-dimensional gas chromatography (GC×GC) in environmental analysis and monitoring. *Anal. Bioanal. Chem.* 386 (4): 1013–1023.

31 Franchina, F.A., Zanella, D., Dubois, L.M., and Focant, J.F. (2021). The role of sample preparation in multidimensional gas chromatographic separations for non-targeted analysis with the focus on recent biomedical, food, and plant applications. *J. Sep. Sci.* 44 (1): 188–210.

32 Matuszewski, B.K., Constanzer, M.L., and Chavez-Eng, C.M. (2003). Strategies for the assessment of matrix effect in quantitative bioanalytical methods based on HPLC-MS/MS. *Anal. Chem.* 75 (13): 3019–3030.

33 Qin, H., Liu, H., Liu, Y. et al. (2023). Recent advances in sample preparation and chromatographic analysis of pharmaceuticals and personal care products in environment. *TrAC Trends Anal. Chem.* 164: 117112.

34 Hollender, J., Schymanski, E.L., Singer, H.P., and Ferguson, P.L. (2017). Non-target screening with high resolution mass. Spectrometry in the environment: ready to go? *Environ. Sci. Technol.* 51 (20): 11505–11512.

35 Mastrangelo, A., Ferrarini, A., Rey-Stolle, F. et al. (2015). From sample treatment to biomarker discovery: a tutorial for untargeted metabolomics based on GC-(EI)-Q-MS. *Anal. Chim. Acta* 900: 21–35.

36 Mazur, D.M., Detenchuk, E.A., Sosnova, A.A. et al. (2021). GC–HRMS with complementary ionization techniques for target and non-target screening for chemical exposure: expanding the insights of the air pollution markers in moscow snow. *Sci. Total Environ.* 761: 144506.

37 Gaida, M., Cain, C.N., Synovec, R.E. et al. (2023). Tile-based random forest analysis for analyte discovery in balanced and unbalanced GC×GC-TOFMS data sets. *Anal. Chem.* 95 (36): 13519–13527.

38 Gaida, M., Stefanuto, P.H., and Focant, J.F. (2023). Theoretical modeling and machine learning-based data processing workflows in comprehensive two-dimensional gas chromatography – a review. *J. Chromatogr. A* 1711: 464467.

39 Stefanuto, P.-H., Smolinska, A., and Focant, J.-F. (2021). Advanced chemometric and data handling tools for GC×GC-TOF-MS. *TrAC Trends Anal. Chem.* 139: 116251.

40 Zanella, D., Focant, J.-F., Hill, J.E., and Stefanuto, P.-H. (2020). Comprehensive gas chromatography–mass spectrometry. In: *Breathborne Biomarkers and the Human Volatilome* (ed. J. Beauchamp, C. Davis, and J. Pleil), 239–251. Elsevier.

2

GC-Suited Sample Preparation Approaches for Omics Studies

Flavio A. Franchina[1], *Damien Pierret*[2], *and Giorgia Purcaro*[2]

[1] *Department of Chemical, Pharmaceutical, and Agricultural Sciences, University of Ferrara, Ferrara, Italy*
[2] *Analytical Chemistry Lab, Gembloux Agro-Bio Tech, University of Liège, Gembloux, Belgium*

2.1 Introduction

Over the last few decades, there has been a clear shift among various disciplines from a reductionist to an interactionist approach. In the field of analytical chemistry, a reductionist approach involves characterizing all the compounds present in a sample to reduce everything to a univariate system, where each compound corresponds to a single target. However, this approach has been proven insufficient to answer increasingly more sophisticated questions. Therefore, scientists have shifted toward a more integrated and interactive approach, also known as "interactionism," which mainly translates into -omics approaches (e.g., metabolomics and proteomics) [1, 2]. The -omics sciences rely on nonspecific signals generated through instrumental signals using advanced analytical instruments, with the ultimate goal of providing information linked to the identity, quality, or quantitative aspects of the sample. This information is then extracted using appropriate data mining methods developed under the umbrella of chemometrics science. Readers are directed to Chapter 5 for a more detailed discussion of the fundamental data mining step. The instrumental signals originate from either a two-dimensional (2D) signal (e.g., absorption spectrum, chromatogram, and kinetic curve) or a three-dimensional (3D) or more-dimensional signal (e.g., fluorescence spectrum, comprehensive 2D chromatogram, and ion-chromatogram). Several techniques can provide an instrumental fingerprint, such as molecular spectroscopy and nuclear magnetic resonance (NMR). However, this book focuses on separative methods, particularly those based on gas chromatography (GC), for which the more specific term "chromatographic fingerprint" is used [3]. As discussed in Chapter 1, it is noteworthy that these chromatographic signals can be processed using either fingerprinting or profiling strategies, thereby providing multilevel information [2, 4]. For the sake of clarity, profiling refers to a semitargeted (e.g., selected chemical classes) or nontargeted approach that aims to obtain detailed information on the qualitative and quantitative distributions of sample components across samples. On the other hand, fingerprinting relies on a high-throughput methodology that is capable of capturing relevant information, but it does not necessarily yield accurate quantitative and qualitative data on sample composition.

When referring to -omics science using GC-based techniques, we naturally narrow down to volatile and semivolatile metabolites [5]. Essentially, we refer to the metabolome, which leads us into the field of metabolomics. As with other -omics techniques (e.g., proteomics and genomics), the goal is to obtain a comprehensive, unbiased, qualitative, and quantitative overview of the metabolomes that participate in the biological functioning of cells or organisms. Nevertheless, the definition of metabolome can be extended to fields of application where the relation with biological functions is not as tight, such as environmental (see Chapter 9), food (see Chapter 10), and petroleum (see Chapter 11) fields. All the different fields are, nonetheless, commonly characterized by a high complexity of sample constituents that can be analyzed using GC-related techniques. Due to this complexity, powerful analytical techniques play a fundamental role, among which comprehensive 2D GC (GC×GC) stands out for its outstanding separation power, flexibility, and increased sensitivity (see Chapter 3).

The focus of the present chapter is to provide an overview of the initial step in the overall analytical workflow, specifically the preparation of samples for GC analysis within the diverse context of omics sciences. Indeed, the first fundamental step is sampling, which depends mainly on the scientific question. Then, sample preparation, which includes extraction, purification, and usually derivatization, follows. A step particularly relevant when studying biological systems is the metabolism quenching. In this scenario, to prepare cells for metabolomics analysis, the quenching step aims to inactivate intracellular enzymes and halt the metabolism, thereby preventing degradation and alterations in sample composition. This step must be fast, as metabolic reactions can occur in milliseconds, and it is often performed by washing the sample (e.g., a cell pellet) in ice-cold organic solvents, which are also stored/frozen before further treatment. Another step relevant for biological matrices is protein precipitation. This step can occur simultaneously with quenching and also involves the use of organic solvents (e.g., acetonitrile and methanol) or salts (e.g., zinc sulfate). It aims at removing proteins that can clog GC inlets and columns, while leaving semivolatile analytes in the supernatant for further processing.

The sample preparation procedures are selected and tuned according to the strategy applied, which can be targeted or nontargeted. The former aims for the quantitative characterization of a limited number of well-selected metabolites, whereas nontargeted analysis is typically applied to cover a broader range of compounds in an unbiased manner, providing a snapshot of the sample's status at the moment of sampling.

In this chapter, the theoretical bases of the most common sample preparation techniques for volatile and semivolatile analysis are presented, with a discussion of the specificity of volatile and semivolatile molecules.

2.2 Sampling Techniques for Volatile Analyte Analysis

The most convenient way to analyze volatile analytes involves collecting the portion above the sample, i.e., its headspace (HS), which can be done using two primary methods: static headspace (SHS) and dynamic headspace (DHS). Both HS sampling methods are solvent-free techniques where the sample is placed in a vial with a space above it, known as the HS.

This section focuses on the sampling techniques for HS analysis, regardless of the type of sample. It provides a general overview and theoretical discussion. In addition to SHS and DHS, the chapter highlights the importance of high-concentration capacity HS (HCC-HS) methods, introduced in the early 1990s, which have largely replaced traditional SHS sampling. As defined by Bicchi et al. [6], HCC-HS serves as a bridge between SHS and DHS, offering the simplicity, speed, automation, and reliability of SHS while also allowing a higher concentration factor, similar to DHS. Since the principles of HCC-HS are similar to those of SHS, but with an extended partition between the HS and the sorbent/polymeric coating, these will be discussed together. A separate section will cover DHS.

2.2.1 Static and High-capacity Headspace Analysis

SHS is a well-established technique, with its introduction dating back to 1939, when it was used as an aerometric method for alcohol determination in water and body fluids, developed by Harger et al. [7]. It was not until 1993 that Zhang and Pawliszyn [7] utilized an enrichment device (known as solid-phase microextraction or SPME), previously employed for extracting organic pollutants from water [8], for trapping the volatiles from the HS. A similar development followed with another enriching device, called stir-bar sorptive extraction (SBSE), which was first used for sampling pollutants in water and subsequently extended to the sampling of HS [8, 9].

The introduction of high-capacity tools has gradually replaced the conventional SHS, thanks to their significant increase in sensitivity and ease of use, particularly for SPME, which can also be fully automated.

Despite the differences in performance, SHS and HCC-HS share the same theory and the practical procedures. A sealed vial is filled with the sample and allowed to rest for a given time at a constant temperature (equilibration time) to enrich the HS with the volatile, according to the equilibrium constant (distribution constant, partition coefficient, and Henry's constant) [10]. A static condition is maintained between the liquid (or solid) and gas phases. The sampling, using a gas-tight syringe or an HCC-HS device, can be carried out at equilibrium or not. In the latter case, the device is exposed to the HS for a certain time (extraction time) before being thermally desorbed into the GC injector.

The analytes are distributed between the gas phase and the condensed phase (sample), producing a partial pressure. The relationship is regulated by the partition coefficient K_{hs}, which is given by Eq. (2.1):

$$K_{hs} = \frac{C_h}{C_s}, \tag{2.1}$$

where C_h is the concentration of the analyte in the HS, and C_s is the concentration of the analyte in the sample. For a solid sample, the adsorption constant (K^A), which characterizes the adsorption process at a constant temperature on a gas–solid interface, should replace the partition constant. Nevertheless, the partition between the sample and the HS can be studied in the same way, regardless of whether the sample is liquid or solid.

Considering C_0 as the initial concentration of the analyte in the sample, Eq. (2.2) describes the mass balance at the equilibrium as follows:

$$C_0V_s = C_hV_h + C_sV_s, \tag{2.2}$$

where C_0 is the initial concentration of the analyte in the sample, C_s and V_s are the concentration and the volume of the sample, and C_h and V_h are the concentration and volume of the HS.

Following a series of rearrangements and substitutions, detailed elsewhere [10, 11], the following relation can be obtained:

$$C_h = \frac{K_{hs}C_0V_s}{K_{hs}V_h + V_s}. \tag{2.3}$$

Therefore, the quantity injected into the GC system is proportional to the initial concentration of the analytes in the sample (C_0). The quantity of analyte effectively in the HS is given by $q_h = C_hV_h$, but, in the case of traditional SHS, the amount transferred to the GC is only a fraction. Like any equilibrium constant, K_{hs} is temperature-dependent, following an exponential relationship.

In the case of HCC-HS techniques, the mass balance equation includes a second equilibrium between the HS and the polymeric coating(s):

$$C_0 V_s = C_h V_h + C_s V_s + C_f V_f, \tag{2.4}$$

where C_f and V_f are the concentration and the volume of the polymeric coating (indicated as f, which stands for the SPME fiber, but valid for any polymeric coating, to distinguish it from the s of the sample).

The distribution constant between the HS and the polymeric coating is defined as a ratio of concentrations or as a product of distribution constants, assuming the effect of moisture is negligible:

$$K_{fs} = \frac{C_f}{C_s} = K_{fh} K_{hs}. \tag{2.5}$$

Moreover, the quantity of analytes trapped in the polymeric coating ($n = C_f V_f$) can be extrapolated as follows:

$$n = \frac{K_{fh} K_{hs} C_0 V_s V_f}{K_{fh} K_{hs} V_f + K_{hs} V_h + V_s} = \frac{K_{fs} C_0 V_s V_f}{K_{fs} V_f + K_{hs} V_h + V_s}. \tag{2.6}$$

This means that the amount of analyte extracted is independent of the location of the polymeric coating, whether it is in HS or immersion mode, as long as V_f, V_h, and V_s are kept constant.

This proportionality can also be expressed as follows:

$$A = \frac{C_0}{K_{hs} + \beta}, \tag{2.7}$$

where A is the chromatographic area, and β is the phase ratio ($\beta = V_h/V_s$) [10].

To maintain direct proportionality between chromatographic area (A) and C_0, the method must operate within the HS linear regime; this condition must be experimentally confirmed, as otherwise, HS saturation may occur [10].

Although the theory is based on thermodynamic equilibrium, sampling is rarely performed at equilibrium. This is particularly true when dealing with complex odor patterns due to the long time required to reach a satisfactory equilibrium for most of the compounds. Despite a possible payback in sensitivity, nonequilibrium sampling provides equally reliable results, although the conditions need to be highly standardized [10, 12–14].

The optimization of HS extraction involves both thermodynamic and kinetic parameters, as well as the preliminary selection of an appropriate extraction device, particularly in the case of polymeric coatings for HCC-HS. Considering the latter, many configurations are now available, as schematized in Figure 2.1. The higher the surface-area-to-volume ratio, the shorter the extraction time and the higher the amount of analyte extracted [16–18].

The chemistry of the coating influences both the selectivity and the extraction sensitivity. A preliminary distinction is between liquid and solid coatings.

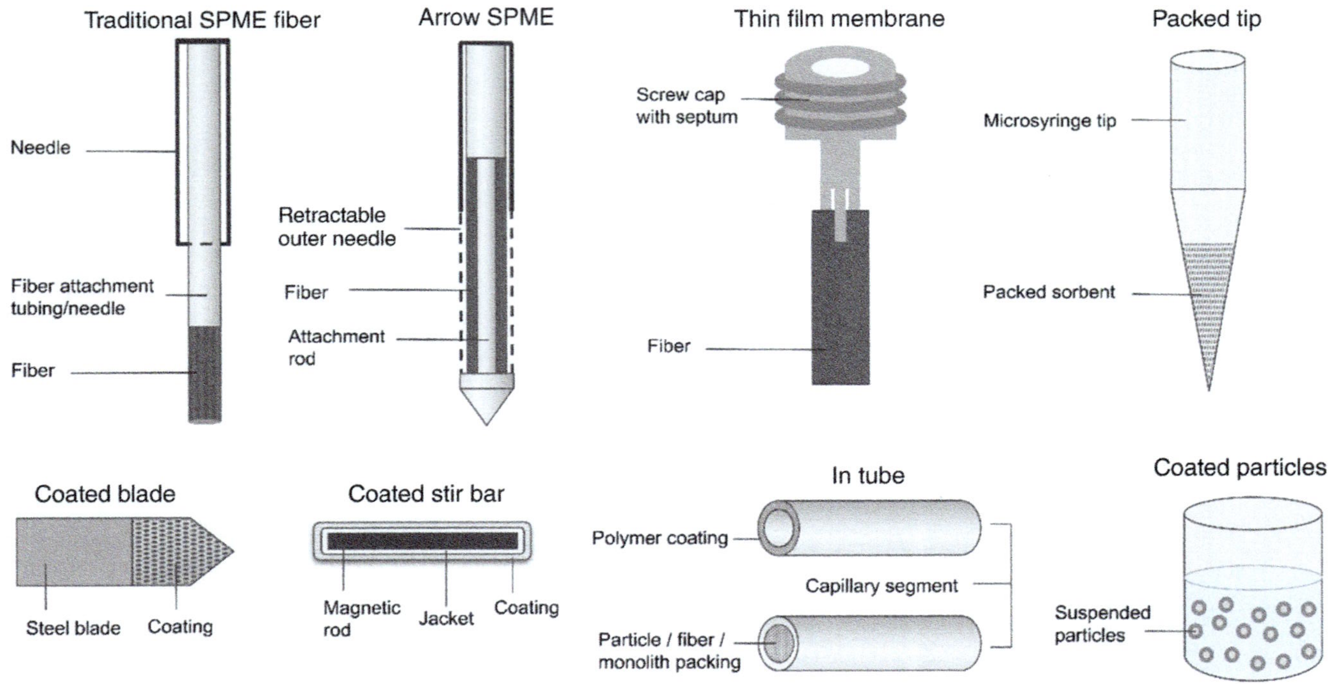

Figure 2.1 Schematic of the main HCC device. Figures not in scale. *Source:* Billiard et al. [15] / MDPI / CC BY 4.0.

Liquid coatings, such as polydimethylsiloxane (PDMS), rely on an absorption mechanism, while solid coatings, made of mixed polymers, like divinylbenzene/carboxen/polydimethylsiloxane (DVB/CAR/PDMS), exhibit an adsorption mechanism, which means that extraction occurs only at the surface of the coating. The adsorption mechanism enables a faster and higher trapping capacity; however, the limited surface area leads to competition for the adsorption site, which can potentially cause a displacement effect [11].

In terms of selectivity, the general rule is "like dissolves like." Apolar coatings, such as PDMS, are preferentially used to extract apolar compounds, while polar coatings, such as polyacrylate (PA), are used for polar compounds. Despite several coatings being available, the DVB/CAR/PDMS coating is by far the most commonly used as it guarantees a broad coverage of compounds characterized by a wide range of physicochemical properties. These characteristics make the DVB/CAR/PDMS coating particularly suitable for nontargeted studies, as often performed using multidimensional chromatographic techniques [11, 19].

In terms of device design, SPME is by far the most widely employed method, although other configurations, such as the SPME arrow or thin-film extraction (TFE), are gaining popularity due to the increased coating volume/surface, which leads to higher sensitivity.

Once the extraction device is chosen, the temperature is the first thermodynamic parameter to optimize, as it directly impacts the distribution constant. Higher temperatures increase Henry's law constant (or other distribution constants, such as the octanol–air distribution constant), facilitating the release of compounds from the matrix and the volatilization of semivolatile compounds [10].

The optimization of the extraction temperature should be based on a deep knowledge of the matrix and consideration of the experimental question. For instance, in the exploration of biological conditions, a 37 °C extraction temperature may be more informative than maximizing the extraction using higher temperatures. Similarly, enzymatic and chemical reactions can be triggered by specific temperatures, altering the volatile profile and creating bias [20, 21].

The same attention must be paid when matrix modifications are considered to increase the extraction yield, such as the addition of salt and pH adjustment. This modification can facilitate the transfer of specific substances to the HS but can also serve to normalize the variability of the sample [22–24].

The presence of macromolecules in the sample also impacts the extractability of molecules due to interactions at the interface [25, 26]. Therefore, the optimized conditions for one matrix may not be extended to others due to the complex interactions involved.

The sample volume and the HS/sample ratio are additional important parameters. The use of excessively high sample volumes, although it may seem beneficial, tends to reduce the extraction of less volatile compounds. Furthermore, if the HS is saturated, the correlation between the chromatographic signal and the analyte concentration in the sample can be compromised, thereby affecting the reliability of the results [27, 28]. Mascrez et al. [28, 29] investigated the effect of saturated HS on cross-sample comparisons by performing multiple-cumulative trapping SPME. The reader is directed toward these references for further details.

In addition to these thermodynamic considerations, the kinetics of extraction play a fundamental role. In practice, this reflects the time required for the analyte to migrate from the sample to the fiber, i.e., the mass transfer process. This process consists of three phases: the sample, the HS, and the fiber (with two interfaces: sample/HS and HS/fiber). The behaviour at the interface can be represented by the two-film model reported in Figure 2.2.

According to the combination of analyte (diffusion coefficient) and sample (viscosity), the main resistance may lie in the liquid-film layer, the gas-film layer, or a combination of both [13]. The mass transfer within the sample itself is typically enhanced through agitation, thereby reducing the liquid-film boundary layer and facilitating the transport of species to the interface. The most volatile

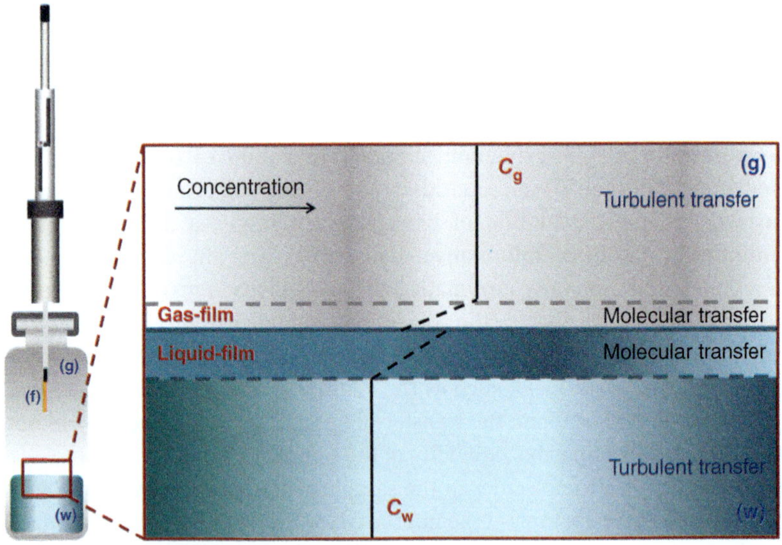

Figure 2.2 Representation of the HS-SPME experimental setup and the two-film model of the gas–liquid interface, where (w), (g), and (f) are the water, gas, and SPME fiber phases; C_g and C_w are the concentrations in the bulk gas and water phases, respectively. *Source:* Psillakis [13] / with permission of Elsevier.

substances rapidly traverse the gas-film layer. Increased temperatures enhance the diffusion coefficient in both phases, thereby reducing viscosity and increasing diffusion in the liquid phase, which facilitates mass transfer to the gas phase and through the HS.

An important additional parameter is the pressure in the HS, which impacts the pre-equilibrium mass transfer, reducing (in the case of subatmospheric pressure) the resistance of the gas-film layer, particularly for molecules for which the evaporation rate is limited by the resistance in this thin layer. The exploration of reduced-pressure conditions started only in 2001 [30]. Another attempt to benefit from the effect of vacuum conditions was made in 2005 [31]. However, the start of a systematic investigation and use of reduced-pressure conditions for HS-SPME (referred to as Vac-HS-SPME) began in 2012 mainly due to the study of Psillakis et al. [13, 32–39].

The ultimate benefit of Vac-HS-SPME is the accelerated extraction of semivolatile compounds, thus obtaining optimized conditions at lower temperatures and shorter extraction times. The extraction kinetics of the most volatile compounds is mainly affected by the liquid-phase resistance, therefore, they are not affected by the pressure conditions. The highest benefit is achieved when Henry's constant locates the evaporation resistance in the gas film. Under these conditions, vacuum reduces that resistance and enables milder conditions, avoiding sample compositional changes due to thermal degradation [13]. A prediction model was established for HS-SPME sampling of water samples based on Henry's law constant, as shown in Figure 2.3.

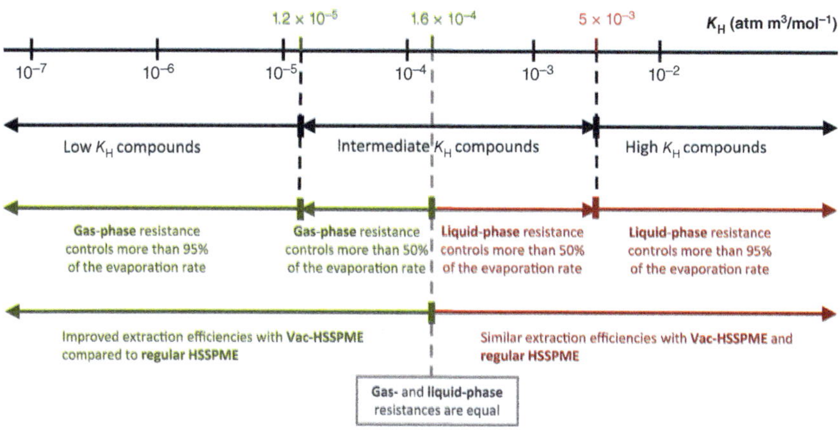

Figure 2.3 Ranges of Henry's law constant associated with the location of the resistance on evaporation rates and the predicted effect of vacuum on HS-SPME sampling of water and water-containing samples. *Source:* Psillakis [13] / with permission of Elsevier.

2.2.2 Dynamic Headspace Sampling

As suggested by the name, DHS implies a dynamic process that can ensure exhaustive extraction [40]. Mixing/stirring, as applied in SHS, although it involves dynamic movements, only affects equilibration speed, not the extraction thermodynamics. The DHS sampling method is also based on adsorptive enrichment and can be performed in either a passive or active sampling mode. In general, both ways can use the same adsorbent materials and the same desorption techniques.

Different from passive sampling, regulated by Fick's first law of diffusion, active sampling is described by a defined air volume forced to pass through a packed tube with one or more sorbents in series. Sampling time can range from a few minutes (active mode) to several weeks (passive mode) [41, 42]. The choice between active/passive sampling depends on the sampling throughput, the required sensitivity, the volatility of the target compounds, and the complexity of the sample matrix.

Such a dynamic flow resembles the chromatographic process, with the stationary phase (i.e., the extraction phase) and the moving mobile phase (i.e., the sample or its vapors). In most applications, the sorbent is typically a porous solid material in which adsorption mechanisms occur with the analytes. Sorbent particles can be placed in either a disk or a tube. The former offers a larger surface area for a higher uptake rate, while the latter (shown in Figure 2.4) can be easily automated with subsequent GC analysis [43].

Different types of adsorbents are available, and their appropriate selection ensures a reliable, robust, and accurate determination of volatile organic compounds (VOCs) [41]. They can be classified based on the type of material, such as inorganic, carbon-based (activated carbons and carbon molecular sieves, graphitized carbon blacks), and porous organic polymers.

The inorganic adsorbents include silica gel, zeolites, and alumina; however, they are highly hydrophilic, which limits their application. Finally, porous organic polymer sorbents are highly pure materials, but generally exhibit limited temperature stability, except for poly(2,6-diphenylphenylene oxide) (Tenax™) and PDMS, which are among the most commonly used sorbents.

Tenax is the most important and widely used polymer material due to its hydrophobic nature and high thermal stability. Due to its low specific surface

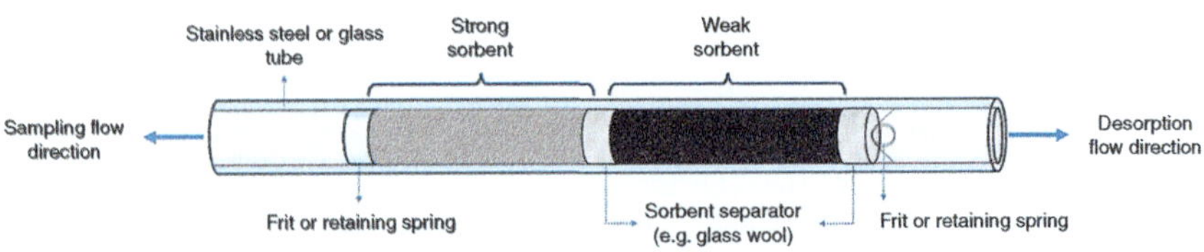

Figure 2.4 Illustration of a dual-bed thermal desorption tube.

area, it is not suitable for sampling highly volatile organics (<4 carbon number). Although Tenax is an excellent adsorbent material, care must be taken in interpreting the data due to some artifacts it tends to form, specifically benzaldehyde, acetophenone, and higher aldehydes (octanal, nonanal, decanal). PDMS is an inert and hydrophobic polymer sorbent that ensures absorption mechanisms, similar to those of chromatography stationary phases [44]. As the energy required in partition processes is lower than that required for adsorption, the analytes can be desorbed at mild temperatures, thus minimizing thermally induced decomposition.

Despite the variety of commercially available sorbents, universal material does not exist [45]. This is the reason why, often, different types of sorbents are packed together to obtain a so-called multibed tube, as illustrated in Figure 2.4. The materials are packed in the direction of increasing adsorbent strength, following the sampling flow. The same result can be obtained by coupling different tubes in series, although this option is far less common. Indeed, it would require multiple analyses for a comprehensive overview of the volatile compounds trapped in the different tubes.

As for the HCC-HS tools, the analytes of interest retained in the sorbent must be desorbed, either by a solvent or a thermal effect. Solvent desorption does not require any dedicated apparatus, but if the solvent is not carefully selected, it may cause degradation of the sorbent or contribute to artifact formation. Moreover, analyte dilution occurs, impacting method sensitivity.

Thermal desorption is generally preferred, especially when using tube geometry, which can be coupled in-line with a GC system. Thermal desorption is solvent-free, minimizing analyte losses and cross-contamination, and maximizing sensitivity, as all trapped analytes can be transferred into the GC column for separation and final determination. Tubes are generally desorbed in back-flush mode, which is particularly important for multibed tubes to guarantee an efficient release of the adsorbed compounds. A high flow should be used to compensate for the slowness of the desorption process, but this is not always compatible with capillary columns or MS coupling. Therefore, in general, a focusing step is used to significantly improve GC separation by refocusing the initial band [46]. Different options can be used to achieve this (e.g., cryotrapping, micro-traps, and 2-step thermal desorption) [43].

Technically, VOCs trapped in a thermal desorption tube can be analyzed only once. In situations where technical replicates or multiple instrument analyses are needed, however, modern thermal desorption systems allow the recollection of the analytes from the GC split flow that would otherwise be lost.

The risk of incomplete desorption exists in case of strongly adsorbed analytes, such as those encountered with activated carbon during the extraction of polar

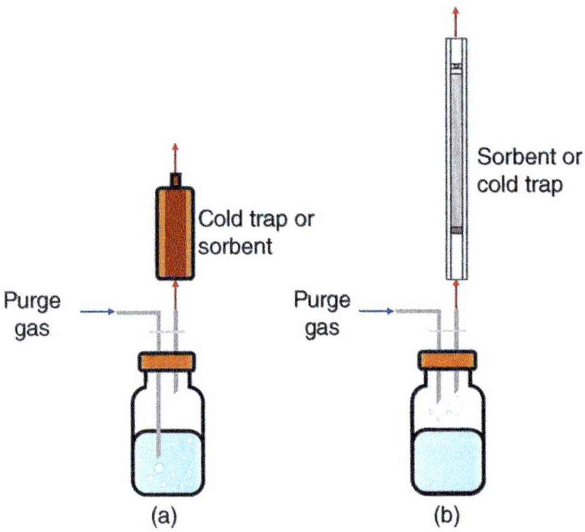

Figure 2.5 Graphical representation of purge-and-trap extraction in (a) and dynamic headspace extraction (b).

compounds, thus proper evaluation and optimization of desorption conditions must be carefully considered.

Another important definition for active sampling is the breakthrough volume or time, which determines the maximum volume of sample that can pass through the trapping device before analytes are no longer fully retained. The specific retention volume depends on the trapping temperature, the presence of other compounds that can compete with the analytes of interest (displacement effect), and the sample humidity, particularly in the case of an adsorbent that is remarkably retentive for water [11]. Last but not least, the sampling flow plays an important role, as it must allow sufficient time for interactions between the analyte and the sorbent surface.

As aforementioned, the tube format is the most common one, as it is more easily coupled online and automated with GC. Based on the setup, purge-and-trap (P&T) or the DHS can be performed (Figure 2.5).

In P&T, an inert gas is bubbled through the liquid sample, favoring the volatilization of volatile and semivolatile molecules, which are subsequently concentrated onto a sorbent or cold trap, as shown in Figure 2.5b. When the purge process is limited to the HS, the approach is referred to as DHS (Figure 2.5b). Also here, the analytes are trapped in a thermal desorption tube.

2.3 Analysis of Semivolatiles Analytes

Semivolatile molecules, often referred to as semivolatile organic compounds (sVOCs), are a broad class of organic compounds characterized by boiling points generally higher than that of water (typically >240 °C) and correspondingly low vapor pressures (ranging from approximately 10^{-14} to 10^{-4} atm). They are less volatile than VOCs but can still readily vaporize from liquid or solid forms at ambient temperatures, allowing for their presence in both gaseous and condensed phases (e.g., adsorbed on particles or surfaces). Examples of sVOCs include polycyclic aromatic hydrocarbons (PAHs), phthalates, pesticides, flame retardants, and polychlorinated biphenyls (PCBs), many of which pose environmental and toxicological concerns.

In the context of biological samples, many important metabolites fall into the semivolatile category due to their polarity and often higher molecular weights, necessitating derivatization to enhance their volatility for GC analysis. Examples of such semivolatile metabolites include amino- and organic acids, sugars (monosaccharides and disaccharides), fatty acids, sterols, and biogenic amines.

A diverse array of sample preparation approaches is employed for these semivolatile molecules, based on the nature of the matrix, the physicochemical properties of the analytes, and the required sensitivity. Based on the physical state of the extractant phase, these can be grouped in liquid-phase extractions (LPE) and solid-phase extractions (SPE). Within LPE, when the sample is a liquid or a solid that can be dissolved in one, liquid–liquid extraction (LLE) is widely utilized, exploiting the differential partitioning of analytes between two immiscible liquid phases [47]. This method is particularly effective for extracting analytes from aqueous samples into an organic solvent, offering good recovery and enrichment for a broad range of semivolatiles.

Classical SPE in cartridge offers a versatile and often more selective alternative to LLE. This utilizes a solid sorbent material to selectively retain analytes from a liquid sample, followed by elution with a small volume of solvent [48]. This technique allows for significant matrix clean-up and analyte concentration and can be tailored to specific analytes by selecting appropriate sorbent chemistries (e.g., reversed-phase, normal-phase, and ion-exchange).

Finally, chemical transformation (derivatization) is often an indispensable step for semivolatile molecules that lack sufficient volatility or thermal stability for direct GC analysis, or for those that do not produce a strong signal with common GC detectors [49]. This process involves chemically modifying the analyte to introduce functional groups that enhance volatility, improve thermal stability, or increase detector response. Common derivatization reactions include silylation, acylation, and alkylation, which transform polar functionalities (e.g., hydroxyl, carboxyl, and amino groups) into more volatile derivatives amenable to GC separation and detection.

The wise selection and optimization of these sample preparation approaches are paramount to achieving reliable and robust analytical results.

It is worth noting that sorbent-based techniques, which belong to the group of SPEs, and which are typically employed for HS sampling of volatile molecules, also enable the access to less volatile molecules when used in direct immersion mode. This mode may require additional steps for interferent removal (e.g., sorbent rinsing for nonvolatile molecule elimination) or a derivatization step directly in the sorbent phase, performed before or during GC injection. Examples of on-fiber SPME derivatization and derivatization in-syringe are more frequent in environmental and food applications. Such hybrid approaches are not discussed in this chapter; instead, the reader is directed to the literature [50–52].

The following sections describe and discuss the most established approaches for preparing semivolatile molecules for GC analysis.

2.3.1 Liquid-phase-based Sampling Techniques

Conventional LLE, also known as solvent extraction, is a foundational technique in chemical separations [47]. It operates on the principle of differential partitioning, where analytes distribute themselves between the sample matrix (the donor phase) and an immiscible extracting solvent (the acceptor phase). This direct contact allows for excellent penetration into the sample, providing comprehensive access to its components.

Its further advancement and use have been hindered by the apparent incompatibility with the recent principles of green chemistry, by the limited possibilities for automation, and the growing and flourishing "competition" of modern SPE techniques.

The push toward miniaturization and environmentally conscious methodologies has spurred the evolution of LLE into liquid-phase microextraction (LPME). This modern approach revitalizes liquid-based methods, aligning them with green chemistry by drastically reducing solvent volumes, often to the microliter scale.

Finally, it is noteworthy that the extraction yield and throughput of these techniques can be further enhanced by applying external energy sources, such as ultrasound or microwaves [53, 54]. These approaches are discussed in detail in the upcoming subsections.

2.3.1.1 Microextractions

LPME is a miniaturized form of sample preparation that operates on the same principles as traditional LLE but with a drastic reduction in solvent volume.

LPME techniques are defined by their small-scale operation and suitability for automation. A key characteristic is the use of an acceptor phase volume that is significantly smaller than the donor phase volume. Consequently, even though the extraction of analytes may be nonexhaustive, this high phase ratio ($V_{sample} \gg V_{extractant}$) results in a very high concentration of the target analytes in the final extract, achieving excellent enrichment factors [55–57]. In LPME, extraction typically occurs from a small amount of a water-immiscible solvent (acceptor phase) into an aqueous sample containing analytes (donor phase).

In LPME, analytes are extracted from a large volume of an aqueous sample (the donor phase) into a tiny volume (typically 1–10 μL) of an immiscible organic solvent (the acceptor phase). This results in a high phase ratio, enabling excellent analyte enrichment even when equilibrium is not achieved. This approach aligns with the principles of green chemistry by minimizing waste and exposure to hazardous solvents [58, 59].

LPME techniques are exceptionally well-suited for preparing samples for GC analysis for several interconnected reasons. Perhaps the most significant advantage is the direct injection; the final micro-extract, which allows for a significant enrichment of analytes, is directly injectable into the GC. This maximizes sensitivity and crucially eliminates any solvent evaporation step, which could cause the loss of volatile analytes. Another key aspect is the careful selection of the extraction solvent. Ideal solvents for this purpose, such as isooctane or undecane, have low volatility at room temperature to prevent evaporation during the extraction but are readily vaporized in the hot GC inlet without interfering with semivolatile analyte peaks. Finally, LPME can be seamlessly coupled with in-situ derivatization. By adding a derivatizing agent to the extraction solvent, polar or nonvolatile analytes can be extracted and chemically modified into GC-amenable derivatives within a single, efficient step [60].

While many variations exist, the technique can be divided into three main categories: single-drop microextraction (SDME), dispersive liquid–liquid microextraction (DLLME), and hollow-fiber LPME (HF-LPME) [61–63].

SDME represents the simplest mode. In this setup, a single microdrop of solvent is suspended directly from a microsyringe needle into the stirred sample. After a predetermined time, the drop, now enriched with analytes, is retracted into the syringe for injection. Its main limitation, however, is the physical instability of the exposed drop [55, 56].

A much faster alternative is DLLME. This method achieves highly efficient extraction by rapidly injecting a mixture of the extraction solvent and a dispersing solvent (such as acetone) into the sample. This action creates a cloudy emulsion of fine droplets, maximizing the surface area for rapid analyte transfer. The phases must then be separated via centrifugation before GC analysis [61, 64].

For enhanced stability and robustness, HF-LPME is often the preferred method. This configuration employs a short, porous polypropylene hollow fiber to contain the acceptor solvent. When submerged in the sample, analytes migrate through the pores into the protected solvent inside. This design effectively prevents solvent loss and avoids the emulsion issues that can sometimes be encountered in DLLME [65, 66].

Each of these modalities serves the same fundamental purpose: to cleanly transfer analytes from a complex aqueous matrix into a concentrated, GC-compatible organic solvent micro-extract.

2.3.1.2 Assisted Techniques

To significantly enhance extraction efficiency beyond conventional methods, accessory physical inputs such as ultrasound, microwaves, or pressure can be applied. These assisted extraction techniques accelerate the transfer of analytes from the sample matrix into the solvent, enhancing the ability of the solvent to penetrate the matrix, thus resulting in substantial reductions in processing time, energy consumption, and solvent volumes, thereby promoting more environmentally friendly, or "green," sample preparation workflows.

Microwave-assisted extraction (MAE) utilizes the heating capabilities of microwave energy to efficiently and rapidly extract compounds. By directly targeting polar molecules within the sample and solvent, microwaves induce molecular rotation and ionic conduction, causing rapid and localized heating that disrupts the sample matrix and enhances the migration rate of target compounds into the extraction phase. This process is significantly more efficient than conventional heating methods, such as conduction and convection [54, 67].

In parallel, ultrasound-assisted extraction (UAE) utilizes high-frequency sound waves to create acoustic cavitation within the solvent. The formation and collapse of microscopic bubbles generate intense localized pressure and temperature gradients, producing powerful shockwaves and micro-jetting effects. This energy violently agitates the sample, fragments the matrix surface, and increases the contact area between the sample and solvent, thereby accelerating mass transfer [53, 68].

Importantly, the benefits of ultrasound and microwaves are not limited to large-scale, conventional procedures, such as LLE. They can also be integrated with the miniaturized liquid-based techniques discussed previously, further boosting the performance of methods like LPME. While these assisted extractions have found widespread use in food and environmental analysis, their application to complex biological samples is a growing area of research, promising to improve throughput and efficiency in clinical and biomedical studies [69, 70].

Pressurized liquid extraction (PLE), also known as accelerated solvent extraction (ASE), is a highly efficient technique for extracting analytes from solid or semisolid sample matrices. The method's efficacy stems from the use of conventional liquid solvents at elevated temperatures and pressures, conditions that significantly enhance the extraction process [71, 72].

Thanks to their enhanced matrix penetration, these approaches are particularly suitable for solid samples and find application in most environmental applications, especially for the target analysis of contaminants (e.g., PCBs, polycyclic aromatic hydrocarbons, and pesticides) [73, 74].

2.3.2 Solid-phase-based Sampling Techniques

SPE is a versatile sample preparation technique based on the principles of chromatography. The method isolates compounds of interest from a liquid or gaseous sample by their interaction with a sorbent; thus, it is particularly suitable and used for target analysis in environmental and food applications [75].

During the process, the sample is passed through the sorbent bed, and analytes are retained on its surface through specific physicochemical interactions, such as hydrophobic, polar, or ion-exchange forces. After the sample is loaded, the sorbent is washed to remove interfering components from the matrix. Finally, the purified analytes are recovered either by elution with a small volume of an appropriate solvent [75].

The downscaling and re-optimization of column properties, specifically for accelerating extraction, quickly gained acceptance and use as a direct alternative to classical LLE. Among the many variants of extraction based on solid sorbents, the advances in miniaturization (μSPE) are significant. μSPE represents a broader category of miniaturized SPE where the sorbent is contained within a small-scale device. This approach often combines the advantages of SPE with the form factor of membrane-based microextractions. For example, sorbent particles can be embedded within a porous membrane sheet, packed into a pipette tip, or contained within a small, permeable bag. This configuration protects the sorbent while allowing it to be immersed directly in the sample, facilitating efficient analyte capture [76–79].

In another configuration, microextraction by packed sorbent (MEPS), the sorbent is embedded directly into the syringe needle. Typically, a small plug containing just a few milligrams of sorbent is packed into the syringe needle. The extraction operates on a “draw-eject” principle, where the sample is repeatedly passed back and forth through the sorbent bed. This cyclic action ensures multiple contacts between the analytes and the sorbent, enhancing extraction kinetics and efficiency. After a wash step, the retained compounds are eluted with a minimal volume of solvent. A key advantage of this syringe-based format is that the eluate can be injected directly into the gas chromatograph, streamlining the entire analytical workflow [80, 81].

The previously listed HCC techniques in Section 2.1 belong to the category of SPEs. As described, when used in the HS, they are amenable to sampling volatile molecules. When used in direct immersion/contact with the sample, the sorbent has access to and can extract less volatile molecules.

2.3.3 Chemical Transformations (Derivatization)

Derivatization is a crucial sample preparation step that chemically transforms analytes to make them suitable for GC analysis [82]. Many compounds, particularly those containing polar functional groups (e.g., —OH, —COOH, —NH_2), are inherently nonvolatile and may interact poorly with the GC system. Derivatization resolves this by converting these active groups into more volatile and thermally stable analogues.

This process serves multiple purposes beyond simply enabling analysis. It can significantly enhance chromatographic separation and improve detection sensitivity, resulting in better peak shapes and lower detection limits. Furthermore, it can be used strategically for sample clean-up or to fractionate complex mixtures. By fundamentally altering the polarity of an analyte, derivatization modifies its interaction with the GC column's stationary phase, thereby optimizing its behavior during separation.

For a derivatization strategy to be analytically valid, two conditions are paramount: the reaction must proceed to completion with a high, reproducible yield, and the resulting derivative products must be stable under analytical conditions. A vast number of derivatization reactions are based on a common principle: the replacement of active hydrogen atoms on polar functional groups (such as —OH, —COOH, —NH_2, and —SH).

Derivatization procedures are applied across various fields to enable or enhance the analysis of specific analytes or a broad range of chemical families. Examples of the most common analytes or classes are fatty acids, chlorophenoxy acid herbicides, steroids, sterols, illicit drugs, organic acids, alcohols, and polymers [83–85]. Chapter 8.1 provides a description of derivatization procedures and examples for biological samples.

Among the most widely employed and robust derivatization chemistries are silylation, acylation, and alkylation. Silylation, particularly using reagents to form trimethylsilyl (TMS) ethers and esters, is extremely common due to its ability to cap hydroxyl and carboxyl groups effectively. Acylation and alkylation reactions serve similar purposes, converting polar functional groups into less polar, more volatile esters, ethers, or amides, thereby ensuring their suitability for GC analysis [83–85].

Depending on the specific scientific objective, derivatization can be carried out at various stages of the analytical workflow (e.g., pre- or post-extraction) and may be applied either to a selected fraction of the sample or to the entire matrix. In a GC-based application, it is always a precolumn process.

Alkylation is a derivatization reaction in which an active hydrogen atom present in polar functional groups, such as hydroxyl (—OH), thiol (—SH), and amino (—NH), is replaced by an alkyl group. This transformation yields derivatives such as (thio/amino) ethers or (thio/amino) esters, depending on the original functional group. The reaction is particularly efficient for amines and carboxylic acids, the latter often being targeted in their free or esterified form [84].

Alkylation is widely employed in profiling studies for metabolomics and food analysis, involving the conversion of free and/or esterified fatty acids (FAs) into their corresponding alkyl esters (AEs). This modification enhances volatility, thermal stability, and chromatographic behavior in GC-based analyses [84, 86].

Although fatty acid methyl esters (FAMEs) are not the optimal derivatives for mass spectrometric structural elucidation, their simplicity and ubiquitous use in general fatty acid analysis have made them a mainstay in GC-MS. The primary limitation of FAMEs lies in the analysis of mono- and di-unsaturated fatty acids, as their electron ionization mass spectra provide little information regarding the location of double bonds. However, such information would be very useful when more complex structures are analyzed. The mass spectra of polyunsaturated fatty acids (PUFAs) and those containing other functional groups, such as hydroxyls or branch points, often yield diagnostic fragmentation patterns. The confident identification of these compounds, however, is greatly enhanced when mass spectra from authentic standards are available for direct comparison [87, 88].

The electron ionization mass spectrum often reveals the molecular ion that, even if with low intensity, provides the invaluable information of the fatty acid's molecular weight. When this information is combined with its corresponding GC retention data (i.e., retention indices), a highly confident peak identification can be achieved. Furthermore, the fragmentation patterns within the mass spectrum can typically be interpreted to deduce the positions of structural features, such as methyl branches and oxygenated functional groups [89, 90].

Also, for alkylation, there does not exist a single reagent that is suitable for every purpose; one must be chosen that best fits the circumstances.

The two main procedures are acid- or base-catalyzed esterification/transesterification. Besides the pH, the most important outcome is the derivatization of FAs present in free forms under the acidic conditions, not achieved under basic conditions. Other practical differences regard timing, temperature, and stability: acid-catalyzed reactions require longer reaction times and higher temperatures compared to basic-catalyzed ones, though the latter are more sensitive to moisture [91, 92].

Among the various derivatization reagents for FAAEs preparation, trimethylsulfonium hydroxide (TMSH) has been widely used in clinical research on different biological samples, such as muscular biopsies, human plasma, and bacterial strains [93]. The TMSH methodology can be directly applied to the bio-sample and/or after isolation of the target analytes (i.e., extraction of phospholipids).

The method may be of value for analyzing nonesterified fatty acids in aqueous solutions, as it eliminates the extraction step that can lead to selective losses of specific components. Due to its simplicity, this approach can be useful for large-batch analysis; however, the limitation of the TMSH method is its insufficient derivatization efficiency for PUFAs [92].

Another popular approach for alkylation in clinical samples involves the use of alkyl chloroformates (e.g., methyl chloroformate), as recently applied for the GC-MS determination of sterols and tocopherols in human serum and amniotic fluid [94], as well as for the GC-MS investigation of more than 150 possible diagnostic amino-carboxylic metabolites in human urine [95]. Such reaction involves the modification of primary and secondary amino groups, as well as carboxyl groups. In this way, amino and nonamino organic acids can be converted into volatile esters and carbamates, and can then be analyzed by GC-MS.

Although this derivatization approach is limited to analytes containing amino and/or carboxyl groups, these represent the majority of intermediates in central carbon metabolism, which are important cell metabolites. In other words, this approach can be useful for amino acid profiling in biological samples. However, chloroformate derivatization is not as well known as other derivatization techniques, and the resulting mass spectra of the derivative forms are still limited in the common databases to support tentative identification [96].

In the acylation process, an acyl group is introduced to an organic, converting an active —H into an ester, a thioester, or an amide. If the organic compound is a carboxylic acid, there will be a loss of the hydroxyl group. Acylation, more suitable for the derivatization of amines, hydroxyls, and thiols, is a derivatization process that prefers an anhydrous environment. It mainly occurs after an extraction procedure to isolate the target analytes. It has been successfully applied to the GC-MS determination of amino acids and their metabolites in various biological samples, including urine, sputum, and blood plasma [97].

Silylation is the most commonly applied derivatization method in metabolomic studies due to its comprehensive coverage of compound classes and relative ease of use [98]. Indeed, the wide choice of silyl reagents available makes silylation suitable for most functional groups containing active hydrogen atoms. To avoid hydrolysis of reagent/derivatized products, care must be taken during this process to ensure that both samples and solvents are in an anhydrous environment [85, 99].

According to the chemical nature of the target analytes, a two-step derivatization process combining consecutive multiple reagents (e.g., methoxylation followed by silylation) is also possible and automatable [100].

The derivatization of metabolite extracts can potentially require lengthy incubation times before injection, which may lead to hydrolysis of the derivatives. Thus, automating the entire process, along with the use of dedicated analytical tools to enhance throughput, is highly desirable.

Moreover, all the derivatization processes can benefit from additional energy sources (e.g., microwaves, ultrasounds), as discussed previously. For example, the use of microwave heating (microwave-assisted derivatization) in combination with common derivatization methods (alkylation, acylation, and silylation) can drastically reduce derivatization time [67, 101].

Additionally, thanks to the advancement of specialized scientific microwave instruments with high-throughput rotors capable of housing GC-ready vials, it is possible to undertake sample derivatization of multiple samples for large-scale omics research.

Finally, it is essential to note that the appropriate selection of GC parts and regular maintenance are necessary to minimize unwanted reactions and surface adsorption phenomena, which can lead to underestimation or overestimation of the metabolites of interest [102]. For example, the presence of active catalytic sites in the liner might indeed generate distorted peaks and contribute to carryover effects, both of which affect the reliability of analyte detection and quantification.

References

1 Fiehn, O. (2002). Metabolomics–the link between genotypes and phenotypes. *Plant Mol. Biol.* 48 (1–2): 155–171.

2 Stilo, F., Bicchi, C., Reichenbach, S.E., and Cordero, C. (2021). Comprehensive two-dimensional gas chromatography as a boosting technology in food-omic investigations. *J. Sep. Sci.* 44 (8): 1592–1611.

3 Cuadros-Rodríguez, L., Ruiz-Samblás, C., Valverde-Som, L. et al. (2016). Chromatographic fingerprinting: an innovative approach for food "identitation" and food authentication – a tutorial. *Anal. Chim. Acta* 909: 9–23.

4 Stilo, F., Bicchi, C., Jimenez-Carvelo, A.M. et al. (2021). Chromatographic fingerprinting by comprehensive two-dimensional chromatography: fundamentals and tools. *TrAC Trends Anal. Chem.* 134: 116133.

5 Franchina, F.A., Zanella, D., Dubois, L.M., and Focant, J. (2021). The role of sample preparation in multidimensional gas chromatographic separations for non-targeted analysis with the focus on recent biomedical, food, and plant applications. *J. Sep. Sci.* 44 (1): 188–210.

6 Bicchi, C., Cordero, C., and Rubiolo, P. (2004). A survey on high-concentration-capability headspace sampling techniques in the analysis of flavors and fragrances. *J. Chromatogr. Sci.* 42 (8): 402–409.

7 Harger, R.N., Raney, B.B., Bridwell, E.G., and Kitchel, M.F. (1950). The partition ratio of alcohol between air and water, urine and blood; estimation and identification of alcohol in these liquids from analysis of air equilibrated with them. *J. Biol. Chem.* 183 (1): 197–213.

8 Tienpont, B., David, F., Bicchi, C., and Sandra, P. (2000). High capacity headspace sorptive extraction. *J. Microcolumn Sep.* 12 (11): 577–584.

9 Bicchi, C., Cordero, C., Iori, C. et al. (2000). Headspace sorptive extraction (HSSE) in the headspace analysis of aromatic and medicinal plants. *J. High Resolut. Chromatogr.* 23 (9): 539–546.

10 Kolb, B. and Ettre, L.S. (2006). *Static Headspace-Gas Chromatography: Theory and Practice*. Wiley.

11 Pawliszyn, J. (2012). *Handbook of Solid Phase Microextraction*. Elsevier Inc.

12 Ouyang, G. and Pawliszyn, J. (2008). A critical review in calibration methods for solid-phase microextraction. *Anal. Chim. Acta* 627 (2): 184–197.

13 Psillakis, E. (2017). Vacuum-assisted headspace solid-phase microextraction: a tutorial review. *Anal. Chim. Acta* 986: 12–24.

14 Snow, N.H. (2024). From gas to gas: fundamentals of static headspace extraction-gas chromatography. *LCGC Int.* 1: 22–25.

15 Billiard, K.M., Dershem, A.R., and Gionfriddo, E. (2020). Implementing green analytical methodologies using solid-phase microextraction: a review. *Molecules* 25: 5297.

16 Helin, A., Rönkkö, T., Parshintsev, J. et al. (2015). Solid phase microextraction Arrow for the sampling of volatile amines in wastewater and atmosphere. *J. Chromatogr. A* 1426: 56–63.

17 Olcer, Y.A., Tascon, M., Eroglu, A.E., and Boyacı, E. (2019). Thin film microextraction: towards faster and more sensitive microextraction. *TrAC Trends Anal. Chem.* 113: 93–101.

18 Reyes-Garcés, N., Gionfriddo, E., Gómez-Ríos, G.A. et al. (2018). Advances in solid phase microextraction and perspective on future directions. *Anal. Chem.* 90 (1): 302–360.

19 Kataoka, H., Lord, H.L., and Pawliszyn, J. (2000). Applications of solid-phase microextraction in food analysis. *J. Chromatogr. A* 880 (1–2): 35–62.

20 Palermo, C., Mentana, A., Tomaiuolo, M. et al. (2024). Headspace solid-phase microextraction/gas chromatography–mass spectrometry and chemometric approach for the study of volatile profile in X-ray irradiated surface-ripened cheeses. *Foods* 13 (3): 416.

21 Rowan, D.D. (2011). Volatile metabolites. *Meta* 1 (1): 41.

22 Fiorini, D., Pacetti, D., Gabbianelli, R. et al. (2015). A salting out system for improving the efficiency of the headspace solid-phase microextraction of short and medium chain free fatty acids. *J. Chromatogr. A* 1409: 282–287.

23 el Manouni el Hassani, S., Soers, R.J., Berkhout, D.J.C. et al. (2020). Optimized sample preparation for fecal volatile organic compound analysis by gas chromatography–mass spectrometry. *Metabolomics* 16 (10): 112.

24 Mills, G.A. and Walker, V. (2000). Headspace solid-phase microextraction procedures for gas chromatographic analysis of biological fluids and materials. *J. Chromatogr. A* 902 (1): 267–287.

25 Fisk, I.D., Boyer, M., and Linforth, R.S.T. (2012). Impact of protein, lipid and carbohydrate on the headspace delivery of volatile compounds from hydrating powders. *Eur. Food Res. Technol.* 235 (3): 517–525.

26 Alam, M.N. and Pawliszyn, J. (2018). Effect of binding components in complex sample matrices on recovery in direct immersion solid-phase microextraction: friends or foe? *Anal. Chem.* 90 (4): 2430–2433.

27 Stilo, F., Cordero, C., Sgorbini, B. et al. (2019). Highly informative fingerprinting of extra-virgin olive oil volatiles: the role of high concentration-capacity sampling in combination with comprehensive two-dimensional gas chromatography. *Separations* 6 (3): 34.

28 Mascrez, S. and Purcaro, G. (2020). Enhancement of volatile profiling using multiple-cumulative trapping solid-phase microextraction. Consideration on sample volume. *Anal. Chim. Acta* 1122: 89–96.

29 Mascrez, S. and Purcaro, G. (2020). Exploring multiple-cumulative trapping solid-phase microextraction for olive oil aroma profiling. *J. Sep. Sci.* 43 (9–10): 1934–1941.

30 Brunton, N.P., Cronin, D.A., and Monahan, F.J. (2001). The effects of temperature and pressure on the performance of Carboxen/PDMS fibres during solid phase microextraction (SPME) of headspace volatiles from cooked and raw turkey breast. *Flavour Fragr. J.* 16 (4): 294–302.

31 Darrouzès, J., Bueno, M., Pécheyran, C. et al. (2005). New approach of solid-phase microextraction improving the extraction yield of butyl and phenyltin compounds by combining the effects of pressure and type of agitation. *J. Chromatogr. A* 1072 (1): 19–27.

32 Psillakis, E., Yiantzi, E., Sanchez-Prado, L., and Kalogerakis, N. (2012). Vacuum-assisted headspace solid phase microextraction: improved extraction of semivolatiles by non-equilibrium headspace sampling under reduced pressure conditions. *Anal. Chim. Acta* 742: 30–36.

33 Psillakis, E., Mousouraki, A., Yiantzi, E., and Kalogerakis, N. (2012). Effect of Henrys law constant and operating parameters on vacuum-assisted headspace solid phase microextraction. *J. Chromatogr. A* 1244: 55–60.

34 Psillakis, E., Yiantzi, E., and Kalogerakis, N. (2013). Downsizing vacuum-assisted headspace solid phase microextraction. *J. Chromatogr. A* 1300: 119–126.

35 Yiantzi, E., Kalogerakis, N., and Psillakis, E. (2015). Vacuum-assisted headspace solid phase microextraction of polycyclic aromatic hydrocarbons in solid samples. *Anal. Chim. Acta* 890: 108–116.

36 Yiantzi, E., Kalogerakis, N., and Psillakis, E. (2016). Design and testing of a new sampler for simplified vacuum-assisted headspace solid-phase microextraction. *Anal. Chim. Acta* 927: 46–54.

37 Glykioti, M.-L., Yiantzi, E., and Psillakis, E. (2016). Room temperature determination of earthy-musty odor compounds in water using vacuum-assisted headspace solid-phase microextraction. *Anal. Methods* 8 (45): 8065–8071.

38 Trujillo-Rodríguez, M.J., Pino, V., Psillakis, E. et al. (2017). Vacuum-assisted headspace-solid phase microextraction for determining volatile free fatty acids and phenols. Investigations on the effect of pressure on competitive adsorption phenomena in a multicomponent system. *Anal. Chim. Acta* 962: 41–51.

39 Orazbayeva, D., Kenessov, B., Psillakis, E. et al. (2018). Determination of transformation products of unsymmetrical dimethylhydrazine in water using vacuum-assisted headspace solid-phase microextraction. *J. Chromatogr. A* 1555: 30–36.

40 Liberto, E., Bicchi, C., Cagliero, C. et al. (2019). Headspace sampling: an "Evergreen" method in constant evolution to characterize food flavors through their volatile fraction. In: *Advanced Gas Chromatography in Food Analysis* (ed. P.Q. Tranchida), 1–37. The Royal Society of Chemistry.

41 Woolfenden, E. (2010). Sorbent-based sampling methods for volatile and semi-volatile organic compounds in air. Part 2. Sorbent selection and other aspects of optimizing air monitoring methods. *J. Chromatogr. A* 1217 (16): 2685–2694.

42 Seethapathy, S., Górecki, T., and Li, X. (2008). Passive sampling in environmental analysis. *J. Chromatogr. A* 1184 (1–2): 234–253.

43 Woolfenden, E. (2021). Thermal desorption gas chromatography. In: *Gas Chromatography* (ed. C.F. Poole), 267–323. Elsevier.

44 Baltussen, E., Sandra, P., David, F., and Cramers, C. (1999). Stir bar sorptive extraction (SBSE), a novel extraction technique for aqueous samples: theory and principles. *J. Microcolumn Sep.* 11 (10): 737–747.

45 Franchina, F.A., Zanella, D., Dejong, T., and Focant, J.-F. (2021). Impact of the adsorbent material on volatile metabolites during in vitro and in vivo bio-sampling. *Talanta* 221, Article 121569. https://doi.org/10.1016/j.talanta.2020.121569.

46 De Poli, M., Chenet, T., Felletti, S. et al. (2025). Sorbent-based sampling with two-stage trapping/desorption coupled to comprehensive two-dimensional gas chromatography and mass spectrometry for terpenoids profiling in cannabis. *Anal. Sci. Adv.* 6 (1): e202400044.

47 Poole, C.F. (2020). *Liquid-Phase Extraction*. Elsevier.

48 Poole, C.F. (2020). *Solid-Phase Extraction*. Elsevier.

49 Moldoveanu, S. and David, V. (2015). The role of derivatization in chromatography. In: *Modern Sample Preparation for Chromatography*, 307–331. Elsevier.

50 Dennenlöhr, J., Thörner, S., and Rettberg, N. (2020). Analysis of hop-derived thiols in beer using on-fiber derivatization in combination with HS-SPME and GC–MS/MS. *J. Agric. Food Chem.* 68 (50): 15036–15047.

51 Deng, C. and Zhang, X. (2004). A simple, rapid and sensitive method for determination of aldehydes in human blood by gas chromatography/mass spectrometry and solid-phase microextraction with on-fiber derivatization. *Rapid Commun. Mass Spectrom.* 18 (15): 1715–1720.

52 Aresta, A., Cotugno, P., and Zambonin, C. (2019). Solid-phase microextraction and on-fiber derivatization for assessment of mammalian and vegetable milks with emphasis on the content of major phytoestrogens. *Sci. Rep.* 9 (1): 6398.

53 Chemat, F., Rombaut, N., Sicaire, A.-G. et al. (2017). Ultrasound assisted extraction of food and natural products. Mechanisms, techniques, combinations, protocols and applications. A review. *Ultrason. Sonochem.* 34: 540–560.

54 Chemat, F., Rombaut, N., Meullemiestre, A. et al. (2017). Review of green food processing techniques. Preservation, transformation, and extraction. *Innovative Food Sci. Emerg. Technol.* 41: 357–377.

55 Yamini, Y., Rezazadeh, M., and Seidi, S. (2019). Liquid-phase microextraction – the different principles and configurations. *TrAC Trends Anal. Chem.* 112: 264–272.

56 Rutkowska, M., Płotka-Wasylka, J., Sajid, M., and Andruch, V. (2019). Liquid–phase microextraction: a review of reviews. *Microchem. J.* 149: 103989.

57 Dugheri, S., Mucci, N., Bonari, A. et al. (2020). Liquid phase microextraction techniques combined with chromatography analysis: a review. *Acta Chromatogr.* 32 (2): 69–79.

58 Pena-Pereira, F., Lavilla, I., and Bendicho, C. (2021). Greening sample preparation: an overview of cutting-edge contributions. *Curr. Opin. Green. Sustainable Chem.* 30: 100481.

59 López-Lorente, Á.I., Pena-Pereira, F., Pedersen-Bjergaard, S. et al. (2022). The ten principles of green sample preparation. *TrAC Trends Anal. Chem.* 148: 116530.

60 Kawaguchi, M., Ito, R., Endo, N. et al. (2006). Liquid phase microextraction with in situ derivatization for measurement of bisphenol A in river water sample by gas chromatography–mass spectrometry. *J. Chromatogr. A* 1110 (1–2): 1–5.

61 Rezaee, M., Assadi, Y., Milani Hosseini, M.R. et al. (2006). Determination of organic compounds in water using dispersive liquid–liquid microextraction. *J. Chromatogr. A* 1116 (1–2): 1–9.

62 Pedersen-Bjergaard, S. and Rasmussen, K.E. (2008). Liquid-phase microextraction with porous hollow fibers, a miniaturized and highly flexible format for liquid–liquid extraction. *J. Chromatogr. A* 1184 (1–2): 132–142.

63 Câmara, J.S., Perestrelo, R., Olayanju, B. et al. (2022). Overview of different modes and applications of liquid phase-based microextraction techniques. *PRO* 10 (7): 1347.

64 Notardonato, I. and Avino, P. (2024). Dispersive liquid–liquid micro extraction: an analytical technique undergoing continuous evolution and development – a review of the last 5 years. *Separations* 11 (7): 203.

65 Madikizela, L.M., Pakade, V.E., Ncube, S. et al. (2020). Application of hollow fibre-liquid phase microextraction technique for isolation and pre-concentration of pharmaceuticals in water. *Membranes (Basel)* 10 (11): 311.

66 Lee, J., Lee, H.K., Rasmussen, K.E., and Pedersen-Bjergaard, S. (2008). Environmental and bioanalytical applications of hollow fiber membrane liquid-phase microextraction: a review. *Anal. Chim. Acta* 624 (2): 253–268.

67 Söderholm, S.L., Damm, M., and Kappe, C.O. (2010). Microwave-assisted derivatization procedures for gas chromatography/mass spectrometry analysis. *Mol. Divers.* 14 (4): 869–888.

68 Panda, D. and Manickam, S. (2019). Cavitation technology – the future of greener extraction method: a review on the extraction of natural products and process intensification mechanism and perspectives. *Appl. Sci.* 9 (4): 766.

69 Khadhraoui, B., Ummat, V., Tiwari, B.K. et al. (2021). Review of ultrasound combinations with hybrid and innovative techniques for extraction and processing of food and natural products. *Ultrason. Sonochem.* 76: 105625.

70 Nayak, N., Bhujle, R.R., Nanje-Gowda, N.A. et al. (2024). Advances in the novel and green-assisted techniques for extraction of bioactive compounds from millets: a comprehensive review. *Heliyon* 10 (10): e30921.

71 Björklund, E., Nilsson, T., and Bøwadt, S. (2000). Pressurised liquid extraction of persistent organic pollutants in environmental analysis. *TrAC Trends Anal. Chem.* 19 (7): 434–445.

72 Richter, B.E., Jones, B.A., Ezzell, J.L. et al. (1996). Accelerated solvent extraction: a technique for sample preparation. *Anal. Chem.* 68 (6): 1033–1039.

73 Primbs, T., Genualdi, S., and Simonich, S. (2008). Solvent selection for pressurized liquid extraction of polymeric sorbents used in air sampling. *Environ. Toxicol. Chem.* 27 (6): 1267.

74 Popp, P., Keil, P., Möder, M. et al. (1997). Application of accelerated solvent extraction followed by gas chromatography, high-performance liquid chromatography and gas chromatography–mass spectrometry for the determination of polycyclic aromatic hydrocarbons, chlorinated pesticides and polychlorinated dibenzo-p-dioxins and dibenzofurans in solid wastes. *J. Chromatogr. A* 774 (1–2): 203–211.

75 Thurman, E.M. and Mills, M.S. (1998). *Solid-Phase Extraction: Principles and Practice*. Wiley.

76 Teixeira, C., Ipatov, A., Carvalho, J. et al. (2023). Miniaturized microscale solid-phase extraction-based module for highly efficient DNA extraction and purification from grapevine samples. *ACS Omega* 8 (35): 31738–31746.

77 Kabir, A. and Samanidou, V. (2021). Fabric phase sorptive extraction: a paradigm shift approach in analytical and bioanalytical sample preparation. *Molecules* 26 (4): 865.

78 He, X., Sun, T., Wang, L., and Jiang, X. (2021). Pipette-tip micro-solid phase extraction based on melamine-foam@polydopamine followed by ultra-high-performance liquid chromatography–quadrupole–time-of-flight mass spectrometry for detection of psychotropic drugs in human serum. *J. Chromatogr. B* 1163: 122499.

79 Rosendo, L.M., Brinca, A.T., Pires, B. et al. (2023). Miniaturized solid phase extraction techniques applied to natural products. *PRO* 11 (1): 243.

80 Abdel-Rehim, M. (2004). New trend in sample preparation: on-line microextraction in packed syringe for liquid and gas chromatography applications I. Determination of local anaesthetics in human plasma samples using gas chromatography–mass spectrometry. *J. Chromatogr. B Analyt. Technol. Biomed. Life Sci.* 801 (2): 317–321.

81 Moein, M.M., Said, R., and Abdel-Rehim, M. (2015). Microextraction by packed sorbent. *Bioanalysis* 7 (17): 2155–2161.

82 Beale, D.J., Pinu, F.R., Kouremenos, K.A. et al. (2018). Review of recent developments in GC–MS approaches to metabolomics-based research. *Metabolomics* 14 (11): 152.

83 Kind, T., Wohlgemuth, G., Lee, D.Y. et al. (2009). FiehnLib: mass spectral and retention index libraries for metabolomics based on quadrupole and time-of-flight gas chromatography/mass spectrometry. *Anal. Chem.* 81 (24): 10038–10048.

84 Blau, K. and Halket, J.M. (1993). *Handbook of Derivatives for Chromatography*. Wiley.

85 Fiehn, O. (2016). Metabolomics by gas chromatography–mass spectrometry: the combination of targeted and untargeted profiling. *Curr. Protoc. Mol. Biol.* 114: 30.4.1.

86 Murrieta, C.M., Hess, B.W., and Rule, D.C. (2003). Comparison of acidic and alkaline catalysts for preparation of fatty acid methyl esters from ovine muscle with emphasis on conjugated linoleic acid. *Meat Sci.* 65 (1): 523–529.

87 Couturier, L.I.E., Michel, L.N., Amaro, T. et al. (2020). State of art and best practices for fatty acid analysis in aquatic sciences. *ICES J. Mar. Sci.* 77 (7–8): 2375–2395.

88 Shibamoto, S., Murata, T., and Yamamoto, K. (2016). Determination of double bond positions and geometry of methyl linoleate isomers with dimethyl disulfide adducts by GC/MS. *Lipids* 51 (9): 1077–1081.

89 Farkas, O., Zenkevich, I.G., Stout, F. et al. (2008). Prediction of retention indices for identification of fatty acid methyl esters. *J. Chromatogr. A* 1198–1199 (1–2): 188–195.

90 van Den Dool, H. and Dec. Kratz, P. (1963). A generalization of the retention index system including linear temperature programmed gas–liquid partition chromatography. *J. Chromatogr. A* 11 (C): 463–471.

91 Ferrara, D., Beccaria, M., Cordero, C.E., and Purcaro, G. (2024). Comprehensive comparison of fatty acid methyl ester profile in different food matrices using microwave-assisted extraction and derivatization methods and comprehensive two-dimensional gas chromatography coupled with flame ionization detection. *Adv. Sample Prep.* 11: 100124.

92 Ostermann, A.I., Müller, M., Willenberg, I., and Schebb, N.H. (2014). Determining the fatty acid composition in plasma and tissues as fatty acid methyl esters using gas chromatography – a comparison of different derivatization and extraction procedures. *Prostaglandins Leukot. Essent. Fatty Acids* 91 (6): 235–241.

93 Müller, K.-D., Husmann, H., Nalik, H.P., and Schomburg, G. (1990). Trans-esterification of fatty acids from microorganisms and human blood serum by trimethylsulfonium hydroxide (TMSH) for GC analysis. *Chromatographia* 30 (5–6): 245–248.

94 Řimnáčová, L., Hušek, P., and Šimek, P. (2014). A new method for immediate derivatization of hydroxyl groups by fluoroalkyl chloroformates and its application for the determination of sterols and tocopherols in human serum and amniotic fluid by gas chromatography–mass spectrometry. *J. Chromatogr. A* 1339: 154–167.

95 Hušek, P., Švagera, Z., Hanzlíková, D. et al. (2016). Profiling of urinary amino-carboxylic metabolites by in-situ heptafluorobutyl chloroformate mediated sample preparation and gas chromatography–mass spectrometry. *J. Chromatogr. A* 1443: 211–232.

96 Smart, K.F., Aggio, R.B.M., Van Houtte, J.R., and Villas-Bôas, S.G. (2010). Analytical platform for metabolome analysis of microbial cells using methyl chloroformate derivatization followed by gas chromatography–mass spectrometry. *Nat. Protoc.* 5 (10): 1709–1729.

97 Kayacelebi, A.A., Knöfel, A.-K., Beckmann, B. et al. (2015). Measurement of unlabeled and stable isotope-labeled homoarginine, arginine and their metabolites in biological samples by GC–MS and GC–MS/MS. *Amino Acids* 47 (9): 2023–2034.

98 Fritsche-Guenther, R., Gloaguen, Y., Bauer, A. et al. (2021). Optimized workflow for on-line derivatization for targeted metabolomics approach by gas chromatography–mass spectrometry. *Meta* 11 (12): 888.

99 Dunn, W.B., Broadhurst, D., Begley, P. et al. (2011). Procedures for large-scale metabolic profiling of serum and plasma using gas chromatography and liquid chromatography coupled to mass spectrometry. *Nat. Protoc.* 6 (7): 1060–1083.

100 Misra, B.B., Upadhayay, R.P., Cox, L.A., and Olivier, M. (2018). Optimized GC–MS metabolomics for the analysis of kidney tissue metabolites. *Metabolomics* 14 (6): 75.

101 Khoomrung, S., Chumnanpuen, P., Jansa-Ard, S. et al. (2012). Fast and accurate preparation fatty acid methyl esters by microwave-assisted derivatization in the yeast *Saccharomyces cerevisiae*. *Appl. Microbiol. Biotechnol.* 94 (6): 1637–1646.

102 Tsochatzis, E.D., Nebel, C., Danielsen, M. et al. (2021). Thermal degradation of metabolites in urine using multiple isotope-labelled internal standards for off-line GC metabolomics – effects of injector and oven temperatures. *J. Chromatogr. B* 1181: 122902.

3

Principles of High-resolution GC and GC×GC

Peter Q. Tranchida[1], *Micaela Galletta*[1], *and Luigi Mondello*[1,2]

[1] *Messina Institute of Technology, c/o Department of Chemical, Biological, Pharmaceutical and Environmental Sciences, University of Messina, Messina, Italy*
[2] *Chromaleont s.r.l., c/o Department of Chemical, Biological, Pharmaceutical and Environmental Sciences, University of Messina, Messina, Italy*

3.1 Introduction

It is generally accepted that gas chromatography (GC), in its various forms, enables the separation of volatile, thermally-stable compounds belonging to a plethora of sample types (*i.e.*, biological, petrochemical, food, and pharmaceutical) of different complexity. If one breaks down such a generic definition and focuses on the terms "separation," "thermally-stable," "volatile compounds," and "complexity," then more detailed information needs to be provided. Now, what is intended by the definition of "separation"? Separation of all the compounds contained in a sample, or better, separation of all the detectable compounds. Co-elution is a very common occurrence in GC (as well as in all forms of chromatography); the percentage of fully-isolated compounds varies greatly, and depends on several factors. For example, the use of a long capillary column (*e.g.*, 100 m) will enable the isolation of a higher number of compounds, compared to a shorter one (*e.g.*, 30 m), at the expense of a lower number of detectable analytes (due to band broadening) and of an extended analysis time. Another frequently used generic term is "volatile compounds." Much better to use the term "GC-amenable," meaning that a compound can pass to the vapor phase inside a hot GC injector and be transferred to the head of the column by the carrier gas (He, H_2, or N_2); here, at a specific temperature, the compound will start its journey along the column. Masses of 700 Da are rarely exceeded in GC. Analyte thermal stability is a further important factor in GC, even though not mandatory; some volatile constituents contained in mixtures may undergo degradation or transformation inside a hot GC injector. Finally, the term "complex" is extremely generic, because samples can be more or less complex, but the extent to which complexity is revealed depends on the GC technique used. The same type of sample analyzed by using packed-column GC, capillary-column GC, and comprehensive two-dimensional (2D) GC (GC×GC) will be revealed at totally different levels of complexity.

3.2 High-resolution GC

The introduction of the fused-silica capillary column marked a revolution in the GC field, overshadowing packed columns in a very short time, and beginning the era of high-resolution GC [1]. Inertness, flexibility, and robustness were the main advantages of fused-silica capillary columns over their glass counterparts. Currently, there is a wide variety of capillary columns in terms of length (5–100 m), internal diameter (0.05–0.53 mm ID), stationary phase thickness (0.05–5.0 μm), and chemistry (*e.g.*, from low to high polarity, measurable through McReynolds constants, among other approaches). The selection of a specific GC column depends on the sample complexity and chemistry, as well as the specific analysis objective (untargeted or targeted). A GC column with dimensions 30 m × 0.25 mm ID × 0.25 μm d_f is a popular first choice and can be considered as a high-resolution GC tool. Among several, there are two main measurable features of a capillary column that must be considered, namely peak capacity (n_c) and selectivity. The peak capacity is a measure of separation power, and it is equivalent to the number of peaks that can potentially be separated in a GC chromatogram, from the end of the void time to that of the temperature program (separation space). Essentially, the peak capacity corresponds to the number of peaks, with a specific average width, that can be placed side-by-side in the separation space, at a specific resolution value (*e.g.*, 1.5). If "normal" operational conditions are applied (gas flow: 1 mL/min; temperature program: 50–300 °C at 3 °C/min), using a 30 m × 0.25 mm ID × 0.25 μm d_f column, then a peak capacity in the range 300–500 can be expected. As a first approximation, peak capacity is proportional to the square root of column efficiency; a 30 m × 0.25 mm ID × 0.25 μm d_f column, operated under optimum gas flow conditions, can generate approximately 120,000 theoretical plates (N). Hence, in this specific case, a hypothetical 120 m × 0.25 mm ID × 0.25 μm d_f column will enable a factor of 2 increase in peak capacity.

Peak capacity is measured by analysing a homologous series of compounds (*e.g.*, alkanes and saturated fatty acid methyl esters) characterized by orderly elution. However, real-world samples are usually composed of a variety of volatile compounds with different molecular weights and chemistries, leading mainly to disordered elution. The main consequence of randomness is that only part of the separation space is usually occupied, and only part of the occupied space relates to single-component peaks. Under the same GC conditions, including column dimensions, the % of occupied separation space, along with the % of single-component peaks, will depend on the stationary phase selectivity. This, despite the fact that elution sequences are always tightly related to analyte vapor pressures (whatever the type of capillary column used). For example, a 100% polyethylene glycol (PEG) stationary phase is characterized by a higher selectivity for the separation of fatty acid methyl esters (FAMEs) compared to a 100% polydimethylsiloxane (PDMS) one. Even so, *cis/trans* FAMEs (*e.g.*, oleic and elaidinic acid methyl esters) undergo separation on a 100% PDMS column, while they co-elute on a 100% PEG one. For an in-depth characterization of a FAME sample, including *cis/trans* composition, highly polar stationary phases such as poly(biscyanopropyl siloxane) or an ionic liquid (IL) are required. In such a respect, Ragonese *et al.* analyzed a standard mixture of 22 *cis/trans* FAMEs (C_{18} group, with 1, 2, and 3 double bonds), and compared the separation performance of a poly(biscyanopropyl siloxane) 100 m × 0.25 mm ID × 0.20 μm d_f column, with one of the same dimensions, containing an IL phase [1,9-di(3-vinyl-imidazolium) nonane bis(trifluoromethyl) sulfonyl imidate] [2]. According to an American Oil Chemists' Society official method (CE 1 h-05), the "determination of the fatty acid composition, including the trans fatty acid isomers of vegetable or non-ruminant animal oils and fats" should be carried out by using the poly(biscyanopropyl siloxane) column. However, it was found that the IL column possessed a superior selectivity, as can be readily observed in Figure 3.1.

Samples composed of FAMEs (vegetable oils, biological samples, etc.) are relatively straightforward to analyze through high-resolution GC because FAMEs are characterized by similar chemical characteristics, with differentiation depending mainly on chain length (usually from C_4 to C_{24}), double bond (DB) number (0–6), position (ω number, isolated, conjugated), and configuration. In addition, the number of detectable FAMEs is usually not high. Therefore, with a peak capacity in the range of 300–500, along with a suitable selectivity, the separation of FAME samples is not usually a real challenge. However, as mentioned previously, the use of a 100 m column is necessary for specific FAME analysis. Real-world samples typically consist of a much higher number of detectable constituents (hundreds, often thousands) with a wide variety of chemistries. In such cases, a peak capacity in the range of 300–500 will be far from sufficient to completely separate all the sample constituents, whatever stationary phase is used. In such a respect, a change in selectivity will alter the elution sequence, improving the separation in some zones of the chromatogram, but deteriorating it in others. In short, the chromatogram profile will change much more than the number of fully-separated compounds.

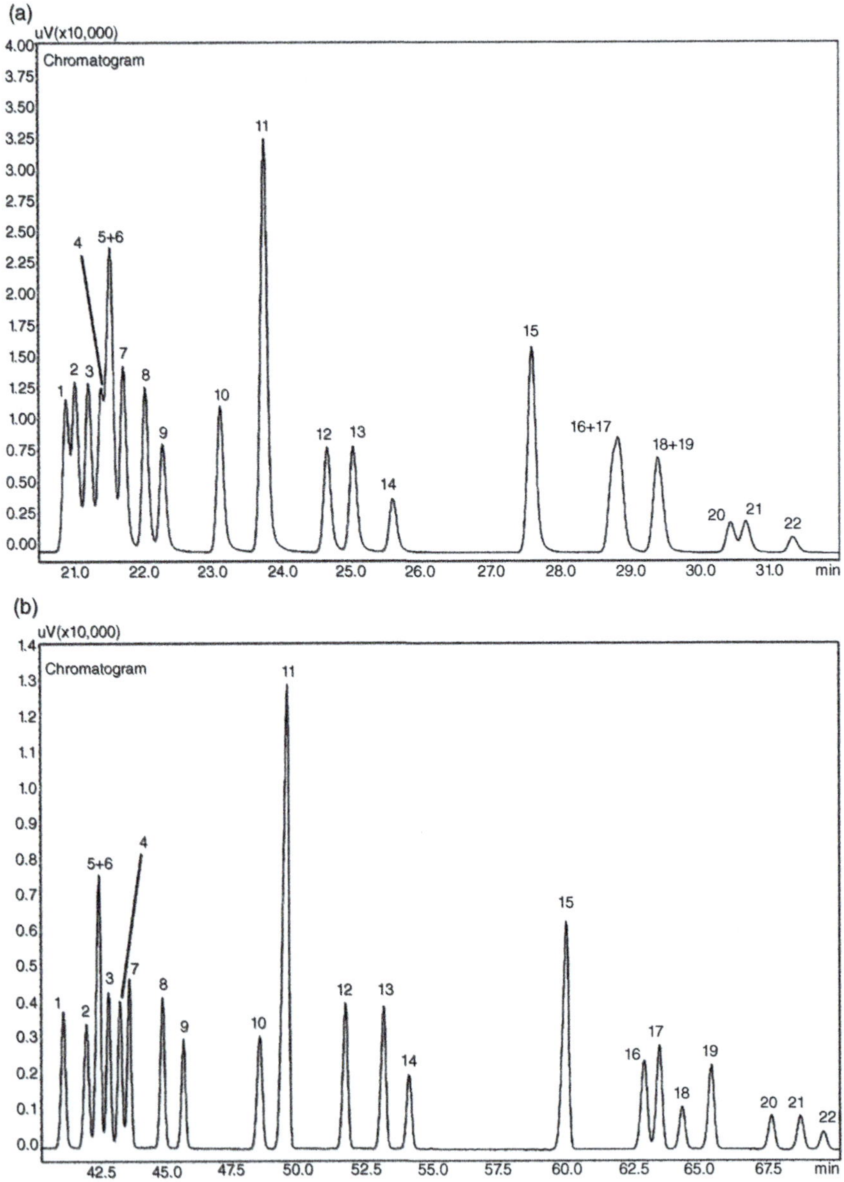

Figure 3.1 Separation of a mixture of 22 *cis/trans* FAMEs on: (a) the poly(biscyanopropyl siloxane) column and (b) the IL one. Peak identification: (1) $C_{18:1\Delta}{}^{6tr}$; (2) $C_{18:1\Delta}{}^{9tr}$; (3) $C_{18:1\Delta}{}^{11tr}$; (4) $C_{18:1\Delta}{}^{12tr}$; (5) $C_{18:1\Delta}{}^{6c}$; (6) $C_{18:1\Delta}{}^{7c}$; (7) $C_{18:1\Delta}{}^{9c}$; (8) $C_{18:1\Delta}{}^{11c}$; (9) $C_{18:1\Delta}{}^{12c}$; (10) $C_{18:1\Delta}{}^{15c}$; (11) $C_{18:2\Delta}{}^{9tr,12tr}$; (12) $C_{18:2\Delta}{}^{9c,12tr}$; (13) $C_{18:2\Delta}{}^{9tr,12c}$; (14) $C_{18:2\Delta}{}^{9c,12c}$; (15) $C_{18:3\Delta}{}^{9tr,12tr,15tr}$; (16) $C_{18:3\Delta}{}^{9tr,12tr,15c}$; (17) $C_{18:3\Delta}{}^{9tr,12c,15tr}$; (18) $C_{18:3\Delta}{}^{9c,12tr,15tr}$; (19) $C_{18:3\Delta}{}^{9c,12c,15tr}$; (20) $C_{18:3\Delta}{}^{9c,12tr,15c}$; (21) $C_{18:3\Delta}{}^{9tr,12c,15c}$; and (22) $C_{18:3\Delta}{}^{9c,12c,15c}$. *Source:* Ragonese et al. [2], with permission of American Chemical Society.

With regard to analyte identification, this is achieved through the use of pure standard compounds, linear retention index (RI) measurements, and mass spectrometry (MS). As will be seen in Chapters 4–6, MS can also be of great help in the identification and quantification of overlapping compounds at the GC column outlet. Mass spectrometry is, in its own right, an analytical dimension, involving the separation of charged species. The various forms of mass analyzers – low-resolution, high-resolution, hybrid, scanning, and non-scanning – as well as different ionization source designs (hard and soft ionization), can greatly boost the GC performance. The combined use of GC and MS forms a powerful two-dimensional technique, providing a wealth of analytical information in the form of a sample-specific fingerprint. GC-MS data can provide knowledge on chemical composition (general profiling) and can enable the differentiation between samples of the same type (*e.g.*, based on geographical origin and type of industrial processing) or the targeted determination of specific compounds (*e.g.*, food contaminants and petrochemical markers).

In many cases, it is not necessary to introduce totally separated compounds to the detector, while in others it is. Such a decision is obviously a choice of the analyst. A brief discussion on the effects of a change in the stationary phase has been made. But how does an increase in column efficiency/peak capacity affect the overall peak resolution? To a first approximation, resolution is proportional to the square root of column efficiency; hence, a $\times 4$ increase in column length will enable a $\times 2$ increase in resolution. The main disadvantages of such an option (as mentioned in the section 3.1) are the extended run time, leading to increased band broadening, and a reduction of analyte detectability. It is also possible to enhance column efficiency/peak capacity by reducing the column diameter (micro-bore columns): $25\,\text{m} \times 0.25\,\text{mm}$ ID $\times\, 0.25\,\mu\text{m}\ d_f$ and $10\,\text{m} \times 0.10\,\text{mm}$ ID $\times\, 0.10\,\mu\text{m}\ d_f$ columns (same phase ratio) can both generate $\approx$100,000 N. In such a manner, high-resolution, faster GC analyses can be performed. However, apart from the limits on the micro-bore column length, due to the high inlet pressure requirements, a further disadvantage is represented by the reduced sample capacity, leading to a lower analyte detectability.

Peak resolution can also be improved by reducing the gradient of the temperature program (*e.g.*, from 3 °C/min to 1 °C/min). Packets of analytes will start to migrate from the head of the column when they only reach a specific temperature. A slower temperature program will result in a higher solubility in the stationary phase during analyte migration. Again, the main disadvantages of such a choice are increased run time, enhanced band broadening, and a decrease in analyte detectability.

One-dimensional (1D) GC, combined with MS (or other types of detection), is still the most commonly used GC technique. Gas chromatography-mass spectrometry, as aforementioned, is a 2D technology: (I) the GC column separates on the basis of analyte boiling points and stationary phase solubilities, delivering a sample elution stream (induced by the temperature program) to the ion source; (II) the mass analyzer separates on the basis of mass-to-charge ratio (m/z) values, with its capability of separation becoming more evident if a soft ionization method is used. For instance, Wang *et al.* analyzed diesel fuel by using GC-MS with field ionization [3]. The GC-MS data were visualized in a 2D format, with the m/z values along a y-axis, and the retention times along an x-axis, leading to the interesting use of the abbreviation GC×MS.

In many cases, 1D GC combined with MS is a sufficient tool, while in others it can prove to be a weak one. The term "weak" obviously relates to both the type of the sample and the result expected.

3.3 Multidimensional GC

The use of multidimensional (MD) GC is a well-known way to increase both peak capacity (separation space) and selectivity. The vast majority of MDGC systems are characterized by two columns with a different stationary phase, forming a 2D instrument. A transfer device, located somewhere between the two columns, enables the transfer of eluate from the first dimension (^{1}D) to the second dimension (^{2}D). Multidimensional GC can be classified into one of two groups: heart-cutting MDGC or GC×GC. Heart-cutting MDGC, which first appeared rather shortly after the invention of GC itself [4], is based on the transfer of selected zones of a ^{1}D chromatogram to the ^{2}D column. So, only a part of the sample is subjected to two sequential separations. The peak capacity of a heart-cutting MDGC method becomes that of the first dimension ($^{1}n_{c}$) plus that of the second dimension ($^{2}n_{c}$), with the latter multiplied by the number of heart cuts. As the present chapter is not focused on the use of heart-cutting MDGC, the reader is directed to the literature for more details [5]. Comprehensive 2D GC, on the other hand, enables the separation of the entire sample on both dimensions, with the overall peak capacity becoming very high, approximately ($^{2D}n_{c}$): $^{1}n_{c} \times {}^{2}n_{c}$.

3.3.1 Comprehensive 2D GC

3.3.1.1 The Invention and Principles

The invention of GC×GC was described in 1991 [6], with it stemming from the research work of J.B. Phillips in prior years [7]. The leap from heart-cutting

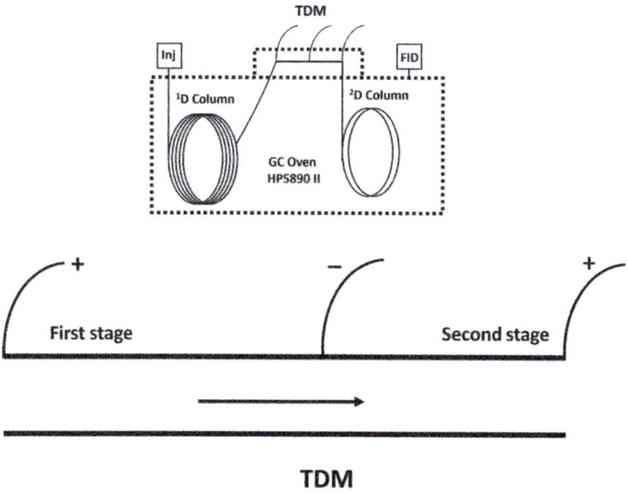

Figure 3.2 The dual-stage thermal desorption modulator. *Source:* Anal. Chem. 1994, v. 66, pp. 3086–3092, with permission of the American Chemical Society.

to comprehensive MDGC was enabled by a brilliant idea and a simple device, the thermal desorption modulator (TDM). The modulator worked like a beating heart, inasmuch that it collected and then released sequential fractions of ^{1}D eluate onto the ^{2}D column, in a continuous manner. A scheme of the TDM is illustrated in Figure 3.2.

The TDM, located outside the GC oven inside a heated chamber, was constructed by coating the initial segment (15 cm) of the ^{2}D column with copper paint. Electrical pulses were applied to the modulator (toward upstream and downstream segments), enabling dual-stage modulation:

- (accumulation I – first stage): a fraction of the ^{1}D eluate is focused at the head of the TDM, which is kept at a lower temperature with respect to the GC oven;
- (release I – first stage): the compressed ^{1}D chromatography band is remobilized through heating induced by the application of a brief electrical pulse (*e.g.*, 18 ms) to the first part of the TDM;
- (accumulation II – second stage): the remobilized ^{1}D chromatography band, after rapidly moving downstream, is compressed a further time at the head of the downstream segment of the TDM;
- (release II – second stage): a second electrical pulse (*e.g.*, 16 ms), applied after a brief period from the first (*e.g.*, 250 ms) and this time directed to the downstream segment of the TDM, enables the release of the compressed ^{1}D chromatography band onto the ^{2}D column (Figure 3.3).

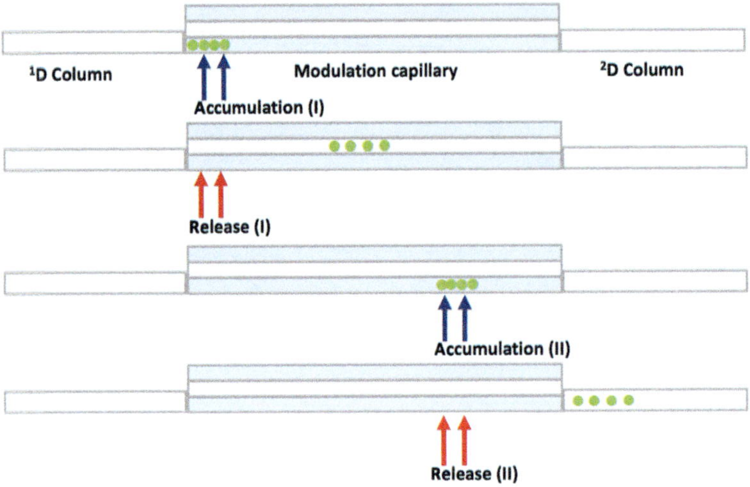

Figure 3.3 Representation of the TDM dual-stage modulation process.

The time between each modulation event is defined as the modulation period (P_M) and should represent the time frame available for each ^{2}D separation. Modulation periods are usually very short (*e.g.*, between 4–8 s) because ^{2}D separations typically occur on short micro-bore column segments (*e.g.*, 1–2 m), even though not always. On the other hand, ^{1}D separations are often performed on a conventional GC column (*e.g.*, 30 m × 0.25 mm ID × 0.25 μm d_f). Even though a wide variety of GC×GC column combinations have been reported, providing different chromatography results, it can be affirmed that ^{1}D peaks are characterized by widths (at the base – w_{b1}) normally in the range 10–20 s, while the ^{2}D peak duration is normally in the range 200–400 ms. The result of eight modulation events, on a ^{1}D peak formed of three co-eluting compounds, is represented graphically in Figure 3.4 (part 1). Ideally, each ^{1}D peak should be modulated 3–4 times; a lower number of modulations could cause a loss of resolution already achieved on the first column (due to remixing during modulation), while an excessively high number of modulations causes a lower analyte detectability (^{1}D peaks are sliced into too many fractions).

A GC×GC chromatogram is a 1D one (the detector is located at the outlet of the second column) formed by a continuous stream of fast ^{2}D separations. However, GC×GC chromatograms are visualized in a planar (2D) manner, by using dedicated software, with ^{1}D and ^{2}D retention times (t_{R1} and t_{R2}) defined along an x-axis and y-axis, respectively. Peaks are characterized by oval/circular shapes, with the intensity visually defined by color and size. GC×GC chromatograms can also be visualized in three dimensions, with peaks taking the form of cones or shark fins, projected along a z-axis (Figure 3.4 – parts 2 and 3). It is

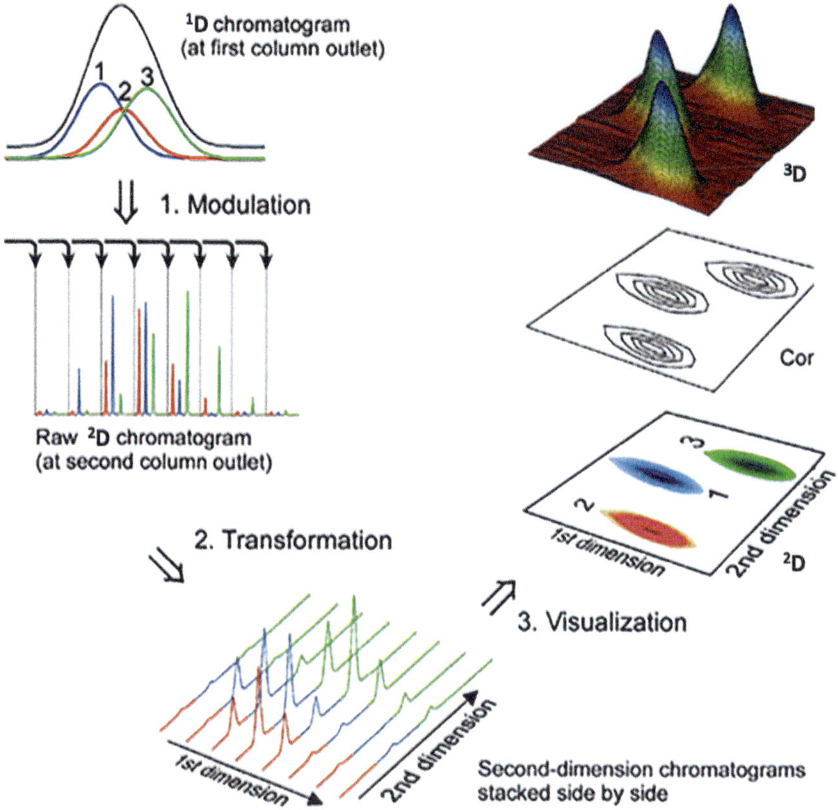

Figure 3.4 Graphical representation of the modulation process (1), and the software visualization process (2 and 3). *Source:* [35] with permission of Elsevier.

common to observe the raw data to evaluate the general peak quality (*e.g.*, fronting or tailing), while it is convenient to observe the 2D chromatogram to evaluate the overall separation performance. How a GC×GC separation performance can be evaluated and improved will be discussed later (Section 3.1.3).

Returning to the concepts of peak capacity and selectivity, it is clear that the use of GC×GC leads to the enhancement of both. Specifically, if the ^{1}D and ^{2}D columns generate peak capacity values of 400 and 20, respectively, then the theoretical $^{2D}n_c$ value becomes 8000. The reasons why theoretical values are never achieved will be given later (Section 3.1.3), even though a 2D separation space can accommodate many more peaks than a 1D one. The increase in selectivity is an obvious consequence of the instrumental configuration. A further advantage, related specifically to thermal modulation, is the considerable increase in analyte detectability, due to modulation focusing.

3.3.2 Modulation Approaches

Modulators can be classified into two groups, namely thermal and flow modulators. The former is further subdivided into cryogenic and non-cryogenic thermal modulators. Flow modulators (FMs) are classified into those that are directly connected to a valve (in-line valve FMs) and those that are indirectly connected to a valve (out-of-line valve FMs). Descriptions of the devices that have made the greatest impact in the GC×GC field (some are still widely used) follow.

3.3.2.1 Thermal Modulation

The TDM, developed by Phillips, was fragile, with frequent occurrences of burnout. Column painting and electrical connections were also delicate processes. Unfortunately, fabrication remained at a research laboratory level, and the device was never subjected to industrial study, development, and production. The TDM is, still today, a potentially simple, effective, and inexpensive form of modulation. The first commercially produced modulator was of a much more complex design/operation and was named the thermal sweeper. It was a moving modulator which never found wide use and soon became obsolete [8]. Both the TDM and the thermal sweeper are examples of non-cryogenic thermal modulators.

The introduction of the first cryogenic modulator (CM), the longitudinally modulated cryogenic system (LMCS), revolutionized the GC×GC field [9]. The LMCS was a moving device, cooled internally through a flow of carbon dioxide, and was initially used to enhance analyte signal-to-noise ratios (*s/n*) [10]. When positioned at the head of a ^{2}D column, its backward and forward movement along the column enabled dual-stage modulation. In general, the use of a CM enables the entrapment of volatiles with a molecular weight $\geq C_{4-8}$, depending also on the type of system used.

Following the introduction of the LMCS, several CMs were developed. Currently, the most popular CMs (and modulators in general) are represented by the loop-type and the quad-jet. Both systems are dual-stage and static and are operated by using liquid N_2 or, alternatively, cryogenic fluid-free refrigeration units capable of cooling down to −90 °C. The quad-jet CM is characterized by two pairs of cold/hot jets located at the head of the ^{2}D column [11], with each pair heating/cooling an upstream and downstream point. Accumulation I (first stage) is achieved through activation of the upstream cold jet, while the downstream hot air jet is activated at the same time. Release I (first stage) and then accumulation II (second stage) are carried out through the simultaneous activation of the upstream hot and downstream cold jets. Release II (second stage) and accumulation I (first stage) are performed through the simultaneous activation of the upstream hot and downstream cold jets. The quad-jet CM, using liquid N_2

in particular, has a high trapping capability (*i.e.*, C_4). Furthermore, the use of a hot jet enables the more efficient release of higher-molecular-weight compounds. The principle of the quad-jet CM differs from that of the LMCS, but is similar to that of the TDM. All three systems, however, are dual-stage – two accumulation and two release steps. The main disadvantage of single-stage thermal modulators (*e.g.*, a single pair of cold/hot jets) is that during the brief release event, along with the focused analytes, there is a high probability that non-focused ^{1}D compounds can also reach the ^{2}D column.

The loop-type CM, proposed shortly after the quad-jet system, is characterized by a single pair of cold/hot jets [12]. Dual-stage operation is enabled by looping a segment of the capillary column (*e.g.*, 1 m) – the delay loop (located between the two dimensions). The loop can also be formed by using the end part of the ^{1}D column or the initial segment of the ^{2}D column. The cooling gas flows continuously and is directed vertically downward onto two overlapping spots of the delay loop (one upstream and the other downstream). Consequently, accumulation I (first stage) and accumulation II (second stage) are performed at the same time. At the end of the accumulation event, the cold jet of gas is diverted by the heating gas for a brief period (*e.g.*, 300 ms). The hot jet is located at 90° with respect to the cold jet, with the gas emitted hitting the same two loop spots.

The extent to which *s/n* values are enhanced when using cryogenic modulation depends on a series of factors, such as ^{2}D analyte-stationary phase solubility, ^{2}D column length, ID and stationary phase thickness, as well as modulation and detector parameters (Sections 3.1.3 and 3.1.4). In general, an increase in *s/n* values by a factor of 10–20 can be expected (compared to 1D GC, using the same ^{1}D column).

The reader is directed to the literature for more detailed information on thermal modulation devices [13].

3.3.2.2 Flow Modulation

Flow modulation represents an interesting alternative to thermal modulation, with both advantages and disadvantages (as will be discussed). The volatility range of compounds amenable to FM GC×GC analysis is the same as for 1D GC. The first example of FM GC×GC was described in 1998 [14], the same year during which CM GC×GC appeared [9]. The FM, an in-line valve one, is illustrated in Figure 3.5. As can be observed, the FM consisted of a diaphragm valve (located inside the GC oven), with four out of six ports connected to the ^{1}D (polar, 4.9 m × 0.53 mm ID × 3.0 μm d_f) and ^{2}D (nonpolar, 0.85 m × 0.18 mm ID × 0.15 μm d_f) columns, to an additional pressure source and to a waste line. Depending on the valve position, the ^{1}D eluate was either directed to waste (position shown in Figure 3.5) or to the ^{2}D column. Specifically, the latter process occurred for 50 ms at the beginning of the modulation period, while after the valve

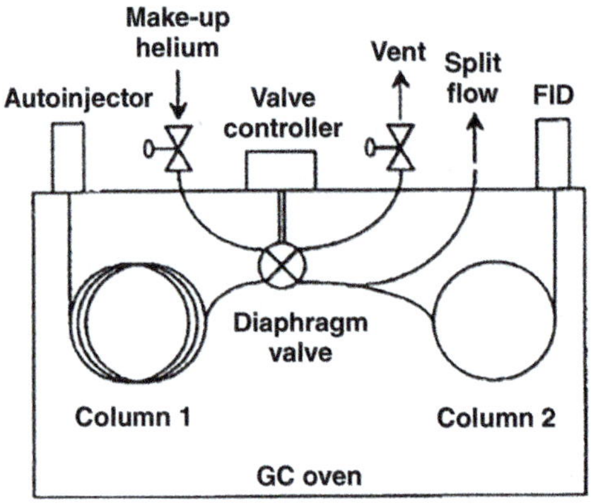

Figure 3.5 Scheme of the first GC×GC flow modulator (in the "waste" position). *Source:* Bruckner et al. [14], with permission of American Chemical Society.

was switched to the waste for 450 ms, leading to a reduced duty cycle (volume of ^{1}D eluate that reaches the ^{2}D column over the volume directed to waste) and a P_M of 0.5 s. While the ^{1}D eluate was directed to waste, the additional pressure source maintained the gas flow in the ^{2}D column. A split line was used to avoid an excessively high gas flow during ^{2}D separations. The generation of high ^{2}D gas flows has often been a disadvantage of FM GC×GC; as will be seen, solutions have been proposed to circumvent such a problem.

The use of a low-duty-cycle FM will lead to the introduction of a narrow eluate band onto the ^{2}D column, which can be considered as a prerequisite for satisfactory high-speed separations. It is obvious, however, that low-duty-cycle FMs will also be characterized by a reduced analyte detectability. This, especially if CMs are considered.

In 2006, Mohler *et al.* described an in-line valve FM that enabled the complete transfer of the eluate between the two analytical dimensions [15]. The system, illustrated schematically in Figure 3.6, consisted of a six-port valve mounted on top of the GC instrument, with only the port side located inside the oven. Using such a configuration, it was possible to perform GC×GC analyses at temperatures up to 275 °C. Observing Figure 3.6, and numbering with 1 the port connected to the ^{1}D column (apolar, 10 m × 0.18 mm ID × 0.25 μm d_f), then proceeding clockwise port 2 is plugged, ports 3 and 6 are bridged by an accumulation loop, port 4 is connected to an auxiliary gas flow source, and port 5 is linked to the ^{2}D column (medium polarity, 3 m × 0.25 mm ID × 0.25 μm d_f). During the accumulation period, ^{1}D eluate entered the loop through the channel linking ports 1 and 6 and was compressed to a certain extent (port 2 is plugged). When the valve passed to the "release" state (represented in Figure 3.6), the content of the loop was rapidly

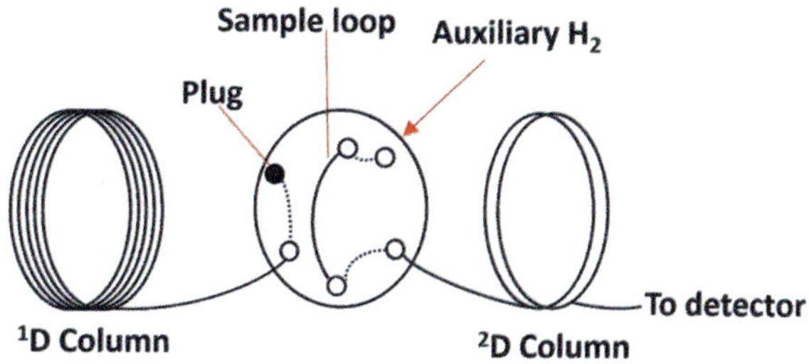

Figure 3.6 Scheme of a high duty cycle, in-line valve GC×GC flow modulator (in the "release" position). *Source:* Mohler et al. [15], with permission of Elsevier.

flushed onto the ^{2}D column by the auxiliary flow. The use of a longer ^{2}D column, compared to those normally used in CM GC×GC, was necessary to accommodate the eluate contained in and extended across the loop. The FM proposed by Mohler *et al.* performed what is defined as reversed fill–flush flow modulation, inasmuch that the directions of loop filling and flushing are opposite.

Within the field of flow modulation, outline valve systems are currently the most commonly used. A breakthrough article appeared in 2006, by Seeley *et al.*, describing a modulator constructed by using two microvolume T-unions, a three-port solenoid valve, and three deactivated fused-silica column segments [16]. The upstream union was linked to the ^{1}D outlet, to the downstream union via a fused-silica segment (15 cm × 0.45 mm ID – accumulation or storage loop), and to the normally closed outlet port of the valve (again, *via* a fused-silica segment). The remaining two ports of the downstream union were linked to the ^{2}D inlet and to the normally open outlet port of the valve (through a fused-silica segment). More specifically, a 10 cm segment of 0.25 mm ID capillary tubing connected the downstream union to a splitter one, with this linked to two polar 5 m × 0.25 mm ID ^{2}D columns. The third port of the valve received gas flow from an auxiliary source. A scheme of the FM is shown in Figure 3.7. During the accumulation period, the valve directed the auxiliary flow to the downstream union, allowing the ^{1}D eluate to enter the loop. At the end of the accumulation period, a pressure pulse of brief duration was directed to the upstream union (normally closed valve position), enabling the release of the eluate band onto the ^{2}D columns. The directions of filling and flushing were the same. The FM proposed by Seeley *et al.* was characterized by a high thermal stability, 100% ^{1}D-^{2}D transfer, and simplicity of construction.

There has been considerable evolution in the field of flow modulation, with different devices currently available. Some derive from the model proposed by Seeley *et al.* in 2006 and are characterized by a high duty cycle [17]. A further FM, defined the multimode modulator, was proposed by Seeley *et al.* in 2018 [18]. A commercial low-duty-cycle FM stemmed from the multimode modulator concept [19]. The reader is directed to the literature for more detailed information on flow modulators [13].

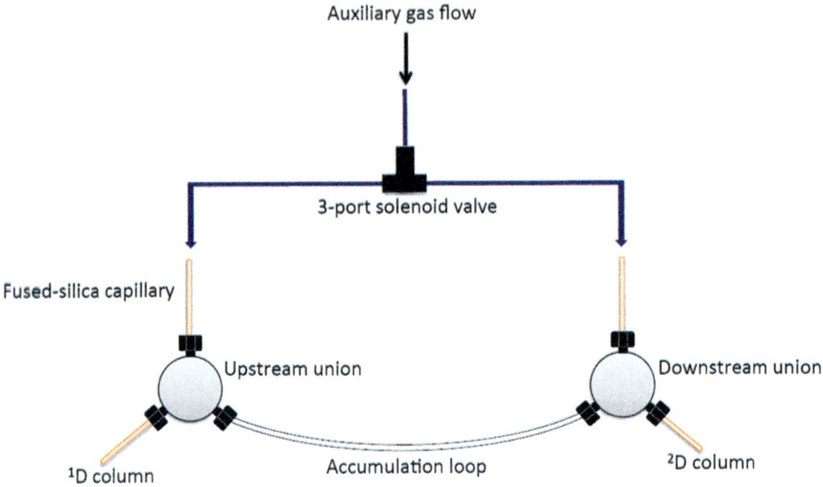

Figure 3.7 Scheme of the FM introduced. *Source:* Adapted from Seeley et al. [16].

3.3.3 Method Optimization

Method optimization in GC×GC is much more demanding compared to 1D GC; apart from injection conditions (which basically do not differ), the operator must select the most appropriate columns, inlet pressure(s), GC oven(s) temperature program(s), and modulation conditions.

3.3.3.1 Stationary Phase Selection

As aforementioned, the use of two GC columns with different stationary phases enhances both selectivity and peak capacity, which are inarguably two advantages of GC×GC over other 1D GC methods. Another benefit relates to the highly organized nature of 2D chromatograms when analyzing a homologous series of compounds (FAMEs, pyrazines, etc.). Such a factor diminishes but certainly does not eliminate the need for mass spectrometry.

Peak capacity depends mainly on the column dimensions, type, and thickness of the stationary phases, and on the modulation process (thermal or flow modulation). However, as in 1D GC, the peak capacity is only in part related to selectivity – *e.g.*, a longer ^{1}D column will improve the overall GC×GC result. Selectivity, on the other hand, is tightly related to the stationary phase combination. Much more than in 1D GC, a change in the selection of the stationary phases can result in a greatly improved selectivity. Column selection is often a case of trial-and-error testing, with a popular initial option, when using cryogenic modulation, being that of a low-polarity (*e.g.*, "5% diphenyl")

30 m × 0.25 mm ID × 0.25 μm d_f ^{1}D and a medium polarity (*e.g.*, polyethylene glycol) 1–2 m × 0.10 mm ID × 0.10 μm d_f ^{2}D. Such a setup is considered the most "orthogonal": packets of overlapping isovolatile analytes (the ^{1}D separation depends strongly on boiling points) are separated in the ^{2}D only based on their different chemistries. It is also true, however, that an inversion of the stationary phases (more polar in the ^{1}D) can often provide a better result. Even so, whatever combination of stationary phases is used, there will always be some correlation between the two sequential separations, due to the dominating role of vapor pressures in GC analyses. Consequently, the upper left part (low t_{R1} and high t_{R2}) and lower right part (high t_{R1} and low t_{R2}) of GC×GC chromatograms are usually characterized by low peak density, or are even empty. On the other hand, the highest peak density is often located along a diagonal elution band (of variable width) crossing the 2D separation space from the lower left side to the upper right one. Apart from correlation, obviously, the ^{1}D and ^{2}D void times contribute to a reduction in peak capacity. However, while analytes will never be present within the ^{1}D void-time space, the same cannot be affirmed for the ^{2}D void-time space. In fact, when analyte t_{R2} values exceed that of the P_M (an occurrence defined as wraparound), then the presence of peaks in the ^{2}D void-time space is rather common. Wraparound is often avoided because it can lead to a loss of the highly organized nature of 2D chromatograms; on the other hand, there are cases in which it can be accepted because it increases the amount of available separation space.

Considering retention times, investigations on prediction in GC×GC, and hence on selectivity, have been reported. For instance, Abraham's solvation model was applied by Seeley *et al.* for the prediction of elution patterns on four column sets, involving 54 analytes [20]. Compared to experimental ^{1}D and ^{2}D retention times, the errors measured were in the 1–5% range. Even though of complicated general use, the research was certainly an interesting GC exercise. The use of RIs is very important in 1D GC analyses, inasmuch that it increases the reliability of analyte identification [21]. Although attempts have been made to measure both ^{1}D and ^{2}D GC×GC RIs, there is no current widespread application [22], the main reason being the difficulty in encompassing analyte t_{R2} values between those of two reference compounds. Therefore, total retention times (t_{R1} plus t_{R2}) are often used to calculate RIs, with the resulting values essentially 1D ones.

The two GC columns can be accommodated in one or two ovens, with the latter option being by far the preferable. If a single oven is used, then the maximum operating temperature will be that of the column with the lower thermal stability, and the ^{2}D analyses temperatures will obviously be related to the ^{1}D elution ones, and are often not ideal. As aforementioned, many GC×GC chromatograms are characterized by diagonal elution bands: if there is limited band expansion across the 2D separation space, with observable insufficient analyte separation, then it is highly probable that the ^{2}D analysis temperatures are too high. In such a case, a slower temperature program could be applied (enabling lower ^{2}D analysis temperatures), which would certainly lead to enhanced ^{2}D resolution values, but also to wider ^{1}D peaks (and also improved ^{1}D resolution). In such an instance, an extension of the modulation period should be reconsidered to avoid oversampling. If, on the other hand, there is an excessive band extension across the 2D separation space, along with an unacceptable wraparound, then the ^{2}D analysis temperatures are certainly too low. The application of a steeper temperature program gradient will lead, on one hand, to higher ^{2}D separation temperatures and on the other, to narrower peaks and reduced resolution in the first dimension. Again, the P_M should be reconsidered (in terms of a reduction) to avoid undersampling. Such issues that arise when using a single oven GC×GC instrument can be easily resolved by using two independent GC ovens.

An extremely common GC×GC concept, of general acceptance, is that separations occur on columns characterized by different stationary phases. The use of the same stationary phase would lead, in principle, to a chromatogram with a diagonal line crossing the separation space from the lower left side to the upper right one, and with no expansion across the second dimension of the separation space. Even so, there have been rare cases in which columns with the same stationary phase have been used successfully. Specifically, Nolvachai *et al.* used the same IL phase in the first and second dimension on a mixture of FAMEs [23]. A certain degree of orthogonality was observed, in essence similar to a polar–apolar separation, *i.e.*, saturated FAMEs eluted at higher ^{2}D retention times, followed by 1 dB, 2 dB FAMEs, etc. The Authors related such "unusual" behavior to a variation of IL selectivity at different temperatures. Delmonte *et al.* used columns with the same highly polar IL phase, linked by means of a capillary containing a layer of Pd (reactor column), again for the analysis of FAMEs [24]. The use of H_2 as mobile phase played a central role, as FAMEs with DBs were hydrogenated to their saturated counterparts (*e.g.*, $C_{18:2\omega6} \rightarrow C_{18:0}$) in the reactor column. Under isothermal and constant flow conditions, FAMEs belonging to the same C-number class (*e.g.*, C_{18}) were transformed to the same FAME (*e.g.*, $C_{18:0}$) prior to ^{2}D analysis, leading to the same ^{2}D retention times. A highly organized GC×GC chromatogram of a fish oil, carried out at an isothermal temperature of 170 °C, is shown in Figure 3.8. As can be seen, the two FAMEs belonging to the class.

C_{24} undergoes wraparound, but it is entirely acceptable because the highly organized nature of the chromatogram is not impaired.

3.3.3.2 Column Dimensions and Gas Flow(s)

Apart from the stationary phase selection, another important aspect relates to the physical dimensions of the columns. Firstly, a distinction must be made between cryogenic (thermal modulation in general) and flow modulation. In the first

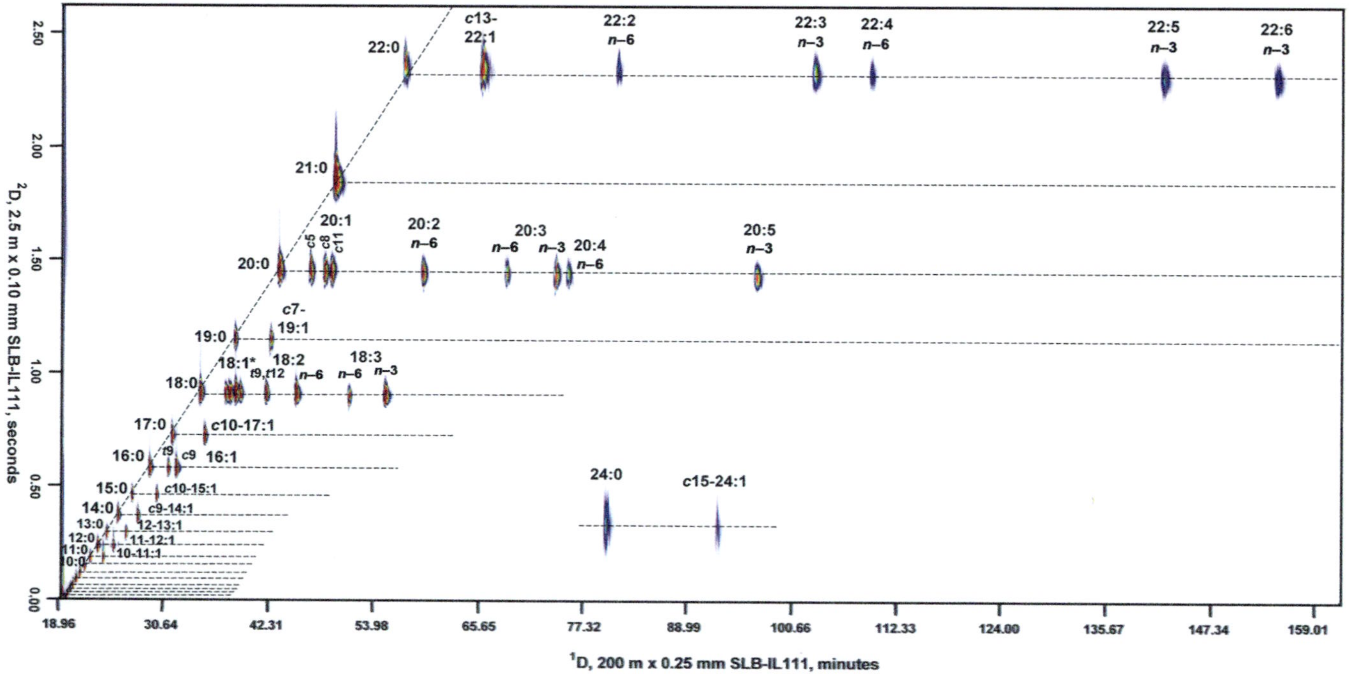

Figure 3.8 On-line hydrogenation GC×GC chromatogram of menhaden oil FAMEs. *Source:* Delmonte et al. [24], with permission of American Chemical Society.

description of a GC×GC analysis, a 21 m × 0.25 mm ID × 0.25 μm d_f polyethylene glycol column and a 1 m × 0.10 mm ID × 0.5 μm d_f nonpolar one were used [6]. In terms of column dimensions, not much has changed to date. In fact, in CM applications, the use of a ^{1}D 30 m × 0.25 mm ID × 0.25 μm d_f column, and a 1 m × 0.10 mm ID × 0.10 μm d_f ^{2}D one, is a common option. However, a direct consequence of using such different IDs is that there will be a great difference in the gas velocities, namely "slow" and close-to-ideal in the ^{1}D (*e.g.*, 20–30 cm/s), and "very fast" and far-from-ideal in the ^{2}D (*e.g.*, 200–300 cm/s). Such gas velocity conditions will generate extremely narrow peak base widths (*e.g.*, 100–200 ms) and can lead to insufficient analyte separation. Going back to the first GC×GC manuscript, a splitter was connected to the junction point between the two analytical dimensions to adjust the ^{2}D gas flow [6]. Such a concept was further investigated by Tranchida *et al.*, who connected a split line to a split valve located on top of the GC oven [25]. Manual adjustment of the split valve enabled regulation of the ^{2}D gas flow and % of ^{1}D eluate directed to waste. A disadvantage of such an approach is the reduction of analyte detectability. Obviously, and as aforementioned, if a dual oven GC×GC system is used, then the temperature programs can be tuned independently, enabling improved resolution.

The great mismatch between the gas velocities becomes less evident when the IDs of the ^{1}D and ^{2}D columns are near one another (*e.g.*, 0.25 mm and 0.18 mm) or are equivalent. In such a respect, the use of a 30 m × 0.25 mm ID + 1–2 m × 0.25 mm ID column combination is an attractive option, especially when using mass spectrometry. Using such a configuration enables conditions of low pressure to extend across the ^{2}D column, leading to higher optimum gas velocities. Furthermore, the sample capacity of a 0.25 mm ID column is also much higher (compared to 0.18 mm and 0.10 mm ones), which will reduce the occurrence of ^{2}D peak overloading.

With regard to FM, a distinction must be made between high and low duty cycle systems. Considering the former, as aforementioned, the model proposed by Seeley *et al.* in 2006 led to the development and introduction of commercial devices. Such FMs are characterized by an accumulation loop/chamber, with a specific volume (Figure 3.7). When using such transfer devices, an ID inversion (with respect to CM) is selected inasmuch that a wider ID column is normally used in the ^{2}D (*e.g.*, 0.25–0.53 mm). Moreover, longer ^{2}D columns are used (*e.g.*, 5–10 m); the reason for such a configuration relates to the fact that rather wide (uncompressed), high-volume chromatography bands are released from the modulator. It is convenient, on the other hand, to use a ^{1}D micro-bore column (*e.g.*, 0.18 mm ID), operated at a limited gas flow, which will equate to that inside the storage loop during the accumulation period. Fine-tuning of the accumulation period (and gas flow) is necessary because if the chromatography band reaches the loop exit too early, then breakthrough will occur. The use of a micro-bore column enables the application of "normal" P_M values (4–6 s). The loop/chamber volume must also be accounted for. Problems related to breakthrough can now be avoided by using the reversed-flow differential flow modulator [17], a modified version of the FM model proposed by Seeley *et al.* [16]. For details, the reader can consult ref. [17].

Optimization of the flushing period (and gas flow) is also important because the loop content must be entirely transferred to the ^{2}D column. It is widely accepted that very brief flushing periods (*e.g.*, 100–150 ms), using very high gas flows (*e.g.*, 20–25 mL/min), are required to adequately transfer the loop content onto the ^{2}D column. Such high flows are not compatible with most MS instruments, and so flow diverting to an additional detector, or to waste, becomes mandatory, causing an inevitable reduction in analyte detectability.

The problems related to high ^{2}D gas flows can be diminished in a simple manner: a flushing period of 100 ms at a gas flow of 25 mL/min is equivalent to one of 400 ms using a flow of 6.25 mL/min. The volume of gas entering the loop – in this case, about 42 μL – needs to be sufficient to expel the loop content onto the ^{2}D column and thus, again, the loop/chamber volume must be accounted for. The simple concept was applied with success in FM GC×GC-MS analyses [26, 27].

Low-duty-cycle FMs, such as the commercial device derived from the multi-mode modulator concept [19], are characterized by the release of very narrow chromatography bands. For such a reason, ^{2}D columns can be altogether similar to those used in CM applications.

3.3.3.3 Modulation

Several details on aspects related to modulation optimization (in particular, flow modulation) have already been provided. Cryogenic modulation (quad-jet and loop-type) is certainly the most efficient form of modulation if peak capacity and analyte detectability are considered. If liquid N_2 is used, then compounds in the C_3–C_{40} range can be analyzed; if, on the other hand, refrigeration units are used, then the volatility range becomes narrower, namely C_{10}–C_{40}. As the analyte molecular weight becomes higher (*e.g.*, C_{30}–C_{40}), an excessively low temperature of entrapment must be avoided (*e.g.*, by reducing the cold gas flow), while a higher reinjection temperature may be needed. The efficient modulation of compounds exceeding C_{40} is difficult, though possible if an ambient temperature cooling gas is used.

The modulation period is tightly related to the average ^{1}D peak base width: considering a common column combination – 30 m × 0.25 mm ID × 0.25 μm d_f + 1 m × 0.10 mm ID × 0.10 μm d_f – then w_{b1} values are normally in the 15–20 s range. In such an instance, a P_M value of 5–6 s range is highly advisable. If a P_M 6 s is applied, with an average ^{2}D peak base width (w_{b2}) of 200 ms and a void time of 1 second, then the ^2n_c reaches a value of 25 (considering no wraparound).

Considering a ${}^{1}n_c$ of 300, then ${}^{2D}n_c$ should reach a value of 7500. The real value, however, is less due to the partial selectivity correlation between the two GC dimensions (certain regions of the chromatograms are not accessible). If a longer ^{1}D column is used (*e.g.*, 100 m), then w_{b1} values will be greater, requiring both an increase in the modulation period and in the length of the ^{2}D column.

3.3.4 Detection

Though various opinions exist, for reliable quantification, a peak profile must be characterized by at least 7 data points above the baseline, equivalent to approximately 10 data points comprising the baseline [28]. Hence, a GC×GC peak with a baseline width of 200 ms will require a detector acquisition frequency of at least 50 Hz. An excessively high acquisition frequency will lead to a reduction of analyte detectability due mainly to an enhancement in high-frequency noise. Furthermore, extra column sources of band broadening must be reduced to a minimum in the presence of such narrow chromatography bands.

Several forms of detection have been used in the GC×GC field, both universal and selective, destructive, and nondestructive. Before 1999 [29] – year of the first paper based on the use of GC×GC-MS (single quadrupole MS – QMS) – the flame ionization detector (FID) was the most commonly used device [30]. The FID possesses many of the requirements for the detection of rapidly-eluting GC×GC peaks: limited internal volume, high acquisition rate and rapid rise time, uniformity of analyte response (for straightforward quantification), low economic cost, reduced necessity of maintenance, and structural robustness.

An alternative to the FID is the helium ionization detector, a universal detector suitable also for compounds characterized by no or a limited FID response. Franchina *et al.* used a helium ionization detector – the barrier discharge ionization detector (BID) – in FM-based applications [31]. It was demonstrated that the BID was characterized by a more restricted linear dynamic range and enhanced analyte detectability compared to the FID, especially for aromatic hydrocarbons (a diesel fuel was subjected to analysis). The FID and BID, however, will provide only a peak area, along with retention times. Chromatograms derived from a GC×GC analysis can be characterized by 100s, if not 1000s of compounds, and so higher levels of analyte information are necessary.

Selective forms of detection have been employed, providing a higher level of information on the presence of specific chemical elements, but none have gained wide use. For example, Khummueng *et al.* used GC×GC with dual detection – electron capture and nitrogen-phosphorus detection – for the analysis of organochlorine, organophosphorus, and *N*-containing pesticides [32]. The sulfur-chemiluminescence and nitrogen-chemiluminescence detectors (SCD/NCD), apart from specificity, have a great deal in common in terms of

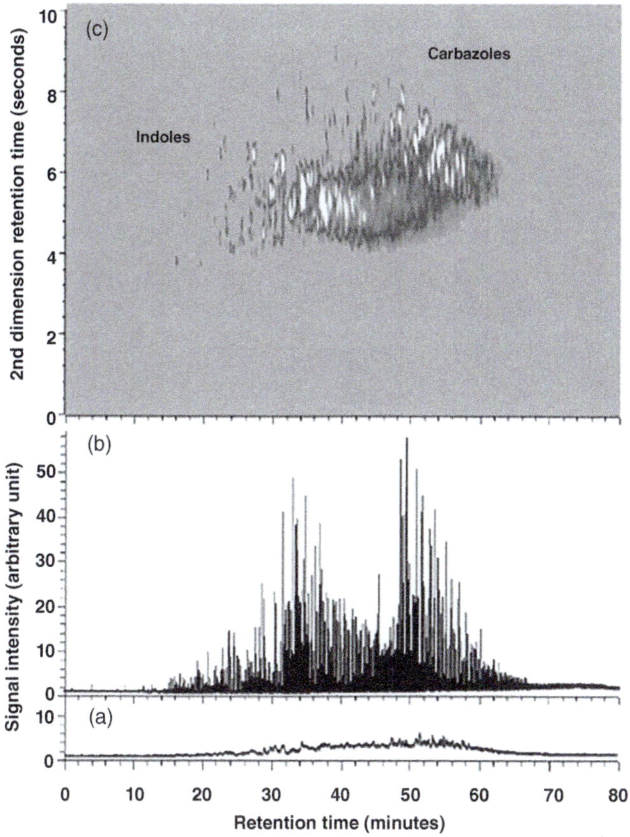

Figure 3.9 (a) GC-NCD chromatogram of a diesel fuel containing 40 ppm of total N; CM GC×GC-NCD untransformed (in b) and 2D (in c) chromatogram of the same sample. *Source:* Wang et al. [33], with permission of John Wiley & Sons.

the reaction process and operational aspects [33, 34]. Wang *et al.* used CM GC×GC-NCD to pinpoint indoles and carbazoles in diesel fuel [33]. The chromatograms shown in Figure 3.9a,b show the capability of cryogenic modulation to enhance analyte detectability (when first using GC-NPD), while that illustrated in Figure 3.9c highlights, in particular, the specificity of the NPD. The detector was operated at an acquisition frequency of 100 Hz and proved its suitability for GC×GC analyses.

Blomberg *et al.* used CM GC×GC-SCD for the analysis of S-containing compounds in petrochemical products [34]. The Authors determined that it was the SCD electronics, at not dead volumes, that had a major effect on band broadening. A modified electrometer resolved such a problem. A highly organized ^{2}D chromatogram, showing the great specificity of the SCD, is reported in Figure 3.10.

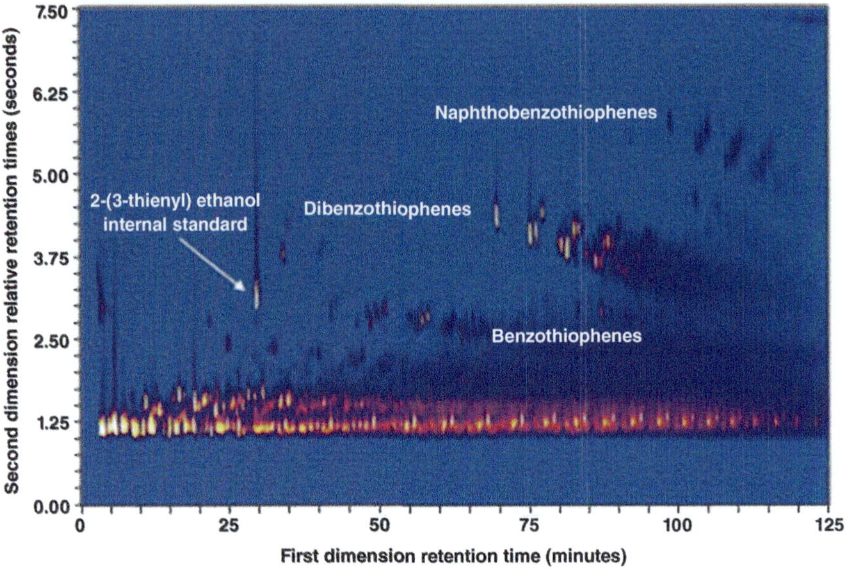

Figure 3.10 CM GC×GC-SCD chromatogram of a Liverpool Bay crude sample. *Source:* Blomberg et al. [34], with permission of Elsevier.

Other forms of selective detection have been used in the field of GC×GC, such as the atomic emission detector and the flame photometric detector [35, 36].

Higher levels of information can be provided by a universal and nondestructive form of detection, such as the ultraviolet detector. In this respect, a vacuum ultraviolet (VUV) detector developed for GC analysis [37] has been used within the context of both CM and FM GC×GC [38, 39]. Such a form of detection can bypass specific MS limitations, such as the identification of isomeric/isobaric compounds characterized by very similar spectra. The deconvolution of overlapping peaks is also possible, along with the possibility to perform pseudo-absolute quantification based on the exploitation of pre-recorded absorption cross-sections of target compounds. Information provided by GC×GC and the VUV detector is entirely orthogonal. Even so, there has been no widespread use of this ^{3}D technology.

The role of mass spectrometry, as for all other forms of column chromatography, has become predominant in the GC×GC field. The first GC×GC-MS paper emphasized the limitations of QMS and that low-resolution time-of-flight (LR TOF) MS would have been an "ideal solution" [29]. At that time, in fact, QMS systems could not reach the required acquisition frequencies of thermal modulation GC×GC. On the other hand, LR TOFMS could easily generate 100s of spectra/s, across a wide mass range, and with consistent spectral profiles (absence of skewing). Nowadays, the situation is completely different because current QMS instrumentation can scan at very high speeds (*e.g.*, 30,000 amu/s), with limited or no skewing [40]. Considering the overall history of GC×GC-MS, LR TOFMS instruments have been by far the most used, followed by QMS. Other forms of MS have been used, albeit to a much lesser extent. Applications describing the use of triple quadrupole MS (QqQMS), high-resolution (HR) TOFMS, Orbitrap MS, QTOFMS, and isotope ratio (IR) MS have been reported and represent a minority [41]. Triple quadrupole MS is characterized by high specificity and analyte detectability and is typically used in targeted analyses (*e.g.*, pesticides in foods), with a 1D GC pre-separation [42]. The use of CM GC×GC – in particular – combined with QqQMS would appear to be an excess in analytical terms. The use of HR MS, i.e., HR TOFMS, Orbitrap MS, again would seem to be excessive: the combination of GC×GC with LR MS covers practically all the necessities for untargeted and targeted analyses. In fact, it is rather difficult to find a sample for which the use of CM GC×GC-HR MS is really justified. Justification, on the other hand, could be found by using FM GC×GC-HR MS (or QqQMS). The same concept can be extended to QTOFMS; in the majority of GC×GC-QTOFMS studies, the Q mass analyzer was used mainly in the "fly through" mode [43, 44]. With regard to isotope ratio mass spectrometry, apart from potential problems related to band broadening (*e.g.*, in the combustion chamber), it must be emphasized that normally only a reduced number of target compounds are monitored. For such a reason, the use of heart-cutting MDGC appears to be the best option [45]. The reader is directed to the literature for more details on GC×GC-MS [41 and references therein], and to Chapters 4–6.

References

1 Dandeneau, R.D. and Zerenner, E.H. (1979). An investigation of glasses for capillary chromatography. *J. High. Resolut. Chromatogr.* 6 (2): 351–356. https://doi.org/10.1002/jhrc.1240020617.

2 Ragonese, C., Tranchida, P.Q., Dugo, P. et al. (2009). Evaluation of use of a dicationic liquid stationary phase in the fast and conventional gas chromatographic analysis of health-hazardous C18 cis/trans fatty acids. *Anal. Chem.* 81: 5561–5568. https://doi.org/10.1021/ac9007094.

3 Wang, F.C.-Y., Qian, K., and Green, L.A. (2005). GC×MS of diesel: a two-dimensional separation approach. *Anal. Chem.* 77 (9): 2777–2785. https://doi.org/10.1021/ac0401624.

4 Simmons, M.C. and Snyder, L.R. (1958). Two-stage gas-liquid chromatography. *Anal. Chem.* 30 (1): 32–35. https://doi.org/10.1021/ac60133a007.

5 Tranchida, P.Q., Sciarrone, D., Dugo, P., and Mondello, L. (2012). Heart-cutting multidimensional gas chromatography: a review of recent evolution, applications, and future prospects. *Anal. Chim. Acta* 716: 66–75. https://doi.org/10.1016/j.aca.2011.12.015.

6 Liu, Z. and Phillips, J.B. (1991). Comprehensive two-dimensional gas chromatography using an on-column thermal modulator interface. *J. Chromatogr. Sci.* 29 (6): 227–231. https://doi.org/10.1093/chromsci/29.6.227.

7 Liu, Z., Zhang, M., and Phillips, J.B. (1990). High-speed gas chromatographic analysis of a simulated process stream using on-column thermal desorption modulation for sample preconcentration and introduction. *J. Chromatogr. Sci.* 28 (11): 567–571. https://doi.org/10.1093/chromsci/28.11.567.

8 Phillips, J.B., Gaines, R.B., Blomberg, J. et al. (1999). A robust thermal modulator for comprehensive two-dimensional gas chromatography. *J. High. Resolut. Chromatogr.* 22 (1): 3–10. https://doi.org/10.1002/(SICI)1521-4168(19990101)22:1<3::AID-JHRC3>3.0.CO;2-U.

9 Kinghorn, R.M. and Marriott, P.J. (1998). Comprehensive two-dimensional gas chromatography using a modulating cryogenic trap. *J. High. Resolut. Chromatogr.* 21 (11): 620–622. https://doi.org/10.1002/(SICI)1521-4168(19981101)21:11<620::AID-JHRC620>3.0.CO;2-#.

10 Kinghorn, R.M. and Marriott, P.J. (1997). Longitudinally modulated cryogenic system. A generally applicable approach to solute trapping and mobilization in gas chromatography. *Anal. Chem.* 69 (13): 2582–2588. https://doi.org/10.1021/ac961310w.

11 Ledford, E.B. Jr., and Billesbach, C. (2000). Jet-cooled thermal modulator for comprehensive multidimensional gas chromatography. *J. High. Resolut. Chromatogr.* 23 (3): 202–204. https://doi.org/10.1002/(SICI)1521-4168(20000301)23:3<202::AID-JHRC202>3.0.CO;2-5.

12 Tranchida, P.Q., Purcaro, G., Dugo, P., and Mondello, L. Modulators for comprehensive two-dimensional gas chromatography. *TrAC Trends Anal. Chem.* 30: 1437–1461. https://doi.org/10.1016/j.trac.2011.06.010.

13 Bahaghighat, H.D., Freye, C.E., and Synovec, R.E. (2019). Recent advances in modulator technology for comprehensive two dimensional gas chromatography. *TrAC Trends Anal. Chem.* 113: 379–391. https://doi.org/10.1016/j.trac.2018.04.016.

14 Bruckner, C.A., Prazan, B.J., and Synovec, R.E. (1998). Comprehensive two-dimensional high-speed gas chromatography with chemometric analysis. *Anal. Chem.* 70: 2796–2804. https://doi.org/10.1021/ac980164m.

15 Mohler, R.E., Prazen, B.J., and Synovec, R.E. (2006). Total-transfer, valve-based comprehensive two-dimensional gas chromatography. *Anal. Chim. Acta* 555: 68–74. https://doi.org/10.1016/j.aca.2005.08.072.

16 Seeley, J.V., Micyus, N.J., McCurry, J.D., and Seeley, S.K. (2006). Comprehensive two-dimensional gas chromatography with a simple fluidic modulator. *Am. Lab.* 38 (9): 24–26.

17 Giardina, M., McCurry, J.D., Cardineal, P. et al. (2018). Development and validation of a pneumatic model for the reversed-flow differential flow modulator for comprehensive two-dimensional gas chromatography. *J. Chromatogr. A* 1577: 72–81. https://doi.org/10.1016/j.chroma.2018.09.022.

18 Seeley, J.V., Schimmel, N.E., and Seeley, S.K. (2018). The multi-mode modulator: a versatile fluidic device for two-dimensional gas chromatography. *J. Chromatogr. A* 1536: 6–15. https://doi.org/10.1016/j.chroma.2017.06.030.

19 Galletta, M., Zoccali, M., Jones, N. et al. (2022). Flow-modulated comprehensive two-dimensional gas chromatography combined with time-of-flight mass spectrometry: use of hydrogen as a more sustainable alternative to helium. *Anal. Bioanal.Chem.* 414 (21): 6371–6378. https://doi.org/10.1007/s00216-022-04086-4.

20 Seeley, J.V., Libby, E.M., Hill Edwards, K.A., and Seeley, S.K. (2009). Solvation parameter model of comprehensive two-dimensional gas chromatography separations. *J. Chromatogr. A* 1216 (10): 1650–1657. https://doi.org/10.1016/j.chroma.2008.07.060.

21 Liberto, E., Cagliero, C., Sgorbini, B. et al. (2008). Enantiomer identification in the flavour and fragrance fields by "interactive" combination of linear retention indices from enantioselective gas chromatography and mass spectrometry. *J. Chromatogr. A* 1195 (1–2): 117–126. https://doi.org/10.1016/j.chroma.2008.04.045.

22 Bizzo, H.R., Brilhante, N.S., Nolvachai, Y., and Marriott, P.J. (2023). Use and abuse of retention indices in gas chromatography. *J. Chromatogr. A* 1708: 464376. https://doi.org/10.1016/j.chroma.2023.464376.

23 Nolvachai, Y., Kulsing, C., and Marriott, P.J. (2015). Thermally sensitive behavior explanation for unusual orthogonality observed in comprehensive two-dimensional gas chromatography comprising a single ionic liquid stationary phase. *Anal. Chem.* 87 (1): 538–544. https://doi.org/10.1021/ac5030039.

24 Delmonte, P., Fardin-Kia, A.R., and Rader, J.I. (2013). Separation of fatty acid methyl esters by GC-online hydrogenation×GC. *Anal. Chem.* 85 (3): 1517–1524. https://doi.org/10.1021/ac302707z.

25 Tranchida, P.Q., Casilli, A., Dugo, P. et al. (2007). Generation of improved gas linear velocities in a comprehensive two-dimensional gas chromatography system. *Anal. Chem.* 79: 2266–2275. https://doi.org/10.1021/ac0618066.

26 Tranchida, P.Q., Franchina, F.A., Dugo, P., and Mondello, L. (2014). Use of greatly-reduced gas flows in flow-modulated comprehensive two-dimensional gas chromatography-mass spectrometry. *J. Chromatogr. A* 1359: 271–276. https://doi.org/10.1016/j.chroma.2014.07.054.

27 Tranchida, P.Q., Franchina, F.A., Dugo, P., and Mondello, L. (2014). Flow-modulation low-pressure comprehensive two-dimensional gas chromatography. *J. Chromatogr. A* 1372: 236–244. https://doi.org/10.1016/j.chroma.2014.10.097.

28 Adahchour, M., Brandt, M., Baier, H.-U. et al. (2005). Comprehensive two-dimensional gas chromatography coupled to a rapid-scanning quadrupole mass spectrometer: principles and applications. *J. Chromatogr. A* 1067 (1–2): 245–254. https://doi.org/10.1016/j.chroma.2004.09.094.

29 Frysinger, G.S. and Gaines, R.B. (1999). Comprehensive two-dimensional gas chromatography with mass spectrometric detection (GC×GC/MS) applied to the analysis of petroleum. *J. High. Resolut. Chromatogr.* 22 (5): 251–255. https://doi.org/10.1002/(SICI)1521-4168(19990501)22:5<251::AID-JHRC251>3.0.CO;2-V.

30 Phillips, J.B. and Beens, J. (1999). Comprehensive two-dimensional gas chromatography: a hyphenated method with strong coupling between the two dimensions. *J. Chromatogr. A* 856 (1–2): 331–347. https://doi.org/10.1016/s0021-9673(99)00815-8.

31 Franchina, F.A., Maimone, M., Sciarrone, D. et al. (2015). Evaluation of a novel helium ionization detector within the context of (low-)flow modulation comprehensive two-dimensional gas chromatography. *J. Chromatogr. A* 1402: 102–109. https://doi.org/10.1016/j.chroma.2015.05.013.

32 Khummueng, W., Morrison, P., and Marriott, P.J. (2008). Dual NPD/ECD detection in comprehensive two-dimensional gas chromatography for multi-class pesticide analysis. *J. Sep. Sci.* 31: 3404–3415. https://doi.org/10.1002/jssc.200800222.

33 Wang, F.C.-Y., Robbins, W.K., and Greaney, M.A. (2004). Speciation of nitrogen-containing compounds in diesel fuel by comprehensive two-dimensional gas chromatography. *J. Sep. Sci.* 27: 468–472. https://doi.org/10.1002/jssc.200301643.

34 Blomberg, J., Riemersma, T., van Zuijlen, M., and Chaabani, H. (2004). Comprehensive two-dimensional gas chromatography coupled with fast sulphur-chemiluminescence detection: implications of detector electronics. *J. Chromatogr. A* 1050 (1): 77–84. https://doi.org/10.1016/j.chroma.2004.07.105.

35 van Stee, L.L.P., Beens, J., Vreuls, R.J.J., and Brinkman, U.A.T. (2003). Comprehensive two-dimensional gas chromatography with atomic emission detection and correlation with mass spectrometric detection: principles and application in petrochemical analysis. *J. Chromatogr. A* 1019 (1–2): 89–99. https://doi.org/10.1016/S0021-9673(03)01301-3.

36 Chin, S.-T., Wu, Z.-Y., Morrison, P.D., and Marriott, P.J. (2020). Observations on comprehensive two-dimensional gas chromatography coupled with flame photometric detection for sulfur- and phosphorus-containing compounds. *Anal. Meth.* 2: 243–253. https://doi.org/10.1039/B9AY00143C.

37 Schug, K.A., Sawicki, I., Carlton, D.D. Jr., et al. (2014). Vacuum ultraviolet detector for gas chromatography. *Anal. Chem.* 86: 8329–8335. https://doi.org/10.1021/ac5018343.

38 Gröger, T., Gruber, B., Harrison, D. et al. (2016). A vacuum ultraviolet absorption array spectrometer as a selective detector for comprehensive two-dimensional gas chromatography: concept and first results. *Anal. Chem.* 88 (6): 3031–3039. https://doi.org/10.1021/acs.analchem.5b02472.

39 Zoccali, M., Schug, K.A., Walsh, P. et al. (2017). Flow-modulated comprehensive two-dimensional gas chromatography combined with a vacuum ultraviolet detector for the analysis of complex mixtures. *J. Chromatogr. A* 1497: 135–143. https://doi.org/10.1016/j.chroma.2017.03.073.

40 Usui, K. and Nagao, Y. (2024). Comprehensive characterization of diesel fuel on GC×GC Utilizing impressive high-speed scan technology of GCMS-QP2050. https://www.shimadzu.com/an/sites/shimadzu.com.an/files/pim/pim_document_file/applications/application_note/24333/an_01-00788-en.pdf.

41 Tranchida, P.Q., Giocastro, B., and Mondello, L. (2022). Multidimensional gas chromatography: Hyphenation with mass spectrometry. *Comprehensive Analytical Chemistry* 96: 93–118. https://doi.org/10.1016/bs.coac.2021.10.004.

42 Lee, J., Kim, L., Shin, Y. et al. (2017). Rapid and simultaneous analysis of 360 pesticides in brown rice, spinach, orange, and potato using microbore GC-MS/MS. *J. Agric. Food. Chem.* 65 (16): 3387–3395. https://doi.org/10.1021/acs.jafc.7b00576.

43 Chin, S.-T., Nolvachai, Y., and Marriott, P.J. (2014). Enantiomeric separation in comprehensive two-dimensional gas chromatography with accurate mass analysis. *Chirality* 26 (11): 747–753. https://doi.org/10.1002/chir.22280.

44 Megson, D., Ortiz, X., Jobst, K.J. et al. (2016). A comparison of fresh and used aircraft oil for the identification of toxic substances linked to aerotoxic syndrome. *Chemosphere* 158: 116–123. https://doi.org/10.1016/j.chemosphere.2016.05.062.

45 Juchelka, D., Beck, T., Hener, U. et al. (1998). Multidimensional gas chromatography coupled on-line with isotope ratio mass spectrometry (MDGC-IRMS): progress in the analytical authentication of genuine flavor components. *J. High. Resolut. Chromatogr.* 21 (3): 145–151. https://doi.org/10.1002/(SICI)1521-4168(19980301)21:3<145::AID-JHRC145>3.0.CO;2-Z.

4

Basic Principles of Mass Spectrometry

Nadine Gawlitta[1], Barbara Giocastro[2], Thomas Gröger[2], Beate Gruber[3], Christopher Rüger[4], Benedikt Weggler[3], and Ralf Zimmermann[4]

[1]*Department of Plant and Environmental Sciences, University of Copenhagen, Frederiksberg, Denmark*
[2]*Chemical Kinetics and Analytics, Institute of Combustion Technology, DLR – German Aerospace Center, Stuttgart, Germany*
[3]*Analytical and Material Science, BASF SE, Ludwigshafen, Germany*
[4]*Institute of Chemistry - Analytical Chemistry at University of Rostock, University of Rostock, Rostock, Germany*

4.1 Key Concepts and Definitions

4.1.1 Getting Started

The first mass spectrum was reported by Thomson in 1913, marking a significant milestone in the development of mass spectrometry. In the late 1950s, with the advent of gas chromatography (GC), the first online gas chromatography-mass spectrometry (GC-MS) experiment was reported. In 1959, Gohlke coupled a GC with a time-of-flight (TOF) mass spectrometer. This initial experiment enabled a mass unit resolution up to a mass of 200 and generated 2000 spectra/s, facilitating the collection of MS spectra for a five-compound mixture separated on a packed column. In his manuscript, Gohlke stated: “Single chromatographic peaks containing two or three components can usually be successfully resolved by a careful examination of several mass spectra obtained at various times during the development of the chromatographic peak.” This statement provided the earliest definition of deconvolution, which is one of the key processing steps for gas chromatography - mass spectrometry. This also shows the unique value and strength of such a coupling, namely the high physical separation of the individual analytes by gas chromatography and the possibility of very selective quantitative and qualitative characterization of the eluent by mass spectrometry. In 1999, Frysinger and Gaines demonstrated the first hyphenation with a comprehensive two-dimensional GC

(GC×GC) system. In their pioneering experiment, a quadrupole MS (QMS) instrument generated a full scan mass spectrum for each resolved peak (over 750 peaks) in a sample of marine diesel fuel. However, full scan spectrum acquisition was limited to 2.43 scans per second. Given that a typical GC×GC peak has a width of about 200 ms at the baseline, the MS acquisition rate was insufficient. To address this challenge, the authors broadened the GC peaks from 200 ms to nearly one second by lengthening the first dimension. This adjustment allowed the QMS to sample each narrow peak with approximately three full scan mass spectra. Modern mass spectrometers coupled to GC×GC systems now meet the MS acquisition frequency requirement (≥50 Hz), enabling the acquisition of a minimum of 10 spectra per peak [1–4].

The advantages of coupling gas chromatography with a mass spectrometer have made it one of the most versatile and powerful analytical techniques for separating, identifying, and quantifying chemical compounds in analytical chemistry. After the elution from gas chromatography, molecules are ionised to create ions, which can be subsequently separated and detected in a mass analyser according to their mass-to-charge ratio (*m/z*). The ionization could be achieved thermally, by electrical fields, or by impacting energetic electrons, ions, or photons. Depending on the transferred energy, ion source geometry, electrical fields, and pressure conditions, the ions can be single or multiply charged, form clusters, or fragment. Given the variety of ionization techniques and mass analysers, the following sections will discuss the most commonly encountered mass spectrometric hyphenations for gas chromatography. It is important to emphasise that mass spectrometry is not a single analytical technique but a highly diverse analytical toolbox with numerous facets. Unlike other analytical techniques, such as various spectroscopic methods, mass spectrometry is a destructive technique, "consuming the analyte". However, the extremely low sample consumption of mass spectrometry makes it the method of choice when other analytical techniques fail because they cannot yield analytical information from nanogram amounts of sample. This capability, combined with the ability of mass spectrometry to provide deep molecular and structural insights into complex organic mixtures, explains the high popularity of GC-MS across almost all scientific and industrial domains [5–8].

4.1.2 The Mass Spectrum and Some Definitions

Results from mass spectrometric analysis are displayed as mass spectra. This visualization shows the *m/z* on the *x*-axis (abscissa) and the detected signal strength on the *y*-axis (ordinate). Signals are commonly referred to as peaks, and the signal strength, or intensity, correlates with the abundance of that ion. Mass spectra using the physical scale of mass per electric charge (SI kg C-1) would be impractical and difficult to handle. Consequently, analytical chemists use the unified atomic mass scale (*u*) instead. The *m/z* is an artificial construct, not a physical unit, used as a dimensionless quantity and converted into the mass of an ion for chemical meaning. In gas chromatography-mass spectrometry, mostly single-charged ions are observed ($z = 1$), meaning the *m/z* scale directly reflects the atomic mass scale, making it easy to interpret. However, depending on the ionization technique, some molecular classes, such as polycyclic aromatic hydrocarbons (PAHs), can be found in higher charge states ($z > 1$). The charge state of an ion may be identified by its isotopic fingerprint, the location, and the abundance profile of the isotopic signals of the respective compound. As mass spectrometry essentially "weighs" ions, the naturally occurring isotopic distributions of the elements are separated into a characteristic pattern called the isotopic fine structure (IFS). For carbon-based organic molecules, the distance between peaks containing only ^{12}C and one ^{13}C isotope can be used to calculate *z*. The difference in mass between these two isotopes is 1.003355 *u*, leading to a difference in the mass spectra location of 1.003355/*z*. The abundance of the isotopic signals can be estimated for rough confirmation or calculated with higher precision for automatic matching. However, calculations considering the correct mass spectrometric parameters, such as peak shape and resolving power, for heavier compounds containing heteroatoms can be highly computationally intensive.

Apart from *u*, other units for *m/z* have historically been used or are still used by some mass spectrometrists. In honour of J. J. Thomson, the unit thomson (Th) is still found and is used equivalently to *m/z*. The *u* is based on the exact mass of the ^{12}C isotope. Until 1961, the *amu* (atomic mass unit) was based on the exact mass of the ^{16}O isotope. Although the deviation seems minor, it is particularly important for high-resolution and mass-accuracy mass spectrometry, and the IUPAC recommends solely using the ^{12}C-normalised scale. In a physical context, mass spectra may also be represented using SI units, which is correct but requires a conversion factor for direct chemical access.

For the ordinate scaling, absolute and relative representations must be distinguished. Absolute scaling displays the read-back of the detector, given as a current (A) or often as an arbitrary unit (a.u.). Most representations normalise signal abundance to 100% relative intensity. The most intense peak of a mass spectrum is called the base peak, while the intact non-fragmented signal of a compound is called the molecular ion, $[M]^{+\bullet}$. Both approaches are often used for normalization. Sometimes, normalization to the total ion count of the displayed spectra or other more specific values, such as subjected concentration, read-back of an added standard, or constant background from the mass spectrometer, can be found. The reader is highly encouraged to check the abundance scaling before comparing mass spectra from different platforms or the literature.

In gas chromatography-mass spectrometry, many different acronyms for ion chromatograms may be found. The most common are the total ion chromatogram (TIC) and extracted ion chromatogram (EIC). Depending on the context, the base peak chromatogram (BPC), which displays the base peak intensity of each spectrum plotted versus the retention time, may also be used [9].

4.1.3 Resolution, Resolving Power, and Accurate Mass

One way to classify mass spectrometers is to categorise them according to their ability to determine and separate the mass of ions. This results in the two categories unit-mass and high-resolution instruments. The ability to separate signals in a mass spectrum is termed mass resolving power (R, dimensionless), which is defined as the smallest difference in m/z that can be separated for a given signal at a given m/z value, whereas mass resolution is defined as $\Delta(m/z)$, commonly given at full width at half maximum (FWHM). This nomenclature is inconsistently used in the literature and often used interchangeably. Adding to the confusion, for older sector field mass spectrometer the peak width, used for calculating the resolution, may be given at 10% peak height. High-resolution mass spectrometry has become a routine technique in many analytical laboratories, but a unified definition convincing all mass spectrometrists is still under debate. The reader is referred to the "Definitions of Terms Relating to Mass Spectrometry" by the IUPAC (2013) [9]. Interestingly, there are also large differences in the notation and usage of these values and terminology between mass spectrometry vendors. Thus, when reporting resolution or resolving power values, one should always give the m/z at which the measurement was made and the methodology by which the value was determined.

$$R = \frac{m}{\Delta m} \tag{4.1}$$

$$= \frac{m/z}{\Delta(m/z)} \tag{4.2}$$

Even though high-resolution and accurate mass are technically closely related parameters and often depend on each other, they should not be confused. A measurement can be performed at high mass resolution (separating the mass information of different ions) but deliver insufficient accurate mass (real mass of the separated ions) results. In principle, high mass resolution allows to separate adjacent signals, whereas accurate mass gives us the ability to derive molecular formula information. The latter concept is based on the mass defect of the elements caused by the mass-to-energy relationship and different energetic arrangements of nucleotides in the atom core. A more detailed description on attributing elemental compositions based on exact mass measurement and the benefits for chemical characterization of unknowns and complex mixtures is given at the end of this chapter. Briefly, the element and isotope dependent mass defect results in unique exact masses for a given elemental composition, which can mathematically be utilized for in-silico calculation of elemental formulae solely based on accurate mass measurement. However, this idealized concept has two major premises: (i) the *m/z* signal has to be separated from other adjacent compound signals, and (ii) the exact mass measurement has to be performed with sufficient accuracy. The first point can be achieved by high-(mass)resolution measurements and/or separating the compounds physically before mass spectrometric detection (e.g., by gas chromatography). Depending on the chemical complexity of the sample, different performance classes of mass spectrometric resolving power are required. These ideas are further outlined in section 4.3.2.5 on "The High Resolution Advantage" [10].

4.1.4 Chromatographic Hyphenation

Chromatographic and mass spectrometric parameters are often considered independently. Therefore, in hyphenated systems, the emphasis is typically on either a detector- or chromatography-focused approach. Commonly, the focus is placed on the mass spectrometer due to its information richness and complexity. However, the quality parameters describing chromatographic peaks have a significant influence on the quality of the mass spectrum, which depends on how the mass spectra are collected and generated during post-processing (Table 4.1).

When obtaining mass spectra, multiple spectra are usually obtained over time intervals or under different conditions. Peak apex spectra refer to the mass spectrum at the highest point of the peak, while mean spectra are obtained by averaging ion intensities at each *m/z* value across all acquired spectra. Mean spectra of a peak can be obtained by averaging all spectra associated with that peak over a defined time interval. (e.g., peak width is 10 seconds and the data acquisition rate is 10 Hz, the mean spectrum of the peak would be obtained by averaging 100 mass spectra.)

4.2 Working Principles and Instrumentation

The second section of this chapter is devoted to the hardware of the mass spectrometer. It follows the molecules as they elute from the gas chromatograph ("The Gas Phase Interface") and discuss the techniques required to first form ions ("Principle of Ion Formation and Ionization Techniques"), then separate and determine the masses of the ions ("The mass analyser"). Finally, the quantity of ions is determined ("Ion Detection and Detectors") and converted into an electronic signal ("Electronic signal processing"), which later results in a

Table 4.1 Non-exhaustive list of chromatographic parameters describing the desired value and their importance for chromatography and mass spectrometry.

Parameter	Desired value	Chromatography	Mass spectrometry
(A)Symmetry	1 for gaussian peaks	Efficient separation and elution of the compounds from the column	Consistent and predictable distribution of ions. Accurate determination of molecular weights and identification of compounds as peak shapes reflects the mass distribution of ions within the peak
Retention time and relative retention time	Consistent across different measurements	Fundamental Parameter for compound identification, quantification, and reproducibility	Complementary Parameter for compound identification and quantification
Peak width	small	Efficient separation	Cleaner mass spectra due to reduced overlap between compounds
Peak area	Consistent across different measurements	Quantitative analysis	Quantitative analysis
Baseline	Stable and well defined	Accurate peak detection	Accurate peak detection
Resolution	~2.5 for baseline separation	Efficient separation	Cleaner mass spectra due to reduced overlap between compounds

chromatogram that combines the chromatographic and mass spectrometric information of the ions.

4.2.1 The Gas Phase Interface

The eluent from a GC system, emerging from the end of a GC column, is analysed by the mass spectrometer in a time-resolved manner, providing both qualitative and quantitative information. The transition between the gas chromatograph and the mass spectrometer is technically achieved through an interface. A key aspect is that the molecules are already in the gas phase, and the analytes are diluted in the carrier gas. Consequently, GC systems predominantly apply gas-phase interfaces, as volatilization of the matrix, such as from a liquid gradient, is unnecessary. The interface's primary role is to ensure an artefact-free transition from the column to the ionization space, where charged ions are formed from uncharged molecules. This ionization preferably occurs under reduced pressure to prevent collision-induced reactions, maximising the mean free path length of the ions. Such low-pressure interfaces, however, result in ions having a high, spatially non-uniform kinetic energy, which must be corrected for accurate mass determination.

Despite the compatibility of GC and MS systems regarding operating temperatures, vapour handling, and small sample requirements (nanograms), challenges arise from the need to reduce the pressure and density of the eluent to the low-pressure regions needed for ionization (10E-7 – 10E-8 bar) and even lower pressures for ion separation in the analyser (10E-8 – 10E-10 bar). In modern GC-MS systems, this pressure transition is managed through a heated interface or transfer line, where the gas expands into a low-pressure region constantly evacuated by pumps such as turbomolecular pumps. Historically, direct interfacing was challenging with older packed column GC-MS systems due to high carrier gas flow rates (up to 30 mL/min) incompatible with low-pressure demands. Specialized interfaces, like jet separators, were used to reduce the gas flow entering the ionization region to within the pump's capacity. Today, capillary GC systems typically operate at flow rates of 1–2 mL/min, compatible with most ionization interfaces. However, GC×GC hyphenation, especially with flow modulation (FM), can be challenging. In flow modulated GC×GC-MS, an auxiliary flow rate (often ≥20 mL/min), much higher than the 1D column flow, is used for rapid re-injection (e.g., 100 ms) of gas-phase contents from accumulation loops onto the 2D column. These high gas flows exceed the pumping capabilities of modern MS systems. Splitting the ^{2}D column effluent to a second detector (e.g., ECD or FID) can aid coupling but may compromise sensitivity. Tranchida et al. demonstrated that optimising flow modulation conditions, such as longer re-injection stages or accumulation loops, can reduce the ^{2}D column flows to 6–8 mL/min, compatible with MS systems, while maintaining acceptable peak shape and capacity. Further reduction to approximately 4 mL/min has been proposed by later studies.

In addition, several other technical aspects have to be considered and technically realized: the interface must operate at elevated temperatures to prevent condensation; dead volumes or loss of the deactivation layer at the column or transfer line must be avoided to prevent peak asymmetry; and high flow rates, which could benefit peak band broadening, must be avoided to maintain adequate vacuum conditions at the interface [11–13].

4.2.1.1 Carrier Gas Consideration

An essential aspect in the application of GC is the choice of carrier gas. When coupling GC to MS, the potential effects on the ionization process within the gas-phase interface must also be considered. Thus, carrier gas selection in GC-MS is critical for optimal analysis, taking into account both chromatographic and mass-spectrometric factors. High purity levels are vital to minimise interference from impurities such as oxygen, hydrocarbons, and moisture, which can cause baseline drift, increase background noise, and disrupt analyte detection. Nitrogen, helium, and hydrogen are commonly used in GC analysis, each providing similar optimal plate height but at significantly different flow rates. Helium has traditionally been preferred for its inertness and efficiency. However, concerns over its cost and sustainability have spurred interest in alternatives like nitrogen and hydrogen. While nitrogen is a potential alternative, it can reduce chromatographic efficiency and extend analysis times due to its narrower optimal range of linear velocity. Additionally, nitrogen's relatively high mass may fall within the range of ion masses relevant for analysis. Generally, ions formed by the ionization of the carrier gas should be technically separated to avoid degrading detector performance. Separation, however, entails a cut-off of smaller masses and possible discrimination of larger, closer masses.

In recent years, there has been increasing interest in using hydrogen as a carrier gas in GC(×GC)-MS, especially in applications requiring high-speed GC separation. With a higher diffusion coefficient and half the viscosity of helium and nitrogen, hydrogen allows for faster GC(×GC)-MS while maintaining similar resolution and chromatographic performance to helium. A study by Galletta et al. compared hydrogen and helium as carrier gases in flow-modulated GC×GC-TOFMS analysis. While hydrogen demonstrated comparable chromatographic resolution to helium, it significantly reduced analysis time by approximately 25%. However, differences in mass spectral profiles were noted, with compound-specific variations. Despite its benefits, hydrogen use can result in drawbacks such as decreased sensitivity, increased background noise, and heightened risk of vacuum instability. Additionally, hydrogen's less inert nature compared to helium can lead to reactions in the ion source, diminishing the response of certain analytes, affecting detection capabilities, and generating ionic background noise, particularly at masses below 200 Da.

If the aim is to switch to hydrogen, the hazard potential must always be assessed in addition to the performance effects. Many manufacturers already certify their MS systems accordingly. However, this is by no means the case for all equipment currently in use. The probability of a crash may seem low, but the potential risks to people and property are enormous [14].

4.2.2 Principle of Ion Formation and Ionization Techniques

Mass spectrometry is based on the separation of charged molecules. The ionization process takes place very close to the point where the eluents exit the chromatographic separation column. The region within the gas phase interface where the ionization is intended is called ion source. In mass spectrometry, various ionization techniques are employed, differing in their instrumental setup and ion generation. One way to classify them is to asseign them to hard and soft techniques. This classification is based on the transferred amount of energy exceeding the energy needed to convert a neutral molecule to an ionised state. Ionization energies (IE) of most organic molecules range from 7–12 eV, and an "overhead" in internal energy is likely to cause fragmentation of the molecule. In gas chromatography-mass spectrometry, electron ionization (EI) at 70 eV is most widely used as a hard ionization technique, and the molecular ion is usually accompanied by several fragment ion peaks at lower *m/z*. To gain a basic understanding of ion formation and fragmentation, a fundamental overview of most common ionization mechanisms for GC-MS will be provided. Electron ionization is initially used as an example. The concepts presented here can then be transferred to other ionization techniques. Ionization generally begins with the formation of ions.

In the case of electron ionization, a high-energy electron passes through a neutral organic molecule, transferring enough of its own kinetic energy to exceed the ionization potential of the molecule. This can lead to the formation of a molecular ion through the removal of an electron. This is the most common ionization process for electron ionization.

$$M + e^- \rightarrow M^{+\bullet} + 2e^- \tag{4.3}$$

However, depending on the energy of the primary electron, the occurrence of double and triple charged ions can also be observed.

After initial ionization and generation of the molecular ion, concurrence and follow-up actions occur. In addition to the desired Ion formations, various other events can result from the interaction between the primary electron and the neutral molecule. As a rule of thumb, with the increase in transferred energy, the frequency and diversity of accompanying processes increase ("hard ionization"). Each of these pathways leads to the formation of characteristic ionized and neutral products, such as molecular ions, fragment ions, multiply charged ions, metastable ions, rearranged ions, and ion pairs.

Regardless of the ionization method, the analyte is in thermal and energetic equilibrium with the inlet system before ionization. Most of the energy of the molecule is determined by its thermal energy. With ionization, the generated ion must internally absorb and distribute much of the energy among its degrees of freedom. The main portion is allocated to the vibrational degrees of freedom of the molecule, as the translational and rotational degrees of freedom cannot contribute significantly. Thus, the ions generated by electron impact are vibrationally excited. Many of these ions possess energies well above the dissociation energies of the molecule. The dissociation of $M^{+\bullet}$ then leads to the formation of a fragment ions (m) and a neutral fragments (n).

$$M^{+\bullet} \rightarrow m_1^{+} + n \tag{4.4}$$

$$\searrow$$

$$m_1^{\bullet} + n^{+} \tag{4.5}$$

There are numerous fragmentation pathways involved. While ionization occurs in the low femtosecond range, fragmentation reactions generally take place within a microsecond. This means that fragment ions may still be formed even after the excited species have already left the ion source [6].

4.2.2.1 Electron Ionization

Electron ionization (EI) is the most employed ionization method in GC-MS. In an EI ion source, a filament (electrical resistance heated wire) emits electrons that are accelerated, directed, and focused to form an ion beam by an electric field. When the eluent from the GC enters the ion source, the electrons from the beam interact with the molecules according to a defined probability. The probability, and consequently the ionization yield, is primarily influenced by the kinetic energy of the electrons and the structure and composition of the molecules. It varies across different compound classes, homologous series, and isomers.

The electrons must have sufficient kinetic energy to remove outer shell electrons from the molecules. This interaction produces additional electrons and a positive radical ion as described above. Due to the high kinetic energy, substantial fragmentation of the molecular ion occurs, resulting in distinct fragment ions and neutral radicals (see Figure 4.1). The extensive fragmentation characteristic of EI classifies it as a "hard" ionization technique. The minimum energy required to ionize a neutral atom or molecule is defined by its ionization energy (IE). Generally, most organic molecules have IEs ranging from 7 to 15 eV. If an electron carries energy close to the IE of a molecule, it must transfer all energy upon collision, resulting in ionization. However, since such collisions are improbable, the ionization efficiency is nearly zero. A slight increase in electron energy can incrementally enhance ionization efficiency. Since EI efficiency is contingent on the intrinsic properties of the ionization process and the ionization

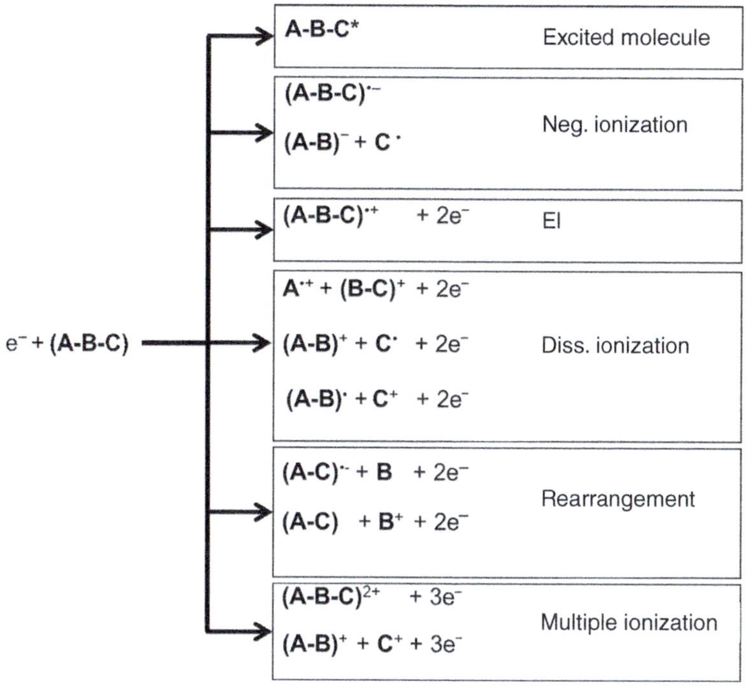

Figure 4.1 Variety of the accompanying processes during electron ionization. Typically, the excitation and negative ionization are processes to be neglected. *Source:* Gross [6] / Springer Science & Business Media.

cross-section of an analyte, each molecule exhibits its own ionization efficiency curve. These curves, plotting ionization cross-section against electron energy, display a consistent type, peaking at approximately 70 eV. The plateau in ionization efficiency around 70 eV for a wide range of organic molecules facilitates reliable fragmentation and reproducible spectra. At lower energies (around 20 eV), the energy transferred during electron-molecule interactions is inadequate for ionization. However, at approximately 70 eV, the de Broglie wavelength of the electrons matches the typical bond lengths in organic molecules (approximately 0.14 nm), optimizing energy transfer and yielding the highest ionization efficiency. Beyond this optimal energy level, the de Broglie wavelength shortens relative to bond lengths, making the molecules effectively "transparent" to the electrons and reducing ionization efficiency. Thus, the ionization cross-section of most organic molecules peaks around 70 eV. Minor variations in electron energy within the 60–80 eV range are negligible. Consequently, mass spectra acquired from different instruments can reliably be compared to one another and to mass spectral libraries.

Even small molecules can exhibit several fragmentation pathways, including primary, secondary, and higher degrees of fragmentation products. The activation energy for these reactions is supplied by the excess energy in EI (70 eV). Ion dissociation is an endothermic process, consuming internal ion energy with each fragmentation. Thus, highly excited ions may undergo multiple consecutive cleavages, while others may experience one or none. However, the typical excess electron energy (70 eV) used for ionization can be reduced by accelerating the electrons to lower kinetic energies, such as 12–15 eV, which is sufficient to ionize most molecules and reduces unwanted fragmentation. In conjunction with low ion source temperatures and/or jet cooling of the eluent, this can further reduce fragmentation. Therefore, using low electron energy has advantages, such as enhanced relative sensitivity of the molecular peak and less overall fragmentation. In general, the fragmentation pattern will be shifted to higher masses. However, low electron energy also results in reduced ionization efficiency, leading to an overall loss of absolute sensitivity. Additionally, low ion source temperatures can cause memory effects from previous samples. Although the weak molecular ion peak may be enhanced, it remains low at 12 eV if it is undetectable at 70 eV. However, the use of very low ionization energies often requires a redesign of the ion generation and cannot be achieved with simple filaments without further focusing and/or acceleration/deceleration [15–17].

4.2.2.2 Chemical Ionization

Chemical ionization (CI) is a also very common soft ionization technique in GC-MS, especially since it could be technically integrated in an EI ion source. The main difference compared to EI is that a reactant gas, which is introduced into the ion source, is primarily ionized by the electron beam. Therefore, the sample molecules are not directly ionized by the electron beam but indirectly by the ions of the reactant gas. This primary ionization of the reactant gas could also be applied by other kinds of ionization techniques, like photoionization (PI), which would result in atmospheric pressure photoionization (APPI). Many different gases (e.g., methane, water, and ammonium) can be used for this technique. In the following, the reaction mechanisms of a neutral molecule with methane as one of the most used reactant gases is demonstrated. Methane is constantly introduced into the ion source, leading to a collision of basically all electrons with methane.

$$CH_4 + e^- \rightarrow CH_4^{+\bullet} + 2e^- \tag{4.6}$$

The radical cation of methane can then undergo fragmentation reactions or further ion-molecule reactions due to the excess and high pressure of neutral methane in the ion source. As a result, initial and secondary reaction stable abundant reagent ions (CH_5^+, $C_2H_5^+$) are formed, which do not further react with neutral methane. These reagent ions are Brønsted acids, which will ionize most organic molecules with sufficient proton affinity by proton transfer.

$$CH_5^+ + M \rightarrow CH_4 + MH^+ \tag{4.7}$$

In contrast to EI, CI usually produces even-electron protonated $[M+H]^+$ molecular ions. Moreover, as less energy is transferred by the ionization process of the analyte and even-electron ions are more stable (than odd-electron ions produced by EI), CI will result in less fragmentation and more abundant molecular ions. Positive ion CI is therefore useful when more intense molecular ions and less fragmentation is desirable. Nevertheless, in some molecules, e.g., saturated alcohols, protonation will occur at the functional group, which results in a loss of H_2O and consequently in very low-abundance molecular ions. An advantage of CI is the possibility of choosing a reagent gas depending on the analyte of interest. For example, if the ionization shall be softer, weaker Brønsted acids can be produced using isobutane instead of methane. Moreover, the addition or substitution of ammonia in the ion source will lead to a proton transfer of more basic compounds, which is of particular interest if amines are the main targets. The sensitivity of CI depends on several experimental parameters, as it can be influenced not only by the electron energy and the applied current, but also by the choice of reactant gas, the reactant gas pressure, and the ion source temperature, which must be stated for better comparability between analyses.

The process described so far results in a positively charged, mainly protonated ion and is therefore called positive ion chemical ionization. Generally, also negative ion chemical ionization could be aimed for and realized. In a CI plasma, positive and negative ions (e.g., $[M+H]^+$ and $[M+H]^-$) are formed simultaneously, and depending on the polarity of the acceleration voltage, either one of the ions is extracted from the ion source. Negative ion chemical ionization (NICI) mass spectra could thus be obtained by deprotonation of acidic analytes like carboxylic acids or by anion attachment. A second process leading to negative ionization results from electron capture, which is a highly sensitive ionization process for many toxic and environmentally relevant substances. High pressure conditions in the CI ion source lead to the collision of primary (produced by the filament) and secondary (produced during the ionization event) electrons until they reach near-thermal energies. Molecules, which contain electronegative substituents and thus have high electron affinities, can very efficiently capture those electrons. Under electron capture conditions, three different ion formation mechanisms, namely resonant electron capture, dissociative electron capture, and ion pair formation, can happen. As the electron-molecule collision rate is much faster than the ion-molecule collisions, this method can be very sensitive for specific types of molecules, such as highly halogenated (especially fluorinated) molecules and other molecules containing electronegative substituents. For example, perfluoroalkanes, which are completely saturated, undergo dissociative electron capture, leading to abundant fragment ions. The high efficiency of electron capture and the high rate of electron-molecule collisions can lead to an increase in sensitivity by a factor of 10^3 in comparison to positive CI.

These soft ionization methods lead to less fragmentation compared to EI. This is especially advantageous if EI cannot provide the molecular ion, which makes the identification of unknowns challenging. Due to less energetic ionization, the molecular ion is usually detectable by softer ionization methods such as CI, enabling confirmation of the molecular mass. In general, the identification and quantification of ions is usually more reliable when the ion has a higher m/z, less background noise, and a greater relative abundance. Hejazi et al. [26] compared soft and hard ionization methods for 16 volatile organic compounds with respect to their ion yields, their potential of identifying molecular and fragment ions, as well as their signal-to-noise (S/N) as qualitative characteristics of the mass spectra. They showed that EI achieved the highest ion yields (10^{-8}) covering less than two decades. Positive CI provided a similar distribution, but on average, lower ion yields by a factor of 10. The ion yields of NICI span comparably wider (3.5 decades) due to high dependencies of ionization cross-sections based on the chemical structure. This can be explained by the working principle of NICI, which is, as previously explained, based on electron capture, and therefore requires chemical structures with high electron affinity, favoring the formation of ions by resonant and/or dissociative electron capture. For such substances, NICI shows very high ion yields, being beneficial for the quantification of distinct compound classes and, thus, a highly selective ionization technique.

EI has several advantages over soft ionization techniques, which can be summarized as follows: EI can deliver structural information, shows highly reproducible mass spectra, which can be compared to large databases, and therefore ease the identification of unknown sample components, and can overall be easily implemented. A combination of hard and soft ionization techniques at the same time would be beneficial for the analysis of complex matrices of volatile compounds in the field of metabolomic research. When the volatility range is more restricted due to the closed ion source design and the ionization of compound classes like aliphatic hydrocarbons is ineffective, EI with low electron energy (e.g., 14 eV) can be a good alternative for the enhancement of the molecular ion [15, 18–29].

4.2.2.3 Photoionization

The interaction of energetic photons with molecules can also result in electrons being removed from neutral molecules, which can then be used to ensure effective ionization of the molecule for mass spectrometry. In general, photoionization processes applied for gas chromatography could be roughly distinguished in either how many photons are involved in the ionization process or in the different pressure regime in which the process takes place. As a consequence of the latter, a further distinction can also be made between direct and indirect ionization. Irrespective of this classification, photoionization is based on the interaction of electrons of a molecule with photons. At resonance, the photon energy is transferred into kinetic energy of electrons, which leads to them assuming a higher energy state. The molecule becomes excited. Ionization takes place when the electron can absorb sufficient energy, thereby exceeding the binding energy to the molecule and entering the continuum state. The threshold value here is the ionization potential of the molecule. While the raising of electrons to bound excited states is bound to specific selection rules, which only allow the absorption of photons of certain energies, an electron in the continuum is not bound to such rules and therefore can have arbitrary kinetic energy.

This explains the universality and non-selectivity of single photon ionization (SPI), in which a single photon transfers enough energy to transport an electron of a molecule directly into the continuum state. Also already mentioned before, most organic molecules have ionization potentials in the 8–12 eV range, while typical background- and carrier gases for gas chromatography need higher energies to become ionized and therefore will not ionize if the light source is adjusted accordingly.

$$AB + h\nu \rightarrow [AB^+]' + e^- \tag{4.8}$$

$$[AB^+]' \rightarrow AB^+ \tag{4.9}$$

$$\rightarrow A + B^+ \tag{4.10}$$

Such energies are in the range of Vacuum Ultraviolet (VUV) light and X-ray. Photons with corresponding wavelength for laboratory environment can be produced by commercially available lamp- and LASER-based light sources. Synchrotron radiation is not an option here, as this is not technically possible in a common GC-MS laboratory. A wide range of different principles and realizations of lamp-based sources are available for use in the laboratory. In most cases, continuous VUV light is emitted, which can be focused by lenses or mirrors. Due to the high energy, however, special materials such as magnesium or lithium fluoride must be used to ensure sufficient transmission. However, the sensitivity is strongly determined by the limited number of usable photons that can be achieved with lamp-based systems. Nevertheless, due to their simple technical realization, they are the most suitable for GC-(SPI)MS applications. Much more intense light (W/m^2) can be achieved by using LASER, but this is at the expense of technical feasibility.

In addition to SPI, a special application of laser-based systems is multiphoton ionization (MPI). In MPI, electrons must firstly be elevated to an intermediate excited state, requiring certain selection rules to be satisfied before they finally reach the continuum state by absorbing a second photon. The MPI process is complex because the photons must match the energy gap to achieve the intermediate state, and a second photon must be absorbed before relaxation occurs. In the context of this chapter, particularly, resonance-enhanced multiphoton ionization mass spectrometry ((REMPI)MS) is considered, where both photons have the same wavelength (one-color or $1+1$ REMPI). The selectivity and limitations of this technique stem from the necessity for the target molecule to have a resonant electronically excitable state prior to the actual ionization step. Due to the low probability of absorbing two photons of the same wavelength and the risk of premature relaxation of the excited state, high light intensities are required, which can only be reliably provided in laboratory settings by using lasers that produce extremely intense light at specific wavelengths.

Both ionization principles (SPI and REMPI) are technically compatible with the design and low pressure regions (10E-6 – 10-E3 mbar) of an EI source and can even be implemented alternately. When selecting a suitable LASER system or the complementary use of different light sources, combined EI/SPI/REMPI sources can also be technically realized. As with EI, the photons are irradiated across the gas inlet. Geometrically, this can be realized in such a way that the ionization takes place at the same point in the ionization chamber so that no further modifications are necessary to extract the ions from the source. However, there is a limitation when using pulsed light sources, as these are generally only compatible with pulsed ion sources, and the pulse rate of the LASER also affects the maximum possible acquisition rate (data points per second). This may be too low for use in fast chromatography.

Photoionization can also be carried out at higher pressures regions up to atmospheric pressure (Atmospheric Pressure Photoionization, APPI). The same light sources are used as for SPI, although in practice, lamps are mainly used again. In addition to the direct ionization of the target molecules as in SPI and MPI, side reactions also take place through the ionization of matrix components. Due to the low free path length of the ions at this pressure, they interact with other charged and uncharged molecules. At higher pressures, these secondary reactions can predominate and can be controlled in a targeted manner using dopants. For coupling with GC systems, mass spectrometers that are originally coupled with liquid chromatography (LC) systems are predominantly used, except that the nebulization step of the liquid matrix can be omitted. More modern (experimental) sources, specially designed for use in conjunction with GC, are primarily aimed at miniaturizing the source and the ionization chamber [30–32].

4.2.2.4 Proton Transfer Reaction Ionization

Proton Transfer Reaction (PTR) is another soft chemical ionization, particularly well-suited for the ionization of volatile organic compounds (VOCs) in the gas phase. This method is widely applied to the analysis of samples from various origins, including plants, human body (breath, blood, etc.), and various anthropogenic sources. The PTR process can be regarded as a specific form of positive ion chemical ionization, which has already been discussed above, although its technical implementation differs. The key advantage of PTR ionization is that it results in minimal fragmentation of the analyte molecules, leading to simpler mass spectra compared to other ionization techniques like electron ionization. Although Proton Transfer Reaction Mass Spectrometry ((PTR)MS)) is commonly known as a direct MS method for the online monitoring of VOCs, this soft ionization technique is also widely used in omics studies and warrants a mention in this context. The combination of (PTR)MS with fast-GC separation has also been reported in the literature.

In PTR, water vapor is first ionized, e.g., by electron ionization to yield the H_2O^+ cation, which then undergoes ion-molecule reactions in the source drift region to form the hydronium ion (H_3O^+). This hydronium ion acts as the primary reagent ion, serving as a proton donor that collides with the neutral analyte molecules (M) to produce protonated species (MH^+):

$$H_3O^+ + M \rightarrow MH^+ + H_2O \tag{4.11}$$

Ideally, the H_3O^+ ion should be the only one reacting with the analyte gas in the drift tube of a (PTR)MS instrument. However, in practice, the presence of unreacted water vapor from the ion source, degassing of the drift tube walls, and the analyte gas itself can lead to the formation of cluster ions of the type $H_3O^+ (H_2O)n$. While H_3O^+ is the most commonly used primary reagent ion, other ions such as $NH_4{}^+$, NO^+, and $O_2{}^+$ can also be employed for PTR ionization, depending on the specific analytical requirements [33].

4.2.3 The Mass Analyser

As we have seen, there are different ways of generating ions that are later separated in the analyser according to their mass-to-charge ratio. While some of the ionization techniques shown are selective for certain classes of molecules, this is not generally the case for the analysers listed below. These are generally compatible with all the ionization techniques mentioned. However, they differ fundamentally in the mass resolution, mass accuracy, and acquisition rate typically achieved. Basically they can also be classified according to whether they scan a specific mass range, i.e., only one mass can be determined from an initial ion packet at a specific time interval, or whether a mass range can be detected (quasi simultaneously). In the following, we will take a closer look at the most common analysers used in conjunction with GC-MS.

4.2.3.1 Quadrupole Mass Analyser

The basic principle of quadrupole mass analyser (QMS) was first described by Paul and Steinwedelin 1953. A decade later, the first prototypes of hyphenated GC-QMS were commercially launched, marking a significant advancement in the GC-MS field. Today, quadrupole mass analysers are widely used in GC-MS hyphenated instruments owing to their cost-effectiveness, compact design, robustness, and the abundance of spectral libraries available in commercial databases. These characteristics position the quadrupole analyser as a popular option for coupling with GC methods across a diverse array of omics-related fields, including metabolomics, lipidomics, and environmental analysis.

While the design of quadrupole instruments may appear simple, the physics behind the mass separation is rather complex and detailed. However, for the scope of this text, understanding the basic principles suffices. A schematic representation of a quadrupole analyser is shown in Figure 4.2. Essentially, quadrupole analysers function as mass filters, sequentially passing ions (one *m/z* at a time) of various *m/z* ratios to the detector. Their basic operation relies on the simultaneous application of ramped direct current (DC) and radio frequency (RF) voltages on a group of parallel rods arranged symmetrically. Opposite rods are connected electrically in pairs. At any given moment, these pairs carry potentials of the same

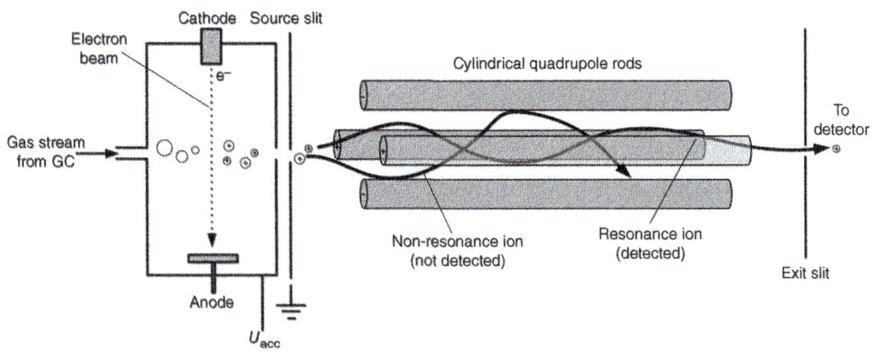

Figure 4.2 Schematic of a Q instrument typically used in GC-(EI)QMS. *Source:* Wittmann [34] / Springer Nature / CC BY 2.0.

magnitude but opposite polarity. By adjusting the voltages on the four rods at a certain frequency, only ions of a specific *m/z* value maintain a stable trajectory through the quadrupole field, reaching the detector, while others collide with the rods, failing to reach the detector. This arrangement induces a mass-dependent oscillatory motion in the ion beam, described by the Mathieu equation.

A quadrupole analyser can operate under two primary acquisition modes: full scan and selected ion monitoring (SIM) mode. In full scan mode, as voltages and frequencies rapidly change over time, the analyser allow to pass ions depending on the time within a user-defined mass range, e.g., a unit mass, generating a full scan mass spectrum. Conversely, in SIM mode, the instrument records the ion current at selected *m/z* values that are characteristic of specific compounds of interest within their expected retention time window, rather than scanning the full mass range. The instrument can switch between different selected *m/z* values during an analysis; however, at any given moment it only monitors one *m/z* value (or a very narrow set of values). If the QMS is operating in SIM mode, the transmission efficiency for ions through the analyser is high compared to other MS systems. This selective approach enables the quadrupole to perform with both high sensitivity and specificity. In SIM mode, the careful selection of monitored ions is essential to balance sensitivity and specificity. Typically, ions of higher abundance provide superior sensitivity, whereas ions unique to the target compound are preferred to maximize specificity. Multiple ions can be selected via SIM; however, increasing the number of ions monitored, reduces both speed and sensitivity. Generally, the "three-ion rule" is followed for each compound if the system operates in SIM: a quantifier ion (I_1) for quantitative determination, and two ions (i_1 and i_2) to confirm the correct peak assignment (qualifier ions). Although SIM mode sacrifices some spectral information, it significantly enhances sensitivity (up to 100 times) and accuracy, making it ideal for quantitative determination of target compounds in complex mixtures. The selection between the full scan and the SIM mode represents a trade-off: one must choose between acquiring full spectra and enhancing sensitivity. However, by alternating between these modes during the elution window, users can optimize the quality and amount of information collected from a single GC-MS run, combining both full spectral data and highly sensitive quantitative information. Moreover, the quadrupole can operate in RF-only mode (i.e., no DC voltage applied), passing all masses through without filtering or only selected bands, serving other purposes, such as ion guides, ion gates, collision cells (often hexapole or octapole) – as seen in analyser like the triple quadrupole (see next section 4.2.3.4).

While widely used in GC-based laboratories, these analysers suffer from some disadvantages. Quadrupole instruments are typically classified as low-duty cycle scanning devices, offering slightly better than unit-mass resolution and a mass accuracy of only around 0.1–0.5 amu. When combined with rapidly eluting GC peaks in methods like GC×GC, the scan rate capability of quadrupole analysers can pose challenges. A typical GC×GC experiment generates narrow effluent pulses lasting around 100 ms or less, necessitating a high scanning rate (≥50 Hz) to sample each narrow-modulated peak with a sufficiently high number of spectra. Due to their limited scan frequency, and the need to scan individual masses in the scan range, quadrupole analysers for many years were primarily used in combination with GC×GC for identification purposes only. The first example of a hyphenated GC×GC-QMS instrument appeared in a study published by Frysinger and Gaines in 1999. The quadrupole used in their study operated in full scan mode (45–350 amu) at a scan rate of only 2.4 Hz. To overcome the slow acquisition scan rate issue, the authors intentionally broadened peaks entering the ion source by lengthening the first dimension, allowing for the acquisition of three spectra per peak, far less than required for a good 2D GC peak reconstruction, though at the cost of reduced peak capacity in the second dimension. It was already obvious that time-of-flight analysers (TOF) are a preferred choice in hyphenated GC×GC-MS platforms, due to their higher data acquisition rate (≥100 Hz) and because all ions in one spectrum are acquired at the same time. Shellie et al. [38] reduced the time necessary to perform a single scanning process by narrowing the mass range (i.e., 41–228 *m/z*), achieving a relatively higher scan rate of 20 Hz. However, this 20 Hz scan rate still resulted in under-sampling of the fast 2D GC peak, indicating the inadequacy of this approach for reliable quantitative analysis. Further attempts to combine quadrupole analysers with GC×GC were made by many authors, resulting in an improvement in the performance of these analysers. Contemporary fast-scanning quadrupole can nowadays reach speeds of up to 20,000 amu/s, allowing for recording wider scan ranges or higher data acquisition rates, meeting the 50-Hz requirement, albeit remaining inherently much slower than TOF instruments (≥100 Hz), and will also suffer from sensitivity issues when operating at high scan rates. Nonetheless, faster scanning has its drawbacks, as the reduced time available for measuring each specific ion may compromise sensitivity, mass accuracy, and mass resolution. To enhance sensitivity, users can either opt for a reduced mass-range or operate in SIM mode. The latter is the recommended choice for quantitative studies, enabling quantitation at the parts-per-billion (ppb) level.

The flexibility to operate in full scan, SIM, or alternate between both modes within the same GC run makes the quadrupole suitable for both targeted and untargeted investigations. Additionally, factors such as robustness and long-term repeatability render the GC×GC-QMS hyphenation potentially appropriate for large-scale, untargeted omics-related studies. In summary, rapid-scanning quadrupoles represent an (economic) valuable alternative to TOF analysers for targeted and untargeted analyses within a limited mass range. Careful monitoring remains essential to address mass spectra skewing, a limiting factor resulting from rapid variations in analyte concentrations during the scanning process. This leads to discrepancies in ion abundances across a chromatographic peak [35–42].

4.2.3.2 Triple Quadrupole Mass Analyser

Triple quadrupoles (QqQ) instruments are tandem mass spectrometry (MS/MS) analysers, mainly used in GC-based applications for targeted approaches. Their configuration includes two quadrupole mass analysers (Q1 and Q3) that function as mass filters, with an RF-only quadrupole (q2) – typically a hexapole or octopole – positioned between them. The q2 functions as a gas-filled collision cell, where ions can undergo low-energy (<100 eV) collision-induced dissociation (CID). This enables MS/MS experiments in which a selected precursor ion is further fragmented providing structural information and enhancing analytical selectivity. Several benchtop QqQ instruments for GC are now commercially available. Owing to their high sensitivity and specificity, QqQ systems are widely regarded as the "gold standard" for quantifying trace-level compounds in complex matrices. Their performance, combined with the cost-effectiveness and compact design compared to other hybrid MS/MS platforms, has driven their widespread adoption as routine benchtop tools in GC-based "omics" studies, including applications in biomedical, food, and environmental analysis.

Ions exiting the first quadrupole, Q1, are accelerated into the collision cell (q2), where they undergo activation via multiple collisions with an inert gas (typically N_2 or Ar), leading to their fragmentation. The third quadrupole (Q3) serves to filter the fragment ions produced in the collision cell (q2), allowing the selection of specific m/z values or a defined range before they reach the detector. Factors such as the nature of the collision gas, its pressure, and the applied collision energy may limit the reproducibility of spectra between QqQ instruments. The flexibility of independently controlling the voltages and acquisition modes of the first (Q1) and third (Q3) quadrupole analysers in a QqQ mass spectrometer is a significant advantage that allows for the selection of various tandem-in-space MS/MS operational modes (Figure 4.3). Considering a precursor ion, M_1^+, which fragments in a product ion, M_2^+, and a neutral loss, N:

$$M_1^+ \rightarrow M_2^+ + N \tag{4.12}$$

A tandem-in-space MS/MS experiment can be classified according to which of these species (precursor, product ion, or neutral loss) is investigated.

The most common acquisition mode, known as multiple reaction monitoring (MRM), operates with both quadrupoles in SIM mode, monitoring several precursor-product ions "reaction channels" or transitions. In MRM mode, Q1 acts as a mass filter, transmitting only the target precursor ion m/z. The selected

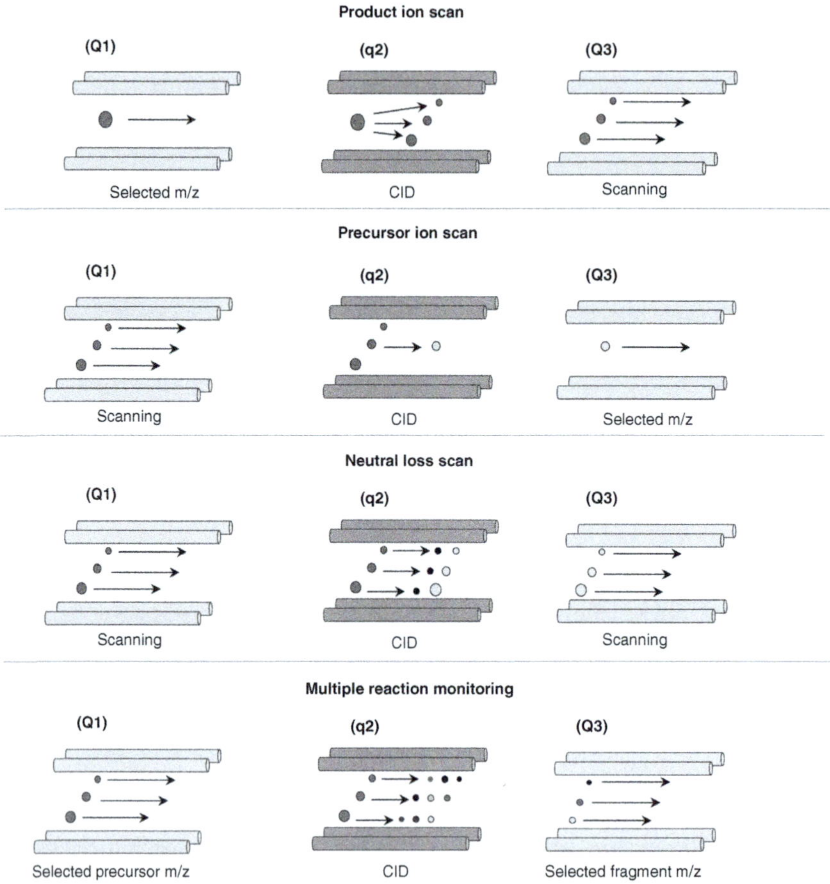

Figure 4.3 Representation of the operational modes in a quadrupole MS/MS analyer.

precursor ions then enter the collision cell (q2), where they collide with the inert gas molecules. Depending on the collision energy applied, the precursor ions will fragment into characteristic product ions. The third quadrupole (Q3) then selects specific *m/z* values corresponding to known fragments of the target compounds. Typically, in MRM experiments, two product ions are isolated and monitored in the Q3- one utilized for quantification purposes (quantifier ion) and the other to confirm correct peak assignment (qualifier ion). The ratio between quantifier and qualifier ion signals should match that of a pure standard to ensure correct peak assignment. Analysing commercial standards of the target compounds is a standard practice to optimize QqQ parameters like collision energy for each MRM transition. Careful optimization of collision-induced dissociation (CID) parameters is crucial, as they can significantly impact the fragmentation pattern, detection sensitivity, and reproducibility of mass spectra. Compared to SIM-QMS, MRM mode offers distinct advantages for quantitative analysis of trace-level analytes. MRM removes most non-target ions in Q1, resulting in reduced background ions and noise, thereby enhancing both selectivity and sensitivity. Like in SIM-QMS, different MRM transitions can be monitored simultaneously. However, parameters such as dwell time (sampling time), chromatographic peak widths, and retention times must be carefully monitored as they impact the number of target analytes that can be monitored in a single "scan". Furthermore, technological evolution in the QqQMS field has advanced greatly, with current systems capable of producing scan/MRM data in a rapid and alternating manner in the same analysis.

In addition to MRM mode, QqQ instruments offer three additional operational modes for MS/MS experiments:

- **Product ion scan (or "daughter scan"):** Q1 operates in SIM mode to select a specific precursor ion while Q3 acquires a full mass spectrum of the fragment ions produced within a defined *m/z* range. This MS/MS mode provides valuable insight into the fragmentation patterns of precursor ions, useful for structural elucidation.
- **Precursor ion scan (or "parent scan"):** the operational modes of Q1 and Q3 are reversed compared to the product ion scan mode. Q3 operates in SIM mode, while Q1 scans within a certain *m/z* range for potential parent ions producing the target fragment ion. This mode is useful for identifying compounds sharing common fragment ions or screening specific chemical classes of compounds.
- **Neutral loss scan:** targets compounds that lose a common neutral species during fragmentation. Both Q1 and Q3 operate simultaneously in scan mode, but they are not in phase: a fixed *m/z* offset is kept between them. This allows for the selective detection of ions undergoing a neutral loss, matching the *m/z* difference between the two analysers. For instance, when coupled with chemical ionization (CI), molecular ions of alcohols lose water, resulting in a neutral loss of 18 mass units. Scanning for this specific neutral loss effectively screens compounds with common functional groups or chemical traits.

Moreover, Q1 or Q3 can be set to RF-only mode when MS/MS analysis is unnecessary, effectively transforming the instrument into a single quadrupole analyser.

The first proof of principle study on GC×GC-QqQMS was published by Poliak et al. [47]. In this study, analyte ionization was performed using supersonic molecular beam (SMB) electron ionization, a soft ionization process also referred to as "cold EI". The study demonstrated how the GC-(SMB)MS approach suffered from considerable peak overlap and unsuccessful MS database matching. In contrast, the increased separation power of GC×GC-(SMB)MS led to a satisfactory database similarity match for the desired compound. However, the authors did not provide any information on the potential performance of a GC-(SMB)MS/MS configuration, which could have offered an interesting comparison within three-dimensional approaches. Is it truly necessary to introduce a fourth dimension to the already three-dimensional GC-QqQMS? This question remains a subject of ongoing debate among scientists. Despite the potential benefits, the use of GC×GC-QqQMS has been limited, mainly because QqQMS instruments typically do not require such high-resolution GC separation [43–47].

4.2.3.3 Time-of-flight Mass Analyser

The concept of a time-of-flight (TOF) mass analyser was first published by Stephens back in 1946. The principle of a TOF mass analyser is quite straightforward. In general, flight differences between ion packages are measured. Ions of a distinct mass to charge ratio (m/z) are introduced into a field-free drift path of known length. Provided similar ionic charge, similar kinetic energy and the ions enter the drift path simultaneously, they will be separated solely on their respective mass. Therefore, lighter ions will arrive at the detector prior to heavier ions.

The correlation of flight time and m/z can be clarified by the following formulas. Ions arriving from a pulsed ion source will be accelerated via a voltage (U) in an electric field before they are introduced into the drift path. When the ion was at rest before entering the electric field, its velocity (v) can be derived from the definition of kinetic energy and can be summarized as follows:

$$v = \sqrt{\frac{2Uez}{m}} \tag{4.13}$$

The time that an ion needs to pass a drift path of a defined length (l) is inversely proportional to the velocity of the ion:

$$t = {}^{l}\!/_{v} \tag{4.14}$$

Combining Eqs. (4.13) and (4.14) and resolving it towards the variable t (time) results in an equation from which the time, an ion needs to pass the drift path, is directly proportional to the square root of its respective m/z:

$$t = \sqrt{\frac{m}{z}} \tag{4.15}$$

Ion packages with the same m/z ratio have some spatial distribution due to the limited extent of ion formation volume and the kinetic energy dispersion induced by, e.g., the ionization process, and instabilities, as well as imprecisions of the electrostatic or pulsed acceleration fields. In a linear TOF system, these processes occur in one axis. Instrumentation and instrumental improvements of TOF systems will be summarized in the next paragraph.

Wiley and McLaren realized in 1955 that the space focus point, at which ions with the same *m/z* show their minimal spatial distribution, is shortly after they leave the ion source. To increase the resolution due to prolongation of the path length for the ions, they enabled the shift of the space focus point further away from the ion source by implementing a second acceleration field. This setup, called two-stage ion source, is still in use and improved the resolution of *linear* TOF systems. Linear means the pulsed extraction of the ions from the source initially takes place in the same direction as the subsequent acceleration and separation of the ions in the field-free drift space. Nevertheless, linear TOF instruments are prone to varying pulse-to-pulse voltages applied in the ion source, which can lead to inconsistent ion acceleration. These inconsistencies in ion acceleration can lead to variations in the flight time and thus reduce mass accuracy and precision. Further instrumental improvements have been established with the introduction of reflection TOF systems. These "reflectrons" are placed in the field-free drift region to enable an active compression and an energy focusing of the ion packages. This was realized for the single-stage ion source with an electrostatic reflector in 1957 and for the two-stage ion source with a two-stage electrostatic reflector in 1973. Principally, an electrical decelerator focuses and reflects ions with a distinct angle towards a detector. Electrodes equipped either with grids (ion transmission up to 95%) or without grids, which can also function as electrostatic lenses, are utilized for the generation of an electric field. The space and energy focusing can achieve a resolution of up to 35,000, and the flight path prolongation a resolution power of 20,000.

In 1989, Dawson and Guilhaus invented the orthogonal acceleration TOF (oaTOF) as another approach to minimize spatial distributions of ions. In oaTOF, an orthogonal inlet system is used to deliver the ions at a 90-degree angle relative to the direction of the drift path of the TOF system (see Figure 4.4). Successive extraction pulses are used to push ions out of this continuous, sharply focused ion beam into the TOF system. This pulsed deflection of the ion beam equals the starting point of the TOF measurement. With this configuration, an improved ion transmission and sensitivity are achieved by minimizing space charge effects. The latest instrumental development in TOF systems is the multireflection TOF, which can achieve a resolution (R) of up to 300,000 in full spectral mode.

$$R = \frac{t}{2\Delta t} \tag{4.16}$$

According to Eqs. (4.1) and (4.15), higher resolution can be reached by drift path prolongation, i.e., an increase in flight time of an ion (t) or by focusing the ion packages, i.e., a decrease in the width of the ion signal (Δt) [50]. A limitation

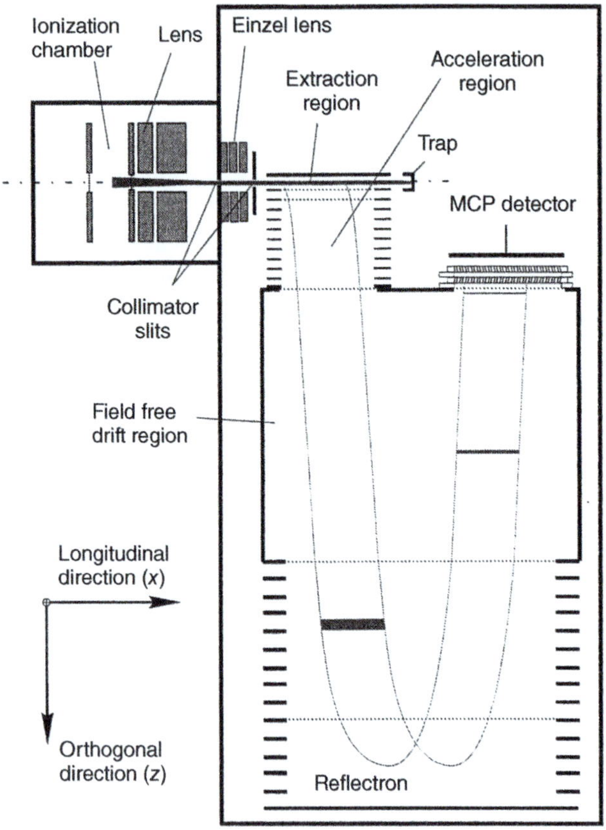

Figure 4.4 Scheme of a reflectron TOF system with an orthogonal inlet. *Source:* Chen et al. [48] / with permission of Elsevier.

is the ion load in the drift region because resolution would suffer from high ion loads and thus charge repulsion. Moreover, ion losses can occur more frequently due to several ion optical devices used for drift path prolongation in the high resolution multireflection TOFMS. The reduced sensitivity can, to some extent, be compensated by higher selectivity and thus less chemical noise. As accurate mass enables the application of a smaller mass window, a better separation of chemical noise and the analyte of interest can be achieved, which finally leads to a higher S/N for high resolution (HR) TOFMS. Moreover, multiplexing (encoded frequent pushing, EFP) was recently made commercially available. In principle, ions are accelerated in an orthogonal manner, which enables pulsing times being faster than the total flight time, producing overlapping mass spectra. These mass spectra can be deconvoluted by mathematical procedures using the EFP process. This improvement is benefiting sensitivity, abundance of the signal, dynamic range, mass accuracy, and noise level of HR TOF measurements.

TOF mass spectrometers are highly advantageous for several applications in the field of analytical chemistry, biology, and biochemistry due to the following reasons: TOF instruments have a wide mass range and thus enable the measurements of small and large molecules which makes them particularly interesting for omics applications, in which small molecules but also larger biomolecules such as peptides and proteins can be of interest. Hyphenation of GC, though, restricts the mass range towards smaller molecular weight (typically up to 600 m/z) and low polarity compounds. An advantage of TOF mass spectrometers is that they can achieve high sensitivity and higher mass accuracy compared to scanning quadrupole MS. Both characteristics are very important for the analysis of complex matrices; high sensitivity enables the detection and quantification of low-abundant analytes (trace-level analysis), high resolution can provide sufficiently accurate mass data which enables the detection of only one or few possible isotopic masses for an analyte and therefore increases the ability for correct compound identification. Nevertheless, the mass resolution that can be reached by TOF measurements is limited to 500,000 (limited spectral range). Therefore, other MS techniques such as Fourier transform ion cyclotron resonance (FT ICR) or Orbitrap need to be employed if higher mass accuracy is required. An important advantage of TOF instruments over other (high resolution) MS systems, though, is the possibility of hyphenation with GC instrumentation (GC, fast GC, GC×GC). Chromatographic separation, especially in fast GC and GC×GC analysis, takes typically only a few seconds per analyte and therefore requires a high acquisition rate to retain the achieved separation. TOF instruments typically acquire mass spectra at acquisition rates of 100–200 spectra/s. HR TOFMS coupled to GC systems are commercially available with a mass resolution of >50,000. Moreover, due to their simple and robust setup and their straightforward hyphenation with GC systems, TOFMS systems are routinely utilized in a broad field of applications, including omics, food, and environmental analysis [26, 49–62].

4.2.3.4 Quadrupole Time-of-flight Hybrid Mass Analyzer

Another hybrid mass spectrometric technique beside QqQ is Quadrupole time-of-flight (QTOF). The technology was developed in the early 1980s but it took until the mid-1990s when the first commercial instrument was introduced to the market by a vendor. Its first adoptions were particularly notable in Liquid Chromatography-Mass Spectrometry (LC-MS), where it did become a standard tool next to other hybrid mass spectrometers like triple quadrupole. The majority of hyphenations with LC were equipped with atmospheric pressure ionization, a process that, in principle, can also be connected to gas chromatographs, but has now been displaced by the ionization techniques mentioned above. The first commercially available QTOF with a low-pressure ionization, such as EI, was only introduced to the market in the 2010s. The ability to combine existing QTOF systems with ionization sources from third-party suppliers has further diversified the field. The technique was quickly accepted by users as it filled a gap that previously common mass spectrometers, such as triple quadrupole and pure TOF systems, could not fill by combining the advantages of both techniques. The essence of QTOF lies in its superior mass accuracy and resolution of the TOF in combination with the capability to apply this advantage for structural determination in MS/MS experiments, while in principle it can also be used for classical TOF or MS/MS applications.

QTOF instrumentation is closely related to triple quadrupole mass spectrometers in their design and general functionality in that only the last quadrupole is replaced by a higher resolution oaTOF. However, there are also systems in which an additional quadrupole q0 is added in front, providing collisional dumping of the ions introduced to the system. Such a QTOF would have the general order q0-Q1-q3-oaTOF (Figure 4.5).

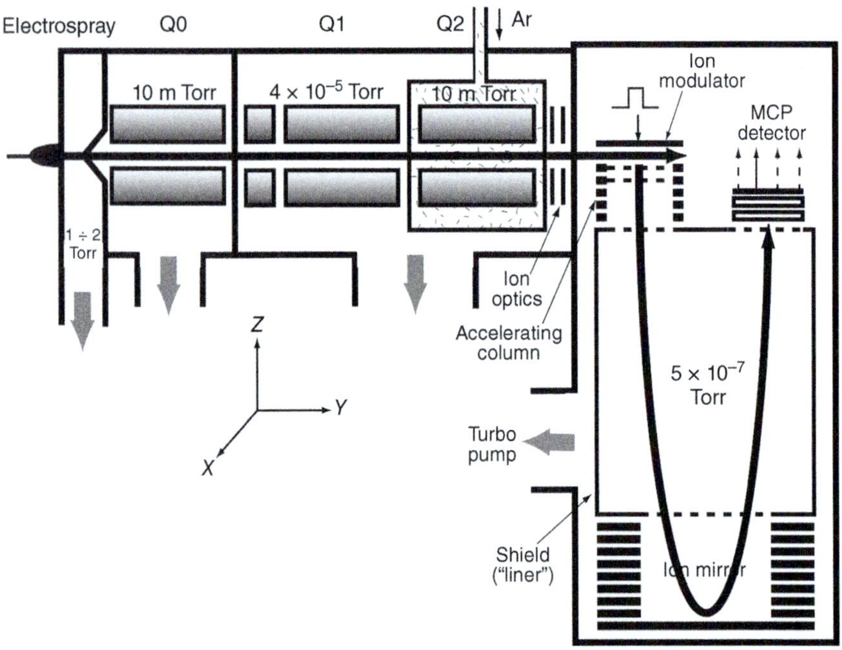

Figure 4.5 Schematic of a QTOF instrument. *Source:* Chernushevich et al. [63] / John Wiley & Sons.

Only Q1 is used as a mass filter. Q0 and q2 act as a collision cell and are generally operated in RF-only mode and set to pass-through mode. If Q1 is also operated in r.f. mode and either no or only very low collision energy is applied, which can lead to further damping of the ion movement and thus higher mass resolution of the TOF, the QTOF operates in TOF-only mode. However, the detectable mass range is not only determined by the TOF, but also by the r.f. applied to the quadrupoles, since the transmission is not constant over the entire mass range, and heavier ions are lost due to poorer focusing. For applications requiring the analysis of ions across a broad *m/z* range, the r.f. voltage can be modulated (either stepped or ramped) during the acquisition to extend the transmission window. However, this broader transmission comes at the expense of the duty cycle; lighter ions are lost when the r.f. voltage is set high, and heavier ions are not effectively focused when the voltage is low, affecting the overall efficiency of ion transmission to the TOF analyzer. The loss of ions could also lead to lower sensitivity. If the system should be used as MS/MS, the collision cell generally operates in the same way as to tripe quadrupole instrumentation, and collision induced dissociation is the main method used to fragment a precursor ion. The degree of fragmentation is determined by the type and pressure of the collision gas as the collision energy and residence time of the ions. Regardless of the hybrid system used, the fragmentation caused by CID can be very dependent on the type of device used, which depends more on the collision cell design than on the subsequent TOF or Quadrupole. CID fragment patterns are often not transferable to other systems, unlike 70 eV electron ionization fragmentation spectra. An additional difference results from the use of the Multichannel plate (MCP) detectors commonly used in TOFs, which decreases the dynamic range compared to a Triple Quadrupole, where, e.g., electron multipliers are used, whereas the noise is increased.

The field of application and the functional modes derived from this result from the advantages offered by the TOF. The focus here is on qualitative aspects, such as structure elucidation or qualitative screening of a matrix, and less on the quantification of individual target compounds. Even if this is not excluded, the literature generally reports better LODs and measurement ranges for triple quadrupoles, but attention must be paid to the comparability of the results, as the systems are not always operated under comparable conditions. In a few cases, the sensitivity of a TOF-based system may be superior to a triple quadrupole due to the selectivity resulting from higher mass resolution. However, this is primarily due to the suppression of the chemical background and not to a higher signal intensity. Especially in terms of ion yield and duty cycle, triple quadrupoles are superior to QTOF systems in MS/MS mode. The typical working mode of a QTOF is the production scan (MS/MS or tandem MS mode), in which the TOF records the occurring fragments according to the CID to a specially selected precursor ion with high mass accuracy and mass resolution. This mode is mainly used to elucidate the structure of a compound. In addition, the full scan mode is of particular importance, as this is where the advantages of a TOF over the otherwise identical QqQ come into play. In this mode, the QTOF records mass spectra over a wide range of *m/z* without selecting or filtering specific ions. This mode is suitable for untargeted analysis and provides a comprehensive overview of all ions present in the sample. In addition to these two modes also common MS/MS modes are possible.

- **Precursor ion scan:** The device searches for precursor ions that lead to a specific product ion. This mode helps identify compounds that produce a common fragment ion, which is useful for studying compounds with similar structures or functional groups.
- **Neutral loss scan:** In this mode, the mass spectrometer identifies precursor ions that lose a neutral fragment with a specific mass during fragmentation. This mode is particularly valuable for identifying classes of compounds that undergo specific fragmentation pathways.
- **Selected reaction monitoring (SRM)/multiple reaction monitoring (MRM):** QTOF instruments, while more common with triple quadrupoles, can also be configured for targeted quantitation by monitoring specific precursor ion transitions, providing high sensitivity and selectivity for the compounds of interest.
- **Precursor ion detection:** A method described as an alternative to precursor ion scanning for LC-MS applications, in which the instrument alternates between low and high collision energy in the collision cell to display both molecular (precursor) ions and their fragment ions. This approach is useful for identifying the potential precursors of a particular fragment ion in a complex sample. However, the significance and applicability will be affected by a highly fragmenting source such as EI.

All previous working methods can be categorized as data-dependent acquisition (DDA), where a specific target, e.g., the fragmentation of a particular precursor ion, is defined. However, QTOF data also opens up a new type of retrospective data analysis. This is based on so-called Data-Independent Data Acquisition (DIA), which is a special form of product ion scan. DIA is a comprehensive approach in which the mass spectrometer runs through a predefined mass range and fragments ions in small windows of this range, regardless of their abundance. This method provides a comprehensive record of all fragmentable ions present in the sample during analysis. This allows researchers to come back to the data set at a later date to search for substances that were not originally in focus or are known from updated databases or new information. This capability is extremely valuable in areas such as toxicology, environmental monitoring, and pharmacokinetics, where the compounds of interest may change over time or comprehensive profiling is required [63–66].

4.2.3.5 Orbitrap Mass Analyzer

Improving the resolving power, mass accuracy, and dynamic range of the mass analyser is central to advancing mass spectrometric analysis. Consequently, researchers and scientists are constantly seeking novel approaches for measuring the m/z of ions with higher precision and sensitivity. The Orbitrap mass analyser, conceived by Alexander Makarov in the late 1990s and commercialized in 2005, represents a breakthrough in high-resolution mass spectrometry. Unlike other mass analysers, such as quadrupole and TOF analysers, the Orbitrap operates by trapping ions in an electrostatic field and then measuring their m/z based on their oscillation frequency. This unique approach has significantly enhanced the precision and sensitivity of mass spectrometry, making the Orbitrap one of the newest and most promising additions to the world of mass analysers. An additional feature of the Orbitrap mass analyser is its ability to perform ion separation and detection simultaneously, which means that no ion detector is required.

At its core, the Orbitrap analyser comprises three electrodes: two outer and one inner electrode. The outer electrodes have the shape of cups opposed to each other and are electrically isolated by a thin dielectric central ring. A spindle-like central electrode holds the trap together and aligns it via nonconductor end-spacers. The electrodes are shaped to create a quadro-logarithmic electrostatic potential. A linear field is formed when voltage is applied between the outer and the central electrodes, allowing for harmonic oscillations along this direction. At the same time, the radial component of the field strongly attracts ions to the central electrode. In principle, the Orbitrap mass analyser is an ion trap without radiofrequency excitation to initiate the motion. Instead, the moving ions are trapped in the constant radial electrostatic field. The electrostatic attraction towards the central electrode is compensated by a centrifugal force arising from the original tangential velocity of the ions.

The initial kinetic energy of the ions is given by the unique concept for injection through a specially machined opening with a compensation deflector electrode in one of the outer electrodes. In contrast to other ion traps, the short burst-like ion injection on a μs to ms time scale eliminates the need for additional excitation. This pulsed ion package injection is realized with an additional ion storage unit. This C-trap is essentially a "flatapole" ion guide that has both a curved central axis and an opening in the electrode faced to the Orbitrap mass analyser. Optimized electric parameters synchronized between C-trap, injection ion optics, and the Orbitrap mass analyser electrodes allow for the stable circular spiral motion ("like planets in the solar system"). After achieving a stable electrostatic field, an image current is recorded on the two outer electrodes acting as receiver plates. The harmonic oscillation movement of the ion packages on elliptical trajectories back and forth between the two outer electrodes results in a sinusoidal signal. This axial motion is independent of the motion around the inner electrode and of all initial parameters of the ions, such as kinetic energy, except their mass-to-charge ratio *m/z*. The harmonic oscillation frequency scales with the square root of the inverse *m/z* with a force constant *k* value according to formula (Eq. (4.17)). The equation also shows that the frequency increases proportional to the square root of the applied voltage U_r and inversely proportional to the scaling of the trap (R_m, R_1, R_2). The digitized image current in the time domain is Fourier-transformed into the frequency domain. Mass spectra in *m/z* are generated by applying internal or external calibration functions.

$$w_z = \sqrt{\frac{k}{m/q}} \tag{4.17}$$

$$= \sqrt{\frac{e}{m/q} \cdot \frac{2 \cdot U_r}{R_m^2 \ln\left(\frac{R_2}{R_1}\right) - \frac{1}{2}\left[R_2^2 - R_1^2\right]}} \tag{4.18}$$

Equation (4.17) and (4.18): Equation for the axial oscillation frequency (w_z) in an Orbitrap mass analyzer; *k* – force constant, *m* – mass of the ion, *q* – charge of the ion, R_m – characteristic radius, R_1 – radius of the central spindle-like electrode, R_2 – maximum radius of the outer barrel-like electrode.

Ions oscillate with kHz frequency, traveling long distances of km in a single acquisition event. Hence, to increase the mean free path, the vacuum requirements of the Orbitrap mass analyser are in the low 10^{-9} to 10^{-10} mbar regime. However, collisions with residual background gas and Coulomb forces between the ion population and imperfect electrical fields cause dephasing of the harmonic motion and signal damping. Consequently, the image current recorded for up to several 100 ms in routine operation and up to several seconds on high-performance platforms is received as free induction decay. Space charge effects, which would be caused by a too-high ion population transferred from the C-trap into the Orbitrap mass analyser, are minimized by deploying an automatic gain control (AGC) concept. AGC measures the total ion flux of the ions with an additional electrode or analyser to calculate optimized C-trap storage times.

In contrast to TOF or QqQ mass spectrometers, the acquisition rate directly depends on the transient length settings, which in turn determines the achievable resolving power *R*. Theoretically, *R* scales linearly with transient length. In practice, longer transient times can lead to improved resolving power up to a certain point, beyond which the benefits plateau or diminish due to the signal damping.

Exemplarily, resolving power settings for the *Q* Exactive series high-field Orbitrap mass analyser range between 17,500 and 140,000 with a scan speed of 13 to 1.5 Hz, respectively. If a continuous switching between positive and negative polarity, which can be beneficial for chemical ionization, with an acquisition rate of 1 Hz, is desired, a resolving power of 35,000 is reached. If used to TOFMS, one must be very careful with these notations, as these values are commonly given for m/z 200. In Orbitrap mass spectrometry, the resolving power is often described as being inversely proportional to the square root of m/z, whereas in TOFMS an almost linear, often in practice slightly increasing R, for increasing m/z is observed. Another important detail is the difference in mass range, particularly the lower start value. Most Orbitrap mass analyzers allow recording of m/z values starting at m/z 50, whereas dedicated GC Orbitrap devices may be able to record from m/z 30 on.

The key characteristics of the Orbitrap mass analyser is the high resolving power (>1e5 at m/z 200), internal mass accuracy (<1 ppm), and high space charge capacity (target AGC settings of >1e5 up to 1e6 charges per scan). The vast majority of the Orbitrap mass analyser portfolio is dedicated to the life-science market for 'omics analysis in the fields of proteomics, lipidomics, glycomics, and metabolomics. Naturally, LC hyphenation with electrospray ionization has been the focal point for the analysis of polar and potentially larger molecules. Here, the rather slow acquisition rate of only several Hz in standard operation mode suits to the separation characteristics of most LC protocols. Gas chromatography with Orbitrap high-resolution mass spectrometric detection of the eluting components is a relatively new technology, which brings a paradigm shift for the annotation of unknowns in non-targeted analysis. However, this technology did not enter the market until 2015 and comes with certain analytical compromises: (i) slower acquisition speeds might be problematic for narrow GC peaks; (ii) different fragmentation patterns that need collision energy adjustment for comparison with other devices; (iii) differences in quantification protocols. A trick to tackle some of these issues is to sacrifice mass resolution in Orbitrap MS/MS mode for the sake of speed. Isotopic pattern accuracy is an important parameter to increase the confidence in annotation for targeted measurements or for finding candidates in comprehensive analysis. Studies demonstrated that GC Orbitrap can be on the same level or even perform better than state-of-the-art QTOFMS.

GC-TOF platforms continue to the be the major workforce in GC-based 'omics analysis. However, with Orbitrap devices reaching the laboratories, there is extensive work performed to address diverse challenges in biomedical research, agriculture or environmental sciences. Readers are referred to excellent review articles summarizing the efforts in untargeted metabolomics. Gas chromatography with high-resolution Orbitrap mass spectrometry, deploying vacuum chemical and electron ionization (CI/EI), has been commercialized on a broad product portfolio covering low tier and top tier platforms starting in 2015. The ion handling is fundamentally different compared to many other GC-MS systems and gas phase chemistry occurring in the C-trap of the hybrid Orbitrap mass analyser has been described. Handling the high-resolution mass spectra can be highlighted as even more important. New software development efforts, implementing the FAIR principles, have been shown to handle the requirements in GC-MS 'omics scale studies. Nonetheless, various bottlenecks and challenges in the domain of GC Orbitrap MS 'omics remain to date, with some of them are: availability of HR MS fragmentation spectral libraries (for EI, NCI/PCI, APPI/APCI), implementation of soft EI (variable electron voltage), and enhanced software capabilities with user-friendly GUI, e.g., for isotope fine structure (IFS) confidence in high throughput annotation.

In the realm of soft ionization, different concepts have been shown, with atmospheric pressure ionization concepts being most promising. Orbitrap mass spectrometry would exceed the scope of this chapter. Finally, exemplary work GC×GC with Orbitrap mass spectrometric detection has been reported. In Fourier transform mass spectrometry (FT MS), resolving power scales inversely with acquisition speed. Thus, the fast acquisition speed required to cope with the high peak capacity and narrow peak widths in GC×GC, results in parametric compromises. A resolving power of 15k was found to be most suitable for delivering good peak shape and distinguish critical ion pairs with close *m/z*. Peak reconstruction reached acquisition rates close to 25 Hz, applicable to most GC×GC protocols.

Fundamentally, the Orbitrap mass analyser belongs to the group of FT MS. Fourier-transform ion cyclotron resonance mass spectrometer (FT ICRMS) completes the ensemble of commercialized FT MS devices. In contrast to the Orbitrap mass analyser, FT ICRMS relies on strong magnetic fields, with sub-ppm spatial homogeneity, ranging from several Tesla up to two pioneering 21 T platforms installed in the US. FT ICRMS delivers unbeaten resolving power and mass accuracy. Together with the capability to hold a high number of charges, resulting in high dynamic range, FT ICRMS is commonly used for the molecular level description of ultra complex mixtures. However, acquisition speeds range from several hundred milliseconds to seconds and are often too slow for high-performance gas chromatography. Thus, this chapter refrains from a detailed description. Excellent reviews on the FT ICRMS technology and handling the complex data are reported elsewhere. There is only limited work on the usage of GC with FT ICRMS detection, often with API. Notable work has been reported on emerging pollutants, plastic pyrolysis oils, fossil petroleum and coal, as well as combustion aerosols [67, 68].

4.2.3.6 Ion Mobility Hyphenation

Ion mobility spectrometry (IMS) is an analytical technique for the gas-phase separation of ions based on the balance of an electrical force and collisions with inert gas (commonly termed "buffer gas") molecules. The applied electric field compels the ions to travel through the buffer gas with a velocity correlated to the mobility (K) of the analyte. The use of IMS in analytical measurements has seen a remarkable surge over the last decades, and GC was one of the first separation techniques coupled to IMS-MS as a front-end device already in 1970 (hybrid IMS-MS instrumentation). Since then, the application range of GC-IMS-MS has expanded, ranging from detecting explosives and volatile compounds in complex mixtures to medical analysis of Chinese medicine and environmental analysis.

Different technical solutions have been developed separating the ions by their mobility either in space or time, resulting in a complex landscape of IMS solutions, with drift tube IMS (DTIMS), traveling wave IMS (TWIMS), trapped IMS (TIMS), differential IMS (DIMS), and field asymmetric IMS (FAIMS) as the most important commercialized devices. 8,2 Each unique instrument platform has distinct advantages and disadvantages, which cannot be addressed here but are discussed in detail elsewhere. IMS allows to separate ions within milliseconds according to their size and shape. Conversion and calibration of the mobility provides access to the collision cross section (CCS). The CCS, denoted in units of square Angstrom ($Å^2$), is an intrinsic property of an ion and can be calculated based on the Mason-Champ equation and experimentally derived with very low error. Combined with the ability to compute these values theoretically, it opens up exciting possibilities for unsupervised prediction of structures in non-targeted analysis or increases the identification confidence in targeted measurements.

Gas chromatographic peak widths in the second domain ideally match IMS separation speed. This makes GC a perfect fit as a front-end separation technique for most IMS-MS techniques. IMS not only seamlessly integrates into most GC-MS approaches without a substantial loss of analytical performance, but it also brings a host of benefits. IMS significantly enhances peak capacity by introducing an additional dimension. This increased separation power, coupled with the improved substance identification through both retention time and drift time, amplifies analytical performance. However, it's important to note that IMS requires a specific pressure of the buffer gas as a working condition, which can complicate the technical implementation for most vacuum ionization mass spectrometers, such as standard electron ionization mass spectrometers. As a result, GC-IMS-MS has often been presented with atmospheric pressure ionization techniques (API). The GC separation prior to the ionization source effectively reduces chemical cross-sensitivities (suppression) and charging phenomena associated with many API sources.

GC-IMS-MS has been depicted as a versatile "Swiss army knife" in gas phase analysis, detecting and identifying volatile and semi-volatile organic compounds in complex mixtures. Modern systems reach detection limits in the ppbv and pptv ranges. Increasing the ion mobility resolving power has been one of the focus technical developments in IMS-MS, with reported values of over 100 for cyclic TWIMS and TIMS devices. With the growing interest in studies of small molecule complex mixtures for various applications, GC-IMS-MS analysis of biological and environmental samples will undoubtedly increase in the coming years. The process can be compared with the development of two different chromatographic separation steps in GC×GC and its subsequent usage for high-performance analytical insights. Artificial intelligence (AI) driven CCS prediction will overcome the limitations of existing databases and calculation-heavy quantum chemistry-based platforms, such as ISiCLE and MOBCAL-MPI [69–78].

4.2.3.7 Mass Calibration

Mass calibration is an essential process to preserve the accuracy of a mass spectrometer to adding the correct m/z value to a detected ion. It is essentially the correction of the mass axis, whereby one or more measured m/z values are compared and corrected with the known true m/z value. In general, such a process can also be referred to as part of a "tuning" of the mass spectrometer and may also include other relevant parameters such as the optimization of the precision & resolution, the correction of relative intensities of the m/z values to each other or a general intensity optimization.

Mass calibration is typically conducted prior to analysis and is relatively stable over an extended period, depending on the mass spectrometer selected. This can and should be verified before and during the measurement by the investigation of known compounds with known spectra. The physical separation of the ions by the mass analyser typically follows a specific physical process, such as the flight time of ions in a TOF analyser, which can be described mathematically. The course of this function can now be fitted ("calibrated") by comparing actual and target values. The selection of calibration substances should be based on the presence of fragments within the mass range of interest for the intended application. This often depends on the ionization technique employed. The most commonly used calibration substance when ion impact ionization is employed is FC-43, which is also known as PFTBA or by its chemical name, perfluorotributylamine. Due to the high degree of fluorination, it is thermally very stable, exhibiting many characteristic fragments over a wide mass range (typically up to $m/z = 614$, however, the molecular ion is usually not detectable) at the typical 70 eV electron ionization. Since fluorine is a monoisotopic element, the mass spectrum is also not very complex in terms of isotopic pattern. In principle, any chemical compound with a known fragment pattern can be used for calibration if enough fragments with sufficient intensity are present to cover the relevant mass range. This may also include compounds that are present as impurities with a sufficiently high intensity, as is often the case with column bleeds. However, in practice, it can happen that the device manufacturer specifies certain fragments. For instance, if additional quality controls, such as detector response or leak tests, are conducted in conjunction with mass calibration, in which the relative response to a reference ion is employed as a criterion, the calibration substance should possess an exceptionally high vapour pressure, enabling direct infusion into the source without the necessity for special thermal treatment.

Mass analysers that are more susceptible to mass drift, such as TOF and particularly high-resolution systems, also permit automatic drift correction to be performed during the analysis. However, this typically only compensates for relative deviations during data acquisition by utilising intensive background signal as a reference. An actual calibration is not performed. Nevertheless, it is often feasible to conduct a post-calibration of individual or all spectra in the chromatogram. For this purpose, one or more existing spectra of known compounds in the chromatogram are employed as a reference. This is particularly pertinent in instances where high mass accuracies are required, such as in the derivation of element compositions.

4.2.4 Ion Detection and Detectors

After the mass spectral analysis of ions and their separation based on different m/z ratios, the ions need to be detected in a reliable and reproducible manner. This applies to all the analysers described above, except the Iontrap system, which can process the oscillation frequency of the ions on their orbit as a signal. Therefore, different detectors are commercially available which generate a signal from incident ions, e.g., by generating secondary electrons. The most common and commercially available detectors for MS are described below. Unfortunately, most detectors are a consumable material and, in many cases, very sensitive if handled or operated incorrectly.

4.2.4.1 Faraday Cup

The Faraday Cup is the simplest detector. It basically consists of an electrode (dynode surface) which emits electrons when hit by an ion beam. The emitted electrons induce a current which is amplified and recorded. The electrode can be of different secondary emitting materials such as metals or semiconductors, for example, BeO, GaP, or CsSb. The ease of secondary electron emission is determined by the electron work function of the respective material. A higher velocity of the impacting particles and a lower electron work function of the surface lead to a higher number of electrons emitted. The Faraday cup is robust but less sensitive than other detectors, as there is only one level of signal amplification, and it can only be used in analogue mode.

4.2.4.2 Secondary Electron Multipliers and Channel Electron Multiplier

Secondary Electron Multipliers (SEMs) are commonly used detectors for ion detection. The working principle is similar to the one of the Faraday cup but extended from one electrode (dynode) to several dynodes operated in series. This configuration is called discrete dynode electron multiplier (DDEMs). These dynodes have increasing potential and thus lead not only to one but several increasing amplifications. More precisely, if an electrode opposite the emissions is held at greater positive potential, the emitted electrons will accelerate towards this electrode and hit its surface, which in turn leads to the emission of even more electrons. Another configuration commonly used is the channel electron multiplier (CEM). The working principle is basically similar to the previous one, but instead of having several dynodes in series, there is only one dynode amplifying the signal through several collisions with the dynode surface. As straight CEMs would be unstable at a gain larger than 10^4 due to the release of positive ions created by the ionization of gas, which can accelerate towards the input signal and randomly contribute to the signal, usually curved ("horn-shaped") CEMs are utilized. This curved design shortens the free path for ion acceleration, which suppresses noise from ion feedback. In both configurations, the ions first pass a conversion dynode and then hit the initial amplification dynode surface, which emits secondary electrons. These electrons are then attracted to the second dynode or into the continuous dynode in which more and more electrons are produced, finally leading to a cascade of electrons. The cascade of electrons produces a current large enough to be detected by a sensitive preamplifier. CEMs are more compact and less expensive compared to DDEMs and, therefore, often found in benchtop instruments. Due to air sensitivity and to prevent discharging due to high voltage, electron multipliers must be operated under vacuum conditions. In general, SEMs can achieve a signal increase of up to 10^6–10^8.

4.2.4.3 Microchannel Plates

Microchannel Plates (MCPs) can be described as an extreme reduction of the size of the previously mentioned channel electron multiplier. These microchannel plates consist of several million small linear channel multipliers aligned together. MCPs are round plates and usually 2–5 cm in diameter when employed in MS instruments. In TOFMS systems, an MCP is usually operated to give an integral output for all entering ions within a certain time interval, e.g., the electron current is detected every nanosecond while ions are arriving at the detector. One MCP can increase the signal of up to 10^4. As this increase is low compared to electron multipliers, commonly two or more MCPs are installed together to gain a signal increase of up to 10^6–10^7.

4.2.4.4 Photomultipliers

In photomultipliers, also known as Scintillation Counters, the first step of detection is similar to the one of the other detectors with an ion beam hitting a dynode and leading to the emission of secondary electrons. In the next step, contrary to the above-mentioned detectors, the electrons are converted into photons before the amplification process. This is realized by a phosphorescent screen, which releases photons when hit by electrons. These photons then strike a photon multiplier which in turn leads to a cascade of photons. An advantage of the use of photons instead of electrons is that the multiplier can be sealed under vacuum, which prevents contamination and extends the lifetime of the detector.

4.2.4.5 Conversion Dynodes

Conversion dynodes are robust electrodes, set to a high potential to attract the ions exiting from the mass analyser. This is especially important for quadrupole instruments, as slow ions exiting from those mass analysers would stop before reaching the detector. Ions impacting the conversion dynode lead to the production of secondary ions or electrons, which can be used for subsequent detection. Conversion dynodes are often used in front of SEM detectors and can also function as a post-acceleration detector [79–88].

4.2.5 Electronic Signal Processing

The primary detector signal is analogue and can have an amplitude of few a microvolts up to 1V, depending on the detector type, and must be converted to a digital signal for further computer-based signal processing and data handling. The individual physical components for signal processing are often arranged on a data acquisition board (DAQ board). Processes which have to be considered are preamplification of the signal, signal conditioning (e.g., de-noising and filtering), digitization, a temporary fast read/write interim storage, and data communication with the computer (e.g., via USB, PCIe, or Ethernet). Timing and time synchronization are also important processes and could be also tasks for the DAQ board, which is closely interlinked with the control of the MS.

The main component of a DAQ system is the analogue-to-digital converter (ADC). The analogue input signal is pre-amplified and converted into a digital signal suitable for data processing. The specifications of the ADC depend strongly on the mass spectrometer used. The main characteristics are sampling rate, bandwidth, and signal resolution. Noticeably short signals, such as those produced by TOF, require a very high sampling rate of gigasamples/second (GS/S = one billion/second), while sampling rates in the megasample range (MS/s = one million/second) are sufficient for quadrupole systems. The sampling rate is usually matched to the maximum frequency of the signals to be recorded. The Nyquist-Shannon theorem is often applied, which states that the sampling rate should be at least twice the maximum frequency. This is imposed by the bandwidth of the ADC. For the different types of MS it is necessary to be able to record frequencies from 1 megahertz (MHz), e.g., for ion trap systems, up to gigahertz (GHz) for TOF systems. The third relevant parameter is the resolution of the ADC. This is the opposite of bandwidth and sampling rate. For TOF systems, ADCs with a resolution of 12 bits are sufficient, which would allow 4096 discrete values (voltage steps of 0.244 millivolts) can be subdivided with a maximum input signal of one volt. Orbitrap systems use 24-bit resolution. The ADC can now, for example, integrate the input signal continuously and store it in discrete, subsequent, and timely equidistant values. In this case, the integration method allows only signals above a certain threshold to be summed or averaged. This can be used to eliminate electronic noise. Alternatively, a non-continuous event-driven (Signal) acquisition would be possible. The non-continuous pulse method only records signals (events) that exceed a certain threshold. It is important to note that a time signal must also be recorded for each event. Pulse Counting is particularly suitable for recording single ion signals or a small number of ions.

Galvanic isolation of the mass spectrometer, including the detector, from the signal processing described above, is also becoming more common. This can be achieved by integrating optocouplers. The ions first hit a scintillator, where a light signal is generated, which is then detected instead of the ions by a multiplier. One of the advantages of this is that the detector and DAQ can be isolated from the high voltages in the MS. In addition, also other sources of interference that could affect an electrical signal can be excluded [5, 89].

4.3 MS Data and Basic Data Handling

GC-MS data are extraordinarily rich in information and belong to the so-called multiway data. If you look at the overall complexity of GC-MS data, it is two-dimensional. The dimensions include retention time (or scan number), *m/z*, and each point in this two-dimensional space represents the intensity of a particular ion (with a particular *m/z*) detected at a particular time. If a GC×GC separation is used, the dimensionality is further increases. In the ideal case, eluting compounds are assigned to a peak. The aim of further processing and handling of the GC-MS raw data is the extraction of qualitative and quantitative information. Qualitative information can simply be the presence of a peak and its retention time. In many cases, however, identification or at least classification based on the retention times and mass spectrometric information obtained is desirable. The quantitative information is usually derived from the height of the peak or integration of the peak area. The third and last section of the chapter is intended to give a brief overview of the basic steps that can be used for this purpose.

4.3.1 Deconvolution of (GC-)MS Signals

The term "deconvolution" emerged in the 1960s, primarily in the context of signal processing and mathematics. It refers to the process of reversing the effects of the superposition or "convolution" of signals (signal processing) or mathematical functions to recover the original signal or function. Chromatographic raw data are usually also a superposition of different signals. The most important signal for the user is the signal associated with the compound of interest, which elute from the chromatographic column onto the detector in the form of a peak. In GC-MS, both quantitative (abundance) and qualitative (mass spectrum, elution time) information can be derived from the peak signal. Possible superimposed signals can be of chemical (column bleed, leaks) or electronic (noise) origin and may also change over time. Often, there are also other compounds that could not be separated chromatographically and now elute simultaneously/closely and overlap (convolute) in the detector signal. In our case, deconvolution refers to the separation of such overlapping chromatographic signals. The underlying process often involves mathematical algorithms. However, some initial parameters must still be specified manually by the user in order for the algorithm to have starting and/or boundary values. Some algorithms and software are described below and discussed very briefly to provide an understanding of the deconvolution process.

A still extremely popular software for deconvoluting GC-MS data to extract "pure" peak information is the Automated Mass Spectral Deconvolution and Identification System (AMDIS). Other classic programs work in a similar way and include the same or similar processing steps. First, the noise of the detector signal is determined in order to identify potential peaks that rise above the noise in the second step. This step also involves determining the start and end points and the APEX, which is usually the highest point of the peak. The actual deconvolution step begins by fitting the individual mass traces to a mathematical function. Mass traces with the same temporal behaviour are assigned to the same source (e.g., peak). By combining the mass traces found with the specified APEX, a peak is constructed from which the qualitative information (mass traces in the form of a mass spectrum as well as the position of the APEX) are derived. This type of deconvolution is still widely used and forms the basis of many one-dimensional GC software. In current software, quantification of the deconvoluted signals is then done by integration of deconvoluted mass traces. However, an important aspect and potential source of error are the assumptions that such a deconvolution approach requires. For example, peak widths or certain shapes often have to be specified.

Another approach would be to exploit the already mentioned multiway structure of a GC-MS data set. In this case, multivariate curve resolution methods are applied used to obtain information about overlapping signals and to calculate eluting peaks. In the case of GC-MS, such a two-way data set is defined in one direction by time (retention time) and in the other direction by the *m/z* information. In such cases, deconvolution benefits from so-called second-order advantages, where peaks do not need to be fully resolved chromatographically, and overlapping signals do not need to be defined beforehand. Common algorithms are Multivariate Curve Resolution Alternating Least Squares (MCR-ALS) or PARAFAC2. While AMDIS-like approaches are already well integrated into manufacturers' software, the fixed integration of multi-curve resolution algorithms is still rare. In the second case, application examples that demonstrate basic applicability for small data sets still predominate and are mostly based on self-programmed algorithms.

The algorithms discussed above are also used in the evaluation of GC×GC data. However, if AMDIS-like algorithms are used, only peaks along the second dimension are initially deconvoluted. In a second step, these partial peaks are then combined into a two-dimensional peak based on certain criteria, such as spectral similarity and elution time in the second dimension. Again, multivariate approaches can be applied directly to the three-way data set. The additional dimension can be used as an additional discriminative feature (e.g., PARAFC3).

For high-resolution data it is even more obvious, and the exact mass is used instead of the nominal mass. However, due to the amount and the complexity of the data, the processing time is sometimes significantly higher and should not be ignored [90, 91].

In the following sections, we will now discuss how to deal with the qualitative and quantitative information that could be determined directly or indirectly from the raw data using deconvolution.

4.3.2 The Qualitative Information

Given the depth and breadth of knowledge required to fully comprehend GC-MS data interpretation, it is no surprise that multiple books have been dedicated solely to this subject. This section aims to provide a concise and accessible overview of the fundamental key concepts in qualitative information extraction from GC-MS data.

While it does not replace the comprehensive understanding gained from extensive study or hands-on experience, it serves as a starting point for those seeking a basic understanding of the principles and techniques involved in qualitative analysis using GC-MS.

4.3.2.1 Interpretation of the Fragmentation Pattern

In mass spectrometry, the preservation or production of a molecular ion is often desired, typically achieved through soft ionization methods. Despite this, (EI)QMS remains the most widely used technique, particularly outside of academia. Considering the prevalence of EI mass spectra in GC-MS analysis, it is worthwhile to consider them as the foundation for interpreting GC-MS data. By understanding the fragmentation patterns and characteristic ions produced in EI mass spectra, researchers get a good starting point for the identification and characterization of compounds. Specific classes of molecules exhibit characteristic ion patterns upon fragmentation. Empirical rules, such as e.g., the even-electron rule, Stevenson's rule, and nitrogen rule, aid in predicting these patterns. These rules guide the interpretation of mass spectra, providing insights into the fragmentation behaviour of different type of molecules. Applying these rules enhances researchers' understanding and interpretation of mass spectra, facilitating compound identification and characterization in analytical studies.

In addition, the presence of characteristic ions and neutral losses can provide valuable insights into specific structural elements within a molecule. However, caution must be exercised when utilizing tables such as Table 4.2, as the suggested interpretations represent only a subset of the numerous possibilities. These tables serve as helpful references, but comprehensive analysis and consideration of all potential fragmentation pathways are essential for accurate interpretation of mass spectra.

4.3.2.2 Scripting and Filtering

The different fragmentation characteristics of molecules and their elution order can be also used to for scripting and filtering. For GC(×GC)-MS data this plays an integral role in qualitative analysis, providing researchers with powerful tools for data interpretation and identification of compounds. Scripting involves using programming languages to automate data processing and analysis of mass spectra and peak data, allowing for the efficient processing of large data sets. It can be used to perform tasks such as peak detection, deconvolution, and spectral library searches. By automating these processes, scripting can significantly reduce the time and effort required for data analysis. Filtering, on the other hand, involves the application of specific criteria to the data set, eliminating unwanted signals and improving the accuracy of the analysis. Filters can be applied based on factors such as peak area, signal-to-noise ratio, retention time, and mass-to-charge ratio. By applying

Table 4.2 Non-exhaustive compilation of characteristic ion fragments and neutral losses: exercise caution in interpretation, as the provided information represents a subset of potential possibilities [92–97].

Ion series	Fragments and neutral losses	Possible information
Carbenium	$[C_nH_{2n+1}]^+$	Alkyl group
Acylium	$[C_nH_{2n-1}O]^+$	Aliphatic aldehydes, ketones, carboxylic acids
Iminium	$[C_nH_{2n+2}N]^+$	Aliphatic amines
Oxonium	$[C_nH_{2n+1}O]^+$	Aliphatic alcohols
Sulfonium	$[C_nH_{2n+1}S]^+$	Aliphatic thiols and thioether
Neutral losses	$[M-1]^+$	Cleavage of H
	$[M-2]^{+*}$	Loss of H_2
	$[M-3]^+$	Loss of H^* and H_2
	$[M-15]^+$	Loss of CH_3
	$[M-17]^{+*}$	Loss of NH_3
	$[M-17]^+$	Loss of OH of tertiary alcohols
	$[M-18]^{+*}$	Loss of HCN
	$[M-27]^{+*}$	Loss of CO/C_2H_4 or N_2
	$[M-29]^{+*}$	Loss of CHO or C_2H_5
	$[M-30]^+$	Loss of H_2CO
	$[M-44]^{+*}$	Loss of CO_2
	$[M-46]^{+*}$	Loss of HCOOH, EtOH or NO_2
	$[M-48]^{+*}$	Loss of SO
	$[M-64]^{+*}$	Loss of SO_2
	$[M-77]^+$	Loss of phenyl
	$[M-91]^+$	Loss of benzyl

these filters, researchers can remove noise and background signals, improving the quality of the data and reducing the risk of false positives. Both scripting and filtering provide possibilities to visualize GC(×GC)-MS data, enabling researchers to efficiently identify and characterize compounds and compound classes in complex mixtures [98, 99].

4.3.2.3 Mass Spectra Libraries

In addition to individual fragments, the entire spectrum of a peak can also be compared with databases to obtain information about the identity of the eluted substance. MS libraries serve as extensive repositories housing the reference mass spectra of identified compounds, facilitating the qualitative analysis of analytes by comparing their unique spectral patterns with those in the library. These databases, including renowned repositories such as the NIST/EPA/NIH Mass Spectral Library and Wiley Registry of Mass Spectral Data, are essential for identifying metabolites and other compounds to unravelling their biological significance. Spectral search algorithms, employing techniques such as spectral correlation, dot product, or cosine similarity, contribute to this identification process by determining the closest matches between the unknown and reference spectra. In Omics research characterized by diverse and intricate samples, MS libraries have become indispensable tools for elucidating the identities of peptides, fatty acids, steroids, and other biomolecules. Spectral search functionalities enable researchers to compare experimental spectra from their samples with those in the library, facilitating the identification process. Qualitative analysis in MS involves the meticulous comparison of unknown compound mass spectra with those in the MS library, leveraging spectral matching algorithms to identify potential matches and ascertain compound identities. The integration of MS libraries and spectral search algorithms significantly enhances the capacity to extract meaningful insights from complex biological samples, emphasizing their paramount role in the qualitative analysis of diverse compounds within the context of omics research. In Omics, the challenges posed by extensive and diverse analytes necessitate the use of expansive and specialized libraries for accurate identification. The accuracy of identification relies on the quality and specificity of the MS library, which requires regular curation and updating. Interpreting MS spectra in omics requires a nuanced understanding of the compound fragmentation patterns and matrix effects, prompting tailored approaches for reliable identification. The accuracy and reliability of spectral searches hinge on the MS library's quality, comprehensiveness, and similarity threshold for matching. However, spectral matching is not always definitive, requiring additional confirmatory techniques, such as retention time matching or fragmentation pattern analysis. Enhancing the integration of MS libraries and spectral search capabilities in GC-MS data facilitates the streamlining of compound identification and elucidates intricate metabolic pathways in the field of omics. This multidisciplinary interplay underscores the significance of robust data-handling strategies for meaningful insights from complex biological samples. MS libraries, whether commercial or custom-built, are continuously updated to enhance identification capabilities, incorporating information beyond spectral matching, such as retention time, molecular weight, and fragmentation patterns, for improved qualitative analysis in omics [5, 6, 100, 101].

4.3.2.4 Seven Golden Rules

In addition to database comparison or calculation of the elemental composition based on exact mass information, rules can also be applied as additional filter. Kind and Fiehn presented in 2008 helpful guidance, referred to as the "Seven Golden Rules" for intelligently restricting the elemental attribution boundaries with the goal of unique sum formulae hits. Firstly, the maximum allowed elemental number is restricted based on spectral database entries, such as Wiley and DNP. Exemplarily for the Wiley database, in a mass range of below 1000 Da, one can expect a maximum of 78 Carbon, 126 Hydrogen, 20 Nitrogen, 7 Oxygen, and 14 sulphur atoms. Obviously, this rule is a rather weak restriction but causes the fewest anomalies, is easy to apply, and supported by millions of database entries. Secondly, we deploy knowledge on chemical bonds. With the LEWIS and SENIOR rules, the sums of the valences are checked and sum formulae not fulfilling these rules discarded. Thirdly, the isotopic pattern is valorised. The fine isotopic pattern of potential hits is calculated and matched with the experimental spectrum. Let us assume a potential sum formula hit containing one Sulphur atom. This candidate can only be true, if the ^{34}S isotope is found with the characteristic *m/z* delta of 1.995796 (^{32}S 31.972072 *u*, ^{34}S 33.967868 *u*) to the monoisotopic mass with an abundance ratio of 95 to 4.2. Naturally, the abundances of the isotopes have to be above the noise level and may vary within experimental and instrumental accuracy boundaries. Here, various automatized workflows have been presented taking the position/existence of the isotopes, their abundance or both information combined into account, often resulting in a filtered matching score. The fourth and fifth restriction rules are limiting further potential hits by filtering elemental ratios, such as hydrogen-carbon-ratio (degree of unsaturation) and oxygen-carbon- or nitrogen-carbon-ratio. Wiley database has shown that 99.99 % of all entries fall within a H/C of 0.2–3.1 and O/C of 0–1.2. Finally, the remaining list of candidates is subjected to an elemental probability (combinations of heteroelements) and trimethylsilyl check (if derivatization has been applied). These "Seven Golden Rules" rapidly allow to substantially lower the number of possible sum formulae. If only one hit remains, the *m/z* signal is attributed to this sum formula. However, if no unambiguity has been reached, the candidate with the lowest mass error or highest matching score (*m/z* deviation and isotopic fingerprint matching) may be taken. Depending on the application field and sample matrix, other aiding rules have been reported using characteristic sum formulae pattern within the samples, such as alkylation (CH_2), oxygenation (O) or larger building blocks, particularly for metabolomics, this concept may allow to drastically increase the assignment confidence based on given biochemical pathways and molecular networks.

4.3.2.5 The High Resolution Advantage

The procedures described above for obtaining qualitative information can be applied to all types of GC-MS data. However, the use of high-resolution mass spectrometers opens up additional possibilities. High-resolution mass spectrometry with exact mass measurement intrinsically allows analytical chemists to move from m/z measurement to chemical knowledge. This chemical information is retrieved by attributing elemental compositions to the m/z signals, either for deciphering fragments in hard ionization or, if the molecular ion is recorded, directly assigning sum formulae to the intact compounds. Structural information and the identity of gas chromatographic signals analysed with mass spectrometric detection is most often based on hard electron ionization. Hereby, spectral library matching can be utilized in an automated manner, i.e., comparing databases, such as NIST, ChemSpider, or Pubchem, with the experimental pattern, finally calculating matching scores. However, these libraries primarily contain nominal mass information from decades of GC-MS research work. Although most of these databases are based on nominally resolved mass spectra, the elemental composition can be used as an additional filter in the search. This is similar to the case for retention indices.

Aside from precisely calibrated mass spectra with ppm-level accuracy, the second prerequisite for converting *our* m/z signals to valuable "chemistry" with high confidence is a spectral separation of the signals of interest. The mass resolving power of the system has to be appropriately chosen for the analytical challenge. Lower complex mixtures or samples fully resolved given the separation power of the gas chromatographic preface, e.g., by comprehensive two-dimensional gas chromatography, can be analysed in-depth on mid-level mass resolving power systems, such as TOF systems. Highly complex mixtures not fully resolved by gas chromatography and with high diversity in the contained heteroelements and chemical functionalities may need analysis on high-resolution mass spectrometric devices.

Starting with the list of calibrated and resolved mass spectrometric regions of interest (features), sum formulae can directly be attributed with various free open-code or commercialized computation software tools. The mass accuracy, given by the system and reassured by comparison of quality control and internal standards with the manufacturer reported values, is set as error width, commonly in the range of 0.5–10 ppm. The fact that accurate mass measurement allows us, analytical chemists, to attribute elemental compositions is based on the mass defect. Different arrangements of neutrons and protons in the core of an atom are causing different energetic states, and consequently, the sum of the masses of the nucleotides is characteristically different to the mass to the core they are forming. On the atomic mass scale [u], the ^{12}C isotope has been defined by the IUPAC as exactly 12 u, whereas other isotopes have exact non-integer values. On this scale, isotopes with exact mass above their nominal weight are referred to as positive mass defect elements, such as ^{1}H, ^{13}C, or ^{14}N, whereas, isotopes with lower exact mass than their nominal integer weight have a negative mass defect, such as ^{16}O and ^{32}S. Hence, the exact mass of any elemental composition is unique and could be resolved by high resolution instrumentation. The mass split notation refers to these difference between two elemental formulae. $^{1}H_{12}$ versus ^{12}C observed in almost all matrices with a high spread in degree of unsaturation results in a difference of 93.9 mDa. A mass split of 36.4 mDa is found for $^{12}C_{1}H_{4}$ versus ^{16}O and 15.3 mDa for $^{12}C_{4}$ versus $^{16}O_{3}$, both frequently observed in dissolved organic matter (DOM), highly oxidized organic aerosols, e.g., secondary organic aerosols (SOA), or biomass pyrolysis oils. More challenging mass splits are given by other heteroelements, such as $^{12}C_{1}H_{2}$ versus ^{14}N, $^{12}C_{2}{}^{1}H_{3}$ versus $^{13}C^{14}N$, $^{14}N_{2}$ versus $^{12}C^{16}O$, and ^{14}N versus $^{13}C^{1}H$, resulting in 12.6, 17, 11.2, and 8.15 mDa, respectively. Challenging for most mass spectrometric platforms utilized in gas chromatographic hyphenation is the 3.4 mDa mass split given by $^{12}C_{3}$ versus $^{32}S_{1}H_{4}$. FT MS and some state-of-the-art time-of-flight mass spectrometer, such as multireflectron and orthogonal acceleration systems, have reached sufficient resolving power for tackling this Sulphur chemistry in the mass spectrometric domain. The latter point is addressed by high accuracy mass measurements in the parts-per-million (ppm) level. Modern FT MS and HR TOFMS routinely result in single-digit ppm mass accuracy. Putting this value in perspective, the measurement of the molecular ion of pyrene ($C_{16}H_{10}{}^{+\bullet}$) with an accuracy of 1 ppm results in 202.077702 ± 0.000202, thus, a deviation of 0.2 mDa. Depending on the allowed chemical diversity (unsaturation and heteroelements), often results in a single mathematically and chemically possible "hit", allowing for elemental formulae attribution with high confidence For low molecular weight constituents ($m/z < 200$) solely containing Carbon, Hydrogen, and Oxygen single-digit ppm mass accuracy is sufficient for unambiguous attribution. Nonetheless, with increasing m/z and heteroatom-content the mathematical possibilities rapidly increase, resulting in a list with potential sum formulae even with high mass accuracy of several ppm. Additional selection criteria, such as those already mentioned in the paragraph Seven Golden Rules, may also be helpful in this regard.

4.3.2.6 Data Visualization of Exact Accurate Data

In the last paragraph of this section, some selected concepts of handling high resolution are presented. In complex mixture analysis, molecular formula attribution easily results in lists of thousands of sum formulae. Thus, intelligent data grouping and visualization approaches have been developed. One of the oldest concepts is the Kendrick mass defect. Historically, the Kendrick mass defect has been defined based on alkylation building blocks (CH_2). The Kendrick mass (KM) is calculated based on the detected accurate mass multiplied by the ratio of the nominal mass of CH_2 and the exact mass of CH_2 (14/14.01565). The Kendrick mass defect (KMD) is retrieved from this value by subtraction from the integer mass. The Kendrick mass defect diagram is visualizing the nominal Kendrick mass on the abscissa and the KMD on the ordinate. Basically, this approach is realigning the data points, resulting in horizontal lines for the building blocks, whereas the ordinate separates other core structures (heteroatoms, aromaticity, etc.). This concept is not limited to CH_2 as building block and has very successfully been deployed for polymer analysis. Beneficially, the Kendrick mass defect visualization does not require attributed data and can be built solely using the accurate mass signals.

Another approach for grouping of the data is the concept of double bond equivalents (DBE). This measure aims to retrieve the number of rings and double bonds (r + d) from a given elemental composition using fundamental valency theory. In the simplest case, restricting the formula to $C_cH_hN_nO_o$ the DBE or r + d can be given as 1 + c–0.5 h + 0.5 n. Exemplarily, the aliphatic compound Decane ($C_{10}H_{22}$) results in a DBE of zero, whereas the aromatic compound Benzene (C_6H_6) directly gives the correct number of four (three double bonds and one ring). Different heteroelements can occur in various oxidation states and bonding situations, which is drastically complicating the calculation for more complex compounds. Moreover, it has to be noted, that DBE and most of the subsequent measures directly extrapolated from the sum formulae are conceptual measures and no direct structural features. Thus, one should always handle these secondary values with care. DBE (ordinate) is most often combined with the carbon number (abscissa) for visualization.

$$\mathrm{RDBE} = \mathrm{C} + \mathrm{Si} - \frac{1}{2}\,(\mathrm{C} + \mathrm{F} + \mathrm{Cl} + \mathrm{Br} + \mathrm{I}) + \frac{1}{2}\,(\mathrm{N} + \mathrm{P}) + 1 \tag{4.19}$$

Elemental-ratios may also be used for data visualization. Here, the concept of the Van Krevelen diagram, originating from coal and petroleum chemistry is used. By retrieving the oxygen-to-carbon (abscissa) and hydrogen-to-carbon (ordinate) value of a geochemical sample, the maturation degree in the catagenesis and metagenesis state can be depicted. This concept has been successfully transferred to the description of complex oxygen-containing molecular mixtures. Hereby, the elemental ratios of each sum formula are represented in the plot, ultimately leading to characteristically scattered data. Depending on the location in this van Krevelen representation, the chemistry of the depicted compounds may be described as carbohydrate-like (high O/C and H/C), lignin-like (intermediate O/C and H/C), or aromatic and condensed aromatic type (low H/C). These chemical class regions in the van Krevelen diagram space allow for easy and rapid representation and grouping of the data. Moreover, characteristic chemical and biochemical pathways, such as the addition/loss of CO (decarbonylation), CO_2

(decarboxylation), H_2O (dehydration), H_2 (dehydrogenation), or other building blocks can be traced.

4.3.3 The Quantitative Information

So far, we have only addressed qualitative aspects that can be obtained from mass spectrometry data to clarify the identity of the matrix. Finally, we will now briefly look at the quantitative aspects so that individual peaks can also be determined in terms of quantity. Quantitative analysis in GC-MS is the process of measuring the abundance of specific compounds in a sample.

4.3.3.1 Quantification

Calibration and standardization are essential for establishing a connection between the instrument's response and the concentration of known standards. In order to ensure precise quantification within specific concentration intervals, a well-defined linear range is necessary. This forms the basis for accurate measurements using analytical methodologies. Calibration curves, which are established through systematic analyses of standard solutions with known concentrations, are crucial tools for assessing linearity. These curves demonstrate the direct proportionality between the detector response, often measured as the peak area or peak height, and analyte concentration. It is crucial to achieve linearity because it ensures the precision and consistency of the method across a range of analyte concentrations. Statistical techniques, including regression analysis, are frequently used to evaluate linearity by calculating the correlation coefficient (R^2). This quantitative assessment provides a measure of how well the calibration curve aligns with the linear model. Internal standards, which are integral to the sample itself, enhance precision by accounting for variations in sample preparation and analysis. This dual approach ensures the reliability of quantitative analyses and supports robust comparisons across diverse omics applications.

Deviations from linearity can arise from various factors, such as matrix effects and inherent instrument limitations, leading to compromised accuracy in quantitative analysis. Saturation in quantitative mass spectrometry analysis occurs when the detector response reaches its maximum, resulting in a plateau in signal intensity despite increasing analyte concentration. This phenomenon can be attributed to overloading the ionization source or detector, thereby hindering the linearity of the response. Saturation poses challenges to accuracy and precision, limits the dynamic range, and may result in the underestimation of high analyte concentrations. To mitigate saturation, optimizing mass spectrometry parameters, such as ionization conditions, and ensuring analyte concentrations fall within the linear range are essential. In cases where saturation occurs, sample dilution or the adoption of alternative analytical techniques with higher dynamic ranges may be necessary for precise quantification. This critical aspect of mass spectrometry data handling underscores the importance of meticulous parameter optimization to maintain linearity and enhance accuracy in quantitative analyses.

Commonly employed in techniques such as LC-MS and GC-MS, the TIC offers a comprehensive view of ion abundance, thereby enabling the quantification of analytes in a sample. Its utility extends to assessing the total ionization efficiency, monitoring changes in ion intensities, and calculating analyte concentrations by comparing ion intensities to known standards. TIC is indispensable for accurate and precise quantification, providing a holistic perspective of ion currents in mass spectrometry experiments. Complementing TIC, Selected Ion Monitoring (SIM) enhances the sensitivity and selectivity in quantitative mass spectrometry analyses. By monitoring specific ions corresponding to target analytes, SIM reduces interference, improves signal-to-noise ratios, and enables the detection and quantification of low-level analytes. SIM is particularly valuable in trace analysis and complex matrices and facilitates targeted and sensitive quantification, offering applications in bioanalysis, drug metabolism studies, and metabolomics. Together, TIC and SIM form a robust foundation for quantitative information in mass spectrometry, providing researchers with the tools needed for nuanced analysis and precise quantification of compounds of interest across various research domains.

The evaluation of data quality is essential for extracting meaningful insights from omics analysis. Rigorous quality control measures, including the assessment of signal-to-noise ratios, peak shapes, and reproducibility, underpin the reliability of quantitative outcomes. This scrutiny is particularly crucial when working with biological samples, where variations and interferences are inherent. Adopting a comprehensive approach to data quality assessment strengthens the foundations of GC-MS data handling, instilling confidence in the reliability and reproducibility of generated results. Established best practices and tips serve as guiding principles for effective GC-MS data handling. These include proper instrument calibration and maintenance, careful sample preparation, appropriate selection and validation of reference standards, rigorous documentation of experimental procedures, and meticulous data management.

4.3.3.2 Isotopic Labelling

Quantitative analysis in mass spectrometry at the pico-nanogram level is often executed using stable isotope-labelled internal standards. Isotopic labelling involves replacing one or more atoms in a molecule with a stable isotope, such as ^{13}C or deuterium. These labelled compounds possess identical or nearly identical chemical properties to the analytes of interest, including ionization efficiency, but can be distinguished based on their mass-to-charge ratio in the mass spectrometer.

They are introduced into the sample at a known concentration before analysis and enable the determination of compound concentrations based on the ratio of labelled to unlabelled analytes. This technique is referred to as isotope dilution. By compensating for variations in sample preparation and injection, ionization efficiency, and instrument response, the isotopic dilution method significantly enhances accuracy and precision, and minimizes errors in the often-challenging matrix of biological samples encountered in omics studies.

The notion of flux, which refers to the rates of metabolic reactions in biological systems, has become a central focus within the field of omics studies. The quantification of metabolite flux provides valuable insight into the dynamic cellular processes and their temporal dynamics. One method for flux analysis is through the use of mass spectrometry, which can measure the turnover rates of isotopically labelled compounds. These compounds are introduced into biological systems, and their incorporation into metabolic pathways is tracked using mass spectrometry. By examining the isotopic labelling patterns of metabolites, flux analysis can reveal information about the flow of metabolites through different pathways and the rates of specific reactions. Flux analysis by mass spectrometry is a powerful tool for understanding metabolic regulation and can be applied in diverse fields such as biochemistry, pharmacology, and systems biology [6, 8, 100].

References

1 Frysinger, G.S. and Gaines, R.B. (1999). Comprehensive two-dimensional gas chromatography with mass spectrometric detection (GC × GC/MS) applied to the analysis of petroleum. *J. High Resolut. Chromatogr.* 22 (5): 251–255.

2 Gohlke, R.S. (1959). Time-of-flight mass spectrometry and gas-liquid partition chromatography. *Anal. Chem.* 31 (4): 535–541.

3 James, A.T. and Martin, A.J. (1952). Gas-liquid partition chromatography: the separation and micro-estimation of volatile fatty acids from formic acid to dodecanoic acid. *Biochem. J.* 50 (5): 679–690.

4 Thomson, J.J. (1921). *Rays of Positive Electricity and Their Application to Chemical Analyses*. Longmans, Green and Company.

5 De Hoffmann, E. and Stroobant, V. (2007). *Mass Spectrometry: Principles and Applications*. Wiley.

6 Gross, J.H. (2006). *Mass Spectrometry: A Textbook*. Springer Science & Business Media.

7 Mondello, L. (2011). *Comprehensive Chromatography in Combination with Mass Spectrometry*. Wiley.

8 Tranchida, P.Q. and Mondello, L. (2019). *Hyphenations of Capillary Chromatography with Mass Spectrometry*. Elsevier.

9 Murray, K.K., Boyd, R.K., Eberlin, M.N. et al. (2013). Definitions of terms relating to mass spectrometry (IUPAC recommendations 2013). *Pure Appl. Chem.* 85 (7): 1515–1609.

10 Murray, K.K. (2022). Resolution and resolving power in mass spectrometry. *J. Am. Soc. Mass Spectrom.* 33 (12): 2342–2347.

11 Franchina, F.A., Maimone, M., Tranchida, P.Q., and Mondello, L. (2016). Flow modulation comprehensive two-dimensional gas chromatography–mass spectrometry using ≈ 4 mL min^{-1} gas flows. *J. Chromatogr. A* 1441: 134–139.

12 Seeley, J.V. et al. (2006). Comprehensive two-dimensional gas chromatography with a simple fluidic modulator. *Am. Lab.* 38 (9): 24.

13 Tranchida, P.Q., Franchina, F.A., Dugo, P., and Mondello, L. (2014). Use of greatly-reduced gas flows in flow-modulated comprehensive two-dimensional gas chromatography-mass spectrometry. *J. Chromatogr. A* 1359: 271–276.

14 Galletta, M., Zoccali, M., Jones, N. et al. (2022). Flow-modulated comprehensive two-dimensional gas chromatography combined with time-of-flight mass spectrometry: use of hydrogen as a more sustainable alternative to helium. *Anal. Bioanal. Chem.* 414 (21): 6371–6378.

15 Kitson, F.G., Larsen, B.S., and McEwen, C.N. (1996). *Gas Chromatography and Mass Spectrometry: A Practical Guide.* Academic Press, San Diego. xii + 381 S. ISBN 0-12-483385-3.

16 Gross, J.H. (2017). Principles of ionization and ion dissociation. In: *Mass spectrometry: A Textbook* (ed. J.H. Gross), 29–84. Springer Science & Business Media.

17 Gross, J.H. (2017). Practical aspects of electron ionization. In: *Mass Spectrometry: A Textbook* (ed. J.H. Gross), 293–324. Springer Science & Business Media.

18 Gross, J.H. (2017). Chemical ionization. In: *Mass Spectrometry: A Textbook* (ed. J.H. Gross), 439–496. Springer Science & Business Media.

19 Hunt, D.F., Stafford, G.C., Crow, F.W., and Russell, J.W. (1976). Pulsed positive negative ion chemical ionization mass spectrometry. *Anal. Chem.* 48 (14): 2098–2104.

20 Yuan, T.-H., Eguchi, A., Tai, C.J. et al. (2022). Comparison of the PCB serum levels among mother-child pairs in areas of Eastern Japan and Central Taiwan. *Sci. Total Environ.* 806: 150272.

21 Cruz, R., Marques, A., Casal, S., and Cunha, S.C. (2019). Fast and environmental-friendly methods for the determination of polybrominated diphenyl ethers and their metabolites in fish tissues and feed. *Sci. Total Environ.* 646: 1503–1515.

22 Worton, D.R., Mills, G.P., Oram, D.E., and Sturges, W.T. (2008). Gas chromatography negative ion chemical ionization mass spectrometry: application to the detection of alkyl nitrates and halocarbons in the atmosphere. *J. Chromatogr. A* 1201 (1): 112–119.

23 Carrizo, D. and Grimalt, J.O. (2006). Rapid and simplified method for the analysis of polychloronaphthalene congener distributions in environmental and human samples by gas chromatography coupled to negative ion chemical ionization mass spectrometry. *J. Chromatogr. A* 1118 (2): 271–277.

24 Budzikiewicz, H. (1981). Mass spectrometry of negative ions. *Angew. Chem. Int. Ed. Engl.* 20 (8): 624–637.

25 Bowie, J.H. (1984). The formation and fragmentation of negative ions derived from organic molecules. *Mass Spectrom. Rev.* 3 (2): 161–207.

26 Hejazi, L., Ebrahimi, D., Hibbert, D.B., and Guilhaus, M. (2009). Compatibility of electron ionization and soft ionization methods in gas chromatography/orthogonal time-of-flight mass spectrometry. *Rapid Commun. Mass Spectrom.* 23 (14): 2181–2189.

27 Mjøs, S.A. (2004). The prediction of fatty acid structure from selected ions in electron impact mass spectra of fatty acid methyl esters. *Eur. J. Lipid Sci. Technol.* 106 (8): 550–560.

28 Wille, S.M., van Hee, P., Neels, H.M. et al. (2007). Comparison of electron and chemical ionization modes by validation of a quantitative gas chromatographic–mass spectrometric assay of new generation antidepressants and their active metabolites in plasma. *J. Chromatogr. A* 1176 (1-2): 236–245.

29 Margolin Eren, K.J., Elkabets, O., and Amirav, A. (2020). A comparison of electron ionization mass spectra obtained at 70 eV, low electron energies, and with cold EI and their NIST library identification probabilities. *J. Mass Spectrom.* 55 (12): e4646.

30 Zimmermann, R. and Hanley, L. (2021). *Photoionization and Photo-Induced Processes in Mass Spectrometry: Fundamentals and Applications*. Wiley.

31 Gunzer, F., Krüger, S., and Grotemeyer, J. (2019). Photoionization and photofragmentation in mass spectrometry with visible and UV lasers. *Mass Spectrom. Rev.* 38 (2): 202–217.

32 Hanley, L. and Zimmermann, R. (2009). Light and molecular ions: the emergence of vacuum UV single-photon ionization in MS. *Anal. Chem.* 81: 4174–4182.

33 Pallozzi, E., Guidolotti, G., Ciccioli, P. et al. (2016). Does the novel fast-GC coupled with PTR-TOF-MS allow a significant advancement in detecting VOC emissions from plants? *Agric. For. Meteorol.* 216: 232–240.

34 Wittmann, C. (2007). Fluxome analysis using GC-MS. *Microb. Cell Factories* 6 (1): 6.

35 Paul, W. and Steinwedel, H. (1953). Notizen: Ein neues Massenspektrometer ohne Magnetfeld. *Z. Naturforsch. A* 8 (7): 448–450.

36 Finnigan, R.E. (1994). Quadrupole mass spectrometers. *Anal. Chem.* 66 (19): 969A–975A.

37 Mathieu, É. (1868). Mémoire sur le mouvement vibratoire d'une membrane de forme elliptique. *J. Math. Pures Appl.* 13: 137–203.

38 Shellie, R.A., Marriott, P.J., and Huie, C.W. (2003). Comprehensive two-dimensional gas chromatography (GC × GC) and GC × GC-quadrupole MS analysis of Asian and American ginseng. *J. Sep. Sci.* 26 (12–13): 1185–1192.

39 Song, S.M., Marriott, P., and Wynne, P. (2004). Comprehensive two-dimensional gas chromatography—quadrupole mass spectrometric analysis of drugs. *J. Chromatogr. A* 1058 (1–2): 223–232.

40 Adahchour, M., Brandt, M., Baier, H.U. et al. (2005). Comprehensive two-dimensional gas chromatography coupled to a rapid-scanning quadrupole mass spectrometer: principles and applications. *J. Chromatogr. A* 1067 (1–2): 245–254.

41 Purcaro, G., Quinto Tranchida, P., Conte, L. et al. (2011). Performance evaluation of a rapid-scanning quadrupole mass spectrometer in the comprehensive two-dimensional gas chromatography analysis of pesticides in water. *J. Sep. Sci.* 34 (18): 2411–2417.

42 Weinert, C.H., Egert, B., and Kulling, S.E. (2015). On the applicability of comprehensive two-dimensional gas chromatography combined with a fast-scanning quadrupole mass spectrometer for untargeted large-scale metabolomics. *J. Chromatogr. A* 1405: 156–167.

43 Regueiro, J., Llompart, M., Garcia-Jares, C., and Cela, R. (2009). Development of a solid-phase microextraction–gas chromatography–tandem mass spectrometry method for the analysis of chlorinated toluenes in environmental waters. *J. Chromatogr. A* 1216 (14): 2816–2824.

44 Rambaud, L., Monteau, F., Deceuninck, Y. et al. (2007). Development and validation of a multi-residue method for the detection of a wide range of hormonal anabolic compounds in hair using gas chromatography–tandem mass spectrometry. *Anal. Chim. Acta* 586 (1–2): 93–104.

45 de Hoffmann, E. (1996). Tandem mass spectrometry: a primer. *J. Mass Spectrom.* 31 (2): 129–137.

46 Franchina, F.A., Machado, M.E., Tranchida, P.Q. et al. (2015). Determination of aromatic sulphur compounds in heavy gas oil by using (low-) flow modulated comprehensive two-dimensional gas chromatography–triple quadrupole mass spectrometry. *J. Chromatogr. A* 1387: 86–94.

47 Poliak, M., Fialkov, A.B., and Amirav, A. (2008). Pulsed flow modulation two-dimensional comprehensive gas chromatography–tandem mass spectrometry with supersonic molecular beams. *J. Chromatogr. A* 1210 (1): 108–114.

48 Chen, Y.H., Gonin, M., Fuhrer, K. et al. (1999). Orthogonal electron impact source for a time-of-flight mass spectrometer with high mass resolving power. *Int. J. Mass Spectrom.* 185-187: 221–226.

49 Stephens, W.E. (1946). A pulsed mass spectrometer with time disaersion. *Phys. Rev.* 69: 691.

50 Gröger, T.M., Käfer, U., and Zimmermann, R. (2020). Gas chromatography in combination with fast high-resolution time-of-flight mass spectrometry: technical overview and perspectives for data visualization. *TrAC Trends Anal. Chem.* 122: 115677.

51 Alikhanov, S. (1957). A new impulse technique for ion mass measurements. *Sov. Phys. JETP* 4 (3): 452.

52 Mamyrin, B. et al. (1973). The mass-reflectron, a new nonmagnetic time-of-flight mass spectrometer with high resolution. *Zh. Eksp. Teor. Fiz.* 64 (1): 82–89.

53 Grix, R., Kutscher, R., Li, G. et al. (1988). A time-of-flight mass analyzer with high resolving power. *Rapid Commun. Mass Spectrom.* 2 (5): 83–85.

54 Dawson, J. and Guilhaus, M. (1989). Orthogonal-acceleration time-of-flight mass spectrometer. *Rapid Commun. Mass Spectrom.* 3 (5): 155–159.

55 Guilhaus, M., Selby, D., and Mlynski, V. (2000). Orthogonal acceleration time-of-flight mass spectrometry. *Mass Spectrom. Rev.* 19 (2): 65–107.

56 Yavor, M.I., Pomozov, T.V., Kirillov, S.N. et al. (2018). High performance gridless ion mirrors for multi-reflection time-of-flight and electrostatic trap mass analyzers. *Int. J. Mass Spectrom.* 426: 1–11.

57 Willis, P., Jaloszynski, J., and Artaev, V. (2021). Improving duty cycle in the folded flight path high-resolution time-of-flight mass spectrometer. *Int. J. Mass Spectrom.* 459: 116467.

58 Hernández, F., Portolés, T., Pitarch, E., and López, F.J. (2011). Gas chromatography coupled to high-resolution time-of-flight mass spectrometry to analyze trace-level organic compounds in the environment, food safety and toxicology. *TrAC Trends Anal. Chem.* 30 (2): 388–400.

59 Guilhaus, M. (1995). *Special feature: Tutorial.* Principles and instrumentation in time-of-flight mass spectrometry. Physical and instrumental concepts. *J. Mass Spectrom.* 30 (11): 1519–1532.

60 Špánik, I. and Machyňáková, A. (2018). Recent applications of gas chromatography with high-resolution mass spectrometry. *J. Sep. Sci.* 41 (1): 163–179.

61 Marney, L.C., Hoggard, J.C., Skogerboe, K.J., and Synovec, R.E. (2014). Methods of discovery-based and targeted metabolite analysis by comprehensive two-dimensional gas chromatography with time-of-flight mass spectrometry detection. In: *Gas Chromatography and Mass Spectrometry: A Practical Guide*, (Methods in Molecular Biology, Vol. 1198, 83–97). New York, NY: Springer New York. https://doi.org/10.1007/978-1-4939-1258-2_6.

62 Wilson, R.B., Hoggard, J.C., and Synovec, R.E. (2012). Fast, high peak capacity separations in gas chromatography–time-of-flight mass spectrometry. *Anal. Chem.* 84 (9): 4167–4173.

63 Chernushevich, I.V., Loboda, A.V., and Thomson, B.A. (2001). An introduction to quadrupole–time-of-flight mass spectrometry. *J. Mass Spectrom.* 36 (8): 849–865.

64 Gushue, J.N. (2013). Principles and applications of gas chromatography quadrupole time-of-flight mass spectrometry. In: *Advanced Techniques in Gas Chromatography-Mass Spectrometry (GC-MS-MS and GC-TOF-MS) for Environmental Chemistry* (ed. I. Ferrer and M.E. Thurman), 255–270. Newnes.

65 Macherone, A. (2013). The future of GC/Q-TOF in environmental analysis. In: *Advanced Techniques in Gas Chromatography-Mass Spectrometry (GC-MS-MS and GC-TOF-MS) for Environmental Chemistry* (ed. I. Ferrer and M.E. Thurman), 471–490. Newnes.

66 Allen, D.R. and McWhinney, B.C. (2019). Quadrupole time-of-flight mass spectrometry: a paradigm shift in toxicology screening applications. *Clin. Biochem. Rev.* 40 (3): 135.

67 Hu, Q., Noll, R.J., Li, H. et al. (2005). The orbitrap: a new mass spectrometer. *J. Mass Spectrom.* 40 (4): 430–443.

68 Zubarev, R.A. and Makarov, A. (2013). Orbitrap mass spectrometry. *Anal. Chem.* 85 (11): 5288–5296.

69 Viehland, L.A. and Mason, E. (1975). Gaseous lon mobility in electric fields of arbitrary strength. *Ann. Phys.* 91 (2): 499–533.

70 Dodds, J.N. and Baker, E.S. (2019). Ion mobility spectrometry: fundamental concepts, instrumentation, applications, and the road ahead. *J. Am. Soc. Mass Spectrom.* 30 (11): 2185–2195.

71 Cohen, M.J. and Karasek, F. (1970). Plasma chromatography™—a new dimension for gas chromatography and mass spectrometry. *J. Chromatogr. Sci.* 8 (6): 330–337.

72 Yin, J., Wu, M., Lin, R. et al. (2021). Application and development trends of gas chromatography–ion mobility spectrometry for traditional Chinese medicine, clinical, food and environmental analysis. *Microchem. J.* 168: 106527.

73 Cook, G.W., LaPuma, P.T., Hook, G.L., and Eckenrode, B.A. (2010). Using gas chromatography with ion mobility spectrometry to resolve explosive compounds in the presence of interferents. *J. Forensic Sci.* 55 (6): 1582–1591.

74 Denawaka, C.J., Fowlis, I.A., and Dean, J.R. (2014). Evaluation and application of static headspace–multicapillary column-gas chromatography–ion mobility spectrometry for complex sample analysis. *J. Chromatogr. A* 1338: 136–148.

75 Reyes-Garcés, N., Gómez-Ríos, G.A., Souza Silva, É.A., and Pawliszyn, J. (2013). Coupling needle trap devices with gas chromatography–ion mobility spectrometry detection as a simple approach for on-site quantitative analysis. *J. Chromatogr. A* 1300: 193–198.

76 Cumeras, R., Figueras, E., Davis, C.E. et al. (2015). Review on ion mobility spectrometry. Part 1: current instrumentation. *Analyst* 140 (5): 1376–1390.

77 Kanu, A.B. and Hill, H.H. Jr., (2008). Ion mobility spectrometry detection for gas chromatography. *J. Chromatogr. A* 1177 (1): 12–27.

78 Parastar, H. and Weller, P. (2023). Towards greener volatilomics: is GC-IMS the new Swiss army knife of gas phase analysis? *TrAC Trends Anal. Chem.* 170: 117438.

79 Wiza, JL (1979). Microchannel Plate Detectors. *Nucl Instrum Methods* 162: 587–601. https://www.sciencedirect.com/science/article/pii/0029554X79907341.

80 Medhe, S. (2018). Mass spectrometry: detectors review. *Chem. Biomol. Eng.* 3 (4): 51–58.

81 Allen, J.S. (1947). An improved electron multiplier particle counter. *Rev. Sci. Instrum.* 18 (10): 739–749.

82 Stanton, H.E., Chupka, W.A., and Inghram, M.G. (1956). Electron multipliers in mass spectrometry; effect of molecular structure. *Rev. Sci. Instrum.* 27 (2): 109–109.

83 Busch, K.L. (2000). The Electron Multiplier. Spectroscopy, June 2000 issue. Kennesaw State University, Digital Commons @ Kennesaw State University. Available at: https://digitalcommons.kennesaw.edu/facpubs/1777/.

84 Kurz, E.A. (1979). Channel electron multipliers. In *Channel Electron Multipliers* (ed. F.I. Scott, Jr.) NASA/George C. Marshall Space Flight Center (NASA Reference Publication 1090).

85 Laprade, B.N. and Labich, R.J. (1994). Microchannel plate-based detectors in mass-spectrometry. *Spectroscopy* 9 (5): 26.

86 Schröder, E. (2013). *Massenspektrometrie: Begriffe und Definitionen*, vol. 260. Springer-Verlag.

87 Daly, N.R. (1960). Scintillation type mass spectrometer ion detector. *Rev. Sci. Instrum.* 31 (3): 264–267.

88 Gross, J.H. (2017). Instrumentation. In: *Mass Spectrometry: A Textbook* (ed. J.H. Gross), 151–292. Cham: Springer International Publishing.

89 Koppenaal, D.W. et al. (2005). MS detectors. *Anal. Chem.* 77 (21): 418A–427A.

90 Du, X. and Zeisel, S.H. (2013). Spectral deconvolution for gas chromatography mass spectrometry-based metabolomics: current status and future perspectives. *Comput. Struct. Biotechnol. J.* 4 (5): e201301013.

91 Kind, T. and Fiehn, O. (2010). Advances in structure elucidation of small molecules using mass spectrometry. *Bioanal. Rev.* 2: 23–60.

92 McLafferty, F. (1993). *Interpretation of Mass Spectra*. University Science Books.

93 Edition, S. and Smith, R.M. (1998). *Understanding Mass Spectra: A Basic Approach*. Wiley.

94 Lee, T.A. (1998). *A Beginner's Guide to Mass Spectral Interpretation*. Wiley.

95 Ham, B.M. (2008). *Even Electron Mass Spectrometry with Biomolecule Applications*. Wiley.

96 Davis, R., Frearson, M.J., and Rose, M.E. (1987). *Mass Spectrometry: Analytical Chemistry by Open Learning*. Wiley.

97 Gross, J.H. (2013). Fragmentation of organic ions and interpretation of EI mass spectra. In: *Mass Spectrometry: A Textbook* (ed. J.H. Gross), 439–496. Springer Science & Business Media.

98 Weggler, B.A., Gröger, T., and Zimmermann, R. (2014). Advanced scripting for the automated profiling of two-dimensional gas chromatography-time-of-flight mass spectrometry data from combustion aerosol. *J. Chromatogr. A* 1364: 241–248.

99 Vogt, L., Gröger, T., and Zimmermann, R. (2007). Automated compound classification for ambient aerosol sample separations using comprehensive two-dimensional gas chromatography–time-of-flight mass spectrometry. *J. Chromatogr. A* 1150 (1–2): 2–12.

100 Holcapek, M. and Byrdwell, W.C. (2017). *Handbook of Advanced Chromatography/Mass Spectrometry Techniques*. Elsevier.

101 Tzanavaras, P.D. (2022). *Derivatization in Analytical Chemistry*. MDPI-Multidisciplinary Digital Publishing Institute.

5

Management and Interpretation of GC and MS Detailed Data

Glaucimar A. P. de Resende[1,2], *Allyson L. R. dos Santos*[1,2], *and Leandro W. Hantao*[1,2]

[1] *Instituto de Química, Universidade Estadual de Campinas, Campinas, SP, Brazil*
[2] *Instituto Nacional de Ciência e Tecnologia em Bioanalítica, Universidade Estadual de Campinas, Campinas, SP, Brazil*

5.1 Fundaments

Separation science encompasses a diverse range of instrumental methods designed to analyze the chemical composition of complex mixtures. Modern advancements in separation modes and column selectivity have made it possible to achieve highly efficient separations, particularly in gas chromatography (GC) for both neutral and ionizable compounds. Technological progress in column design has led to faster analyses with significantly improved peak capacities compared to what was achievable in the 1990s, a period marked by the introduction of groundbreaking methods, such as the comprehensive two-dimensional GC (GC×GC) technique developed by Liu and Phillips.

In their pioneering study, Liu and Phillips introduced a system where two chromatographic columns, each with distinct selectivity, were connected sequentially. The interface between the columns, referred to as a modulator (initially named on-column thermal desorption modulation), acted as a conduit that multiplexed the chromatographic data from both dimensions, maintaining the separation profiles of each column and allowing for detection using a single detector. This innovative concept quickly captured the interest of the scientific community and set the stage for a series of continuous enhancements, resulting in a diverse array of modulation methods utilizing either thermal (cryogen-based and cryogen-free) or flow modulation techniques.

The steady growth in publications on GC×GC over the years (Figure 5.1) illustrates the technique's expanding influence and adoption. Today, GC×GC stands as a leading method for extracting detailed chemical information from volatile and

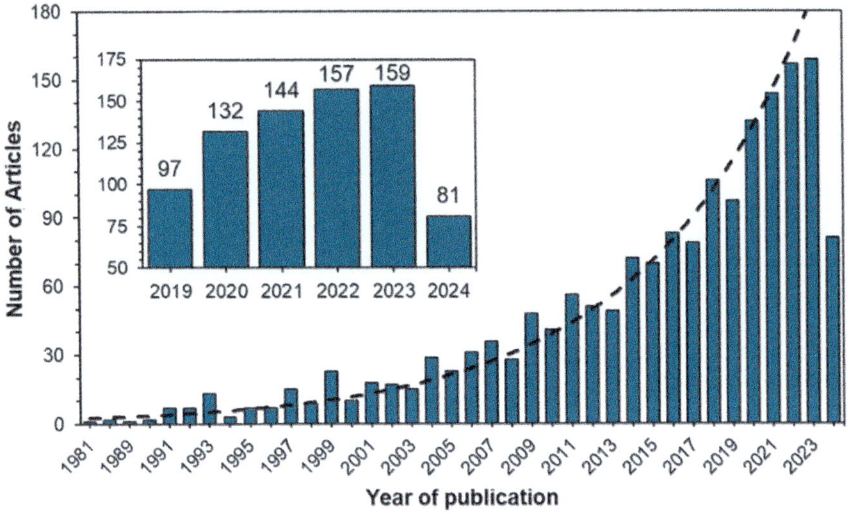

Figure 5.1 Number of papers published over the years, as registered in July of 2024 on the Web of Science collection, using the keywords "gas chromatography" and "chemometrics".

semivolatile complex mixtures. Its application extends beyond chromatography, playing a transformative role in sample preparation and, most notably, data analysis. As a result, GC×GC is now viewed not just as a tool, but as a foundational component for developing new analytical strategies in separation science.

Additionally, the integration of data processing tools, such as deconvolution algorithms and chemometric methods, has improved GC×GC workflows for untargeted analysis. These innovations have facilitated the identification and quantification of numerous compounds in a single run, overcoming the limitations of conventional GC and making it possible to explore chemical compositions in greater depth. The enhanced resolution of GC×GC, when combined with modern data mining techniques, allows researchers to uncover hidden patterns and relationships within datasets, paving the way for new discoveries in diverse scientific fields.

Untargeted analysis through the use of GC×GC serves as a compelling example to highlight the major advancements within the field. By utilizing GC×GC for the analysis of volatile and semivolatile organic compounds (SVOCs), unprecedented peak capacities have been achieved, substantially minimizing the occurrence of overlapping peaks. Moreover, modern software solutions such as ChromaTOF®, ChromSpace®, and GC Image® now incorporate deconvolution algorithms that can extract clear chromatographic and spectral profiles from complex datasets, thereby improving the tentative identification of unknowns. As a result, the wealth of chemical information generated by GC×GC can be overwhelming and poses a challenge for data processing and mining. While traditional one-dimensional (1D) GC workflows can manage up to about 30 peaks manually, GC×GC routinely produces hundreds of integrated peaks, necessitating the use of automated processing techniques and advanced chemometric methods. For instance, studies like that of Bean et al. [1] successfully identified over 500 metabolites from bacterial volatiles using solid-phase microextraction (SPME) coupled with GC×GC-MS.

While data processing efficiency is crucial, the ability to extract chemically relevant information is equally significant. The number of peaks that can be detected with GC×GC is far higher compared to 1D-GC, resulting in a much denser information profile, particularly for trace compounds that would otherwise be obscured by more dominant peaks in the first dimension. Furthermore, advancements in related fields have provided additional benefits to GC×GC-based untargeted analysis.

A key concept in GC×GC is understanding the data structure, which is inherently pixel-based. Each pixel serves as the fundamental unit of intensity in a chromatogram or mass spectrum. For instance, a gas chromatography-flame ionization detector (GC-FID) operating at 100 Hz over a 5-minute experiment will generate around 30,000 pixels, each representing a specific intensity value. Higher acquisition rates, such as 200 Hz, double the resolution, resulting in even more detailed chromatographic profiles. Effective reconstruction of these profiles requires an adequate number of pixels per peak, typically estimated at 10 pixels for conventional reconstruction and 17 pixels for multi-channel deconvolution. Likewise, GC×GC datasets are structured as matrices of pixel intensities (Figure 5.2). For example, a GC×GC-FID experiment with a modulation period of 6 seconds over 5 minutes will yield a 600 by 50 matrix, where the matrix dimensions correspond to the number of pixels required to reconstruct the second- and first-dimension profiles, respectively. Adjustments to acquisition rates or modulation periods can further refine the temporal resolution, thereby affecting the peak reconstruction quality in both dimensions.

Standard parameters, such as retention time and peak area, remain central to data extraction regardless of the dimensionality of the chromatogram. Typically, the peak area is the primary metric used for GC data processing. A conventional workflow involves processing the chromatographic pixels to eliminate noise, baseline drift, and contaminants, followed by integration of the analyte peaks and construction of a global peak table for multivariate analysis. While conventional processing strategies rely on peak tables derived from chromatographic areas, newer approaches employ the pixel itself as the primary input, as seen in tile-based methods like the Fisher's ratio, which are now available in ChromSpace® and ChromaTOF Tile®. Alternative workflows that bypass peak detection and rely directly on pixel-based processing are gaining popularity, as

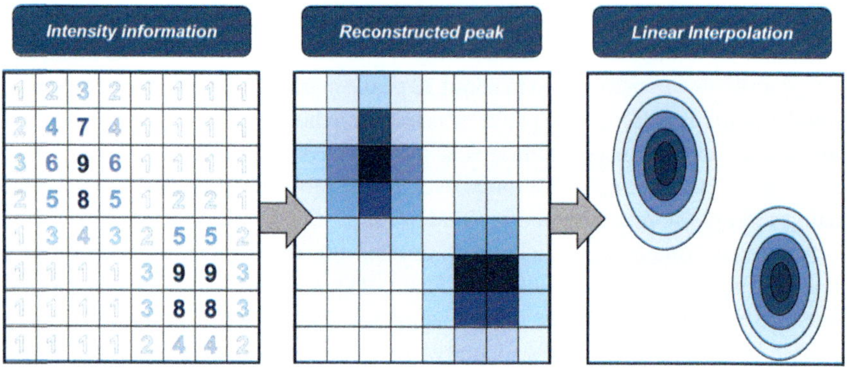

Figure 5.2 GC×GC chromatogram formed by the intensity information acquired, translated into reconstructed peaks in the pixel format and the classical format formed by the linear interpolation.

they are less sensitive to low-intensity or distorted peaks. However, pixel-based data handling is computationally intensive, often requiring specialized software environments like MATLAB®.

In this chapter, we demonstrate how the integration of GC(×GC)-MS with advanced chemometric techniques has improved the way complex chemical data are processed and interpreted. By leveraging multivariate statistical methods, such as principal component analysis (PCA), partial least squares (PLS) regression, and machine learning algorithms, researchers can extract valuable insights from intricate datasets that would otherwise remain hidden. This synergy between GC(×GC)-MS and chemometrics not only enhances compound identification and classification but also enables the discovery of subtle chemical differences, making it a pivotal approach for untargeted analysis in various fields. As these methodologies continue to evolve, the combination of GC(×GC)-MS and chemometrics is poised to drive new analytical strategies, providing a deeper understanding of complex chemical systems.

5.2 GC-MS Applications

GC is an essential and well-developed analytical technique with its origin in 1952 by James and Martin [2]. Their work was essential for the separation of volatile compounds and based on the liquid-phase partition chromatography research of Martin and Synge from 1941 [3, 4]. The GC history is rich, with remarkable milestones [5] in which the use of mass spectrometers (MS) hyphenated with GC was one of them. First reported in 1959 by Gohlke [4], GC-MS has the biggest advantage of having both selectivity from GC and structural elucidation from MS.

The actual GC-MS systems have evolved a lot since their first report. Common mass analyzers include single quadrupole (QMS), triple quadrupole (QqQ), time of flight (TOF), and Orbitrap systems. Regardless of the MS type, raw data acquired from chromatographic data is also rich in information.

The chromatographic data will dictate the selectivity nature of the technique. The chromatogram, plotted by retention time versus intensity, represents the diversity of volatile compounds present in the sample characterized by the peaks. Besides volatile and semivolatile analytes, these compounds must be thermally stable or, even in the case of lower volatility, have the capacity of being derivatized. The separation occurs based on the equilibrium of compounds into the stationary phase present in the columns (Figure 5.3). This equilibrium is established between the stationary phase and the mobile phase (the inert carrier gas, e.g., H_2 and He). Selectivity directly influences peak resolution, where resolution values higher than 1.5 represent a fully separated peak. Retention time is described as the time in which an analyte takes since the injection of the analyte until it reaches the detector. The retention time (t_R) represents the position in which a peak is present in the chromatogram, and as for consequence the analyte. The efficiency of separation is an important characteristic to express a good method, and it can be expressed by the number of plates, plate height, and peak capacity.

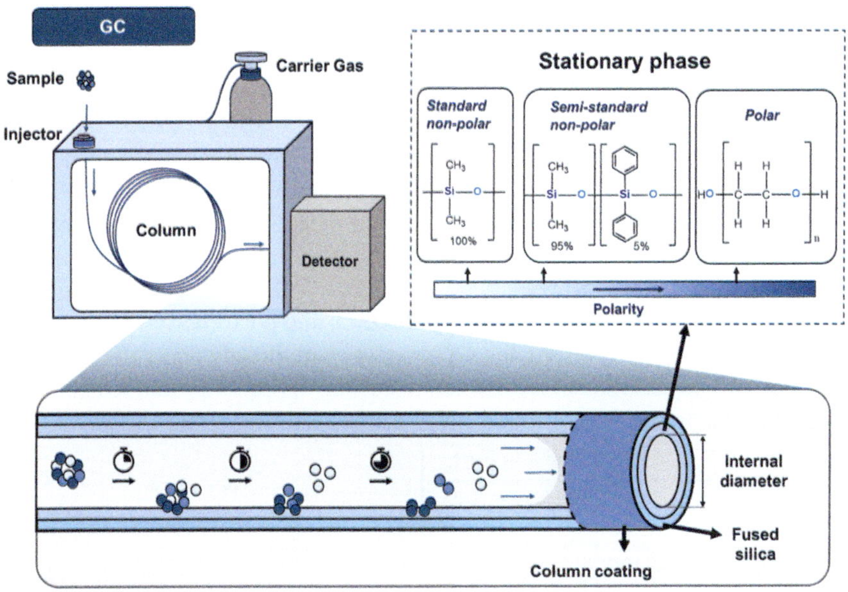

Figure 5.3 Diagram of a GC and its components such as injector, carrier gas, column structure, detector, and the most common structures present in the stationary phase.

An essential step for any quantitative or statistical analysis to obtain a clean, structured, and reliable output is the treatment of raw data. This information needs to be processed, aligned, and/or filtered.

The processing step involves conversion of files from the instrument into a standardized format, e.g., mzXML files. This step also includes the cleansing of the data to obtain the final chromatogram. Remotion of noise and correction of baseline are commonly applied to the data and are sometimes included into instrument software. Also, algorithms for peak detection, deconvolution, and identification from the MS spectral information can be applied.

The elution of compounds in GC generally happens with minor variations in the retention time. A strategy to minimize time variations is the alignment, correcting the peak time-shift in the chromatograms. In this way, a final matrix with all samples can easily identify their compounds having the same retention time. Strategies for alignment will be better discussed in the section on data treatment. Filtering can help in peak analysis, removing no significative features from the samples. The purpose of normalization is to correct systematic variations caused by the instrument, injection, and matrix effect, among other effects. The general output of the processed data will be a matrix or feature table with samples as rows, which can facilitate the data analysis.

Regarding the structural elucidation from MS, there are different compound identification strategies. The most common is the matching information from mass spectral libraries to the compound's mass spectrum.

To enhance confidence, the retention index (RI) is used as the second identification of compounds by GC-MS. Its principle is based on the normalization of a compound's retention time relative to a series of linear n-alkanes, making a more robust and transferable identification. In practical terms, analyte identity assumes a comparison between the calculated RI and MS libraries (e.g., NIST and Wiley). As the equilibrium depends on the type of stationary phase, three phases are most found in these libraries. As represented in Figure 5.3, these phases are standard nonpolar (100% polydimethylsiloxane, PDMS), semistandard nonpolar (95% PDMS and 5% phenyl methyl-polysiloxane), and polar (polyethylene glycol, PEG, also known as Wax). As the GC program will directly affect the RI, there are two ways to calculate it: isothermal by Kovats [6] (Eq. (5.1)) and temperature programming from the description of van den Dool/Kratz [7] (Eq. (5.2)):

$$I_x = 100n + 100\left[\frac{\log t_x - \log t_n}{\log t_{n+1} - \log t_n}\right], \tag{5.1}$$

$$I_x = 100n + 100\left[\frac{t_x - t_n}{t_{n+1} - t_n}\right], \tag{5.2}$$

where I_x is the retention index, n is the number of carbons from the linear n-alkane before the calculated compound, t_x is the retention time of the compound, while

t_n and t_{n+1} are, respectively, the retention time from the previous and posterior linear *n*-alkanes.

In some cases, the identifications by RI will not be available. For example, this may occur when using a nonreported stationary phase, when compounds are not described in the library, or when the library for the specific type of MS is not available. In this way, high-resolution mass spectrometry (HR MS) can provide accurate mass and isotopic patterns, which can provide an unequivocal molecular formula. For complex samples, the total ion chromatograms (TIC) can be replaced by extracted ion chromatograms (EIC), filtering the desired fragments and improving the desired identification.

An approach to standardize the confidence of the identification has been established by the Metabolomics Standards Initiative (MSI) [8–10]. Reporting the identification levels of MSI fortifies data transparency and reproducibility, being crucial to the literature analysis. The levels can be described from 1 to 4 [10], where level 1 is the best scenario of identification, where a standard reference can confirm the measurements by MS, MS/MS, and retention time matching. The levels can be described as follows:

Level 1. Confirmed structure: Achieved by matching at least two orthogonal properties to an authentic chemical standard in the same experimental conditions.

Level 2. Putatively annotated compounds: Identification occurs based on the analysis results with high-mass spectral library matching and, ideally, with a match of the literature retention index.

Level 3. Putatively characterized compound classes: It happens when a class of compounds can be suggested by the mass spectrum and/or evidence as the elution region. That happens when the identification cannot predict the precise chemical structure but the class behavior.

Level 4. Unknown compounds: This scenario happens by the detection of a compound, while its abundance can be measured. In this way, compounds can be expressed by their fragments and/or retention index.

5.3 GC×GC-MS Applications

One of the biggest evolutions related to GC since the development of open tubular capillary columns is the comprehensive two-dimensional GC first reported in 1991 by Liu and Philips [11]. This technique is based on the principle of using a second column, connected to the first by a component called modulator aiming for better separation of compounds. As in the GC-MS, analytes must be volatile or semivolatile, and thermally stable or, in the case of lower volatility, have the capacity of being derivatized. The first column, also known as the first dimension (^{1}D), works similarly to the traditional GC, separating the compounds based on the interaction along the column. Next, the analytes are focused and sent to the second column rapidly and sequentially by the modulator. This part is fundamental and plays a critical role in the high separation capacity of GC×GC by conducting fractions of samples regularly to the second column and by focusing on the analytes. The second column, or second dimension (^{2}D), is usually shorter than the ^{1}D and further separates the analytes coming from the modulator. Finally, the sample fractions are analyzed by the detectors, and the data are provided for analytical processing. Among the range of advantages of using GC×GC, when compared to the classical GC, it offers better sensitivity, selectivity, an increase in the peak capacity, and detectability as a good response in fingerprint studies.

To enhance results and to improve the analytical capabilities of the GC×GC technique, it is important to comprehend its components. Starting with the column settings, there are different approaches described in the literature. GC×GC and heart-cut (H/C) multidimensional GC (GC-GC) are two major classes of column setup. While GC×GC has the presence of a modulator for comprehensive evaluation of samples, the H/C MDGC has the toggle valve (or Deans switch), which allows part of the ^{1}D fraction to be transferred into a second column at certain times and for a specific duration. As illustrated in Figure 5.4, the configurations of columns, as the elements present in the techniques and their data acquisition, are shown for GC, GC-GC, and GC×GC. For this reason, the H/C MDGC is normally used in targeted studies as the GC×GC has better potential for untargeted analysis. In some cases, the separation improvement can be achieved by the inclusion of a third (^{3}D) or even a fourth dimension (^{4}D). For example, the comprehensive three-dimensional GC (GC^3 or GC×GC×GC) is a system with 3 columns [12, 13], while a parallel comprehensive two-dimensional GC (2GC×2GC) has 4 columns [14]. The stationary phases of the settled dimensions are normally different. They may be constituted by PEG, ionic liquids, chiral phases, and copolymers of PDMS modified with diphenylsiloxane or cyanopropylsiloxane.

Another key component that continually evolves to enhance hardware configuration is the modulator. Modulators are classified into two main categories: thermal and pneumatic. The first application of a modulator in GC×GC was introduced by Liu and Phillips using a thermal modulator. This approach was demonstrated through a proof-of-concept separation of a 14-compound solution and the analysis of coal liquids [11]. The modulation system has naturally evolved through various studies [15], leading to the development of technologies such as cryogenic modulation [16], advancements in flow modulation [17], solid-state modulators (SSM) [18], and longitudinally modulated cryogenic systems (LMCS) [16, 19]. The modulation process determines the modulation period (P_M), which defines the chromatographic "slices" – or, in other words, the time intervals in the second dimension. This period is directly related to the reconstruction of the raw chromatogram into the conventional GC×GC chromatogram. Figure 5.5

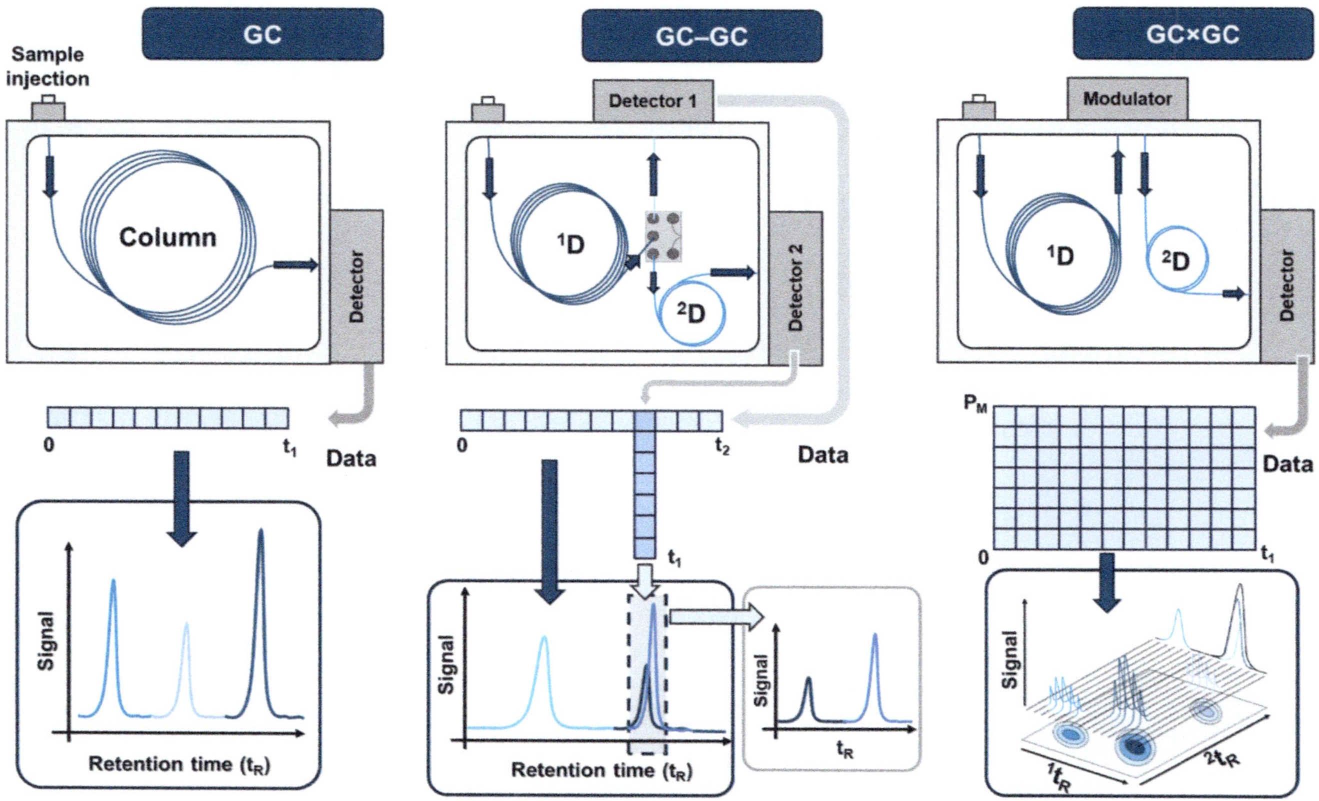

Figure 5.4 GC, GC-GC, and GC×GC configuration and data acquisition expressing the hardware configuration, the format of data, and how the chromatograms are represented. In the case of the modulator representation in GC×GC, its presence can be external or internal to the oven.

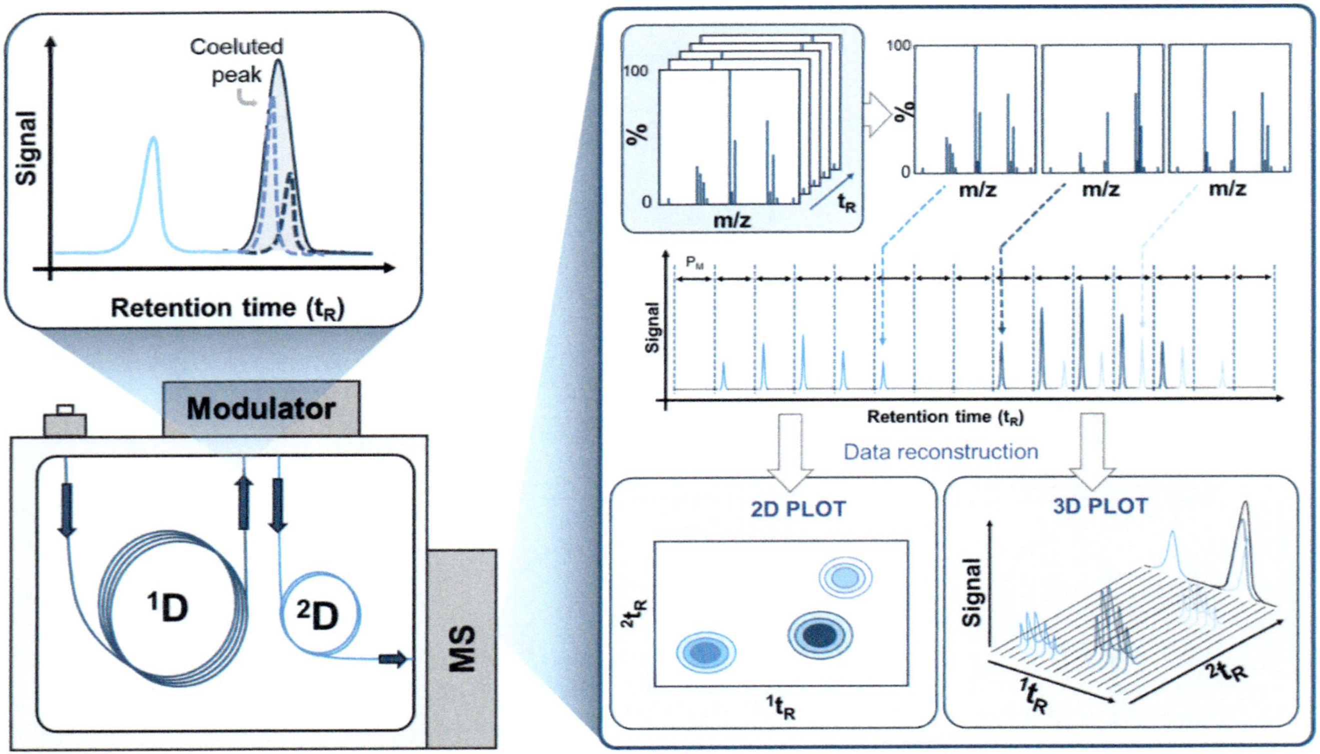

Figure 5.5 Representation of GC×GC chromatogram reconstruction, showing a 1D GC chromatogram in the top left, which would be generated in the absence of a modulator. On the right, the figure displays a GC×GC chromatogram in its raw format, along with ^{2}D and ^{3}D plots, where an MS spectrum is associated with each retention time (t_R).

illustrates how the acquired data, initially formatted as a classical GC chromatogram, can be transformed into the final plot. The graphical representation shows the chromatogram recorded before modulation and the chromatograms obtained after the modulation process. The final chromatogram contains MS spectra linked to each retention time, and the reconstruction considers P_M as the total time in the second dimension, represented as ^{2}D and ^{3}D plots. During this reconstruction, P_M determines how the data matrix will be folded, and other factors such as time drift and signal-to-noise ratio (*S/N*) influence the quality of the resulting data.

The last and perhaps most critical element for data acquisition is the detector. In GC×GC, the application of a detector requires a high acquisition rate, as the signals generated in the second dimension typically range from less than 100 to 500 ms, with a preferred frequency of up to 200 Hz [20]. The peaks produced by the second column are narrow, necessitating fast detectors with an appropriate frequency. Before discussing the various types of detectors, two key questions arise: how to choose a detector and how do they influence the results? The first question can be complex, but it is primarily addressed by understanding the research purpose, sample type, and available equipment. In essence, the choice of detector depends on a combination of factors, including the nature of the matrix, sample preparation, analytical goals, economic constraints, and overall feasibility. The second question relates to how detectors influence the signal, as they can also increase the data dimensionality. Detectors are generally classified into two categories: single-channel and multichannel. Single-channel detectors provide a scalar response, as seen with FID, nitrogen-phosphorus detectors (NPD), electron capture detectors (ECD), photoionization detectors (PID), and mass spectrometers (MS) operating in selected ion monitoring (SIM) mode or tandem mass spectrometry (MS/MS). In contrast, multichannel detectors are spectroscopic and add a new dimension to the data [21]. Examples include MS in scan mode, Fourier transform infrared spectroscopy (FTIR), and vacuum ultraviolet spectroscopy (VUV).

GC×GC is considered an ideal tool for untargeted analysis and MS plays a crucial role in a wide range of applications. Several advancements can be achieved, such as enhanced selectivity, improved separation, more accurate compound identification, and, in some cases, the discovery of new analytes. Various types of mass spectrometers can be coupled with GC×GC for analyte investigation, including single transmission quadrupole (QMS), time-of-flight (TOFMS), and Orbitrap MS. One challenge with using MS in conjunction with GC×GC is the acquisition frequency, as some QMS instruments have lower acquisition rates than required. TOFMS addresses this issue by achieving acquisition rates of up to 500 Hz and accommodating both low and high mass resolutions. The Orbitrap MS quickly gained attention for its high-resolution accurate mass capabilities, leading to its integration into GC and GC×GC instruments [22]. Its implementation in these systems has made various high-performance configurations commercially available [23]. GC×GC-MS produces continuous data primarily for quantitative purposes, which can also be processed for qualitative analysis, such as compound identification and clustering, depending on the application. However, coupling GC×GC with HR MS often results in large file sizes – sometimes exceeding 3 GB – posing challenges for model construction and data handling [24, 25].

GC×GC-MS offers various configurations, enabling the collection of diverse and versatile datasets. Along with these distinct data types, the large volume of information acquired can be challenging to analyze. This is where statistical strategies and chemometrics become essential. From a data analysis perspective, GC×GC-MS provides a wealth of possibilities and avenues for new developments. As illustrated in Figure 5.4, the column setup influences how the information is organized: in GC, data are represented in a vector arrangement; in GC-GC, a larger vector arrangement is used along with selected slices as a matrix; and in GC×GC, data are structured as a matrix. In essence, the organization of the datasets defines the peak capacity (n), where for classical GC, it depends on the peak capacity of a single column (n_1); for GC-GC, it is the sum of the dimensions ($n_t = n_1 + n_2$) from the first and second columns; and for GC×GC, it is a geometric increment expressed as the multiplication of dimensions ($n_t = n_1 \times n_2$). These relationships hold true for single-channel detectors and single samples, but the dimensions of the data can expand with an increase in the number of samples or the use of multichannel detectors. As shown in Figure 5.5, data can adopt multiple structures, where the first row represents results for single samples, the second row shows the samples' dataset, and the last row provides examples for each column configuration. Analysis can be performed using univariate and multivariate methods. In univariate analysis, a single sample set is used, with the peak area serving as an example. In contrast, multivariate analysis involves datasets with dimensions greater than two. When dealing with high-order datasets – exceeding three dimensions – multiway strategies are required to accommodate the complexity of the data's geometry. Mathematical tools such as parallel factor analysis (PARAFAC) and multiway principal component analysis (MPCA) are well-suited for these scenarios. Figure 5.6 also illustrates the geometric terminology for data tensors up to the fourth order and how samples add new dimensions to the structure. For data acquired with single-channel detectors (e.g., FID), both GC and GC×GC produce vector and matrix representations, respectively. Other examples include GC-MS in scan mode and GC×GC with a single-channel detector, both of which can be visualized as a matrix for single-sample data. Data cubes and hypercubes can be represented by GC×GC-MS and GC^3-MS, respectively, when using MS in scan mode.

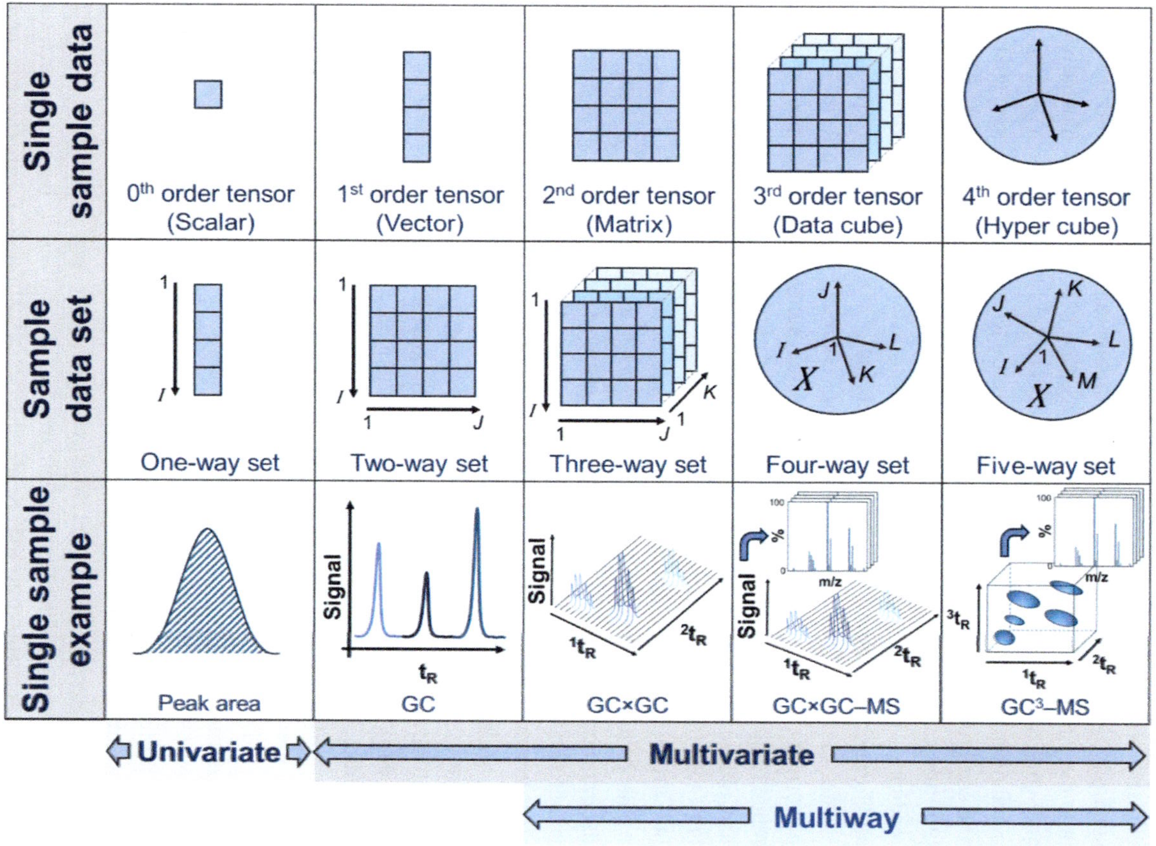

Figure 5.6 Data arrangement and structure for different order tensors, including sample datasets and an example related to the GC system.

Recently, several comprehensive review papers have been published, compiling information on GC×GC and chemometric tools available in the literature. For a more detailed understanding of specific applications of GC×GC and chemometrics, we recommend referring to these reviews. They cover research topics such as Chinese medicines [26], ^{2}D chromatography and spectroscopy [27], new tools for untargeted analysis [28], the use of GC×GC-MS in bioanalytical and clinical metabolomic investigations [29], GC-MS and GC×GC-MS-based metabolomic studies [30], the application of GC-GC and GC×GC for food analysis and contaminant detection [31], chemometric and data tools for GC×GC-TOFMS [20], advancements in "omics" sciences [24], applications in cosmetics, food, and medicines containing essential oils [32], chromatographic fingerprinting [33], and forensic and compliance studies in fuels [34].

Due to its effectiveness in evaluating complex samples and its strong performance in nontargeted and fingerprint analysis, GC×GC has a wide range of applications. Research areas include "omics" sciences [35], biomedical and pharmaceutical research [36], environmental studies [37, 38], fuel analysis [12, 39], forensic and toxicological investigations [40, 41], polymer research [42, 43], and the study of food, flavors, and fragrances [44, 45].

The primary objective of omics sciences is to comprehensively study large datasets across various fields, and GC×GC is an excellent technique for profiling volatile and semivolatile compounds in samples. Within omics research, fields such as metabolomics and petroleomics benefit significantly from GC×GC-MS. These studies analyze the metabolic profiles of organisms and complex mixtures of organic constituents in petroleum samples, respectively. Additionally, some specialized areas of metabolomics have emerged, leveraging the advantages of this technique and receiving unique designations, such as lipidomics [46–48], foodomics [49–51], and volatilomics [52, 53]. Examples of health-related studies using GC×GC-MS and chemometrics include the analysis of colorectal cancer biomarkers in human serum [46], detection of tuberculosis-associated compounds on human skin [36], urinary metabolites in patients with type 2 diabetes mellitus [54], and volatile compounds in the breath of lung cancer patients [55]. Plant characterization studies have also been reported, including the profiling of *Varronia curassavica* Jacq. [56], hops (*Humulus lupulus* L.) [50], Hainan camellia (*Camellia oleifera* Abel.) oil [44], and various tea varieties (*Camellia sinensis*) [49]. Food research is another key area where GC×GC finds extensive applicability, particularly in the analysis of beverages such as wines [57–62], beers [62, 63], peach juices [64], Chinese Baijiu [65–67], and coffee products [45, 68–70]. With a clear understanding of the setup, data acquisition, format, and sample applications, the next step is to explore the acquired data.

5.4 Data Treatment

The data analysis process is complex and begins even before experimental results are obtained. A summary of the data analysis workflow is shown in Figure 5.7. As illustrated, the process starts by defining the research goal, which is further refined by scientific questions and limitations. Additionally, the randomization and injection plans are crucial for generating reliable data. Multivariate statistical strategies, such as Design of Experiments (DoE), can optimize results using a variety of tools. DoE is a central part of chemometrics and includes methodologies such as response surface methodology (RSM), full and fractional factorial designs, central composite design (CCD), and Plackett–Burman designs. While this chapter does not delve deeply into experimental planning, we recommend consulting a review on DoE in chemometrics [71] and a detailed description of its history and use in chemometrics and analytical chemistry [72].

The knowledge of prior information will guide the methods applied to the samples. In targeted studies, the focus is on specific compounds, aiming for their identification and quantification. Conversely, untargeted or nontargeted analyses involve the comprehensive study of a sample to extract unknown information, detect, and identify compounds. The choice between these strategies should be made before the experimental phase, which includes defining the type of normalization (e.g., internal standard) and selecting the appropriate model type (e.g., classification or regression). These aspects will be discussed in more detail later in the text.

After acquiring GC×GC-MS results, the first step is data pretreatment, which includes signal preprocessing, alignment, and data preprocessing. Following these steps, modeling – or in other words, chemometric methods – can be performed. These methods are categorized into unsupervised and supervised approaches. While unsupervised methods rely solely on the acquired data for exploratory analysis, supervised methods require external information to guide machine learning models. Figure 5.7 illustrates examples of unsupervised models for dimensionality reduction (PCA, MPCA, and WPCA) and clustering (HCA and ^{2}D-HCA), as well as supervised univariate models (Fisher's ratio and t-test) and supervised multivariate models for classification (PLS-DA, OPLS-DA, and DD-SIMCA) and regression (PLS).

Throughout each step, data verification is essential to ensure the consistency of the methods used. Checking for missing values is crucial, as the presence of such gaps can introduce early issues that may compromise or mislead the results. Although the occurrence of missing values is common, researchers now have tools to address this issue. Missing data points can arise due to the absence of certain compounds, instrumental errors, or sample preparation issues. These situations are typically classified into three mechanisms: (i) missing at random (MAR), (ii)

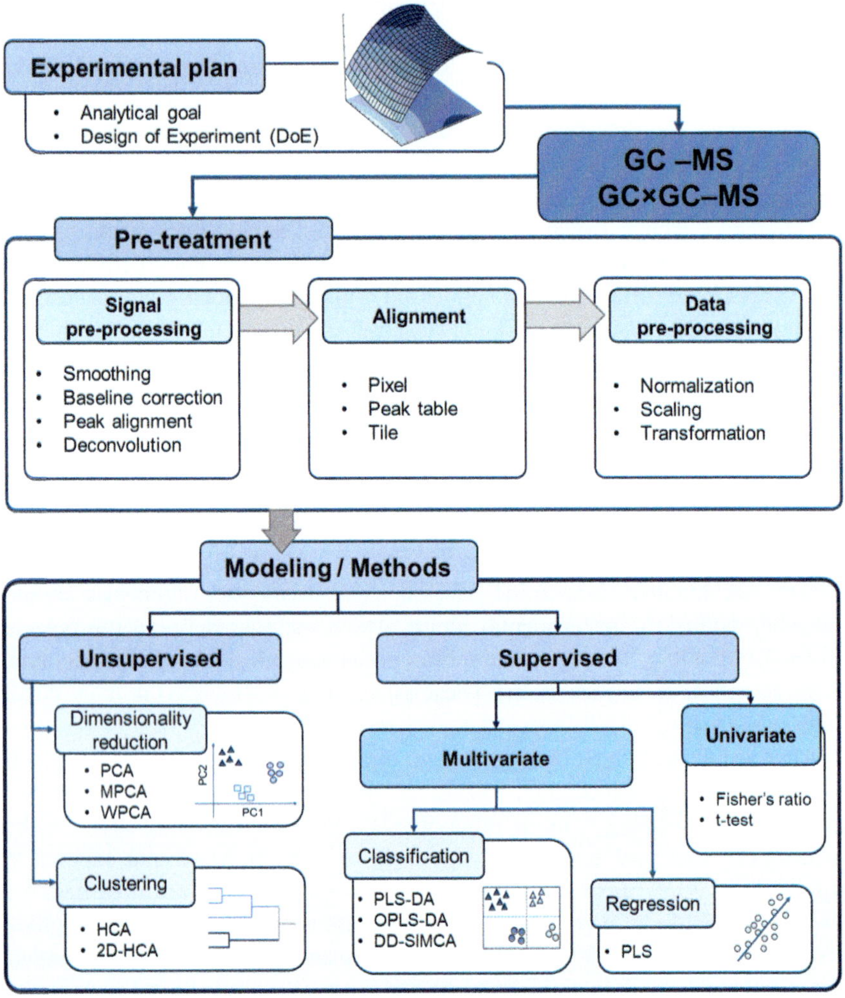

Figure 5.7 Workflow of data analysis applied to GC×GC-MS.

missing completely at random (MCAR), and (iii) missing not at random (MNAR). MAR is linked to dependencies on other detected values, often caused by factors like deconvolution or alignment issues. MCAR occurs when entire data points are missing due to random errors, such as sample preparation problems or instrument malfunctions. MNAR, on the other hand, pertains to values missing due to inherent properties, such as limits of detection or quantification [73, 74]. Some authors suggest strategies to address missing data, such as excluding rows or columns associated with missing values if they are not critical or estimating these values using various techniques. Estimation methods may include calculating the mean value of neighboring data points, weighting using k-nearest neighbor (kNN), iterative methods based on PCA, or even interpolation [75, 76]. These strategies must be applied with caution, as inappropriate use can negatively impact results, necessitating a thorough understanding of the data and validation of the chosen approach.

A relevant example of missing data management in GC×GC applied to metabolomics involved identifying primary types of missing values in real sample datasets of bacterial (n = 258) and fungal (n = 78) samples [74]. The study effectively described and demonstrated the missingness issues in the datasets, which were attributed to a mix of MAR and MNAR. They also introduced an R package (MetabImpute) designed to assist with data preprocessing steps. The following section will further describe the pretreatment approaches applied to GC×GC-MS data.

5.4.1 Pretreatment

This step involves processing the raw data before any data handling procedures are applied to correct common variations in sample chromatograms. The main objective is to eliminate undesirable variations and contaminants that may arise from experimental conditions. The pretreatment process follows the sequence illustrated in Figure 5.7, starting with signal preprocessing, followed by alignment, and concluding with data preprocessing. Typical deviations present in raw chromatograms include noise, baseline drift, time shifts, and co-elution, as shown in Figure 5.8. To address these issues, the signal must be refined using specific strategies before proceeding to further data organization steps. Signal preprocessing methods include noise smoothing, baseline correction, peak alignment, and deconvolution, all of which are typically implemented using chromatographic acquisition software as the data are reconstructed.

In practical terms, GC×GC-MS data are generated by the instrument and reconstructed using various commercial software solutions. A review in Ref. 20 detailed commonly used software packages, including Analyzer Pro® (SpectralWorks Ltd, Runcorn, Cheshire, UK), Canvas® (J&X Technologies, Shanghai, China), ChromSpace® (SepSolve Analytical Ltd, Hampton, Peterborough, UK), ChromSquare® (Shimadzu Corporation, Kyoto, Japan), ChromaTOF™ (LECO Corporations, St. Joseph, MI, USA), GasPedal® (Decodon Software UG, BioTechnikum Greifswald, Germany), GC Image™ (Zoex Corporation, Lincoln, NE, USA), and the open-source platforms Guineau [77] and OpenChrom® (Lablicate GmbH, Hamburg, Germany). These software solutions provide tools for signal preprocessing, including smoothing, baseline correction, GC×GC chromatogram reconstruction, peak identification using MS spectra or RI, and deconvolution.

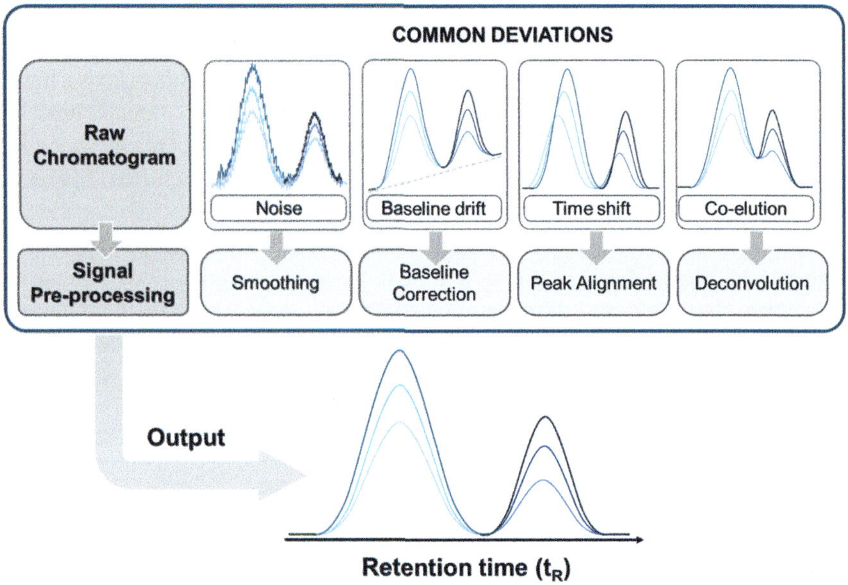

Figure 5.8 Common deviations in signal pretreatment.

One of the most common deviations in GC×GC-MS data is noise, which arises due to signal fluctuations caused by the detection process under high acquisition frequencies. Smoothing algorithms generally improve data quality by increasing the signal-to-noise ratio (S/N). Unfolded GC×GC-MS data typically use the same smoothing strategies as conventional GC [20]. A commonly applied method is the Savitzky–Golay filter [78], developed in 1964, which uses linear least squares to enhance data quality without altering the signal trend. This technique is widely used in chemometrics. For instance, it has been applied in studies of getter substrates by GC×GC-TOFMS [79], the decomposition of GC×GC-TOFMS data through partial modulation in the negative pulse mode [80], and ion filter-based data reduction [81]. Another smoothing technique is the 3-point Gaussian filter, which Oliva, Cheung, and Perrault used in their GC×GC-QMS/FID data analysis of standard mixtures [82].

Baseline fluctuations are also common in GC×GC-MS and are typically caused by low-frequency noise from the detector, complex sample matrices, variations in sample injection, column bleeding, and changes in setup parameters such as temperature and flow [20, 24]. In signal preprocessing, unfolded chromatograms can apply baseline correction methods from conventional GC. These include the asymmetric least squares algorithm and iterative Bayesian regularized artificial neural networks (iBRANN). The asymmetric least squares algorithm corrects the baseline by adapting Whittaker smoothing to estimate the baseline trend [83].

This method was used to develop an R-package (RGCxGC toolbox) designed for GC×GC-MS data processing [84]. The iBRANN tool, developed in 2016, uses machine learning-based artificial neural networks (ANNs) for baseline adjustment in both ^{1}D and ^{2}D data, making it applicable to GC×GC [85]. Other preprocessing tools reported in the literature include blank chromatogram subtraction, local minima value (LMV) detection, trilinear decomposition, and singular value decomposition (SVD) [62].

As mentioned previously, GC×GC has a higher peak capacity than conventional GC, leading to improved compound separation. However, some peaks still experience co-elution, requiring deconvolution. As a third-order tensor, GC×GC data necessitate multiway strategies such as parallel factor analysis (PARAFAC and PARAFAC2), which will be further explained in Section 5.4.5.

Chromatographic alignment addresses the inherent time shifts in chromatography. While 1D GC experiences shifts in only one dimension, GC×GC increases complexity with shifts occurring in a second dimension. Misalignment can lead to false results, such as the apparent absence of an analyte. Alignment approaches are categorized into table-based, pixel-based, and tile-based methods [28], as illustrated in Figure 5.9.

The table-based approach is the most commonly used method for alignment. It involves reducing the data into a batch sample table containing information on analyte identification and quantification. MS identification is typically based on spectrum matching using libraries such as NIST, RI, and standard sample injections. Each chromatogram is analyzed individually under standardized conditions (e.g., P_M and ^{2}D shift) and a defined signal-to-noise (S/N) ratio. The match value for mass spectra identification is recommended to be as high as possible (ideally close to 1000), but the criteria for compound identification will depend on the specific study. The final table includes all identified analytes with their retention times ($^{1}t_R$ and $^{2}t_R$), peak areas, and relevant analytical conditions. The quality of the peak table is directly related to the chromatographic separation, and the retention times and identities of the compounds must be carefully verified throughout the entire process. One example of the table-based approach is the HS-SPME-GC×GC-QMS analysis of 137 compounds in hop samples [50]. The authors constructed the table using peak areas obtained from Canvas® software, under conditions of P_M 6 s, S/N 5, ^{2}D shift of −4.29, NIST11 as the library, MS match values >700, and RI ± 20. Due to the high number of compounds, including terpenes and esters, this approach was time-consuming.

The pixel-based approach operates on every individual data point, where each pixel in the GC×GC chromatogram represents a variable [86]. As shown in Figure 5.9, chromatograms are treated in an unfolded format, preserving the raw data and maintaining their original fidelity [28]. One advantage of this method is that chemometric results can be represented as a GC×GC image,

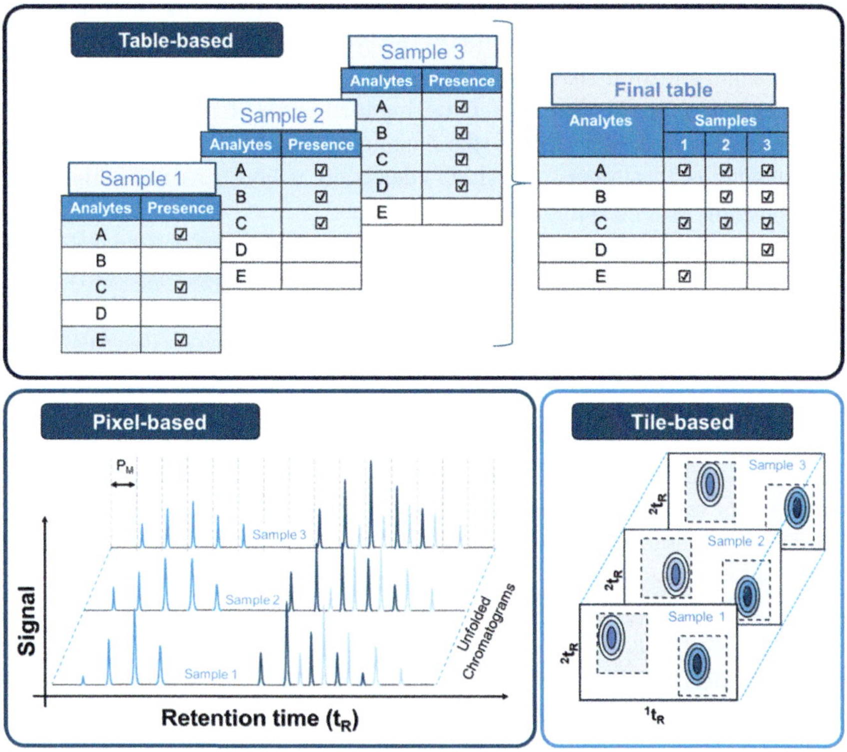

Figure 5.9 Alignment strategies for GC×GC-MS data.

closely resembling a chromatogram, which aids in chemical interpretation. However, the pixel-based approach has two main drawbacks: it is computationally intensive and sensitive to false positives. This method is not recommended for large datasets due to its high processing demands and the need for advanced computational setups. False positives can arise from detector noise and time shifts [28]. These issues can be minimized by selecting specific mass channels (*m/z*) or regions of interest (ROI) within a GC×GC-MS system, thereby reducing the file size. Several algorithms assist in alignment, such as the ^{2}D correlation optimized warping algorithm (^{2}DCOW) [87] and iCoshift [88], along with newer strategies like pixel-by-pixel retention time shift alignment [89]. The literature offers several examples of these strategies: ^{2}DCOW has been applied to fingerprint micropollutants in Copenhagen freshwater sediment samples using GC×GC-HR MS [38], forensic analysis of diesel oil spills in the environment using GC×GC-HR MS [90], and getter substrate analysis using GC×GC-TOFMS [70]. Meanwhile, iCoshift has been utilized in studies of allergens in perfumes using GC^3-MS [91] and in produced water waste from petroleum exploration [37].

The tile-based methodology is a valuable tool for GC×GC data, as it operates by averaging ^{2}D chromatograms across both dimensions. Binning strategies identify tiles that encompass the peak widths, represented as rectangles in the chromatogram (Figure 5.9). The primary benefits include addressing misalignment issues, improving the S/N ratio, and reducing data size before further analysis, thus enhancing processing speed and performance [28]. The main drawback is the potential for false positives, making it essential to carefully select tile sizes and confirm peak locations within the bins. Two examples of this strategy in GC×GC-MS studies include the use of tile-based PCA and Fisher's ratio approaches for the analysis of ignitable liquid residues in a forensic context [92] and the evaluation of postmortem VOCs produced by bacterial species using tile sizes of 10 s (^1t_R) and 0.7 s (^2t_R) in a GC×GC-QMS/FID system [40].

After alignment, the final step of pretreatment is data preprocessing, which involves correcting the signal intensity. This step is crucial for large datasets as it enables a more accurate evaluation of the features affecting the samples. Before data preprocessing, it is essential to visualize the data and assess the presence of missing values [93]. The main data preprocessing strategies include normalization, scaling, centering, and transformation. Figure 5.10 illustrates the objectives of each method, along with equations for scaling, centering, and transformation. The rationale for data preprocessing lies in the variations introduced by fluctuations in the chromatographic system, which directly impact peak areas [20]. More detailed descriptions of each pretreatment method will be provided in the next section.

Normalization adjusts peak areas that are typically affected by sample composition or the analytical procedure. The use of an internal standard (IS) can mitigate these variations by spiking a known, constant amount of an analyte that is not normally present in the sample. This approach compensates for fluctuations in the chromatogram. The internal standard can also be isotopically labeled but must maintain similarity to the target compound. Consequently, I.S. methods are typically applied in targeted analyses and are difficult to configure in complex samples containing diverse classes of compounds [20]. For untargeted analyses, an alternative is total area normalization, which standardizes the chromatographic area by dividing the peak area of an analyte by the total chromatographic area sum, representing the result as a percentage. This method assumes that the samples have a similar number of analytes and response factors, which is not always the case [20, 34].

Another normalization method gaining attention in GC×GC is probabilistic quotient normalization (PQN) [94]. This approach corrects results based on the individual influence of analytes rather than the entire chromatogram, as is assumed by classical normalization methods. PQN partially adjusts the chromatogram according to differential concentrations, while integral normalization

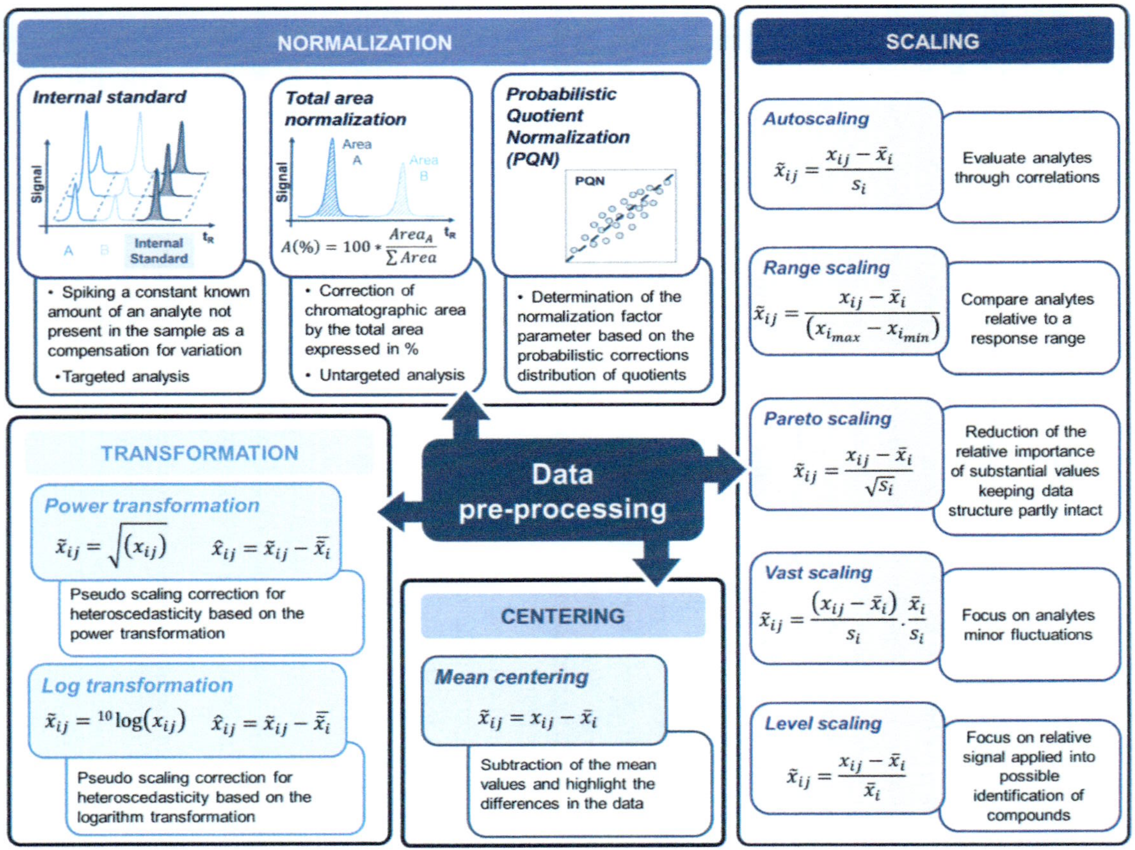

Figure 5.10 A schematic representation of data preprocessing strategies applied to analytes using GC×GC-MS.

considers signals in relation to dilution factors. PQN has been widely used in GC×GC-MS studies, such as the discrimination of decaffeinated and regular coffee [45], tuberculosis detection through human breath analysis [52], and identifying ignitable liquid residues in wildfire incidents [95].

Nonlinear data transformations are applied to reduce heteroscedasticity using mathematical strategies such as logarithmic and power transformations. These transformations also help improve the symmetry of skewed distributions and convert multiplicative relationships into linear ones [96]. Biological samples often exhibit complex multiplicative variable relationships that require transformation into a linear format. Logarithmic transformations convert multiplicative effects into additive forms, but care must be taken to avoid using this method on data containing zeros or with large standard deviations (s_i) due to the limitations of the log function. In contrast, power transformations are less affected by small values, and their use can be tailored to specific data sets [96].

The centering pretreatment strategy, also known as mean centering, aims to correct data by subtracting the mean value of each sample ($\tilde{x}_{ij} = x_{ij} - \underline{x_i}$, where $\tilde{x}_{ij}$ is the centered value, x_{ij} is the raw data point, and $\overline{x_i}$ is the mean value). The main advantage is a reduction in multicollinearity, which attributes a new zero to the dataset while maintaining homoscedasticity. If heteroscedasticity is present, the transformation strategies discussed earlier can be applied.

Scaling is another preprocessing strategy that adjusts data by dividing the mean-centered values by a scaling factor, aiming to balance the contribution of each variable to the overall model. This method has been widely used in metabolomics studies involving GC×GC-TOFMS [20] and other data preprocessing techniques [96]. Figure 5.10 outlines the goals, advantages, and disadvantages of five scaling tools: autoscaling, range scaling, Pareto scaling, vast scaling, and level scaling, which are further categorized based on data dispersion and size measurement. Autoscaling, range scaling, and Pareto scaling have been explored in GC×GC-TOFMS metabolomics applications [20].

Autoscaling evaluates analytes through correlations using the standard deviation (s_i) as the scaling factor ($\tilde{x}_{ij} = \frac{[x_{ij} - \underline{\underline{x_i}}]}{s_i}$, where s_i is the standard deviation value). With this approach, the data are examined based on correlations, and s_i is standardized to one. However, a major drawback is the inflation of errors, although it equalizes the importance of analytes. Pareto scaling, similar to autoscaling, uses the square root of s_i as the scaling factor ($\tilde{x}_{ij} = \frac{[x_{ij} - \underline{\underline{x_i}}]}{\sqrt{s_i}}$). This method reduces the relative importance of large values while keeping the data structure partly intact. It is recommended to carefully evaluate its use, as it is highly sensitive to large fold changes.

In biological studies, range scaling compares analytes relative to their response range ($\tilde{x}_{ij} = \frac{[x_{ij} - \underline{\underline{x_i}}]}{[x_{ij\,\max} - x_{ij\,\min}]}$). The scaling factor corresponds to the difference between the maximum and minimum values, giving equal importance to the analytes (e.g., metabolites). Disadvantages include error inflation and sensitivity to outliers. Vast scaling focuses on minor analyte fluctuations by multiplying the autoscaled value by the coefficient of variation (CV) ($\tilde{x}_{ij} = \frac{[x_{ij}-\bar{x}_i]}{s_i}.\frac{\bar{x}_i}{s_i}$). Its name, VAriable STability scaling, reflects its emphasis on variables with low variation. This approach enhances robustness in both supervised and unsupervised preprocessing paths, but should not be used for groups with significant induced variations.

For scaling based on average values, level scaling utilizes the mean as the scaling factor ($\tilde{x}_{ij} = \frac{[x_{ij}-\bar{x}_i]}{\bar{x}_i}$). It is applied when seeking to identify potential compounds (e.g., biomarkers), although this method is also prone to error inflation and sensitivity to outliers.

5.4.2 Model Construction

After preprocessing, the data are ready for statistical evaluation. As depicted in Figure 5.7, there are two main types of modeling: unsupervised and supervised. In unsupervised modeling, the focus is on identifying patterns in unlabeled data. This strategy primarily involves dimensionality reduction and clustering methods, exemplified by PCA and Hierarchical Cluster Analysis (HCA), respectively. In contrast, supervised modeling utilizes labeled data, where the method "supervises" the learning process, as seen in machine learning algorithms for classification and regression. Classical examples of classification methods include PLS-discriminant analysis (PLS-DA) and soft independent modeling of class analogy (SIMCA). For regression, the most commonly used technique is PLS regression.

In supervised modeling, the model's performance depends on data splitting, cross-validation, and external validation. Starting with sampling selection, specific algorithms are used to divide the data into calibration and validation sets, also known as training and test sets. This step is crucial to ensure the unbiased and correct structure of the models. Traditional algorithms like Kennard–Stone [97], Duplex [98], D-optimal [99], and Kohonen Neural Network [100] are typically used for this purpose [101].

Cross-validation is a key step in model construction, functioning as an internal validation procedure. It defines the number of latent variables (LVs) and reduces prediction error. In practical terms, it involves the following steps: ordering the samples, removing a subset, building a model with the remaining data, and then testing the model on the removed subset [102]. This strategy provides information on model stability and highlights variations that require special attention, typically using different sample subsets [101]. Depending on the data structure, sample type, and analytical technique, four main cross-validation

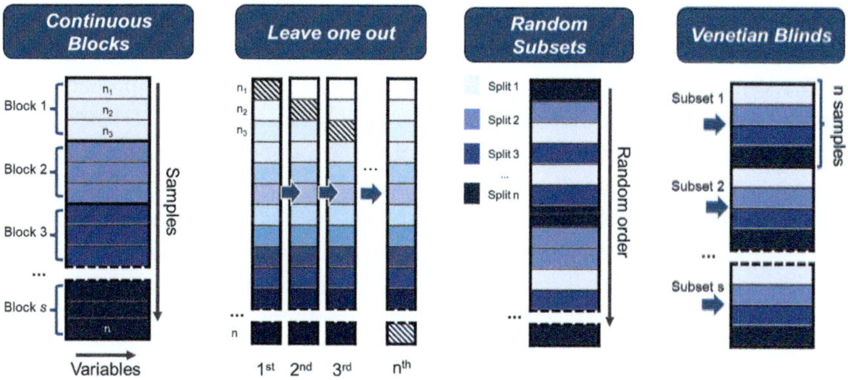

Figure 5.11 Representation of the continuous blocks, leave one out, random subsets, and venetian blinds strategies of cross-validation.

tools are commonly used: continuous block, leave-one-out, random subsets, and Venetian blinds [102, 103]. As illustrated in Figure 5.11, these four cross-validation methods are schematically represented, showing the order and selection criteria for resampling. The lines from top to bottom indicate the order of samples up to the last sample (n), while the variables are listed in each row. Each color represents a subset used in the cross-validation process, with samples arranged according to each specific strategy.

The continuous block strategy divides the dataset into blocks containing an equal number of samples up to the final block (s). In this method, a block is excluded from the model during each sampling round, then reinserted, and the next block is excluded until all blocks have been used. Since the blocks need to be representative, the organization of data must be carefully considered to ensure randomness, thereby minimizing errors. In summary, continuous blocks require a large dataset and careful attention to the ordering of samples to avoid biased results [102, 103].

The leave-one-out strategy removes one sample at a time from the dataset in each round of cross-validation. As the name suggests, one sample is excluded, and the remaining data are used to build the model, similar to the continuous block strategy. This approach is typically used when the number of samples is small ($n < 20$) and is not affected by sample order. Although it is a simple modeling strategy, if a larger sample size is used, the cross-validation error can be underestimated [102, 103].

The random subsets strategy involves dividing the dataset into multiple subsets by randomly selecting n samples for each subset. This approach is robust in terms of data organization and is not limited by sample size. However, the random selection can cause issues with reproducibility, and in some cases, may result in the presence of unusual patterns or behaviors [102, 103].

The Venetian blinds strategy is similar to the continuous block approach, but instead of using blocks, a single sample from each subset is selected. Essentially, the selection follows a regular interval based on the number of samples allocated to each subset. This method allows for flexibility in using both ordered and randomly arranged data [102, 103].

The fundamentals and guidelines for constructing a reliable model have been reviewed, providing a basis for choosing the appropriate statistical model for each data class. Table 5.1 outlines the chemometric methods applied to GC×GC-MS,

Table 5.1 Some chemometric tools applied to GC×GC-MS present in the literature.

		Application example	
Statistical analysis method	**Abbreviation**	**Doi**	**References**
Dimensionality reduction			
Multiple factor analysis	MFA	10.3390/foods13010018	[57]
Multiway principal component analysis	MPCA	10.3390/separations6040046	[63]
Principal component analysis	PCA	10.3390/metabo14040237	[50]
Principal coordinate analysis	PCoA	10.1016/j.chroma.2024.465017	[40]
Weighted principal component analysis	WPCA	10.1021/acs.est.8b05238	[90]
Clustering			
Hierarchical clustering analysis	HCA	10.3390/foods13010018	[57]
Two-dimensional hierarchical clustering analysis	^{2}D-HCA	10.1016/j.scitotenv.2018.08.317	[104]
Univariate			
Analysis of variance	ANOVA	10.3390/separations11020058	[105]
Duncan's multiple test	Duncan's MT	10.1016/j.jfca.2024.106398	[64]
Fisher's exact test	—	10.1016/j.jchromb.2023.123937	[36]
Fisher's ratio	*F*-ratio	10.3390/metabo13070828	[106]
Least significant difference	LSD	10.1016/j.jhazmat.2022.130407	[42]
Pearson's correlation	—	10.1021/acsomega.0c05359	[107]

Table 5.1 (Continued)

Statistical analysis method	Abbreviation	Application example: Doi	Application example: References
Student's T-test	T-test	10.3390/molecules27061806	[45]
Tukey's honestly significant difference	Tukey's HSD	10.3390/foods13010018	[57]
Discrimination/ Classification			
Artificial neural network	ANN	10.1016/j.fpsl.2022.100978	[108]
Data-driven soft independent modelling of class analogy	DD-SIMCA	10.1007/s00216-022-04428-2	[109]
Decision tree	DT	10.1016/j.fpsl.2022.100978	[108]
Factor analysis	FA	10.1016/j.fpsl.2022.100978	[108]
Independent component analysis	ICA	10.1016/j.fpsl.2022.100978	[108]
k-nearest neighbor	kNN	10.3390/foods13101504	[66]
Linear discriminant analysis	LDA	10.1016/j.fpsl.2022.100978	[108]
Orthogonal partial least squares discriminant analysis	OPLS-DA	10.1016/j.fbio.2023.103403	[110]
Partial least squares discriminant analysis	PLS-DA	10.3390/ijms24119614	[46]
Randon forest	RF	10.3390/molecules27061806	[45]
Support vector machine	SVM	10.1016/j.fpsl.2022.100978	[108]
Two-way orthogonal partial least squares discriminant analysis	O2PLS-DA	10.1016/j.aca.2024.342555	[58]
Regression			
Binary logistic regression	BLR	10.1016/j.jchromb.2023.123937	[36]
Partial least squares regression	PLS	10.3390/foods13010018	[57]
RReliefF analysis	RRelieF	10.1016/j.chroma.2023.463920	[39]
Curve resolution/Deconvolution			
Multivariate curve resolution – alternating least square	MCR-ALS	10.1016/j.foodchem.2024.138690	[65]
Parallel factor analysis	PARAFAC	10.1007/s10337-020-03863-6	[91]
Trilinear deviation ratio	TDR	10.1016/j.chroma.2019.460368	[111]

including their abbreviations and categories. The papers referenced in this table were identified in the Web of Science database from 2019 up to July 2024. The five categories presented are integrated into the workflow shown in Figure 5.7 and will be detailed in their respective sections later in this chapter.

Further details on GC×GC-MS data analysis using chemometric tools are summarized in Table 5.2, which lists research in this field in chronological order, starting from the most recent papers down to those published in 2019. The table categorizes the studies by sample type, research objective, chromatographic system and MS type, data processing software, statistical methods used, and corresponding references. In addition to MS hyphenated systems (e.g., TOFMS and QMS), complementary analytical techniques such as FID [40, 54, 56] and olfactometry (O) [49, 113] have been employed. Regarding software tools, MATLAB (MathWorks, Natick, MA, USA) is among the most frequently used platforms, employing various toolboxes such as the PLS Toolbox (Eigenvector Technologies, Manson, WA, USA), Statistics and Machine Learning, Curve Fitting 8.1.1, Neural Network Fitting, Decision Tree (DT), MexStandalone-v0.02, and LIBSVM version 3.25. Other software solutions utilized in data analysis include ChromCompare+ (SepSolve Analytical, Peterborough, UK), Microsoft Excel (Microsoft, Redmond, WA, USA) with the XLSTAT add-on (Lumivero, Denver, CO, USA), R Statistical Software (R Foundation for Statistical Computing, Vienna, Austria), SPSS (SPSS Inc., Chicago, IL, USA), Unscrambler X 10.4 (CAMO ASA, Oslo, Norway), SIMCA (Umetrics, Umea, Sweden), TBtools [119], OriginPro (OriginLab Corporation, Northampton, MA, USA), MetaboAnalyst 5.0 (Xia Lab, McGill University, Montréal, QC, Canada), Python (Python Software Foundation, Wilmington, DE, USA), SAS 9.4 (SAS Institute, Cary, NC, USA), ChromaTOF Tile Software (LECO, St. Joseph, MI, USA), and LAV Software version 2.2.1 (Gesellschaft für Analytische Sensorsysteme mbH, Germany).

5.4.3 Univariate Strategies

According to the parsimony principle, the simplest approach is often the best option for data analysis. In this context, multivariate statistics can sometimes be replaced by univariate and classical statistical analyses. These are powerful tools that can effectively describe the desired responses in GC×GC-MS techniques, including ANOVA, t-tests, and Fisher's ratio. The use of ANOVA has been reported in the literature since 2011, with examples including studies of Pinot Gris and Pinot Noir wines [57], heat-sterilized peach juice [64], feng-flavored Baijiu (a distilled Chinese beverage) [66], essential oil from *Varronia curassavica* Jacq. [56], and water samples from Alberta's oil sand wells [104]. Similarly, the

Table 5.2 Overview of some papers published from 2019 until the moment of the writing using chemometrics and GC×GC-MS.

Sample type (number of samples)	Aim	Chromatographic system (modulator)	Processing software	Statistical strategies	References
Packages for wildfire investigations ($n = 10$)	Evaluation of the effectiveness of different packaging types in inhibiting cross-contamination by analyzing the relative target compounds	GC×GC-TOFMS (flow)	ChromCompare + V2.1.7, and Excel 365	ANOVA, PCA, tile-based *F*-ratio, RF and kNN	[92]
Pinot Gris ($n = 7$) and Pinot Noir Wines ($n = 7$)	Characterization of the chemical and sensory profile of Pinot Gris and Pinot Noir wines available in the market	GC×GC-TOFMS (Flow)	XLSTAT add-on for Excel	ANOVA, Tukey's HSD, HCA, MFA and PLS	[57]
Postmortem mammalian tissues ($n = 175$)	Study of VOC produced by five postmortem bacterial species in a period of five days by headspace and gas chromatography	GC×GC-QMS/ FID (flow)	ChromCompare + v. 1.4 and R statistical software	tile-based *F*-ratio, PCA and PCoA	[40]
Heat-sterilized peach juice ($n = 4$)	Characterization and comparison of VOC in samples of heat-sterilized peach juice by GC-MS, GC×GC-MS and aroma assessment by QDA	GC×GC-TOFMS (thermal)	SPSS versus 25.0 and Unscrambler x. 10.4	ANOVA, Duncan's multiple tests and PLS	[64]
Xinomavro red wine ($n = 24$)	Investigation of wine aging applying SPME Arrow GC×GC-MS coupled with chemometric tools applied to VOC markers	GC×GC-QMS (thermal)	SIMCA 14.1	HCA, PCA and O2PLS-DA	[58]
Chinese Baijiu ($n = 75$)	Development of a workflow using GC×GC data by different chemometrics approaches through an untargeted analysis of volatiles	GC×GC-TOFMS (thermal)	EI-TOFMS software and MATLAB R2020b	MCR-ALS, CARS-PLS and PLS-DA	[65]

(Continued)

Table 5.2 (Continued)

Sample type (number of samples)	Aim	Chromatographic system (modulator)	Processing software	Statistical strategies	References
Feng-flavored Baijiu ($n = 30$)	Study of the differences and correlation between sensorial analysis and volatile profile of aged Baijiu beverages	GC×GC-TOFMS (thermal)	SPSS 22.0 and SIMCA-P 14.1	kNN, ANOVA, PCA, PLS-DA and PLS	[66]
Hops ($n = 5$)	Determination of the comprehensive profile of volatile metabolites in hops by HS-SPME-GC×GC-MS and the chemotyping of five different species	GC×GC-QMS (thermal)	MATLAB 7.13 with PLS Toolbox 6.5	PCA e HCA	[50]
Organo-phosphorus nerve agents ($n = 18$)	Analysis of organic impurity profile of class G and V weapons in three groups of samples by GC×GC	GC×GC-TOFMS (thermal)	—	HCA, PCA, ANOVA and OPLS-DA	[41]
Sub-bituminous coal ($n = 12$)	Analysis of two low-rank coals by different extracts using GC×GC-TOFMS	GC×GC-TOFMS (thermal)	Origin 2021b	PCA and HCA	[112]
Hainan camellia oils and their adulteration solutions ($n = 138$)	Compositional analysis of flavor VOC from Hainan oils and establishment of adulteration with different oils by HS-SPME-GC×GC	GC×GC-QTOFMS (thermal)	SIMCA 14.1 and SPSS 22.0	PCA and OPLS-DA	[44]
Serum ($n = 64$)	Identification of colorectal cancer markers by low- and high-resolution GC×GC-TOFMS	GC×GC-TOFMS (thermal)	MetaboAnalyst 5.0 and Python 3.9.7	ANOVA, PCA, HCA, PLS-DA and RF	[46]

Arabica coffee ($n = 56$)	Determination of the volatile fingerprint of the potato taste defect in roasted arabica coffee using HS-SPME-GC×GC-TOFMS and chemometric tools	GC×GC-TOFMS (thermal)	MATLAB 2019b with PLS Toolbox 8.9	t-test, MCR-ALS, PCA and PLS	[68]
Aerospace fuels ($n = 58$)	Implementation of tile-based variance ranking as an improvement to PLS modeling data of aerospace fuels analyzed by GC×GC-TOFMS	GC×GC-TOFMS (thermal)	MATLAB 2019b with PLS Toolbox 8.9	RReliefF, PLS and PARAFAC	[39]
Produced water from petroleum exploration ($n = 75$)	Evaluation, optimization, and analysis of an alternative method aiming the classification of produced water in accordance with Brazilian regulation (CONAMA)	GC×GC-QMS (flow)	MATLAB R2021a	DD-SIMCA	[37]
Breast milk ($n = 12$)	Determination of volatiles in the storage of breast milk by two different sample preparations (SPME and DHS)	GC×GC-O-MS (thermal)	SIMCA v.14	ANOVA and OPLS-DA	[113]
Tea (*Camellia sinesis*) ($n = 18$)	Analysis of green, oolong, and black tea using HS-SPME, GC-IMS, and GC×GC-O-MS for the identification of their odorants by means of sensorial evaluation and chemometrics	GC×GC-O-MS (thermal)	SIMCA 14.1 and TBtools	PCA and OPLS-DA	[49]
Human skin ($n = 38$)	Targeted and untargeted analysis of a noninvasive approach of skin VOCs for tuberculosis diagnosis	GC×GC-TOFMS (thermal)	SPSS 25 and SAS 9.4	Fisher's exact test, ANN, BLR and PCA	[36]

(Continued)

Table 5.2 (Continued)

Sample type (number of samples)	Aim	Chromatographic system (modulator)	Processing software	Statistical strategies	References
Pooled human feces ($n = 24$)	Development of a human stool reference material for metabolomics studies applying analysis of volatiles metabolites in aqueous and lyophilized human feces from omnivore and vegan diets	GC×GC-TOFMS (thermal)	MATLAB 2020a with PLS Toolbox 9.1	F-ratio and PCA	[106]
JP-4 jets fuel ($n = 12$)	A proof-of-concept study aiming the use of cuboid-base F-ratio analysis using GC^3–TOFMS data from 38 analytes spiked into jet fuel	GC^3–TOFMS (^1M: thermal and ^2M: flow)	MATLAB 2019b	Cuboid-based F-ratio and tile-based F-ratio	[12]
Grass carp ($n = 5$)	Establishment and improvement of a new method for the detection of volatiles present in grass carp samples by HS-SPME-GC×GC	GC×GC-QMS (thermal)	MATLAB 7.0 and SIMCA 14.1	OPLS-DA and PCA	[110]
rPET flakes ($n = 57$)	Development of a HS-SPME-GC×GC-QTOFMS untargeted screening method for the detection of volatile contaminants present in recycled poly (ethylene terephthalate) associated with chemometric tools for authenticity evaluation of recycling areas by region	GC×GC-QTOFMS (thermal)	SIMCA 14.0 and SPSS 21 software	OPLS-DA, PCA and LSD	[42]

Poly (ethylene terephthalate) (PET) for food packaging ($n = 105$)	A novel discrimination method by DSI on the semivolatile organic compounds of recycled and virgin PET packages using different chemometric tools	GC×GC-QTOFMS (thermal)	SIMCA 14.1, SPSS 25.0 and MATLAB, neural net fitting toolbox, decision tree toolbox, MexStandalone-v0.02, LIBSVM toolbox version 3.25	PCA, OPLS-DA, LDA, ANN, DT, RF, SVM, FA, ICA and HCA	[108]
Fish oil ($n = 32$)	Evaluation and development of a GC×GC analysis method for fish oil supplements authentication using chemometric modeling DD-SIMCA and its figures of merit	GC×GC-QMS (flow)	MATLAB R2021a	OPLS-DA and DD-SIMCA	[109]
Diploic yeast strain ($n = 24$)	Development of a computational method for the determination of cycling yeast metabolites by the untargeted analysis of 24 time-point samples using GC×GC-MS and statistical tools	GC×GC-TOFMS (thermal)	ChromaTOF Tile Software version 101, MATLAB 2019b with PLS Toolbox	Tile-based *F*-ratic analysis and PCA	[114]
Pacu fish ($n = 20$)	Metabolomic studies of pacu fish raised in rice-fish farming and tank-raised using tile-based Fisher's ratio analysis	GC×GC-TOFMS (thermal)	MATLAB v 2019a, in-house tile-based *F*-ratio software [115]	*F*-ratic, PARAFAC and PCA	[35]

(Continued)

Table 5.2 (Continued)

Sample type (number of samples)	Aim	Chromatographic system (modulator)	Processing software	Statistical strategies	References
Regular and decaffeinated coffee ($n = 16$)	Discrimination of decaffeinated and regular coffee by their volatiles using HS-GC×GC, machine learning, and chemometric tools	GC×GC-TOFMS (flow)	R version 4.0.2 and MetaboAnalyst 5.0 in-house code for MATLAB R2019b	T-test, ANOVA, PCA, HCA, PLS-DA and RF	[45]
Fractions of hydrothermal petroleum extracts ($n = 34$)	Determination of multimolecular organic changes in the apolar fraction of petroleum extracts present in sediments taken along a transect originated in the Catedral Hill vent mound (Guaymas Basin, Gulf of California) by GC×GC coupled with MPCA and HCA	GC×GC-QTOFMS (thermal)	MATLAB 7.12 R2011a with PLS Toolbox 8.5.2	MPCA and HCA	[116]
Nonfuel coal ($n = 11$)	Determination of pyrolysis products originated from bituminous and reference coals by means of py-GC×GC-TOFMS and chemometrics	py-GC×GC-TOFMS (thermal)	IBM SPSS 19	Pearson's correlation, HCA and PCA	[107]
Type 2 diabetes mellitus (T2DM) patients($n = 86$)	Investigation of the relations between sedentary (SED)-time in patients with T2DM by means of the metabolic profile and biological pathway linked with the urinary system	GC×^{2}GC-MS/FID (thermal)	R statistical software 3.2.2	PCA and OPLS-DA	[54]

Essential oil from plant *Varronia curassavica* Jacq. ($n = 9$)	Investigation of the metabolomic profile, effect of seasonality, and number of anti-inflammatory components present in hydrodistillated essential oils of three genotypes of *Varronia curassavica* Jacq.	GC×GC-FID/QMS (flow)	XLSTAT-2020 version	ANOVA, PCA and HCA	[56]
Allergens in perfumes ($n = 2$)	Development of a new GC3 setup and analysis of allergens in perfumes using chemometric tools	GC^3–TOFMS (^{1}M: thermal and ^{2}M: flow)	MATLAB 2017a with PLS Toolbox 8.21 and N-way toolbox version 3.31	PARAFAC	[91]
Fortress channel ($n = 16$) and lake water ($n = 21$)	Nontarget screening characterization of water from a lake and channel of an urban area of Copenhagen (Denmark) using GC×GC-HR MS	GC×GC-HR MS (thermal)	MATLAB R2017b	Pixel-based PCA and WPCA	[38]
Diesel fuel ($n = 6$)	Development of a more automatable approach using Fisher's ratio analysis and GC×GC-TOFMS data	GC×GC-TOFMS (thermal)	MATLAB R2018a and in-house developed tile-based *F*-ratio software [115, 117]	Tile-based Fisher ratio, S-ratio and PARAFAC	[118]
Diesel oil spills ($n = 32$)	Analysis and determination of diesel oil spills in the context of environmental investigations by a new method of GC×GC-HR MS and ^{2}D chemometric analysis of selected ion chromatograms (^{2}D-CHEMSIC)	GC×GC-HR QTOFMS (thermal)	MATLAB R2016a and icoshift v. 3.0 algorithm [88]	WPCA	[90]

(Continued)

Table 5.2 (Continued)

Sample type (number of samples)	Aim	Chromatographic system (modulator)	Processing software	Statistical strategies	References
Patients with active pulmonary tuberculosis ($n = 32$) and symptomatic but with but without tuberculosis ($n = 18$)	Analysis of exhaled human breath from patients living in an endemic tuberculosis zone, including positive cases, smokers, and HIV positive subjects by GC×GC-MS and chemometrics	GC×GC-TOFMS (thermal)	R v3.4.3	RF, SVM with a linear kernel, PLS-DA, PCA and HCA	[52]
Water from Albertas' oil sand wells ($n = 4$)	Profiling resolved components of individual naphthenic acids and their temporal and spatial variation at Syncrude Canada's Sandhill Fen reclamation site	GC×GC-TOFMS (thermal)	SPSS 17.0.0 and MetaboAnalyst 3.0	ANOVA and ^{2}D-HCA	[104]
Analytes test mixture ($n = 21$) in different chromatographic settings ($n = 4$)	Experimental validation using TDR as a predictive tool for PARAFAC performance, and an informative guide related to determining separation conditions for quantitative PARAFAC	GC×GC-TOFMS (thermal)	MATLAB R2016b with MATLAB curve fitting toolbox 8.1.1 and PLS Toolbox 8.2.1	TDR and PARAFAC	[111]
Fermented beer with different yeast ($n = 2$)	Application of HS-SPME-FM-GC×GC-MS and bottom-up approach for data mining applying MPCA in the bioaromatized beer fermented sample with different yeasts	GC×GC-QMS (flow)	MATLAB R2017b with PLS Toolbox 7.5 B	MPCA	[63]

t-test has been applied to samples of roasted Arabica coffee to identify off-flavors [68] and for discriminating between decaffeinated and regular coffee [45].

Since its first application in GC×GC-MS [120], Fisher's ratio (*F*-ratio) has demonstrated its effectiveness as a univariate supervised method. The benefits of this technique include its ability to select significant features in both parametric and nonparametric approaches and to identify compounds in nontargeted analyses, which highlights its powerful utility. This procedure is based on the one-way ANOVA, where the hypothesis tests the ratio of between-class variance (σ^2_{bc}) to within-class variance (σ^2_{bc}) [28], as defined in Eq. (5.3). The output of this method is a ranked list, with analytes ordered by their *F*-ratio values, indicating their potential to show distinct behavior. The values in the "hit list" range from zero to infinity, reflecting the significance of variance in the statistical analysis. In chromatographic data, accurate alignment is crucial for obtaining reliable *F*-ratio results, and this can be achieved using the three alignment strategies described earlier in this chapter. If the data acquisition conditions are not optimized, a high number of false responses may appear at the top of the hit list. This can manifest as both false negatives and false positives – where false negatives mean that analytes with chemical significance are not highlighted by the *F*-ratio analysis, and false positives indicate features incorrectly identified as chemically distinct [121],

$$F - \mathrm{ratio} = \frac{\sigma^2_{\mathrm{bc}}}{\sum \sigma^2_{\mathrm{wc}}}. \tag{5.3}$$

Consequently, *F*-ratio has been widely applied in various research fields [12, 35, 105, 106, 114, 118]. One example is the use of a tile-based alignment approach combined with *F*-ratio in the analysis of wildfire investigation samples using the GC×GC-TOFMS technique [105]. In this study, ANOVA was also employed in a univariate context, while PQN normalization was applied alongside random forest (RF) and k-nearest neighbor (kNN) methods for class prediction. The results assessed the individual profiles and classes of VOCs in the samples, demonstrating the effectiveness of the packaging in minimizing cross-contamination. The experiments confirmed the integrity of the samples, despite the packaging containing ignitable liquid residues. Nylon-based materials showed the best performance for packaging, while triple-layered approaches did not present significant differences. The authors concluded that a thermally sealed bag made of nylon, combined with a metal can, was the most effective packaging solution.

Another example of *F*-ratio application, shown in Table 5.2, is the metabolic profiling of human fecal samples from individuals following vegan and omnivorous diets [106]. The study included samples from eight different individuals provided by the NIST institute, which were analyzed using GC×GC-TOFMS for aqueous and lyophilized volatile analysis. This complex sample was investigated to develop a stool metabolomic reference procedure. The total useful peak area, number of peaks, and total peak area were evaluated, and the *F*-ratio selection was based on values exceeding 300 and presence in at least 66.67% of the samples. A higher number of peaks were observed in samples from omnivorous diets, while storage conditions had a more significant impact on the metabolic profile. PCA was used for pattern recognition, aiding in the understanding of sample variations based on storage, diet, and their interactions. The next section will further discuss the application of pattern recognition in data analysis.

5.4.4 Pattern Recognition

The primary goal of unsupervised methods is to identify patterns in the data and extract meaningful information. As previously mentioned, there are two main classes of unsupervised methods: dimensionality reduction and clustering. GC×GC-MS, like other techniques, generates large datasets that can obscure patterns, which can be revealed through unsupervised methodologies.

One of the most widely used dimensionality reduction tools, and a foundation for other methods, is PCA. The core principle of PCA is to simplify the data structure by creating variables known as principal components (PCs) through the algebraic decomposition of the data matrix (X). This process performs linear combinations, resulting in a reduced number of significant PCs. The number of PCs typically ranges from 1 to a value that adequately represents the significant variance in the model. In other words, PC1 captures the largest variance in the data, followed by PC2, and so on, until all significant variance is accounted for [122]. To elucidate the mathematical process and its algebraic structure, Figure 5.12 provides a schematic representation of PCA. The data matrix is decomposed into the product of a score matrix (T) and a loading matrix (P), plus the residuals (E) found in the model. The scores represent the projection of samples in the new PC space, while the loadings – defined by the cosine of the angle between the PCs and the original variables – indicate the magnitude of contribution of each PC. Several algorithms are available for PCA modeling, with SVD being the most commonly used. To build a reliable PCA model, removing outliers and selecting the appropriate number of PCs is essential. For more detailed information, the interested reader is referred to other sources [102, 122–124].

Various dimensionality reduction models based on PCA are utilized in GC×GC-MS data analysis. Examples include principal coordinate analysis (PCoA) [40], MPCA [63, 116], and weighted principal component analysis (WPCA) [38, 90], as listed in Tables 5.1 and 5.2. A schematic representation of MPCA is shown in Figure 5.12. MPCA is a multiway method where the data tensor (X) is decomposed into the sum of the outer product (Kronecker product, represented as "⊗") of score vectors (tr) and loading matrices (Pr), along with the residual tensor. In this context, the first mode (I) corresponds to samples, the

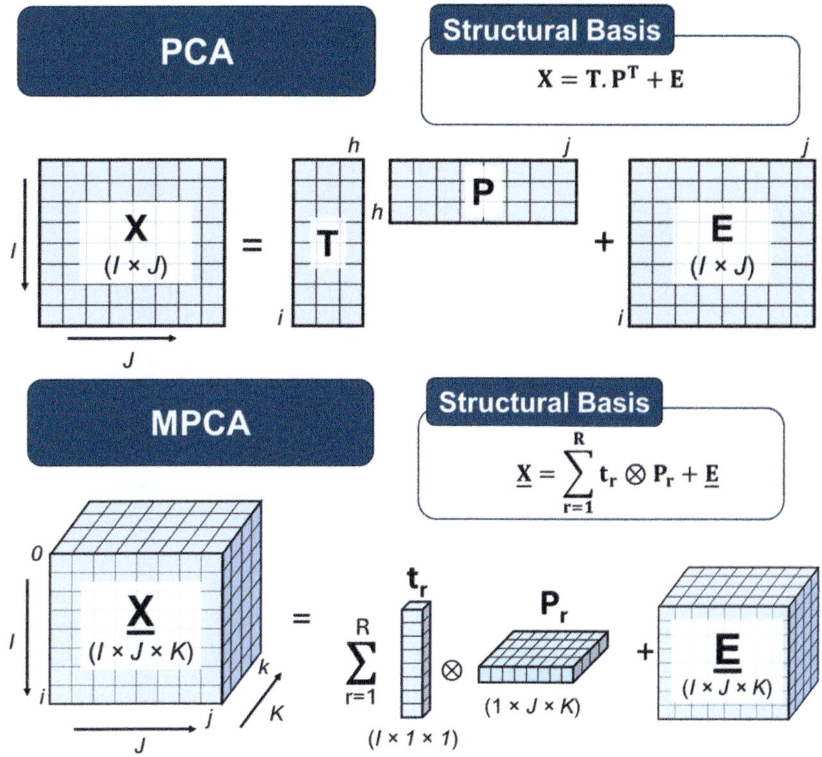

Figure 5.12 Representation and structural basis for the construction of PCA and MPCA models.

second mode (J) to retention time, and the third mode (K) can represent either the ^{2}D retention time or m/z values in the case of GC-MS data.

A representation of PCA results is illustrated in Figure 5.13. The graph shows score plots for PC1 versus PC2 and PC1 versus PC3, using three data groups (A, B, and C) as an example. As noted, PC1 captures the highest variance, and its capacity to distinguish group C is evident. This group is clearly separated in both score plots, where PC1 is plotted on the *x*-axis. The underlying reason for this separation can be investigated by examining the loading plot, which helps attribute the analytical significance of the sample features to the model's performance. An alternative way to present the PCA graph is through a biplot, which simultaneously displays both scores and loadings on the same graph. In the top score plot (PC1 versus PC2), PC2 is represented on the *y*-axis, and it reveals a distinct pattern where this component can indicate a variable specific to group A. Additionally, the separation of group B is visible in PC1 versus PC2, characterized by lower PC1 values and higher PC2 values. In this model, PC3 (the third component, shown on the *y*-axis of the second

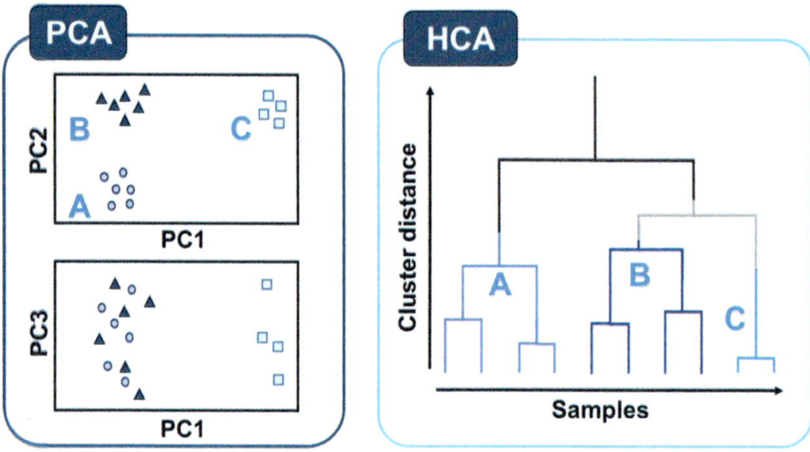

Figure 5.13 Results of the sample classes A, B, and C by the score plots of PCA and HCA dendrogram.

score plot) does not effectively separate the groups, resulting in a mix of samples in this dimension. While PC3 may contain some chemical information relevant to the samples, it does not significantly contribute to the group patterns observed in the model, indicating that two components are sufficient to describe this particular characteristic.

As shown in Table 5.2, PCA is frequently reported in a wide range of applications in GC×GC-MS. In the context of metabolomic studies, PCA has been applied to investigate metabolites in hop species [50], biomarkers of colorectal cancer in human serum [46], skin compounds as markers of tuberculosis [36], and pacu fish metabolites from rice farms [35]. A more specific example involved using both PCA and PCoA to analyze volatiles produced by bacteria in postmortem mammalian tissues [40]. In this study, the volatile profiles of *Bacillus subtilis*, *Ignatzschineria indica*, *Ignatzschineria ureiclastica*, *Vagococcus lutrae*, and *Curtobacterium luteum* were monitored from day zero to day four using GC×GC-QMS/FID. Procrustes analysis was also applied to maximize similarity between different ordinations by rotating one ordination to match another. Results were presented using PCA plots based on Euclidean distance and PCoA plots using Bray–Curtis dissimilarity. Although individual volatile profiles for each bacterium showed slight overlap, Procrustes analysis revealed that while both PCA and PCoA provided similar results, PCoA was a better tool for this dataset. The control sample displayed a clear profile in the diagrams, with time acting as a key variable influencing sample configuration. The study concluded that multivariate data visualization was simplified and provided a solid comparison of different dimensionality reduction approaches.

Among unsupervised clustering methods, HCA is one of the most commonly applied to chemical data. Pattern recognition and exploratory analysis are achieved by constructing clusters based on sample similarities. The samples are compared to each other and their relative distances in HCA space are used to form clusters, which are then visualized in a dendrogram. Various factors must be considered when building an HCA model, such as the choice of distance measures (e.g., Euclidean, Manhattan, and Mahalanobis), clustering algorithm (agglomerative or divisive), and the grouping strategy [103, 122]. A classical HCA result is depicted in Figure 5.13, where the data are presented in a two-dimensional tree-like diagram. A well-defined cluster is indicated by shorter branches that represent distinct groups, whereas long branches suggest less clear group formation. In Figure 5.13, group C shows the highest similarity among samples, evidenced by the shortest cluster distances, while it is more closely related to group B than to group A, as indicated by the cluster's formation order.

HCA is often combined with a heatmap to display cluster similarities visually, with HCA represented along the axes. An example of HCA in the context of lipidomics and metabolomics was reported in a proof-of-concept investigation of lipids and metabolites in human serum as markers of colorectal cancer [46]. Various statistical methods were used, including ANOVA, PCA, HCA, PLS-DA, and RF. The study analyzed 64 human serum samples divided into 3 groups: 20 adenocarcinoma, 23 adenoma, and 21 control samples using GC×GC-TOFMS. This research identified 40 lipids spanning six different classes, with HCA results presented as both a dendrogram and a heatmap. The heatmap incorporated the eight most significant attributes, resulting in a dendrogram that categorized the serum samples into their respective groups. The clustering potential of the groups was evident from the selected fatty acid profiles using Pearson's correlation and Ward's algorithm, with the adenocarcinoma group showing the most dissimilar pattern. This methodology also distinguished the stages of disease progression based on long-chain polyunsaturated fatty acids (PUFA), specifically ω-3 and ω-6. Omega-3 PUFA was higher in control groups, whereas omega-6 PUFA was more prevalent in adenocarcinoma samples. The study successfully demonstrated the potential of lipidomics in identifying biomarkers for disease differentiation.

5.4.5 Curve Resolution/Deconvolution

Returning to two of the biggest and most common challenges in chromatography – complex data with more than three dimensions and the co-elution of analytes – curve resolution and deconvolution methodologies are particularly useful. Multiway methods use tensors, where in GC×GC-MS data, it is common to encounter dimensions such as ${}^1t_R$, ${}^2t_R$, m/z, and samples. Deconvolution is sometimes referred to as "mathematical chromatography" due to its algebraic capability to extract valuable information for separating target compounds. Techniques such as parallel factor analysis (PARAFAC) and multivariate curve resolution–alternating least squares (MCR-ALS) are among the most important tools for addressing these issues.

Similar to PCA, PARAFAC is also a decomposition method, first reported in 1970, and is also known as canonical decomposition (CANDECOMP) [125, 126]. To facilitate understanding, Figure 5.14 presents a schematic illustration of the structural basis of PARAFAC and its algebraic pathway. The figure depicts a three-way decomposition of a data tensor (X) into three matrices (A, B^T, and C^T), along with the residuals (E). This decomposition results in trilinear components or triads, as shown on the bottom left side of the figure [127]. The mathematical framework of PARAFAC involves the Kathri–Rao product (represented by "|⊗|"), and a simple description of the triad calculations is provided on the bottom right side of the figure.

PARAFAC offers several advantages, including uniqueness, interpretability, symmetry, and noise separation, and various constraints can be applied during

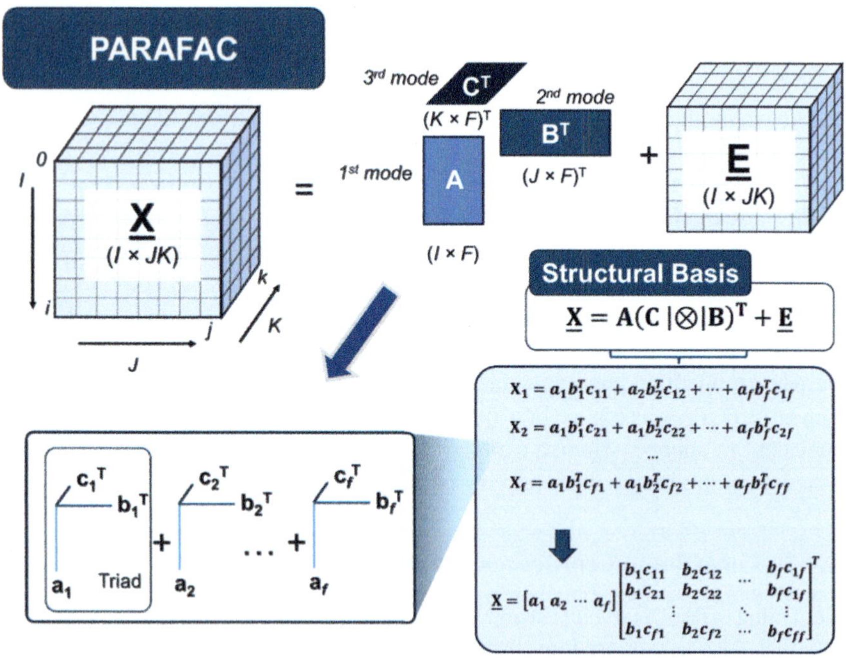

Figure 5.14 Representation and structural basis for the construction of a PARAFAC model.

model construction. For more in-depth details on PARAFAC and its variations, such as PARAFAC2, Bro's (1998) thesis [127] is an excellent reference for understanding their theoretical foundations.

In MCR-ALS, the ALS algorithm enables the incorporation of constraints, which enhances the chemical interpretability of the samples (e.g., nonnegativity constraints for concentration values). The MCR method works by mathematically decomposing a data matrix (D). Figure 5.15 illustrates the operational mechanism using a GC×GC-MS dataset as an example. The first step involves unfolding a tensor into row-wise augmented arrays to convert the tensor into a matrix, represented by modulations known as matrix "folds" or "slices." The structural basis of MCR-ALS is expressed as the multiplication of a contribution matrix (C) by the response matrix (S) (e.g., m/z ratio), with the addition of residuals (E).

GC×GC-MS data are well-suited for deconvolution strategies, as demonstrated in Figure 5.16, which illustrates both PARAFAC and MCR-ALS methodologies. In PARAFAC, the data are decomposed into a product of matrices corresponding to the ^{1}D retention time, ^{2}D retention time, and m/z ratio, resulting in the desired components. In MCR-ALS, the data are first unfolded into a retention time matrix, transposed, and then the analytes A and B are characterized by their m/z "spectrograms" (ST).

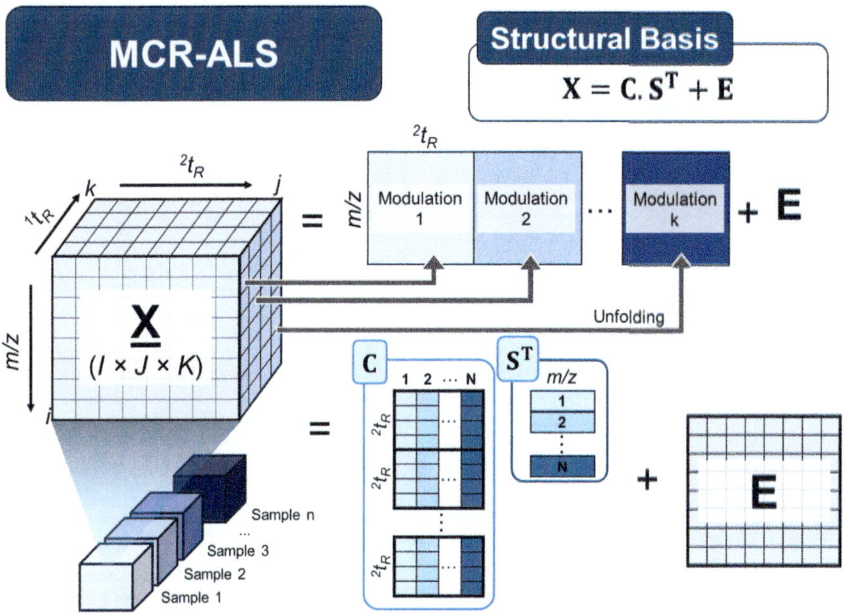

Figure 5.15 Representation and structural basis for the construction of the MCR-ALS model.

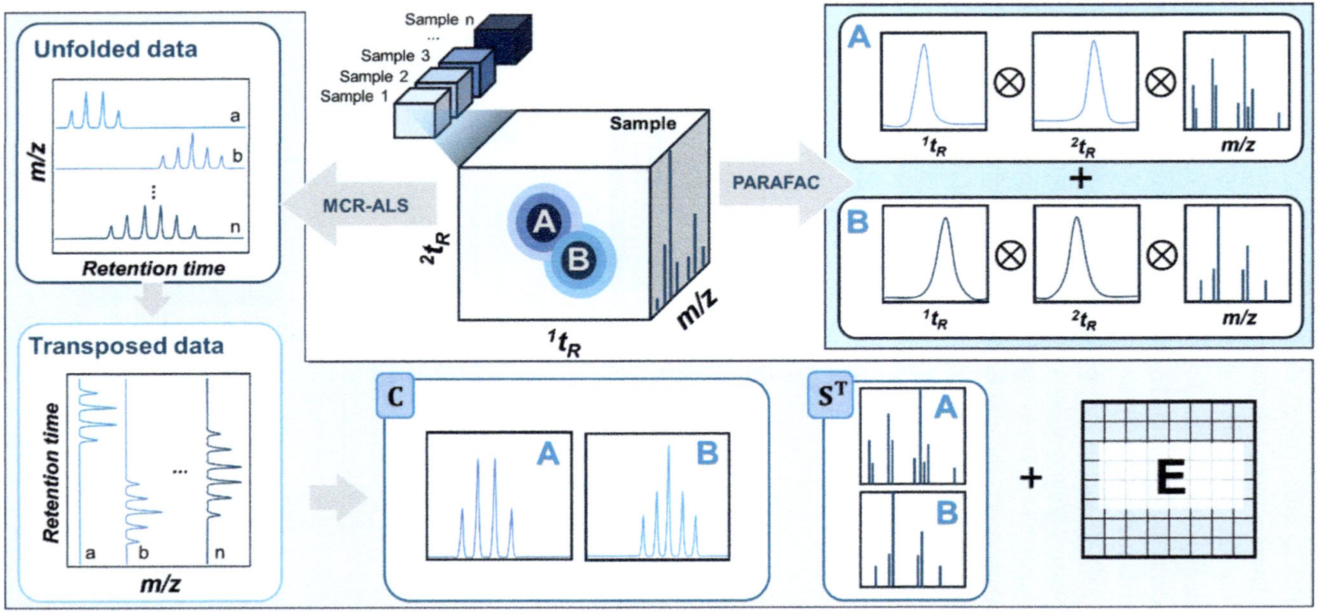

Figure 5.16 Application of MCR-ALS and PARAFAC models to GC×GC-MS data.

A recent study [128] explored various chemometric tools for metabolomic analysis of *Daphnia magna* samples using GC×GC-TOFMS, including bilinear MCR-ALS, synchronized and non-synchronized trilinear MCR, PARAFAC, and PARAFAC2. Classical metrics such as lack of fit (LOF) and explained variance (R^2) were used to compare the statistical performance of each model, providing a clear example of how analysts can evaluate and select the most appropriate model. Another valuable reference for understanding this class of methods was developed by Berrier [121], who offers a detailed explanation of data handling in GC×GC and a deeper insight into deconvolution methods using PARAFAC and MCR-ALS.

An example from Table 5.2 is the screening analysis of markers in the distilled beverage Baijiu using GC×GC-TOFMS [65]. MCR-ALS, combined with competitive adaptive reweighted sampling partial least squares (CARS-PLS), enabled the identification of key compounds in the samples. The authors implemented data compression, achieving a reduction of over 1000-fold without compromising data quality. PLS-DA was used for discriminant analysis (DA), resulting in 30 potential compounds after concentration analysis and outlier detection, while PLS regression was applied to the dataset. The markers were identified based on their mass spectra obtained through the MCR-ALS strategy. The authors validated the significance of their findings by highlighting the relevant compounds contributing to the flavor profile of Chinese Baijiu.

5.4.6 Regression

In analytical chemistry, as in many other fields, quantitative models are crucial for determining the amount of a specific variable. From a multivariate perspective, several variables are used to build calibration models and establish relationships for the properties measured. Among the regression strategies, multiple linear regression (MLR), principal component regression (PCR), and PLS regression are widely reported in the literature [122].

PLS regression is a popular method in multivariate calibration. It is based on the decomposition of a data matrix (X), resulting in a product of loadings (P^T, Q^T) and scores (T, U), along with residuals (E, F) that represent the nonessential information for modeling, as described in Eq. (5.4). The decomposition process is carried out simultaneously for two data blocks, targeting the highest correlation with the dependent variable (Y) and maximizing the covariance between the data matrix (X) and y. This dependent variable, represented in Eq. (5.5), defines PLS as a supervised method, requiring an additional set of data for model development. In Eq. (5.5), q denotes the regression vector associated with T, as seen in Eq. (5.4):

$$X = T.P^T + E, \tag{5.4}$$

$$Y = U.Q^T + F. \tag{5.5}$$

Since the two data blocks are affected simultaneously by variance, the principal components (PCs) differ, as some of their orthogonality is lost. These components are referred to as LVs for the decomposed factors. Another important aspect to note is that the LVs are not arranged in descending order of explained variance, as in PCA [122].

If the vector "*y*" is used as the dependent variable, PLS is referred to as PLS1. When the model requires more than one variable for the dependent set, the "*y*" vector can be replaced by a matrix (*Y*), and in this case, the methodology is called PLS2. Various algorithms can be employed for this method, such as NIPALS [129]. The removal of outliers and the appropriate selection of LVs are critical for constructing a meaningful regression model. If fewer LVs are selected than necessary, valuable information is excluded, resulting in an underfit model. Conversely, selecting too many LVs leads to overfitting, where redundant variance is included, making the model accurate only under the limited conditions defined during its construction [122].

Another reason for the broad applicability of PLS as a regression model is the use of informative vectors that provide chemical meaning within the model. A commonly used approach in recent years is the variable importance in projection (VIP) scores, which evaluate the significance of each variable as chemical information based on absolute values using the weights in the model's projection [102, 122].

Other regression methodologies such as binary logistic regression (BLR) [36] and RReliefF analysis [39] have also been applied to GC×GC-MS data, as reported in Tables 5.1 and 5.2. Different calibration methodologies can follow either linear or nonlinear approaches. For example, while linear PLS expresses the relationship between *X* and *y* in a linear manner, nonlinear PLS or kernel PLS assumes a nonlinear relationship [130]. To evaluate the quality of a model, metrics such as root mean square error of cross-validation (RMSECV), root mean square error of calibration (RMSEC), root mean square error of prediction (RMSEP), systematic error (Bias), and percentage error of prediction (%RE) are commonly used to assess model reliability [101, 122, 130].

An application example of PLS regression using GC×GC-MS data can be found in the study of aroma profiles in peach juice samples and the effects of thermal sterilization [64]. The research investigated the impact of three different heat sterilization treatments on the volatile profile changes in flat peach juice using GC-MS, GC×GC-TOFMS, and sensory analysis. Among the 175 compounds identified, 26 were classified as aroma-active substances based on their relative odor activity values (rOAVs) being higher than one. The goal of PLS regression was to correlate sensory attributes ($n = 6$: cooked, fresh, grassy, off-odor, peach-like, and sweet), aroma profiles (26 compounds), and sterilization methodologies ($n = 3$). The model included two components, PC1 and PC2, which accounted for 100% of the variance in the aroma-active compounds matrix (*X*) and 91% in the juices and sterilization matrix (Y). The authors described the model as effective, identifying relationships between the compounds and the sensory profile. The PLS model's effectiveness was demonstrated using a correlation loading plot and VIP values higher than one in descending order for the following compounds: hexanal, (*E*)-2-hexenal, acetaldehyde, 2-methyl-butanal, (*Z*)-3-hexenyl acetate, and (*Z*)-3-hexenal.

5.4.7 Classification/Discrimination

Depending on the research goal, qualitative strategies may be necessary. In chemometrics and machine learning, classification methods encompass a broad range of approaches, as outlined in Table 5.1. These methods aim to classify or discriminate new objects into predefined classes established by the constructed model. Thus, systematic model construction relies on a supervised training strategy, selecting compounds present in specific regions of the multivariate space [122]. Typically, before building a prediction model, pattern recognition is applied to assess data behavior. The diversity of classification methods arises from the need to represent class separation within the multivariate space and characterize the prediction power based o7n sample characteristics. To illustrate their applicability in GC×GC-MS, two studies comparing different discriminant methods are presented in the following paragraphs.

In the first study [108], SVOCs present in recycled and virgin forms of poly(ethylene terephthalate) (PET) were analyzed. Since PET is used in food packaging, careful consideration is needed because the presence of contaminants in recycled PET could pose health risks. To discriminate between new and recycled PET samples, 105 decontaminated PET samples were analyzed using direct sample introduction (DSI) and GC×GC-QTOFMS. The comparison involved various classification methods, including orthogonal partial least squares discriminant analysis (OPLS-DA), linear discriminant analysis (LDA), ANN, DT, RF, and support vector machine (SVM). The SVOCs were examined from an untargeted perspective, revealing a total of 267 compounds for discrimination. The first chemometric method applied was OPLS-DA along with nonparametric tests. OPLS-DA uses orthogonal signal correction (OSC) and PLS to remove irrelevant information from the X and Y matrices. It showed a high predictive capability, with a Q^2 value of 0.937 and an explained variance of Y (R^2Y) of 0.971, indicating that the model was well-predicted and well-fitted (as good prediction models typically have Q^2 and R^2Y values higher than 0.5). A total of 46 markers were selected based on VIP scores greater than 1 and the Kruskal–Wallis test ($p < 0.05$). Of these, 14 compounds were classified as "high-level markers" with VIP > 2 and $p < 0.05$. The other discriminant methods were tested using three datasets: all discriminated compounds, the 46 markers, and the 14 high-level markers. The prediction accuracy of the calibration and validation sets was tested using random, systematic, and stratified sampling methodologies. OPLS-DA produced the best results, achieving 100% prediction accuracy for all three datasets. For ANN, DT, LDA, RF, and SVM, prediction accuracy for the high-level markers set was approximately 95%. This study demonstrated robust results for discrimination and highlighted its potential use for detecting PET adulteration.

The second study compared different classification models to evaluate machine learning strategies for GC×GC data [131]. The dataset comprised volatile compounds from six wine varieties, acquired using headspace-SPME (HS-SPME) and analyzed by GC×GC-TOFMS. The classification methods compared included DT, DA, ensemble methods, SVM, and kNN. Leave-one-out cross-validation was applied across various models, totaling 17 methods in the study. The best methods achieved 96–100% accuracy for different class pairs (e.g., pairwise comparison of wine varieties, all wine samples, production year, and wineries). The paper provided an extensive evaluation of discrimination methodologies, demonstrating the performance of each method for various variables in wine samples.

5.4.8 Validation

After going through all these steps, the ultimate goal for an analyst is to have reliable parameters for a well-adjusted model. After its construction, figures of merit (FOM) are used to validate the model and numerically represent how well it performs [101, 124, 132]. These metrics can take different forms depending on whether the model is used for qualitative or quantitative supervised analysis. For example, in the context of cross-validation, the root mean square error of cross-validation (RMSECV), defined by Eq. (5.6), plays a key role when selecting the number of LVs in PLS-DA. The optimal number of LVs is determined by identifying the lowest RMSECV value or finding a threshold in the RMSECV versus LV plot [122]:

$$\text{RMSECV} = \sqrt{\frac{\sum_{i=1}^{n} (y_i - \hat{y}_i)^2}{n_c}}, \tag{5.6}$$

where y_i is the reference value, $\hat{y}_i$ is the predicted value, and n_c is the sample number in the calibration set.

Analogous to RMSECV, the root mean square error of calibration (RMSEC) is calculated using calibration samples, while the root mean square error of prediction (RMSEP) is computed using validation samples to evaluate the model after the calibration step. RMSEC, RMSEP, and RMSECV are key parameters for assessing the performance of supervised models.

In qualitative approaches, the predictive capability is evaluated through metrics that quantify the accuracy of classification. In a binary prediction system, the

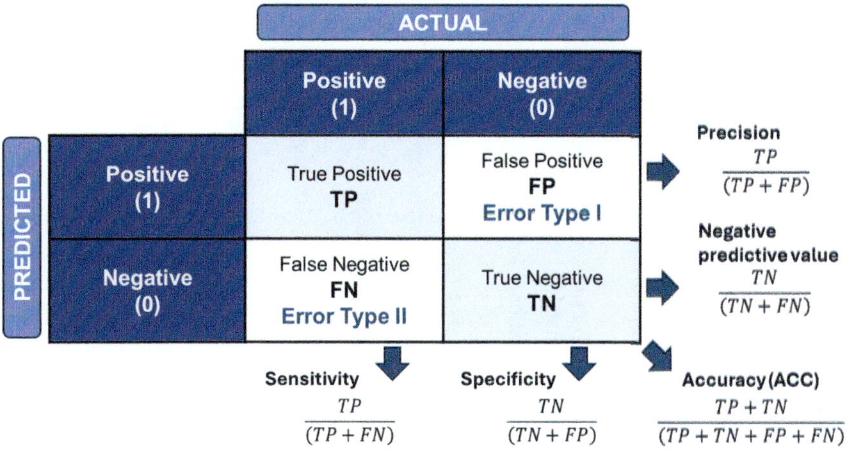

Figure 5.17 Confusion matrix representation and the equations of precision, negative predictive value, sensitivity, specificity, and accuracy.

possible outcomes are represented by 1 for positive predictions and 0 for negative predictions. The model's class prediction performance is visualized using a confusion matrix, as shown in Figure 5.17 [133].

If samples from class 1 are correctly classified into the predicted class, they are categorized as true positives (TP). Similarly, if samples from class 0 are correctly predicted, they are considered true negatives (TN). Two types of errors can occur in this matrix: Type I errors, characterized by false positives (FP), where class 0 samples are incorrectly predicted as class 1; and Type II errors, characterized by false negatives (FN), where class 1 samples are incorrectly predicted as class 0. Figure 5.17 also provides equations for FOM such as precision, negative predictive value, sensitivity (ST), specificity (SP), and accuracy (ACC). In addition to these metrics, false positive rate (FPR) and false negative rate (FNR), defined by Eqs. (5.7) and (5.8), can also be calculated using the values from the confusion matrix [122, 133],

$$\mathrm{FPR} = \frac{\mathrm{FP}}{\mathrm{FP} + \mathrm{TN}}, \tag{5.7}$$

$$FNR = \frac{FN}{FN + TP}. \tag{5.8}$$

Precision metrics used in qualitative validation include accordance (ACO) and concordance (CON) [134], as defined by Eqs. (5.9) and (5.10), respectively. ACO is used to estimate the probability that identical samples from the same batch or laboratory will yield consistent results. In contrast, CON estimates the probability that

Table 5.3 Figures of merit, their expression, and information applied to quantitative analytical validation.

Figure of merit	Expression	Information
Sensitivity (SEN)	$\frac{1}{\|\|b\|\|}$	Gives the partial response of an analytical variable responsible for an increment of a unit of analyte concentration
Analytical sensitivity (SEN_A)	$\frac{SEN}{\delta x}$	A relative response of sensitivity correcting variations from the utilized analytical technique
Variance of the prediction errors ($V(PE_i)$)	$\left(\frac{1}{n_c} + h_i\right) MSEC$	Calculation of an estimative confidence interval for unknown samples
Limit of detection (LoD)	$3.3\delta x\|\|b\|\| = 3.3\delta x\frac{1}{SEN}$	Multivariate calculation of LoD
Bias	$\sum_{i=1}^{n_v}\frac{(y_i^{ref}-\hat{y}_i)}{n_v}$	Amount of the systematic errors in a model
Standard deviation of validation (SDV)	$\sqrt{\frac{\sum[(y_i^{ref}-\hat{y}_i)-bias]^2}{n_v-1}}$	Student's t test estimative of statistical bias significance
Residual prediction deviation of calibration (RPD_{cal})	$\frac{SD_{cal}}{RMSECV}$	Natural variation ratio related to the samples and $RMSE_{CV}$
Residual prediction deviation of validation (RPD_{val})	$\frac{SD_{val}}{RMSEP}$	Natural variation ratio related to the samples and $RMSE_P$

$\|\|b\|\|$ = magnitude of the vector of regression coefficients; δx = estimative of instrumental noise; n_c = number of calibration samples; h_i = leverage; MSEC = mean square error of calibration; y_i^{ref} = reference value; $\hat{y}_i$ = predicted value; n_v = number of validation samples; SD_{cal} = standard deviation of reference values for calibration set; SD_{val} = standard deviation of reference values for validation set.

samples will produce the same results across different batches and laboratories,

$$\text{ACO} = \frac{\{k(k-1) + (n-k)(n-k-1)\}}{n(n-1)}, \tag{5.9}$$

$$\text{CON} = \frac{2[k(k-n.b)] + n.b(n.b-1) - \text{ACO}[n.b(n-1)]}{n^2.b(b-1)}, \tag{5.10}$$

where k means the concordant results, n means the sample number present in the set, and b is the number of laboratories and batches.

In quantitative methods, validation involves a variety of FOM. A summarized version of some key metrics, along with their expressions and additional information, is provided in Table 5.3. A more detailed explanation of these figures can be found in other sources [122].

5.5 Conclusions and Perspectives

Data analysis is an essential step in achieving reliable results. With advancements in computational instrumentation, along with the information-rich datasets generated by GC(×GC)-MS, the strategic extraction of significant chemical information has become feasible through chemometrics. This chapter outlines a brief methodology for accessing chemical insights. Selecting the appropriate method is crucial, and when paired with the adequate data analysis strategy, it enables the development of representative and predictive models. Depending on the desired outcomes, various unsupervised and supervised models can be applied to GC(×GC)-MS data, as demonstrated throughout the chapter. Naturally, the evolution of new methodologies and applications is closely linked to advances in computational tools and analytical instrumentation, with new algorithms being continuously developed to facilitate data analysis [135–137]. In the context of HR MS, such large datasets pose new challenges and have yet to be addressed, highlighting the need for improved chemometric workflows.

References

1. Bean, H.D., Hill, J.E., and Dimandja, J.-M.D. (2015). Improving the quality of biomarker candidates in untargeted metabolomics via peak table-based alignment of comprehensive two-dimensional gas chromatography–mass spectrometry data. *J. Chromatogr. A* 1394: 111–117.
2. James, A.T. and Martin, A.J.P. (1952). Gas-liquid partition chromatography: the separation and micro-estimation of volatile fatty acids from formic acid to dodecanoic acid. *Biochem. J.* 50 (5): 679–690.

3 Martin, A.J.P. and Synge, R.L.M. (1941). A new form of chromatogram employing two liquid phases. *Biochem. J.* 35 (12): 1358–1368.

4 Gohlke, R.S. (1959). Time-of-flight mass spectrometry and gas-liquid partition chromatography. *Anal. Chem.* 31 (4): 535–541.

5 Bartle, K.D. and Myers, P. (2002). History of gas chromatography. *TrAC Trends Anal. Chem.* 21 (9–10): 547–557.

6 Kováts, E. (1958). Gas-chromatographische charakterisierung organischer verbindungen. Teil 1: retentionsindices aliphatischer halogenide, alkohole, aldehyde und ketone. *Helv. Chim. Acta* 41 (7): 1915–1932.

7 van Den Dool, H. and Kratz, P.D. (1963). A generalization of the retention index system including linear temperature programmed gas—liquid partition chromatography. *J. Chromatogr. A* 11: 463–471.

8 Cuykx, M., Rodrigues, R.M., Laukens, K. et al. (2018). in vitro assessment of hepatotoxicity by metabolomics: a review. *Arch. Toxicol.* 92 (10): 3007–3029.

9 Schymanski, E.L., Jeon, J., Gulde, R. et al. (2014). Identifying small molecules via high resolution mass spectrometry: communicating confidence. *Environ. Sci. Technol.* 48 (4): 2097–2098.

10 Sumner, L.W., Amberg, A., Barrett, D. et al. (2007). Proposed minimum reporting standards for chemical analysis. *Metabolomics* 3 (3): 211–221.

11 Liu, Z. and Phillips, J.B. (1991). Comprehensive two-dimensional gas chromatography using an on-column thermal modulator interface. *J. Chromatogr. Sci.* 29 (6): 227–231.

12 Schöneich, S., Cain, C.N., Sudol, P.E., and Synovec, R.E. (2023). Enabling cuboid-based fisher ratio analysis using total-transfer comprehensive three-dimensional gas chromatography with time-of-flight mass spectrometry. *J. Chromatogr. A* 1708: 464341.

13 Ledford, E.B., Billesbach, C.A., and Zhu, Q. (2000). GC3: Comprehensive three-dimensional gas chromatography. *J. High Resolut. Chromatogr.* 23 (3): 205–207.

14 Yan, D., Tedone, L., Koutoulis, A. et al. (2017). Parallel comprehensive two-dimensional gas chromatography. *J. Chromatogr. A* 1524: 202–209.

15 Tranchida, P.Q., Purcaro, G., Dugo, P. et al. (2011). Modulators for comprehensive two-dimensional gas chromatography. *TrAC Trends Anal. Chem.* 30 (9): 1437–1461.

16 Kinghorn, R.M. and Marriott, P.J. (1998). Comprehensive two-dimensional gas chromatography using a modulating cryogenic trap. *J. High Resolut. Chromatogr.* 21: 620–622.

17 Bueno, P.A. and Seeley, J.V. (2004). Flow-switching device for comprehensive two-dimensional gas chromatography. *J. Chromatogr. A* 1027 (1–2): 3–10.

18 Luong, J., Guan, X., Xu, S. et al. (2016). Thermal independent modulator for comprehensive two-dimensional gas chromatography. *Anal. Chem.* 88 (17): 8428–8432.

19 Beens, J., Dallüge, J., Adahchour, M. et al. (2001). Moving cryogenic modulator for the comprehensive two-dimensional gas chromatography (GC×GC) of surface water contaminants. *J. Microcolumn Sep.* 13 (3): 134–140.

20 Stefanuto, P.-H., Smolinska, A., and Focant, J.-F. (2021). Advanced chemometric and data handling tools for GC×GC-TOF-MS. *TrAC Trends Anal. Chem.* 139: 116251.

21 Zavahir, J.S., Nolvachai, Y., and Marriott, P.J. (2018). Molecular spectroscopy – information rich detection for gas chromatography. *TrAC Trends Anal. Chem.* 99: 47–65.

22 Zubarev, R.A. and Makarov, A. (2013). Orbitrap mass spectrometry. *Anal. Chem.* 85 (11): 5288–5296.

23 Peterson, A.C., Hauschild, J.-P., Quarmby, S.T. et al. (2014). Development of a GC/Quadrupole-Orbitrap mass spectrometer, part I: design and characterization. *Anal. Chem.* 86 (20): 10036–10043.

24 Pollo, B.J., Teixeira, C.A., Belinato, J.R. et al. (2021). Chemometrics, comprehensive two-dimensional gas chromatography and "omics" sciences: basic tools and recent applications. *TrAC Trends Anal. Chem.* 134: 116111.

25 Pollo, B.J., Alexandrino, G.L., Augusto, F., and Hantao, L.W. (2018). The impact of comprehensive two-dimensional gas chromatography on oil & gas analysis: recent advances and applications in petroleum industry. *TrAC Trends Anal. Chem.* 105: 202–217.

26 Yang, X., Zeng, P., Wen, J. et al. (2023). *Gain deeper insights into traditional Chinese medicines using multidimensional chromatography combined with chemometric approaches*, vol. 16, 27–41. *Chin Herb Med.*

27 Sudheeshna, M., Malarvannan, M., Kumar, K.V. et al. (2023). ^{2}D chromatography and ^{2}D spectroscopy in analytical chemistry: an overview. *J. Anal. Chem.* 78 (9): 1213–1230.

28 Trinklein, T.J., Cain, C.N., Ochoa, G.S. et al. (2023). Recent advances in GC×GC and chemometrics to address emerging challenges in nontargeted analysis. *Anal. Chem.* 95 (1): 264–286.

29 Zaid, A., Khan, M.S., Yan, D. et al. (2022). Comprehensive two-dimensional gas chromatography with mass spectrometry: an advanced bioanalytical technique for clinical metabolomics studies. *Analyst* 147 (18): 3974–3992.

30 Feizi, N., Hashemi-Nasab, F.S., Golpelichi, F. et al. (2021). Recent trends in application of chemometric methods for GC-MS and GC×GC-MS-based metabolomic studies. *TrAC Trends Anal. Chem.* 138: 116239.

31 Nolvachai, Y., Amaral, M.S.S., and Marriott, P.J. (2023). Foods and contaminants analysis using multidimensional gas chromatography: an update of recent studies, technology, and applications. *Anal. Chem.* 95 (1): 238–263.

32 Rasheed, D.M., Serag, A., Shakour, Z.T.A., and Farag, M. (2021). Novel trends and applications of multidimensional chromatography in the analysis of food, cosmetics and medicine bearing essential oils. *Talanta* 223: 121710.

33 Stilo, F., Bicchi, C., Jimenez-Carvelo, A.M. et al. (2021). Chromatographic fingerprinting by comprehensive two-dimensional chromatography: fundamentals and tools. *TrAC Trends Anal. Chem.* 134: 116133.

34 Sudol, P.E., Pierce, K.M., Prebihalo, S.E. et al. (2020). Development of gas chromatographic pattern recognition and classification tools for compliance and forensic analyses of fuels: a review. *Anal. Chim. Acta* 1132: 157–186.

35 Schöneich, S., Ochoa, G.S., Monzón, C.M., and Synovec, R.E. (2022). Minimum variance optimized Fisher ratio analysis of comprehensive two-dimensional gas chromatography / mass spectrometry data: study of the pacu fish metabolome. *J. Chromatogr. A* 1667: 462868.

36 Makhubela, P.C.K., Rohwer, E.R., and Naudé, Y. (2023). Detection of tuberculosis-associated compounds from human skin by GCxGC-TOFMS. *J. Chromatogr. B* 1231: 123937.

37 Castiblanco, J.E.B., Ferreira, V.H.C., Teixeira, C.A., and Hantao, L.W. (2024). Classification of produced water samples using class-oriented chemometrics and comprehensive two-dimensional gas chromatography coupled to mass spectrometry. *Talanta* 268: 125343.

38 Lübeck, J.S., Alexandrino, G.L., and Christensen, J.H. (2020). GC×GC–HRMS nontarget fingerprinting of organic micropollutants in urban freshwater sediments. *Environ. Sci. Eur.* 32 (1): 78.

39 Cain, C.N., Ochoa, G.S., and Synovec, R.E. (2023). Enhancing partial least squares modeling of comprehensive two-dimensional gas chromatography time-of-flight mass spectrometry data by tile-based variance ranking. *J. Chromatogr. A* 1694: 463920.

40 Furuta, K., Byrne, J., Luat, K. et al. (2024). Volatile organic compounds produced during postmortem processes can be linked via chromatographic profiles to individual postmortem bacterial species. *J. Chromatogr. A* 1728: 465017.

41 Wang, J., Lu, X., Zhang, Z. et al. (2024). Application of chemical attribution in matching OPNAs-exposed biological samples with exposure sources- based on the impurity profiles via GC×GC-TOFMS analysis. *J. Chromatogr. A* 1718: 464718.

42 Dong, B., Wu, X., Wu, S. et al. (2023). Occurrence of volatile contaminants in recycled poly(ethylene terephthalate) by HS-SPME-GC×GC-QTOF-MS combined with chemometrics for authenticity assessment of geographical recycling regions. *J. Hazard. Mater.* 445: 130407.

43 Wu, S., Wu, X., Li, H. et al. (2022). The characterization and influence factors of semi-volatile compounds from mechanically recycled polyethylene terephthalate (rPET) by combining GC×GC-TOFMS and chemometrics. *J. Hazard. Mater.* 439: 129583.

44 Zheng, X., Zhang, S., Ai, B. et al. (2024). Adulteration identification of Hainan camellia (*Camellia oleifera* Abel.) oil based on comprehensive two-dimensional gas chromatography and quadrupole time-of-flight mass coupled with chemometrics spectrometry. *Int. J. Food Sci. Technol.* 59 (4): 2305–2317.

45 Zou, Y., Gaida, M., Franchina, F.A. et al. (2022). Distinguishing between decaffeinated and regular coffee by HS-SPME-GC×GC-TOFMS, chemometrics, and machine learning. *Molecules* 27 (6): 1806.

46 Bhatt, K., Orlando, T., Meuwis, M.-A. et al. (2023). Comprehensive insight into colorectal cancer metabolites and lipids for human serum: a proof-of-concept study. *Int. J. Mol. Sci.* 24 (11): 9614.

47 Bhatt, K., Dejong, T., Dubois, L.M. et al. (2022). Lipid serum profiling of boar-tainted and untainted pigs using GC×GC–TOFMS: an exploratory study. *Meta* 12 (11): 1111.

48 Wenk, M.R. (2005). The emerging field of lipidomics. *Nat. Rev. Drug Discov.* 4 (7): 594–610.

49 Lin, Y., Wang, Y., Huang, Y. et al. (2023). Aroma identification and classification in 18 Kinds of teas (*Camellia sinensis*) by sensory evaluation, HS-SPME-GC-IMS/GC×GC-MS, and chemometrics. *Foods* 12 (13): 2433.

50 Resende, G.A.P., Amaral, M.S.S., Botelho, B.G., and Marriott, P.J. (2024). Comprehensive two-dimensional gas chromatography–mass spectrometry as a tool for the untargeted study of hop and their metabolites. *Meta* 14 (4): 237.

51 Valdés, A., Álvarez-Rivera, G., Socas-Rodríguez, B. et al. (2022). Foodomics: analytical opportunities and challenges. *Anal. Chem.* 94 (1): 366–381.

52 Beccaria, M., Bobak, C., Maitshotlo, B. et al. (2018). Exhaled human breath analysis in active pulmonary tuberculosis diagnostics by comprehensive gas chromatography-mass spectrometry and chemometric techniques. *J. Breath Res.* 13 (1): 16005.

53 Llambrich, M., Brezmes, J., and Cumeras, R. (2022). The untargeted urine volatilome for biomedical applications: methodology and volatilome database. *Biol. Proced. Online* 24 (1): 20.

54 Benetti, E., Liberto, E., Bressanello, D. et al. (2020). Sedentariness and urinary metabolite profile in type 2 diabetic patients, a cross-sectional study. *Meta* 10 (5): 205.

55 Pesesse, R., Stefanuto, P.-H., Schleich, F. et al. (2019). Multimodal chemometric approach for the analysis of human exhaled breath in lung cancer patients by TD-GC×GC-TOFMS. *J. Chromatogr. B* 1114–1115: 146–153.

56 Facanali, R., Marques, M.O.M., and Hantao, L.W. (2020). Metabolic profiling of *Varronia curassavica* Jacq. terpenoids by flow modulated two-dimensional gas chromatography coupled to mass spectrometry. *Separations* 7 (1): 18.

57 Darnal, A., Poggesi, S., Longo, E. et al. (2023). Decoding the identity of Pinot Gris and Pinot Noir wines: a comprehensive chemometric fusion of sensory (from dual panel) and chemical analysis. *Foods* 13 (1): 18.

58 Kalogiouri, N.P., Manousi, N., Ferracane, A. et al. (2024). A novel headspace solid-phase microextraction arrow method employing comprehensive two-dimensional gas chromatography–mass spectrometry combined with chemometric tools for the investigation of wine aging. *Anal. Chim. Acta* 1304: 342555.

59 Koljančić, N., Gomes, A.A., and Špánik, I. (2023). A non-target geographical origin screening of botrytized wines through comprehensive two-dimensional gas chromatography coupled with high-resolution mass spectrometry. *J. Sep. Sci.* 46 (19).

60 Lukić, I., Carlin, S., and Vrhovsek, U. (2022). Utility of comprehensive GC×GC gas chromatography in finding varietal markers among volatile compounds in non-aromatic red wines. *Agronomy* 12 (10): 2512.

61 Perestrelo, R., Silva, C., Gonçalves, C. et al. (2020). An approach of the madeira wine chemistry. *Beverages* 6 (1): 12.

62 Zhang, P., Piergiovanni, M., Franceschi, P. et al. (2023). Application of comprehensive ^{2}D gas chromatography coupled with mass spectrometry in beer and wine VOC analysis. *Analytica* 4 (3): 347–373.

63 Paiva, A.C., Simões Oliveira, D., and Hantao, L.W. (2019). A Bottom-up approach for data mining in bioaromatization of beers using flow-modulated comprehensive two-dimensional gas chromatography/mass spectrometry. *Separations* 6 (4): 46.

64 Huang, J., He, W., Wang, P. et al. (2024). Characterization of aroma profiles in flat peach juice post-thermal sterilization: insights from comprehensive two-dimensional GC with time-of-flight mass spectrometry (GC×GC-TOF-MS), sensory analysis, and chemometric approaches. *J. Food Compos. Anal.* 133: 106398.

65 Li, M., Zhao, Z., Zhang, Y. et al. (2024). Chemometrics combined with comprehensive two-dimensional gas chromatography-mass spectrometry for the identification of Baijiu vintage. *Food Chem.* 444: 138690.

66 Ren, J., Li, Z., and Jia, W. (2024). Key aroma differences in volatile compounds of aged Feng-flavored Baijiu determined using sensory descriptive analysis and GC×GC–TOFMS. *Foods* 13 (10): 1504.

67 Song, X., Wang, G., Zhu, L. et al. (2021). Comparison of two cooked vegetable aroma compounds, dimethyl disulfide and methional, in Chinese Baijiu by a sensory-guided approach and chemometrics. *LWT* 146: 111427.

68 Cain, C.N., Gaida, M., Stefanuto, P.-H. et al. (2024). Investigating sensory-classified roasted arabica coffee with GC×GC-TOFMS and chemometrics to understand potato taste defect. *Microchem. J.* 196: 109578.

69 Eggermont, D., Spadafora, N.D., Aspromonte, J., and Purcaro, G. (2023). Unraveling the impact of the capsule material on the aroma of brewed coffee by headspace analysis using a HiSorb probe followed by reverse fill/flush flow modulation GC×GC-MS. *Anal. Bioanal. Chem.* 415 (13): 2511–2521.

70 Lopes, G.R., Petronilho, S., Ferreira, A.S. et al. (2021). Insights on single-dose espresso coffee capsules' volatile profile: from ground powder volatiles to prediction of espresso brew aroma properties. *Foods* 10 (10): 2508.

71 Hibbert, D.B. (2012). Experimental design in chromatography: a tutorial review. *J. Chromatogr. B* 910: 2–13.

72 Brereton, R., Jenson, J., Lopes, J. et al. (2017). Chemometrics in analytical chemistry—part I: history, experimental design and data analysis tools. *Anal. Bioanal. Chem.* 409: 5891–5899.

73 Rubin, D.B. (1976). Inference and missing data. *Biometrika* 63 (3): 581–592.

74 Davis, T.J., Firzli, T.R., Higgins Keppler, E.A. et al. (2022). Addressing missing data in GC×GC metabolomics: identifying missingness type and evaluating the impact of imputation methods on experimental replication. *Anal. Chem.* 94 (31): 10912–10920.

75 Picornell, A., Oteros, J., Ruiz-Mata, R. et al. (2021). Methods for interpolating missing data in aerobiological databases. *Environ. Res.* 200: 111391.

76 Walczak, B. and Massart, D.L. (2001). Dealing with missing data. *Chemom. Intell. Lab. Syst.* 58 (1): 15–27.

77 Castillo, S., Mattila, I., Miettinen, J. et al. (2011). Data analysis tool for comprehensive two-dimensional gas chromatography/time-of-flight mass spectrometry. *Anal. Chem.* 83 (8): 3058–3067.

78 Savitzky, A. and Golay, M.J.E. (1964). Smoothing and differentiation of data by simplified least squares procedures. *Anal. Chem.* 36 (8): 1627–1639.

79 Davis, J.T., Beaux, M.F., and Freye, C.E. (2023). Evaluation of different getter substrates using two-dimensional gas chromatography with time of flight mass spectrometry. *J. Chromatogr. A* 1689: 463760.

80 Gough, D.V., Schöneich, S., and Synovec, R.E. (2020). Chemometric decomposition of comprehensive two-dimensional gas chromatography time-of-flight mass spectrometry data employing partial modulation in the negative pulse mode. *Talanta* 210: 120670.

81 Adutwum, L.A., Kwao, J.K., and Harynuk, J.J. (2021). Unique ion filter—a data reduction tool for chemometric analysis of raw comprehensive two-dimensional gas chromatography-mass spectrometry data. *J. Sep. Sci.* 44 (14): 2773–2784.

82 Oliva, D.J., Cheung, C., and Perrault, K.A. (2023). Fisher ratio feature selection by manual peak area calculations on comprehensive two-dimensional gas chromatography data using standard mixtures with variable composition, storage, and interferences. *Anal. Bioanal. Chem.* 415 (13): 2575–2585.

83 Peng, J., Peng, S., Jiang, A. et al. (2010). Asymmetric least squares for multiple spectra baseline correction. *Anal. Chim. Acta* 683 (1): 63–68.

84 Quiroz-Moreno, C., Furlan, M.F., Belinato, J.R. et al. (2020). RGCxGC toolbox: an R-package for data processing in comprehensive two-dimensional gas chromatography-mass spectrometry. *Microchem. J.* 156: 104830.

85 Mani-Varnosfaderani, A., Kanginejad, A., Gilany, K., and Valadkhani, A. (2016). Estimating complicated baselines in analytical signals using the iterative training of Bayesian regularized artificial neural networks. *Anal. Chim. Acta* 940: 56–64.

86 Furbo, S., Hansen, A.B., Skov, T., and Christensen, J.H. (2014). Pixel-based analysis of comprehensive two-dimensional gas chromatograms (color plots) of petroleum: a tutorial. *Anal. Chem.* 86 (15): 7160–7170.

87 Zhang, D., Huang, X., Regnier, F.E., and Zhang, M. (2008). Two-dimensional correlation optimized warping algorithm for aligning GC×GC–MS data. *Anal. Chem.* 80 (8): 2664–2671.

88 Tomasi, G., Savorani, F., and Engelsen, S.B. (2011). icoshift: An effective tool for the alignment of chromatographic data. *J. Chromatogr. A* 1218 (43): 7832–7840.

89 Zushi, Y., Gros, J., Tao, Q. et al. (2017). Pixel-by-pixel correction of retention time shifts in chromatograms from comprehensive two-dimensional gas chromatography coupled to high resolution time-of-flight mass spectrometry. *J. Chromatogr. A* 1508: 121–129.

90 Alexandrino, G.L., Tomasi, G., Kienhuis, P.G.M. et al. (2019). Forensic investigations of diesel oil spills in the environment using comprehensive two-dimensional gas chromatography–high resolution mass spectrometry and chemometrics: new perspectives in the absence of recalcitrant biomarkers. *Environ. Sci. Technol.* 53 (1): 550–559.

91 Ferreira, V.H.C., Hantao, L.W., and Poppi, R.J. (2020). Consumable-free comprehensive three-dimensional gas chromatography and PARAFAC for determination of allergens in perfumes. *Chromatographia* 83 (5): 581–592.

92 Boegelsack, N., Walker, J., Sandau, C.D. et al. (2023). Cross-contamination of ignitable liquid residues on wildfire debris—detection and characterization in matrices commonly encountered at wildfire scenes. *Separations* 10 (9): 491.

93 Gromski, P., Xu, Y., Kotze, H. et al. (2014). Influence of missing values substitutes on multivariate analysis of metabolomics data. *Meta* 4 (2): 433–452.

94 Dieterle, F., Ross, A., Schlotterbeck, G., and Senn, H. (2006). Probabilistic quotient normalization as robust method to account for dilution of complex biological mixtures. Application in 1H NMR metabolomics. *Anal. Chem.* 78 (13): 4281–4290.

95 Boegelsack, N., Sandau, C., McMartin, D.W. et al. (2021). Development of retention time indices for comprehensive multidimensional gas chromatography and application to ignitable liquid residue mapping in wildfire investigations. *J. Chromatogr. A* 1635: 461717.

96 van den Berg, R.A., Hoefsloot, H.C., Westerhuis, J.A. et al. (2006). Centering, scaling, and transformations: improving the biological information content of metabolomics data. *BMC Genomics* 7 (1): 142.

97 Kennard, R.W. and Stone, L.A. (1969). Computer aided design of experiments. *Technometrics* 11 (1): 137–148.

98 Snee, R.D. (1977). Validation of regression models: methods and examples. *Technometrics* 19 (4): 415–428.

99 Wu, W., Walczak, B., Massart, D.L. et al. (1996). Artificial neural networks in classification of NIR spectral data: design of the training set. *Chemom. Intell. Lab. Syst.* 33 (1): 35–46.

100 Rajer-Kanduč, K., Zupan, J., and Majcen, N. (2003). Separation of data on the training and test set for modelling: a case study for modelling of five colour properties of a white pigment. *Chemom. Intell. Lab. Syst.* 65 (2): 221–229.

101 Westad, F. and Marini, F. (2015). Validation of chemometric models – a tutorial. *Anal. Chim. Acta* 893: 14–24.

102 Wise, B.M., Gallagher, N.B., Bro, R. et al. (2006). *Chemometrics Tutorial for PLS_Toolbox and Solo*. Wenatchee, WA, USA: Eigenvector Research, Inc.

103 Fulgêncio, A.C.C., Resende, G.A.P., and Botelho, B.G. (2022). Chemometrics fundamentals and its application in infrared spectroscopy. In: *FTIR Spectroscopy: Advances in Research and Applications* (ed. A.S. França and L.S. Oliveira), 1–26. Nova Science Publishers.

104 Bowman, D.T., Warren, L.A., McCarry, B.E., and Slater, G.F. (2019). Profiling of individual naphthenic acids at a composite tailings reclamation fen by comprehensive two-dimensional gas chromatography-mass spectrometry. *Sci. Total Environ.* 649: 1522–1531.

105 Boegelsack, N., Walker, J., Sandau, C.D. et al. (2024). Cross-contamination of ignitable liquid residues on wildfire debris—effects of packaging and storage on detection and characterization. *Separations* 11 (2): 58.

106 Nam, S.L., Carrillo, K.T., de la Mata, A.P., and Harynuk, J.J. (2023). Untargeted metabolomic profiling of aqueous and lyophilized pooled human feces from two diet cohorts using two-dimensional gas chromatography coupled with time-of-flight mass spectrometry. *Meta* 13 (7): 828.

107 Yin, H., Lu, J., Liu, G. et al. (2021). Application of chemometrics for coal pyrolysis products by online py-GC×GC–MS. *ACS Omega* 6 (5): 3763–3770.

108 Hao, T.-Y., Xu, X., Lin, Q.-B. et al. (2022). Rapid discrimination of recycled and virgin poly(ethylene terephthalate) based on non-targeted screening of semi-volatile organic compounds using a novel method of DSI/GC×GC-Q-TOF-MS coupled with various chemometrics. *Food Packag. Shelf Life* 34: 100978.

109 Lima, R.A.M., Ferraz, S.M.M., Cardoso, V.G.K. et al. (2023). Authentication of fish oil (omega-3) supplements using class-oriented chemometrics and comprehensive two-dimensional gas chromatography coupled to mass spectrometry. *Anal. Bioanal. Chem.* 415 (13): 2601–2611.

110 Zhao, G., Yuan, Y., Zhou, H. et al. (2023). Determination of volatile compounds in different parts of grass carp using GC×GC-MS combined with chemometrics. *Food Biosci.* 56: 103403.

111 Prebihalo, S.E., Pinkerton, D.K., and Synovec, R.E. (2019). Impact of comprehensive two-dimensional gas chromatography time-of-flight mass spectrometry experimental design on data trilinearity and parallel factor analysis deconvolution. *J. Chromatogr. A* 1605: 460368.

112 Xu, H., Fan, X., Zhao, M.-H. et al. (2024). Exploring the molecular characteristics of organic matter in low-rank coals using GC×GC/TOF-MS plus data mining. *J. Anal. Appl. Pyrolysis* 181: 106605.

113 Guo, K., Zhang, Y., Zhang, H. et al. (2023). Odor changes in breast milk during different storage temperatures and times using GC×GC-O-MS. *Food Res. Int.* 168: 112792.

114 Mikaliunaite, L. and Synovec, R.E. (2022). Computational method for untargeted determination of cycling yeast metabolites using comprehensive two-dimensional gas chromatography time-of-flight mass spectrometry. *Talanta* 244: 123396.

115 Marney, L.C., Christopher Siegler, W., Parsons, B.A. et al. (2013). Tile-based Fisher-ratio software for improved feature selection analysis of comprehensive two-dimensional gas chromatography–time-of-flight mass spectrometry data. *Talanta* 115: 887–895.

116 Dalzell, C.J., Ventura, G.T., Walters, C.C. et al. (2021). Hydrocarbon transformations in sediments from the Cathedral Hill hydrothermal vent complex at Guaymas Basin, Gulf of California – a chemometric study of shallow seep architecture. *Org. Geochem.* 152: 104173.

117 Reaser, B.C., Wright, B.W., and Synovec, R.E. (2017). Using receiver operating characteristic curves to optimize discovery-based software with comprehensive two-dimensional gas chromatography with time-of-flight mass spectrometry. *Anal. Chem.* 89 (6): 3606–3612.

118 Ochoa, G.S., Prebihalo, S.E., Reaser, B.C. et al. (2020). Statistical inference of mass channel purity from Fisher ratio analysis using comprehensive two-dimensional gas chromatography with time of flight mass spectrometry data. *J. Chromatogr. A* 1627: 461401.

119 Chen, C., Chen, H., Zhang, Y. et al. (2020). TBtools: an integrative toolkit developed for interactive analyses of big biological data. *Mol. Plant* 13 (8): 1194–1202.

120 Pierce, K.M., Hoggard, J.C., Hope, J.L. et al. (2006). Fisher ratio method applied to third-order separation data to identify significant chemical components of metabolite extracts. *Anal. Chem.* 78 (14): 5068–5075.

121 Berrier, K.L., Prebihalo, S.E., and Synovec, R.E. (2020). Advanced data handling in comprehensive two-dimensional gas chromatography. *Sep. Sci. Technol.* 12: 229–268.

122 Sena, M.M., Almeida, M.R., Braga, J.W.B., and Poppi, R.J. (2017). Multivariate statistical analysis and chemometrics. In: *Spectroscopic Methods in Food Analysis*, 1e (ed. A.S. Franca and L.M.L. Nollet), 273–314. Boca Raton, FL: CRC Press.

123 Komsta, Ł., Heyden, Y.V., and Sherma, J. (2018). *Chemometrics in Chromatography*, Chromatographic Science Series. Boca Raton: CRC Press.

124 Brereton, R.G., Jansen, J., Lopes, J. et al. (2018). Chemometrics in analytical chemistry—part II: modeling, validation, and applications. *Anal. Bioanal. Chem.* 410 (26): 6691–6704.

125 Harshman, R.A. (1970). Foundations of the PARAFAC procedure: models and conditions for an "explanatory" multimodal factor analysis. *UCLA WPP* 16 (1): 1–84.

126 Carroll, J.D. and Chang, J.-J. (1970). Analysis of individual differences in multidimensional scaling via an n-way generalization of "Eckart-Young" decomposition. *Psychometrika* 35 (3): 283–319.

127 Bro, R. (1998) Multiway analysis in the food industry. Models, algorithms and applications. Universiteit van Amsterdam.

128 Izadmanesh, Y., Garreta-Lara, E., Ghasemi, J.B. et al. (2017). Chemometric analysis of comprehensive two dimensional gas chromatography–mass spectrometry metabolomics data. *J. Chromatogr. A* 1488: 113–125.

129 Wold, H. (1975). Path models with latent variables: the NIPALS approach. In: *Quantitative Sociology* (ed. H.M. Blalock, A. Aganbegian, F.M. Borodkin, et al.), 307–357. Academic Press.

130 Brendel, R., Schwolow, S., Rohn, S., and Weller, P. (2020). Comparison of PLSR, MCR-ALS and Kernel-PLSR for the quantification of allergenic fragrance compounds in complex cosmetic products based on nonlinear ^{2}D GC-IMS data. *Chemom. Intell. Lab. Syst.* 205: 104128.

131 Reichenbach, S.E., Zini, C.A., Nicolli, K.P. et al. (2019). Benchmarking machine learning methods for comprehensive chemical fingerprinting and pattern recognition. *J. Chromatogr. A* 1595: 158–167.

132 Olivieri, A.C., Faber, N.M., Ferré, J. et al. (2006). Uncertainty estimation and figures of merit for multivariate calibration (IUPAC Technical Report). *Pure Appl. Chem.* 78 (3): 633–661.

133 Tiwari, A. (2022). Supervised learning: from theory to applications. In: *Artificial Intelligence and Machine Learning for EDGE Computing* (ed. R. Pandey, S.K. Khatri, N.K. Singh, and P. Verma), 23–32. Academic Press.

134 Langton, S.D., Chevennement, R., Nagelkerke, N., and Lombard, B. (2002). Analysing collaborative trials for qualitative microbiological methods: accordance and concordance. *Int. J. Food Microbiol.* 79 (3): 175–181.

135 Yang, X., Sima, Y., Luo, X. et al. (2024). Analysis of GC×GC fingerprints from medicinal materials using a novel contour detection algorithm: a case of Curcuma wenyujin. *J. Pharm. Anal.* 14 (4): 100936.

136 Molenaar, S.R.A., Mommers, J.H.M., Stoll, D.R. et al. (2023). Algorithm for tracking peaks amongst numerous datasets in comprehensive two-dimensional chromatography to enhance data analysis and interpretation. *J. Chromatogr. A* 1705: 464223.

137 Gaida, M., Franchina, F.A., Stefanuto, P.-H., and Focant, J.-F. (2022). Top-down approach to retention time prediction in comprehensive two-dimensional gas chromatography–mass spectrometry. *Anal. Chem.* 94 (49): 17081–17089.

6

System Suitability and Quality Control in GC-MS and GC×GC-MS

Anaïs Rodrigues, Djulia Bensaada, Jean-François Focant, and Pierre-Hugues Stefanuto

Organic and Biological Analytical Chemistry Group, MolSys, University of Liège, Liège, Belgium

6.1 Introduction

Gas chromatography-mass spectrometry (GC-MS) and comprehensive two-dimensional (2D) GC×GC-MS are powerful analytical techniques used for identifying and quantifying compounds in complex mixtures [1]. Ensuring high-quality data in these methods is critical for accurate and reliable results. Quality assurance (QA) and quality control (QC) play a pivotal role in achieving this.

Reproducibility of results is a cornerstone of QA. Regular calibration and tuning in GC-MS confirm consistent detector response and ion source performance. For GC×GC-MS, the large data volume necessitates robust QC to ensure software algorithms for peak alignment and integration produce consistent results across runs. QC also helps detect instrumental errors early, such as column degradation, ion source contamination, or detector malfunctions, which can compromise separation efficiency, ionization efficiency, and peak intensities.

QC methods for targeted GC-MS and GC×GC-MS are well established in the literature and guidelines linked to regulated methods. Accuracy and reliability of results are fundamental, particularly in applications such as environmental monitoring, food safety, and clinical diagnostics, where incorrect results can lead to misinterpretation and costly consequences. Most targeted omics workflows have established QC protocols, using reference materials (RMs) to monitor the analytical workflow from sampling to data interpretation.

In GC-MS, single-dimension systems require validation for peak identification and quantification, while the added complexity of GC×GC-MS demands QC to maintain the integrity of second-dimension chromatography without loss of resolution or peak misalignment. In GC×GC-MS, the QC needs to consider the enhanced resolution and sensitivity, ensuring that complex multidimensional separations are reproducible and accurate. On the detector side, response factors need to be properly monitored to ensure consistency.

With MS, extra parameters need to be carefully monitored. For example, the ability to distinguish between compounds with similar molecular weights. High mass accuracy is crucial for correct molecular annotation, and maintaining a strong signal-to-noise ratio ensures the detectability of trace-level analytes. Finally, all these steps need to take place on a regular basis to ensure consistency over time.

6.1.1 Challenges in Non-targeted Analysis

The main challenge in the field of omics is the shift toward nontargeted analytical methods. The rapid advancements in instrument capabilities, particularly in resolution and sensitivity, have allowed researchers to investigate complex mixtures without predefining targets. Nontargeted screening offers immense potential for generating knowledge with a limited number of analytical runs. However, ensuring robustness and reliability in nontargeted methods has become a major challenge [2–5].

Unlike targeted methods, QC protocols for nontargeted screening are not well defined. Researcher initiatives, such as the metabolomics standards initiative and the Metabolomics QA and QC Consortium (mQACC), have been working to establish guidelines that provide guidance in the field [6, 7]. In high-throughput omics environments, QC minimizes batch-to-batch variability, ensuring consistent data quality. Well-documented QC procedures provide traceable records for audits and QA, enhancing credibility in research, allowing study replication, and ensuring compliance with regulatory environments. From an economic and efficiency perspective, identifying and addressing issues early through QC avoids expensive instrument downtime, wasted samples and reagents, and the need for time-consuming reanalysis of samples.

6.2 QA/QC Systems for Untargeted Metabolomics

In a study conducted in 2020, the mQACC examined QA/QC practices among 23 laboratories using liquid chromatography coupled to MS (LC-MS) for untargeted metabolomics [7]. Participants represented academia, industry, and government. The objective was to document common practices, identify discrepancies, and guide best practices in the metabolomics community.

Key findings included:

- **QA Practices:** 68% of laboratories implemented standard operating procedures (SOPs), while 82% documented equipment maintenance and calibration. Additionally, training programs were established in 55% of the laboratories.
- **QC Practices:** All participants employed system suitability checks, and 96% used pooled QC samples to ensure reproducibility. Internal standards were incorporated by 91% of laboratories, primarily for assessing mass accuracy and sensitivity. However, only 36% of labs included technical replicates of experimental samples, highlighting a gap in ensuring experimental reliability.
- **Data Management:** 71% of laboratories maintained formal data storage processes, while 55% archived their data in public repositories.

Despite these efforts, several challenges remain. Practices such as long-term reference standards and formal audits are still underutilized. Greater adoption of QA/QC training and cross-laboratory comparison exercises is needed to enhance standardization. Best practices should be tailored to accommodate diverse workflows while promoting harmonization across the field.

6.2.1 Reference Materials for Untargeted Metabolomics

A 2022 review by the mQACC emphasizes the development and use of RMs to standardize untargeted metabolomics and lipidomics [8]. The primary goal is to ensure the reproducibility, comparability, and reliability of results across studies and laboratories.

RMs play a crucial role in QA/QC by facilitating data standardization and enabling cross-study comparisons. These materials include both matrix-based and synthetic substances designed for calibration, validation, and routine QC. Current efforts focus on developing plasma-based RMs, such as National Institute of Standards and Technology (NIST) SRM 1950 and materials from the MetQual program [8, 9]. Researchers are also working to expand RMs into diverse biological matrices, including tissues, saliva, and feces, as well as synthetic chemical mixtures for system suitability assessments [10].

Despite their benefits, the adoption of RMs remains limited due to high costs and the absence of unified best practices. Broader accessibility, sustainability, and increased educational outreach are necessary to encourage their widespread use. Greater community collaboration is needed to establish harmonized methodologies and conduct interlaboratory studies. Improvements in metadata transparency and adherence to FAIR data principles will further enhance global data usability and drive standardization efforts in metabolomics and lipidomics research [11].

6.2.2 Lessons from Mass Spectrometry-Based Proteomics

In 2017, the data quality task group of the Metabolomics Society surveyed QA/QC practices in the global metabolomics community [12]. They received responses from 97 participants representing 84 institutions.

Key findings included:

- **Training Deficiencies:** 65% of respondents lacked in-house training programs, and 74% were not required to pursue continuous education.
- **SOP Utilization:** 76% of labs used SOPs, but most lacked written protocols for QA review or independent evaluations.
- **QC Practices:** While sample blanks and randomization were common, the use of standard RMs (SRMs) was limited (21%).
- **Interlaboratory Comparisons:** Only 33% participated in such studies, despite interest in standardization efforts.
- **Laboratory Accreditation:** 89% of labs had not pursued accreditation.

Recommendations emphasize the need for improved training, widespread adoption of SOPs, better use of QC materials, and incentivization of QA/QC practices. Establishing global QA/QC frameworks and fostering community collaboration were identified as priorities.

A separate study conducted within the core for life alliance aimed to establish a harmonized framework for QC in MS-based proteomics. By involving eight European core facilities over four years, this research enhanced reproducibility, comparability, and data integrity across laboratories. The findings highlighted intra- and interlaboratory variability with performance differences attributed to equipment setups, method configurations, and instrument usage patterns. Collaborative insights from shared data facilitated discussions that improved system tuning and operator practices, reducing variability. These recommendations could also be applied to both GC and GC×GC-MS analysis.

In this chapter, we will provide a summary of guidelines and initiatives for applications of GC-MS and GC×GC-MS in metabolomics. Our main focus will be directed at nontargeted screening since targeted QA/QC procedures are already well established. These are generally broad analytical methods, with many of the standard procedures originating from the UK's Standing Committee of Analysts, international bodies such as the European Committee for Standardization (CEN), and the International Organization for Standardization (ISO) also publish comparable standard methods that are equally applicable [13].

6.3 System Suitability Testing in GC-MS

6.3.1 Definition and Purpose

System suitability testing (SST) ensures that an instrument is performing at its specifications before sample analysis begins. It is a critical component of QC, verifying that chromatographic parameters meet predefined criteria to guarantee reliable and reproducible results. SST is particularly important in regulated environments where compliance with analytical performance standards is required to ensure data integrity. SST helps in:

- Identifying and correcting instrument malfunctions before analysis.
- Ensuring analytical methods produce consistent and reproducible results.
- Preventing errors that could lead to inaccurate data interpretation.
- Complying with regulatory requirements and maintaining laboratory accreditation.

A well-defined SST protocol includes specific acceptance criteria based on historical performance data and established guidelines. If any parameter falls outside acceptable limits, corrective actions such as recalibration, column replacement, or system maintenance must be taken before proceeding with sample analysis.

6.3.2 Instrument Maintenance and Troubleshooting

The following section may seem trivial, but it represents an essential aspect of the QA strategy. With the increasing complexity of hyphenated instrumental setups, a rigorous maintenance schedule must be established.

The first point is the proper use of built-in system monitoring tools, which come with most of the mass spectrometer software. Key metrics are detector voltage, vacuum pressure, resolution, intensity, and relative intensity of key ions. These reports are usually automatically generated, but there is no proper longitudinal monitoring. A simple spreadsheet-based system can be implemented to avoid and detect longitudinal drift. Such tracking allows for proper planning of maintenance activities, such as ion source cleaning and GC and MS consumable replacements.

6.3.3 Key Parameters to Monitor in GC-MS

To ensure robust analytical performance, several key parameters must be routinely monitored during SST:

- **Retention Time Reproducibility:** Consistency in retention times ensures accurate compound annotation. Variations may indicate column degradation, flow rate inconsistencies, or temperature fluctuations. The utilization of retention index methods allows the minimization of retention time variations. Nevertheless, the stability of these metrics still needs to be properly monitored in order to use them during annotation [14, 15].

- **Peak Resolution and Shape:** Well-resolved, symmetrical peaks indicate proper column performance and separation efficiency. Poor peak shapes, such as tailing or fronting, can signify issues with injection techniques, sample preparation, or column contamination.
- **Signal-to-Noise Ratio (S/N):** A high S/N ratio is essential for detecting trace-level analytes. A low ratio may suggest detector issues, contamination, or baseline instability.
- **Ion Source Stability and Mass Calibration:** Consistent ionization efficiency and accurate mass calibration ensure reliable mass spectral data. Drift in calibration may lead to misidentification of analytes, requiring frequent tuning and recalibration of the mass spectrometer. Mass spectrometers are generally used in omics nontargeted methods. This chapter will focus only on these detectors.

SST should be performed at regular intervals, including before each analytical batch, after maintenance procedures, and whenever significant changes in instrument performance are observed (see Section 6.4). By integrating SST into routine workflows, laboratories can enhance data quality, improve instrument longevity, and maintain compliance with regulatory standards. To properly monitor all those key parameters, standardized reference mixtures are required.

6.3.4 Standardized References for SST in GC-MS

6.3.4.1 Synthetic Chemical Mixtures

There are some specific reference mixtures for key applications in GC-MS (e.g., fatty acid methyl esters [FAMEs] and polycyclic aromatic hydrocarbons [PAHs]). However, for a nontargeted approach, SST can be performed using general and simpler solutions. A mix of GC-compatible chemicals is commercially available. For example, the RESTEK Mega mix catalog contains 20 references, containing from 16 to 75 compounds depending on applications and solvents. The first evaluation mixture was developed by Grob in 1978 and contains 10 compounds selected across different chemical and volatility classes. The Grob mix was first developed as a standardized quality test for glass capillary columns in GC [16]. This proposed test mixture can be used to determine key properties like separation efficiency, adsorption behaviors, and film thickness, providing quantitative information to evaluate columns for specific analytical purposes. Grob tests use a single chromatographic run with a standardized test mixture under temperature-programmed conditions, making results comparable across different column setups. By focusing on adsorption phenomena and practical test conditions, the methodology offers a pragmatic solution to optimize column performance for diverse applications.

The paper outlines a comprehensive, standardized quality test for glass capillary columns designed to address key challenges in GC [16]. Historically, GC column evaluation focused heavily on separation efficiency, often neglecting adsorption characteristics and film properties, which are critical for reliable and versatile performance. This test represents a significant advance, providing a robust methodology for evaluating and comparing columns with diverse stationary phases.

Key features of the proposed test include:

- **Single Chromatographic Run:** The test simplifies column evaluation by extracting all necessary data in one run, eliminating the need for multiple iterations under varying conditions.
- **Standardized Conditions:** By controlling variables like carrier gas flow rate and temperature programming, the test ensures reproducibility and comparability of results across laboratories.
- **Quantitative Information:** Unlike earlier qualitative approaches, this test quantifies adsorption of hydroxyl and aldehyde functions, separation efficiency, acid-base behavior, and film thickness, providing actionable insights for column optimization.

The Grob mix, designed to include compounds with varied functional groups, reveals a column's capacity for separating polar and nonpolar analytes, while the use of temperature-programmed runs ensures effective elution across different film types. Adsorption phenomena, both reversible and irreversible, are identified, ensuring the reliability of results even for complex samples. Notably, the test accounts for variables, like peak shape, retention times, and quantitative peak areas, offering a comprehensive picture of column performance.

This method addresses pressing issues in GC, such as the need for better control of adsorption processes and enhanced column robustness. Its application facilitates the optimization of column properties, improving their suitability for specialized analytical tasks, from environmental monitoring to complex organic analyses. The inclusion of standardized testing conditions and the focus on actionable metrics make this protocol a valuable tool for both column manufacturers and end-users.

Currently, the Grob mix still generates valuable insight for SST. Routine monitoring of key chromatographic parameters extracted from the mix in specific conditions allows maintaining strict control on all aspects of the analytical method, from injection to detection [17].

6.3.5 Internal and External Standards

In a targeted workflow, internal and external standards are largely used to obtain accurate quantification, establish the figure of merit, and monitor method fidelity. For specific applications, especially for regulated methods, isotopically labeled standards are used in order to generate isotopic dilution quantification. More information on these approaches can be found in the order work [18, 19].

These strategies are difficult to translate into nontargeted screening methods. Indeed, the large number of compounds present in the sample makes the selection of standards complicated. A priori knowledge of the sample can help, but other strategies need to be implemented. These strategies are described below.

6.3.5.1 Biological and Matrix-based Reference Materials

In the context of omics, synthetic chemical mixtures are usually not enough to ensure quality. Anything happening before the injection can have a significant impact on the generated data. Indeed, sampling conditions, sample preparation protocols, storage conditions and thawing conditions will have a critical impact on the quality of the results. For this specific task, SRM must be used when available.

In metabolomics, the NIST SRM 1950 is the most commonly used for serum/plasma analysis. This SRM has been developed by the NIST, in partnership with the National Institutes of Health, to aid technological advancements in metabolomics. Known as SRM 1950—Metabolites in Human Plasma, this RM contains concentrations of metabolites that reflect typical levels in adult human plasma. The plasma used to produce SRM 1950 was collected from male and female donors, with an ethnic distribution chosen to mirror the diversity of the US population [9]. Because metabolomics research varies widely in both methods and objectives, SRM 1950 was designed for general use across the field rather than for specific applications and marked the first RM specifically created for use in metabolomics studies [9].

Based on the success of SRM1950, the NIST decided to extend its library of biological matrix SRM. In August 2020, they initiated the Gut Microbiome Metabolomics Interlaboratory Program to achieve a consensus-based characterization of potential human whole stool RMs [10]. This multiplatform metabolomic interlaboratory study examined a new candidate RM for omics applications using GC(×GC)-MS, but also LC-MS and NMR (nuclear magnetic resonance) [10]. The Gut Microbiome SRM is now available at NIST.

6.3.5.2 Pooled Quality Control Mixture

As the list of available SRMs is still short, pooled QC samples play a critical role in ensuring system suitability and analytical reliability in metabolomics studies [20]. A pooled QC mixture is obtained by mixing an aliquot of each sample included in the study, generating an "average" sample that contains all the characteristics from the batches [21]. This technique is particularly useful to account for variability at every step of the analytical process. In nontargeted metabolomics, where large-scale studies involve hundreds to thousands of samples, systematic variations can arise from multiple factors, including instrument drift, sample preparation inconsistencies, and batch effects [21]. These variations can arise from multiple sources, such as instrument drift, batch effects, and sample preparation inconsistencies [21]. Pooled QC samples help address these challenges by providing a representative mixture of all study samples, allowing for the detection of systematic errors and enabling data normalization [22]. Unlike single-sample QCs, pooled QCs capture the full complexity of the dataset, making them particularly useful in untargeted analyses where metabolite compositions are highly variable. They are widely used in metabolomics, clinical studies, and environmental monitoring to assess analytical precision, correct for technical variations, and improve comparability across studies [21]. Advanced data correction techniques, such as QC-based signal correction algorithms, further enhance the reliability of pooled QCs by minimizing batch effects and ensuring reproducibility. As omics applications continue to grow in scale and complexity, incorporating pooled QC samples into analytical workflows remains one of the best practices for achieving high-quality, consistent, and comparable results [4]. A summary of different types of pooled QC samples and their associated advantages and limitations obtained from Broadhurst et al. [4] is presented in Table 6.1. It is worth noting that sample campaign logistics sometimes prevent the inclusion of all samples in the pooled QC. Partial pooling is then usually applied, but sampling randomization and consistency need to be guaranteed.

6.4 Advanced System Suitability for GC×GC-MS

6.4.1 Unique Considerations in GC×GC-MS

The increased chromatographic resolution obtained with GC×GC-MS makes it the method of choice for nontargeted screening of small metabolites in an omics context. However, it presents additional parameters to monitor compared to traditional one-dimensional GC-MS. Due to the added modulation process and the 2D signal reconstruction, advanced SST is crucial for ensuring reproducible and high-quality analytical results.

6.4.1.1 Modulation Efficiency

The modulator in GC×GC-MS plays a key role in trapping and re-injecting analytes from the first column into the second column at precise time intervals. Any inefficiency in this process can lead to loss of resolution or incomplete transfer of compounds, affecting quantitative accuracy and peak integrity [23]. Most of the current nontargeted workflows using GC×GC-MS are based on thermal modulators. However, the constant development of flow modulator technology makes them more and more compatible with omics applications.

Table 6.1 A summary of different types of pooled quality control samples and their associated advantages and limitations.

Option	Type of pooled QC sample	Preparation method	Typical sample types	Comments
1	A pooled QC sample created from ALL of the biological test samples	A small aliquot of each biological test sample is pooled into a single QC sample, followed by sample processing of aliquots of the pooled sample using the same approach as for the biological test samples	Biofluids where adequate volumes are available, including urine and plasma/serum	(i) Applied for biofluids where suitable sample volumes are available for all samples; (ii) the most representative pooled QC sample for a biological study
2	A pooled QC sample created from a representative SUBSET of the biological test samples	A small aliquot of a representative subset of biological test samples is pooled into a single QC sample, followed by sample processing of aliquots of the pooled sample using the same approach as for the biological test samples	Biofluids where adequate volumes are available, including urine and plasma/serum	(i) Applied for biofluids where suitable sample volumes are available for some samples; (ii) applied when data acquisition of samples collected in beginning of project is started before all samples have been collected; (iii) the most representative pooled QC sample for a biological study with the exception of option 1; (iv) uses smaller sample volumes for a subset of samples; (v) can be applied to prepare a representative pooled QC sample for each biological class where the composition of different samples in different biological classes is very different

3	A pooled sample created from the same sample type but from a DIFFERENT BIOLOGICAL SOURCE	A small aliquot of each biological sample acquired from a different biological source is pooled into a single QC sample, followed by sample processing of aliquots of the pooled sample using the same approach as for the biological test samples	Any sample type	(i) Applied for biofluids where small sample volumes are available; (ii) typically applied when insufficient sample volume is available to prepare a pooled QC sample; (iii) the metabolites present, their concentration, and the sample matrix are not an average of the biological test samples
4	A pooled sample created from the PROCESSED SAMPLE SOLUTIONS for all or a representative subset of biological test samples	A small aliquot of the processed sample solution from all of the biological test samples is pooled together, and sub-aliquots of this pooled sample are placed in autosampler vials/96-well plates for analysis	Cellular and tissue-based samples	(i) Most representative pooled sample for cellular and tissue samples; (ii) represents variation introduced during data acquisition and raw data processing; variation associated with sample processing is not assessed
5	An ARTIFICAL QC SAMPLE created with authentic chemical standards and a dummy sample matrix	Authentic chemical standards for as many metabolites as is achievable or representing as many metabolite classes as possible are dissolved in an artificial sample matrix (e.g., saline)	Biofluids, including tears and breath samples	(i) Provides a measure of data quality but at its lowest representative level; (ii) not all metabolites present in the biological test samples will be present in this pooled QC sample; (iii) the biological matrix is not accurately represented

Source: Broadhurst et al. [4] / Springer Nature / CC BY 4.0.

Thermal modulators

- Thermal stability is another critical aspect, as temperature fluctuations can alter analyte retention and separation. Modulators typically operate at elevated temperatures, and inconsistencies in heating cycles can impact chromatographic performance. Proper monitoring of modulator temperature settings is necessary to ensure robust performance.
- The choice of cooling gas (e.g., liquid nitrogen and carbon dioxide) in accordance with the sample type is critical, as it influences modulation efficiency. Insufficient cooling can lead to peak broadening and reduced separation [24].

Flow modulators

- Consistency of the flow in the different sections of the modulator is crucial for repeatability of the modulation process. Any variation will affect peak sampling and splitting between the different detectors or the waste line [25–27].
- The type of flow modulator will also be dependent on the sample type and analyte concentrations.

Retention time alignment across dimensions

- Unlike conventional GC-MS, where time locking is able to account for most of the variations, GC×GC-MS relies on retention time stability across two dimensions to accurately classify and quantify analytes. Variations in retention times between runs can significantly impact peak matching and data reliability.
- Alignment issues can arise due to slight fluctuations in carrier gas flow, temperature gradients, or modulator timing inconsistencies. Implementing internal retention markers and rigorous calibration protocols helps maintain retention time consistency.
- Automated software tools for peak alignment, such as correlation optimized warping or dynamic time warping, can further improve reproducibility across samples by adjusting for small retention time shifts without compromising resolution [28, 29].

Advanced system suitability tests tailored for GC×GC-MS should incorporate validation of modulation efficiency, peak shape integrity, and precise retention time synchronization across two dimensions. Regular assessment of these parameters ensures high-quality, reproducible data for complex sample analyses.

In order to monitor those key parameters, GC×GC-specific standardized mixtures have to be used.

6.4.2 Standardized Test Mixtures for GC×GC

6.4.2.1 Gasoline: Used to Monitor Separation Quality

During its early development in the 1990s, GC×GC has been mostly used for petroleum applications [25, 30]. Based on the extensive applications developed in this field, gasoline and diesel have been the "go-to option" to compare separation performances in specific conditions [31]. Nowadays, GC×GC applications have extended from petroleum to food and biological matrices. The historical use of gasoline needs to be backed up with other standardized mixtures [32].

6.4.2.2 Phillips Mix: Developed to Evaluate GC×GC Performance

Tailored for characterizing GC×GC columns, the Phillips mix, a standardized test mixture developed to honor Professor John B. Phillips, encompasses a diverse range of homologous compounds, alkanes, alkenes, alkynes, alcohols, aldehydes, ketones, FAMEs, and more, providing a wide range of volatility and polarity profiles [33].

Unlike the Grob mix, traditionally used for one-dimensional GC, the Phillips mix is tailored for the unique complexities of GC×GC, addressing issues like modulation, stationary phase interactions, and column coupling. The study by Dimandja and colleagues highlights the mixture's ability in assessing chromatographic performance, optimize operational parameters, and troubleshooting column setups. The research further demonstrates how different column combinations yield distinct chromatographic fingerprints, offering practical insights for GC×GC practitioners.

The Phillips mix serves as a benchmark for normalizing GC×GC operations and advancing the technique's applicability across industries, paving the way for tailored mixtures for specific applications like environmental analysis or flavor and fragrance studies.

6.4.2.3 Century Mix: A 100-Compound Reference Mixture for Standardizing Column Sets

To go further, Dimandja et al. recently introduced the Century Mix (CM), a 100-compound reference mixture designed to standardize the classification and performance evaluation of column sets in GC×GC [32]. GC×GC enhances analyte separation by coupling two stationary phases with differing polarities, classified as forward, reversed, nonorthogonal, or hybrid orthogonality. Forward orthogonality, where the secondary column is more polar than the primary, remains predominant, though reversed configurations are gaining traction. The CM extends earlier efforts, like the Phillips mix, and integrates compounds relevant to both 1D GC and GC×GC to link methodologies. A 2D chromatogram of the CM injected on a GC×GC system equipped with a normal set combination is presented in Figure 6.1.

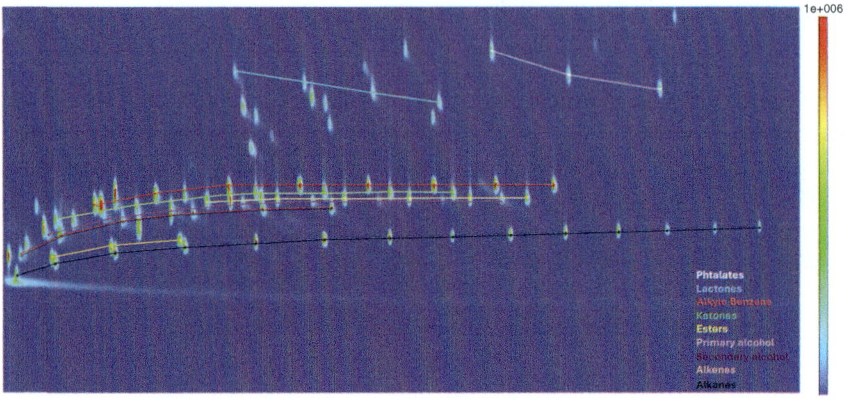

Figure 6.1 A 2D chromatogram of the Century Mix injected on a GC×GC system equipped with a normal set combination. Homologous series are highlighted: Phthalates, lactones, alkylbenzene, ketones, esters, primary & secondary alcohols, alkenes, and alkanes.

This work outlines standardized conditions for operating GC×GC systems and highlights how the CM facilitates the evaluation of column orthogonality and separation efficiency. Predictive chromatograms, derived from retention indices (RIs) of the compounds, allow researchers to explore new column combinations before experimental trials. This approach underscores the importance of orthogonality in achieving efficient separations, while recognizing the nuanced effects of column selectivity. Future efforts will focus on building databases of RIs and refining predictive modeling to optimize GC×GC method development.

There are already two studies using the CM as SST in the context of nontargeted screening.

6.4.2.4 Prehistoric Adhesives Characterization: Used in GC×GC-TOFMS for Identifying Organic Residues in Archaeological Contexts

This first case study demonstrating SST effectiveness focuses on the characterization of hafting adhesives found on prehistoric tools using comprehensive 2D GC coupled with time-of-flight MS (GC×GC-TOFMS) [34]. The goal is to enhance nondestructive analytical techniques for identifying organic residues in archaeological contexts. In this study, the authors used the CM as a chemical reference standard to optimize and evaluate the instrumental conditions. By using the CM, the study ensures that the analytical method is robust and capable of detecting a broad spectrum of organic molecules relevant to ancient adhesives.

The CM was utilized to

- Assess instrument performance, ensuring high chromatographic resolution and sensitivity.

- Validate the reproducibility of retention times and mass spectral accuracy, crucial for distinguishing between different adhesive mixtures.
- Provide a structured GC×GC contour plot, allowing the annotation of specific compound classes based on their chromatographic behavior.

This reference standard played a critical role in demonstrating that the method could effectively analyze archaeological adhesives without requiring destructive sample preparation. The use of CM strengthened the study's ability to differentiate between resin-based adhesives, beeswax mixtures, and other plant-derived substances.

The findings suggest that GC×GC-TOFMS validated through CM analysis is a powerful tool for noninvasive screening of archaeological residues, providing a pathway for further biomarker studies without altering fragile specimens.

6.4.2.5 Forensic Decomposition Analysis: Applied to Monitor Volatile Organic Compounds in Tissue Decomposition

This second case study presents a comprehensive approach for monitoring human tissue degradation, with a particular focus on volatile organic compounds (VOCs) emitted during decomposition using the CM as a reference standard [35].

Role of CM in the study:

- Instrumental calibration and performance assessment: CM was used to evaluate and optimize GC×GC-TOFMS conditions, ensuring high separation efficiency and analytical precision.
- Annotation and comparison: The mixture helped establish elution patterns for decomposition-relevant compounds, facilitating accurate detection of VOCs in decomposing tissue samples.
- Baseline for decomposition VOC analysis: The CM was additionally spiked with compounds typically associated with decomposition, enabling a controlled reference for VOC evolution tracking over time.

By leveraging the CM, the study effectively demonstrates how VOC profiles differ among human tissues, offering forensic insights into cadaveric decomposition and post-mortem interval estimation. The method's robustness, validated through CM, enhances forensic applications where noninvasive, high-resolution chemical analysis is critical.

6.4.3 In-house QC Mixture: Case of QC39

When standardized mixtures are not available for QC purposes, GC×GC users can prepare in-house QC mixtures tailored to their specific research needs. However, the lack of widely available commercial standards remains a significant limitation, underscoring the need for further development in this area. As a proof-of-concept, the OBiAChem laboratory from the University of Liège in Belgium developed the QC39 mixture, a non-commercial QC solution composed of 39 compounds derived from a combination of commercially available standard mixes and individual standards. The composition of QC39 is shown in Figure 6.2 and detailed in Table 6.2. A representative chromatogram from a liquid injection of QC39, analyzed using a GC×GC-TOFMS system with cryogenic modulation, is presented in Figure 6.3.

QC39 is currently employed to monitor instrument stability, with injections performed routinely before and after instrument use, tuning, maintenance, and other critical procedures. To assess instrument performance, QC charts were generated to monitor both retention times and peak areas of each compound. Evaluation thresholds were set using one, two, and three standard deviations. A longitudinal QC chart spanning a year, shown in Figure 6.4, illustrates area scores for all 39 compounds, providing a reliable indication of instrument fluctuations and enabling timely corrective actions to prevent compromised data quality.

While this approach demonstrates practical utility, it also highlights the broader opportunity for future development of standardized QC mixtures and advanced metrics tailored to GC×GC systems. Standardization across laboratories and platforms, supported by robust, commercially available QC materials and performance indicators, remains a critical frontier for enhancing reproducibility and confidence in GC×GC analyses.

This QC chart is used to monitor instrument variations over time. In this particular case, most data points remain within limits, suggesting a stable process, though a few values near or beyond the -2SD and -3SD lines may indicate anomalies requiring further analysis. The data fluctuates around the mean, but no clear upward or downward trend suggests an immediate issue. If significant deviations beyond control limits appear consistently, a deeper investigation would be necessary to determine the potential cause.

The use of an in-house QC mixture allows a tailored composition that fits the need. However, it still presents some limitations. First, the generation of different commercial solutions and individual standards is generating batch-to-batch variation in the preparation. Moreover, the available concentration ranges are dependent on the starting solutions. Ideally, in-house solutions need to rely on a robust preparation SOP and should be used concomitantly with standard solutions.

Table 6.2 Detailed list of the composition of the QC39 mixture for all 39 compounds.

#	Compound	CAS	Reference	#	Compound	CAS	Reference
1	1-ethyl-3methyl-cyclopentane	3726-47-4	Individual Standard	**21**	Menthone	3391-87-5	Individual Standard
2	2-Hexanone	591-78-6	Individual Standard	**22**	Menthofuran	494-90-6	Individual Standard
3	p-xylene	106-42-3	Individual Standard	**23**	Borneol	464-45-9	Cannabis Terpenes Mix B
4	Cyclohexanone	108-94-1	Individual Standard	**24**	2,6-dimethylaniline	87-62-7	Grob Test Mix
5	beta-inene	18172-67-3	Cannabis Terpenes Mix B	**25**	Menthol	89-78-1	Individual Standard
6	6-methyl-5hetpen-2-one	110-93-0	Individual Standard	**26**	alpha-Terpineol	98-55-5	Cannabis Terpenes Mix B
7	n-Decane (C10)	124-18-5	Grob Test Mix	**27**	Geraniol	106-24-1	Cannabis Terpenes Mix B
8	3-Carene	498-15-7	Cannabis Terpenes Mix B	**28**	Methyl Caprate	110-42-9	Grob Test Mix
9	p-Cymene	99-87-6	Cannabis Terpenes Mix B	**29**	1-tetradecene	1120-36-1	Individual Standard
10	Limonene	5989-27-5	Cannabis Terpenes Mix B	**30**	Dicylcohexylamine	101-83-7	Grob Test Mix

(Continued)

Table 6.2 (Continued)

#	Compound	CAS	Reference	#	Compound	CAS	Reference
11	Eucalyptol	470-82-6	Individual Standard	**31**	Methyl Undecanoate	1731-86-8	Grob Test Mix
12	g-Terpinene	99-85-4	Cannabis Terpenes Mix B	**32**	beta-Caryophyllene	87-44-5	Cannabis Terpenes Mix B
13	Aceptophenone	98-86-2	Individual Standard	**33**	1-pentadecene	13360-61-7	Individual Standard
14	1-Octanol	111-87-5	Grob Test Mix	**34**	Methyl Laurate	111-82-0	Grob Test Mix
15	Terpinolene	586-62-9	Cannabis Terpenes Mix B	**35**	cis-Nerodiol	40716-66-3	Cannabis Terpenes Mix B
16	Undecane	1120-21-4	Grob Test Mix	**36**	beta-Eudesmol	473-15-4	Cannabis Terpenes Mix B
17	Linalool	78-70-6	Cannabis Terpenes Mix B	**37**	Phytol	150-86-7	Cannabis Terpenes Mix B
18	Nonanal	124-19-6	Grob Test Mix	**38**	2,3-Butanediol	19132-06-0	Grob Test Mix
19	2,6-Dimethylphenol	576-26-1	Grob Test Mix	**39**	2-Ethylhexanoic acid	149-57-5	Grob Test Mix
20	Camphor	464-48-2	Cannabis Terpenes Mix B				

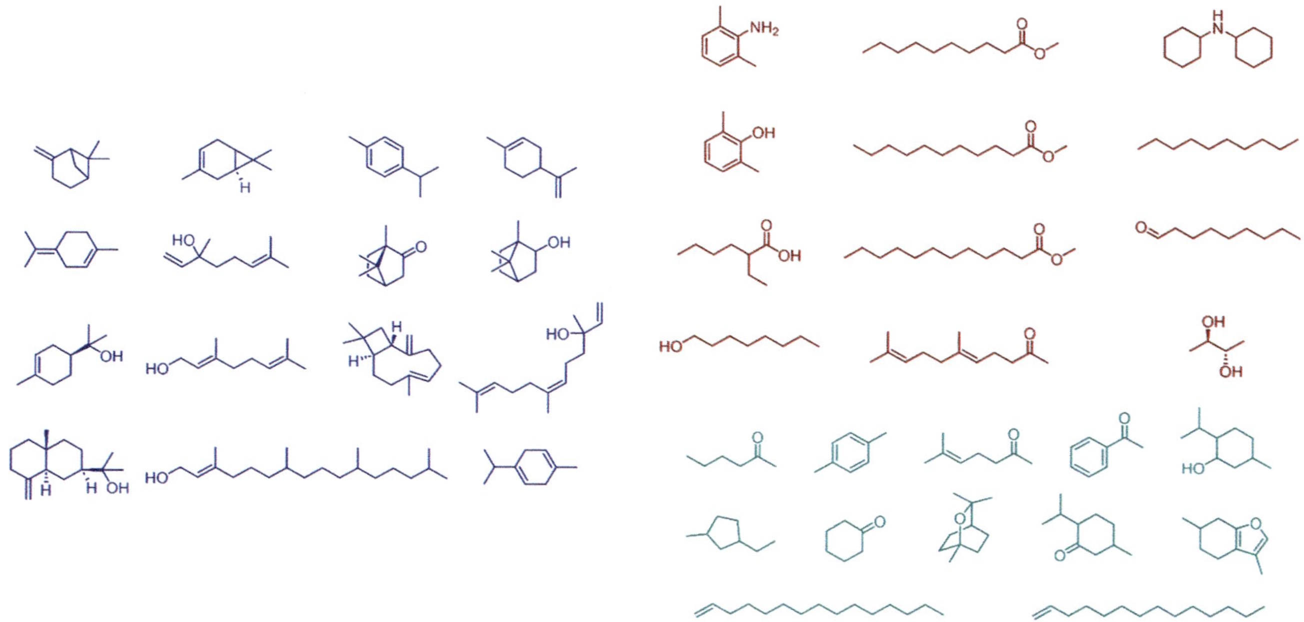

Figure 6.2 Composition of the QC39 mixture. The mixture is composed of 39 compounds coming from two different commercialized mixtures: Cannabis terpenes mix B obtained from Merck (ref. CRM40937) represented in blue, Grob Test Mix obtained from RESTEK (ref. 35000) represented in red, and 12 additional compounds obtained from individual standards, represented in green.

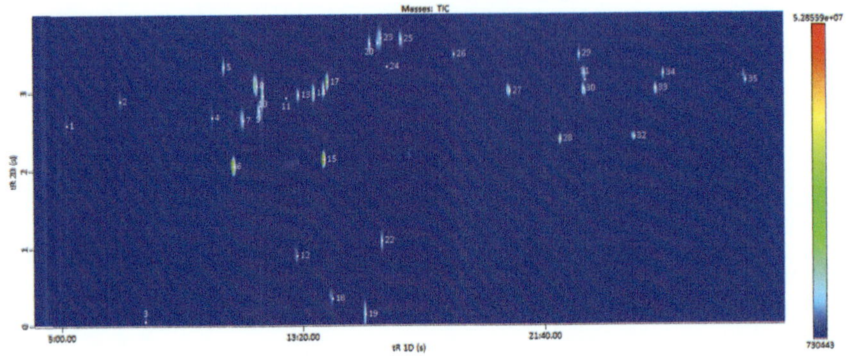

Figure 6.3 GC×GC-TOFMS chromatogram (quad-jet modulator) of the QC39 using a normal column set combination ([1]D: Rxi-5SilMS [30 m, 0.25 mm ID, 0.25 μm] and [2]D: Rxi-17SilMS [2 m, 0.25 mm ID, 0.25 μm]). The modulation period was 4 s, the injection volume was 1 μL in split mode (split ratio of 10) of the 1 ppm solution.

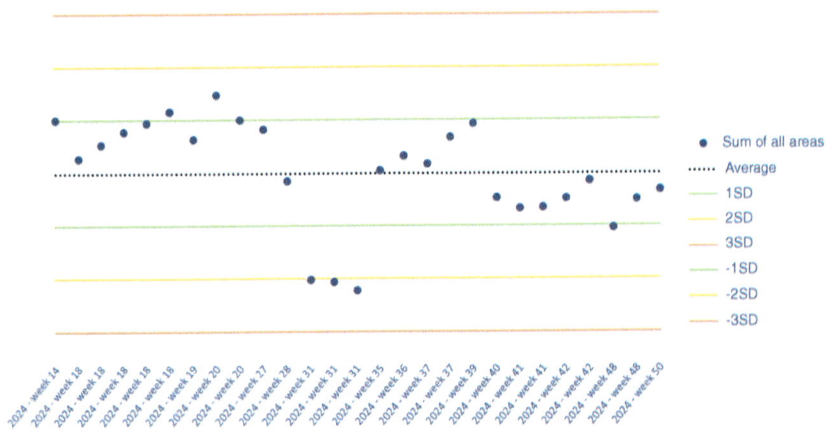

Figure 6.4 Follow-up QC chart with the score of all areas from 39 compounds of QC39. The blue dots represent the "Sum of all areas" obtained for all 39 compounds measured across different periods in 2014, showing how the process metric evolved. A central dotted blue line marks the average value, with additional lines indicating one, two, and three standard deviations (SD) above and below this mean. The green lines correspond to ±1SD, the yellow lines to ±2SD, and the orange lines to ±3SD, which typically define control limits in statistical process control.

6.5 Quality Control Strategies for Non-targeted Screening in Omics

6.5.1 Statistical QC Tools

6.5.1.1 QC Charts

QC charts are still the most important tools to monitor analytical workflow stability and robustness over time. QC charts can be used to monitor the key GC and GC×GC parameters listed above for different standards. There are several tools and rules to use QC charts efficiently. They can be found in key references [36–38]. Nevertheless, classical QC charts are univariate and usually treated independently. This single dimension is usually not adapted for nontargeted screening, for which factors are usually interconnected.

6.5.1.2 Multivariate QC Charts – Introduction to Multivariate Tools

Castiblanco et al. explored the development and application of multivariate QC charts within the context of analytical chemistry, particularly in monitoring 3D-printed microfluidic devices for improved sample preparation [39]. This work aimed to enhance the selectivity of urea crystallization for separating linear hydrocarbons from branched and cyclic hydrocarbons by leveraging 3D-printed microfluidic devices. The reproducibility of these analytical devices was validated using principal component analysis (PCA) and Hotelling's T^2 charts, ensuring consistency in the crystallization process. Additionally, multivariate control charts were implemented to monitor batch-to-batch quality, confirming the reliability of the devices for untargeted petroleum analyses.

The study achieved improved selective trapping of linear hydrocarbons (n-C13 to n-C17) and FAMEs compared to manual methods. It also enhanced reproducibility by reducing variability in sample preparation. Furthermore, the approach demonstrated potential for automation and seamless integration with GC-MS systems.

Multivariate QC charts were employed to ensure the quality and consistency of device manufacturing across multiple batches. These charts also facilitated real-time monitoring of key process parameters influencing crystallization, including flow rates and solvent compatibility. This study underscored the value of integrating multivariate QC charts with advanced manufacturing technologies to optimize analytical workflows. The proposed method has significant implications for petroleum analysis, offering a robust and environmentally friendly alternative for sample preparation in industries requiring high-throughput and reliable data.

6.5.1.3 Sequence Design and QC Monitoring

Following the recommendation of the mQACC, to avoid batch-to-batch and run-to-run variation during analysis, it is recommended to build specific QC charts for study follow-up in four steps, as presented in Figure 6.5 from Bhatt et al. [40].

In this workflow, following the recommendation of the mQACC, four steps are represented:

- Identification assurance, checked for carry over (Step 1)
- Baseline for QC chart (Step 2)
- Randomized sample analysis with QC samples (Step 3)
- Repeating step 1 to check the system stability (Step 4)

The first step is to ensure identification assurance. This involves checking for carryover, with standards and reference standards grouped separately and separated by system blank. Alkanes are used to calculate and verify RIs, while FAMEs standards are used for identification assurance. The chosen reference standard is NIST SRM 1950, commonly used for serum monitoring. A sample blank is also included to check for background signals.

The second step involves building a baseline for the QC chart. This baseline consists of 10 QC samples of identical composition and concentration, injected in groups of 5 with a method blank in between. This step is essential for monitoring instrument variation over the duration of the study.

The third step is the injection of study samples, which may be distributed over several days depending on the dataset. In this case, the injection schedule spans from Day 1 to Day 7, indicating a staggered injection plan. Samples are injected randomly in groups of five, with a QC sample injected at the beginning and end of each run, and after every five samples. A system blank is injected at the beginning and end, as well as after every 10 samples.

The final step is a repetition of the first step to assess system stability.

This structured layout reflects a well-controlled experimental design, where standards and samples are consistently measured over time to ensure analytical accuracy and reliability. The main limitation of such a system lies in the use of SRMs, which may not be available for all sample types (e.g., breath, air, and dust). Additionally, careful selection of standards and QC samples is critical and has to be discussed in the scope of the study.

6.5.2 A Concrete Example of Quality Control QC Strategy for HPLC-MS and GC-MS-Based Metabonomic Analysis

In this study, Gas Chromatography-Mass Spectrometry (GC-MS and HPLC-MS have been applied to metabolomic analysis of plasma and human urine

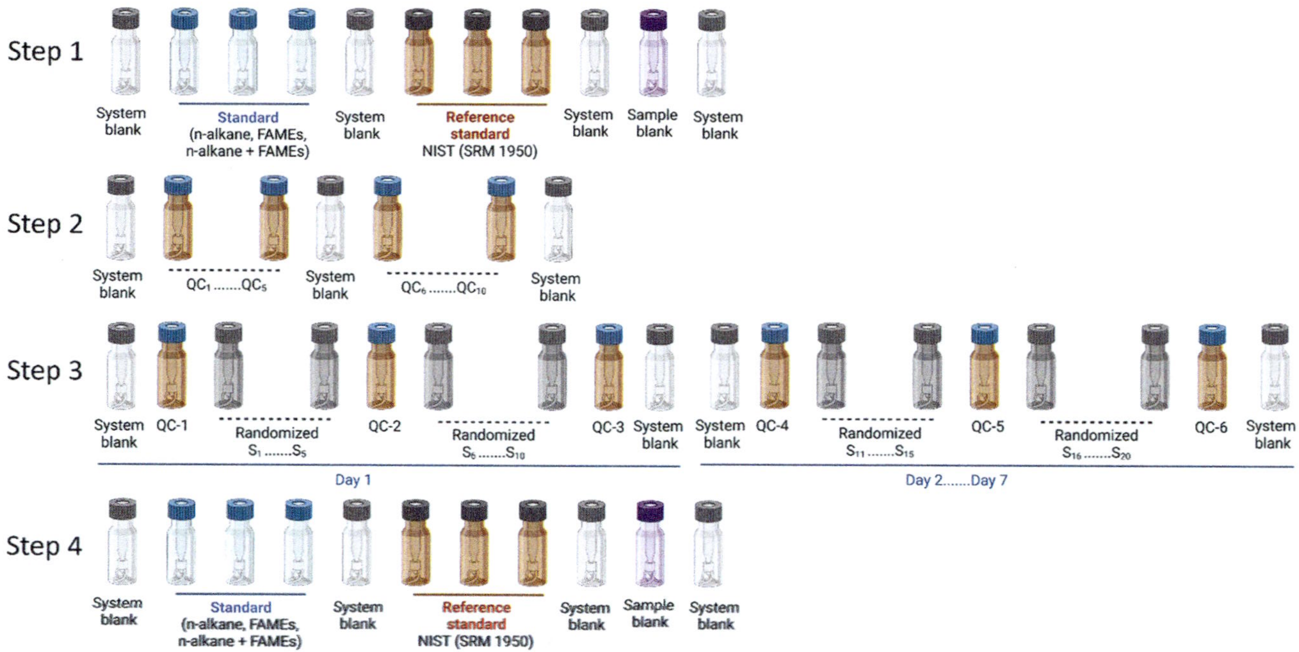

Figure 6.5 Sequence of sample vials used in an analytical testing workflow. The vials are color-coded and categorized into different types, including Standard (blue vials with blue caps), Reference Standard (brown vials with black caps), and additional sample vials with varying shades, including a purple-capped vial. *Source:* Bhatt et al. [40] / MDPI / CC BY 4.0.

samples, ensuring robust and reproducible results [37]. QC strategies have been implemented through the establishment of pooled QC sample protocols, allowing continuous monitoring of analytical performance throughout runs. Sample preparation has included protein precipitation, evaporation, and derivatization, with double derivatization (methoxylamine hydrochloride and MSTFA) applied for GC-MS workflows to enhance detection capabilities.

Analytical procedures have been conducted using capillary GC-MS systems with DB5 columns, utilizing temperature gradients from 85 to 320 °C to optimize compound separation. HPLC-MS methods have been developed for reversed-phase gradient separations, incorporating UPLC systems with C18 columns and formic acid-acetonitrile gradients to ensure high-resolution metabolite profiling. Chromatographic retention shifts and mass spectrometer sensitivity have been continuously monitored to maintain data integrity.

To support metabolomic data analysis, PCA has been employed for quality assessment, ensuring QC clustering and analytical stability. Potential biomarkers have been evaluated through statistical validation, reducing the likelihood of analytical artifacts. Best practices have been implemented, including GC system conditioning protocols to minimize early run variability and HPLC-MS system suitability tests to confirm method performance before sample analysis. Additionally, data rejection criteria have been established to maintain high-quality metabolomic datasets for further interpretation.

This approach integrates methodological precision, QA, and statistical validation, supporting reliable biomarker discovery in metabolomic research.

6.6 Chemometric Tools for Quality Assurance in Annotation

Compounds annotation represents the biggest challenge for the reproducibility of nontargeted studies. Being able to correctly annotate a compound of interest is a key element for discovery investigations. To improve the robustness of this step, several guidelines have been published by scientists' associations, mainly coming from the Metabolomics Society [4, 8, 12]. In this quest, the combination of chromatographic annotation metrics, such as RIs, with electron ionization-specific fragmentation is the gold standard for small molecules.

In order to achieve a faster and more robust annotation workflow, recent advancements in machine learning allow for enhanced compound annotation and peak annotation in GC-MS, reducing false identifications and improving classification accuracy.

This section summarizes the main metrics used for compound annotation with GC and GC×GC-MS.

6.6.1 Role of Retention Indices and Spectral Libraries in Compound Annotations

The accurate annotation of solutes in GC is fundamental for obtaining reliable analysis results, particularly in nontargeted metabolomics and other "omic" technologies. RIs play a crucial role in compound annotation by providing a standardized and reliable measure of retention behavior in GC. Unlike retention time (t_R), which varies based on operating conditions such as temperature, column type, and carrier gas flow, RIs offer consistency across different systems. Several RI systems exist, each using specific reference compounds for calculation: Van Den Dool and Krat RI (alkanes), Lee RI [41] (PAHs), and Fiehn RI [42] (FAMEs). Since RIs depend only on a compound's chemical structure and the stationary phase used, they provide a more robust tool for compound annotation [14]. The latest NIST 23 database includes 492,000 retention index values (Kovats and Lee) for 153,000 compounds [43]. Compared to NIST 17, this represents an increase of 40,000 compounds (10%).

Known unknowns are compounds with data in the database but unknown to the investigator. For these compounds, RI, when paired with mass spectral data, helps identify them by providing a consistent reference for their chromatographic behavior. However, for unknown unknown compounds with no prior data in databases, predicting both the RI and mass spectra becomes essential. Machine learning methods are increasingly being used to predict RI and spectral data, improving the annotation of both known and unknown compounds.

A deep learning-based predictive model, named DeepReI [44], was developed for accurate prediction of gas chromatographic RIs using Simplified Molecular Input Entry System as input, achieving median absolute errors of 16.4 and 16.0 units for validation and test sets, respectively. The model is available through an R package and a user-friendly graphical interface at https://github.com/TomasVrzal/DeepReI.

6.6.2 Spectral Libraries

Reference electronic mass spectral libraries and databases are essential tools for compound annotation. They assist in QC by providing a reliable reference for comparing acquired mass spectra to known spectra. Various spectral databases are available, including general-purpose commercial databases like NIST 17, as well as specialized libraries for metabolomic applications such as the Golm Metabolome Database (GMD) [45], Human Metabolome Database (HMDB) [46], and FiehnLib. The FiehnLib libraries contain 2212 electron impact mass spectra and RIs for over 1000 primary metabolites, covering various classes of compounds such as lipids, amino acids, sugars, and organic acids[42]. These spectra, obtained through GC-MS, include methoximated and trimethylsilylated metabolites. Compared to the public Golm Metabolite Database, the FiehnLib libraries are 68% larger and offer greater spectral diversity, including unique compounds not found in the commercial NIST05 database.

6.6.3 Machine Learning Approaches for Better Annotation with Spectral Libraries

Machine learning approaches, particularly deep learning techniques, have shown great potential in improving the accuracy of compound annotation when using spectral libraries. A notable example is the work by Matyushin et al., who developed a deep learning-based approach to improve compound annotation and peak annotation in GC-MS by enhancing library search accuracy [47]. Traditional library search algorithms often return incorrect results, even when the correct compound is present. The authors introduce a deep learning ranking method using a convolutional neural network (CNN) for better comparison of mass spectra. This approach outperforms traditional methods, reducing the fraction of incorrect answers by 9–23% at rank-1. The model was tested using spectra from the GMD, HMDB, and FiehnLib. Another study by Wong et al. addresses the challenge of identifying novel psychoactive substances (NPS) using GC-MS data [48]. With the increasing structural diversity of NPS, traditional detection methods struggle to keep up. To overcome this, machine learning models, specifically artificial neural networks, CNN, and balanced random forests, were trained and evaluated using 891 GC-MS spectra. The models were able to classify seven classes, including six NPS types (cathinone, cannabinoids, phenethylamine, piperazine, tryptamines, and fentanyl) and unrelated compounds, achieving a high macro-F1 score of 0.9 across cross-validation folds.

6.7 Future Trends and Innovations

The ballistic rise of AI-based tools has a significant impact on science, including analytical chemistry. From here, it is not difficult to imagine the emergence of AI-assisted workflow in SST and QAQC systems. In addition, instrument providers are currently promoting smart software to assist analysts, such as the Analytical Intelligence platform from Shimadzu. This section presents some perspectives on developments in the direction of AI-based QA/QC.

6.7.1 Automation in System Suitability Testing and Quality Control Workflows

Automation reduces manual errors, improving reproducibility and efficiency in QC. AI-based tools will help the development of automated tasks, from data interpretation to reporting and smart archiving. In addition, large language models, such as ChatGPT, will allow interfacing numerical output with text-based communication and reporting.

6.7.2 Case I: AI and Machine Learning for Real-Time Monitoring

Image recognition techniques are a good example of the integration of artificial intelligence for real-time monitoring, as demonstrated by Bi et al. and Stilo et al., who use the technique for food characterization purposes [49, 50]. An important aspect of the studies involved the use of image recognition techniques for analyzing GC-MS fingerprints, which has significant implications for QC purposes. By leveraging machine learning models, such as CNNs, the study transformed complex instrumental data into structured fingerprint images. This method allows automated, consistent evaluation of food composition and sensory attributes, reducing reliance on subjective human olfactometry and minimizing errors due to human variability.

This image-based fingerprinting approach has notable benefits for QC applications as it enhances accuracy in identifying chemical markers and classifying materials. It also enables large-scale analysis, making it an efficient tool for ensuring product integrity and compliance with quality standards. The integration of machine learning in GC-MS fingerprint recognition represents a major advancement in analytical chemistry, offering a scalable solution for real-time monitoring and automated decision-making in industrial and research settings. This aspect could easily be extended to GC×GC using the entire 2D chromatogram for image recognition.

6.7.3 Case II: Advances in Hardware and Software

Recently, Aigensberger et al. investigated the optimization of data processing and QC in untargeted metabolomics, focusing on reproducibility, reliability, and the minimization of false positives in MS analyses [51]. This study highlights a comparison of four major workflows: XCMS, compound discoverer, MS-DIAL, and MZmine. It evaluates data treatment strategies, including missing value imputation, transformation, scaling, and filtering. The focus is on detecting true positive signals while reducing noise and false positives.

Key QC findings reveal that missing value imputation predominantly arises from low-abundance compounds or absent peaks. A 10%-minimum imputation with variance is the most effective approach, preserving data integrity for downstream analysis. Data transformation and scaling methods, such as logarithmic, square root, and cube root transformations, improve normality in data distribution. Among these, auto-scaling emerges as the most reliable for multivariate analyses, ensuring uniform feature significance. Filtering techniques using low blank ratio thresholds and QC filtering enhance data quality, while overly stringent filters risk excluding vital features. Moderate thresholds offer an optimal balance.

Regarding workflow insights, XCMS is the most comprehensive but requires manual reintegration and blank filtering for noise reduction. MS-DIAL is closest to manual integration, providing consistent detection of added compounds. MZmine offers reliable integration with robust parameter optimization and effective QC workflows. CD, however, presents high baseline peaks that require additional external blank subtraction.

Finally, this work highlights the importance of adaptable QC frameworks tailored to specific datasets, ensuring robust metabolomic insights. It also provides a comparative resource for choosing suitable workflows and processing strategies in untargeted metabolomics.

6.8 Conclusion

In this chapter, we aimed to explore the essential role of QC in GC-MS and comprehensive 2D GC×GC-MS. We discussed key concepts such as SST, the importance of standardized test mixtures, statistical QC tools, and emerging machine learning approaches for data analysis and QA. These elements collectively ensure the reliability, accuracy and reproducibility of analytical results, which are critical for applications ranging from environmental monitoring to forensic and metabolomics studies.

The impact of QC on analytical confidence and reproducibility cannot be overstated. By implementing robust QC measures, laboratories can detect and address instrumental errors early, minimize batch-to-batch variability, and enhance compliance with regulatory standards. The use of standardized test mixtures, such as the Grob mix for GC-MS and the CM for GC×GC-MS, provides a benchmark for instrument performance, facilitating cross-laboratory comparability. Additionally, statistical tools like PCA and Hotelling's T^2 charts improve QC monitoring, ensuring consistent data integrity over time. Moreover, advancements in artificial intelligence and machine learning are further enhancing QC processes by automating peak alignment, pattern recognition, and outlier detection.

To fully harness the benefits of QC, laboratories must adopt the best practices and standardized methodologies. We encourage researchers and industry professionals to integrate routine system suitability testing, leverage standardized QC mixtures, and utilize advanced chemometric tools for data validation. Collaboration within the scientific community is essential to develop universally accepted QC frameworks and expand the availability of RMs. By committing to these best practices, the field of GC-MS and GC×GC-MS can continue to advance, ensuring high-quality analytical outcomes that support scientific discovery and regulatory compliance.

References

1 Mondello, L., Cordero, C., Janssen, H.-G. et al. (2025). Comprehensive two-dimensional gas chromatography–mass spectrometry. *Nat. Rev. Methods Primers* 5 (1): 7.

2 Dudzik, D., Barbas-Bernardos, C., García, A., and Barbas, C. (2018). Quality assurance procedures for mass spectrometry untargeted metabolomics. A review. *J. Pharm. Biomed. Anal.* 147: 149–173.

3 Viant, M.R., Ebbels, T.M.D., Beger, R.D. et al. (2019). Use cases, best practice and reporting standards for metabolomics in regulatory toxicology. *Nat. Commun.* 10 (1): 3041.

4 Broadhurst, D., Goodacre, R., Reinke, S.N. et al. (2018). Guidelines and considerations for the use of system suitability and quality control samples in mass spectrometry assays applied in untargeted clinical metabolomic studies. *Metabolomics* 14 (6): 72.

5 Bouhifd, M. (2015). Quality assurance of metabolomics. *ALTEX* 32 (4): 319–326.

6 Sumner, L.W., Amberg, A., Barrett, D. et al. (2007). Proposed minimum reporting standards for chemical analysis. *Metabolomics* 3 (3): 211–221.

7 Evans, A.M., O'Donovan, C., Playdon, M. et al. (2020). Dissemination and analysis of the quality assurance (QA) and quality control (QC) practices of LC–MS based untargeted metabolomics practitioners. *Metabolomics* 16 (10): 113.

8 Lippa, K.A., Aristizabal-Henao, J.J., Beger, R.D. et al. (2022). Reference materials for MS-based untargeted metabolomics and lipidomics: a review by the metabolomics quality assurance and quality control consortium (mQACC). *Metabolomics* 18 (4): 24.

9 Phinney, K.W., Ballihaut, G., Bedner, M. et al. (2013). Development of a standard reference material for metabolomics research. *Anal. Chem.* 85 (24): 11732–11738.

10 Cruz, A.K., Alves, M.A., Andresson, T. et al. (2024). Multiplatform metabolomic interlaboratory study of a whole human stool candidate reference material from omnivore and vegan donors. *Metabolomics* 20 (6): 125.

11 Wilkinson, M.D., Dumontier, M., Aalbersberg, I.J.J. et al. (2016). The FAIR Guiding Principles for scientific data management and stewardship. *Sci Data* 3 (1): 160018.

12 Dunn, W.B., Broadhurst, D.I., Edison, A. et al. (2017). Quality assurance and quality control processes: summary of a metabolomics community questionnaire. *Metabolomics* 13 (5): 50.

13 Nixon, S.C., Rees, Y.J., Gunby, J.A., and Lack, T.J. (1997). European freshwater monitoring network design. ISBN: 92-9167-024-3.

14 Bizzo, H.R., Brilhante, N.S., Nolvachai, Y., and Marriott, P.J. (2023). Use and abuse of retention indices in gas chromatography. *J. Chromatogr. A* 1708: 464376.

15 Zellner, B.d.'A., Bicchi, C., Dugo, P. et al. (2008). Linear retention indices in gas chromatographic analysis: a review. *Flavour Fragr. J.* 23 (5): 297–314.

16 Grob, K., Grob, G., and Grob, K. (1978). Comprehensive, standardized quality test for glass capillary columns. *J. Chromatogr. A* 156 (1): 1–20.

17 Perrault, K.A., Stefanuto, P.-H., Stuart, B.H. et al. (2015). Reducing variation in decomposition odour profiling using comprehensive two-dimensional gas chromatography. *J. Sep. Sci.* 38 (1): 73–80.

18 Maier, B. and Vogeser, M. (2013). Target analyte quantification by isotope dilution LC-MS/MS directly referring to internal standard concentrations – validation for serum cortisol measurement. *Clin. Chem. Lab. Med.* 51 (4): 833–837.

19 Ryan, M.J., Grant-St James, A., Lawler, N.G. et al. (2023). Comprehensive lipidomic workflow for multicohort population phenotyping using stable isotope dilution targeted liquid chromatography-mass spectrometry. *J. Proteome Res.* 22 (5): 1419–1433.

20 Broeckling, C.D., Beger, R.D., Cheng, L.L. et al. (2023). Current practices in LC-MS untargeted metabolomics: a scoping review on the use of pooled quality control samples. *Anal. Chem.* 95 (51): 18645–18654.

21 Dunn, W.B., Wilson, I.D., Nicholls, A.W., and Broadhurst, D. (2012). The importance of experimental design and QC samples in large-scale and MS-driven untargeted metabolomic studies of humans. *Bioanalysis* 4 (18): 2249–2264.

22 Ahmed, W., Wilkinson, M., and Fowler, S.J. (2024). Generating pooled quality control samples of volatile organic compounds. *J. Breath Res.* 18 (4): 041004.

23 Bahaghighat, H.D., Freye, C.E., and Synovec, R.E. (2019). Recent advances in modulator technology for comprehensive two dimensional gas chromatography. *TrAC Trends Anal. Chem.* 113: 379–391.

24 Boswell, H., Chow, H.-Y., and Gorecki, T. (2020). Basic multidimensional gas chromatography – Chapter 4. In: *Modulators*, 89–140. Elsevier.

25 Snow, N.H. (2020). *Basic Multidimensional Gas Chromatography.* Academic Press.

26 Dubois, L.M., Perrault, K.A., Stefanuto, P.-H. et al. (2017). Thermal desorption comprehensive two-dimensional gas chromatography coupled to variable-energy electron ionization time-of-flight mass spectrometry for monitoring subtle changes in volatile organic compound profiles of human blood. *J. Chromatogr. A* 1501: 117–127.

27 Wilde, M.J., Cordell, R.L., Salman, D. et al. (2019). Breath analysis by two-dimensional gas chromatography with dual flame ionisation and mass spectrometric detection – method optimisation and integration within a large-scale clinical study. *J. Chromatogr. A* 1594: 160–172.

28 Pierce, K.M., Wood, L.F., Wright, B.W., and Synovec, R.E. (2005). A comprehensive two-dimensional retention time alignment algorithm to enhance chemometric analysis of comprehensive two-dimensional separation data. *Anal. Chem.* 77 (23): 7735–7743.

29 Gamble, S.N., Granger, C.O., and Mannion, J.M. (2024). gcxgclab: an R package for two-dimensional gas chromatography preprocessing and analysis. *Anal. Chem.* 96 (43): 17059–17064.

30 Liu, Z. and Phillips, J.B. (1991). Comprehensive two-dimensional gas chromatography using an on-column thermal modulator interface. *J. Chromatogr. Sci.* 29 (6): 227–231.

31 Stefanuto, P.-H. and Focant, J.-F. (2020). Basic multidimensional gas chromatography – Chapter 3 columns and column configurations. In: *Separation Science and Technology*, vol. 12, 69–88. Academic Press.

32 Dimandja, J.-M.D., Archer, J., Casanova, J. et al. (2024). Development of a standardized protocol for the classification of column sets in comprehensive two-dimensional gas chromatography. *LCGC Int.* 1: 12–20.

33 Dimandja, J.-M.D., Clouden, G.C., Colón, I. et al. (2003). Standardized test mixture for the characterization of comprehensive two-dimensional gas chromatography columns: the Phillips mix. *J. Chromatogr. A* 1019 (1–2): 261–272.

34 Perrault, K.A., Dubois, L.M., Cnuts, D. et al. (2018). Characterization of hafting adhesives using comprehensive two-dimensional gas chromatography coupled to time-of-flight mass spectrometry. *Sep. Sci. Plus* 1 (11): 726–737.

35 Dubois, L.M., Stefanuto, P.-H., Perrault, K.A. et al. (2019). Comprehensive approach for monitoring human tissue degradation. *Chromatographia* 82 (5): 857–871.

36 Food and Drug Administration, Center for Drug Evaluation and Research (CDER) (2001). FDA Guidance for industry, bioanalytical method validation.

37 Sangster, T., Major, H., Plumb, R. et al. (2006). A pragmatic and readily implemented quality control strategy for HPLC-MS and GC-MS-based metabonomic analysis. *Analyst* 131 (10): 1075.

38 Schubert, R., Pesch, R., Pfaff, H., and Schroeder, E. (1993). An integrated concept of automated instrument control, quality assurance and data reporting for dioxin/furan analysis. *Chemosphere* 27 (1–3): 25–32.

39 Castiblanco, J.E.B., Quero, R.F., Oliveira, A.M.d. et al. (2021). Improving selective channel occlusion of complex hydrocarbons and fatty acid methyl esters in urea crystals by using an expendable 3D-printed microfluidic device for sample preparation in untargeted petroleomics. *Anal. Chim. Acta* 1160: 338425.

40 Bhatt, K., Orlando, T., Meuwis, M.-A. et al. (2023). Comprehensive insight into colorectal cancer metabolites and lipids for human serum: a proof-of-concept study. *Int. J. Mol. Sci.* 24 (11): 9614.

41 Lee, M.L., Vassilaros, D.L., and White, C.M. (1979). Retention indices for programmed-temperature capillary-column gas chromatography of polycyclic aromatic hydrocarbons. *Anal. Chem.* 51 (6): 768–773.

42 Kind, T., Wohlgemuth, G., Lee, D.Y. et al. (2009). FiehnLib: mass spectral and retention index libraries for metabolomics based on quadrupole and time-of-flight gas chromatography/mass spectrometry. *Anal. Chem.* 81 (24): 10038–10048.

43 NIST23: Updates to the NIST Tandem and Electron Ionization Spectral Libraries.

44 Vrzal, T., Malečková, M., and Olšovská, J. (2021). DeepReI: deep learning-based gas chromatographic retention index predictor. *Anal. Chim. Acta* 1147: 64–71.

45 Kopka, J., Schauer, N., Krueger, S. et al. (2005). GMD@CSB.DB: the Golm metabolome database. *Bioinformatics* 21 (8): 1635–1638.

46 Wishart, D.S., Jewison, T., Guo, A.C. et al. (2012). HMDB 3.0—the human metabolome database in 2013. *Nucleic Acids Res.* 41 (D1): D801–D807.

47 Matyushin, D.D., Sholokhova, A.Y., and Buryak, A.K. (2020). deep learning driven GC-MS library search and its application for metabolomics. *Anal. Chem.* 92 (17): 11818–11825.

48 Wong, S.L., Ng, L.T., Tan, J., and Pan, J. (2023). Screening unknown novel psychoactive substances using GC–MS based machine learning. *Forensic Chem.* 34: 100499.

49 Stilo, F., Alladio, E., Squara, S. et al. (2023). Delineating unique and discriminant chemical traits in Brazilian and Italian extra-virgin olive oils by quantitative 2D-fingerprinting and pattern recognition algorithms. *J. Food Compos. Anal.* 115: 104899.

50 Bi, K., Zhang, D., Qiu, T., and Huang, Y. (2019). GC-MS Fingerprints profiling using machine learning models for food flavor prediction. *PRO* 8 (1): 23.

51 Aigensberger, M., Bueschl, C., Castillo-Lopez, E. et al. (2025). Modular comparison of untargeted metabolomics processing steps. *Anal. Chim. Acta* 1336: 343491.

Section II

Omics Applications

7

In vivo and *in situ* Analysis of Plant Volatilome: Strategies, Challenges, and Applications for Understanding Biogenic Volatile Organic Compound Emissions

Cecilia Cagliero and Carlo Bicchi

Department of Drug Science and Technology, University of Turin, Turin, Italy

7.1 Introduction

The biogenic volatile organic compounds (BVOCs) emitted from plants are an important biosensor diagnostic of the changes that take place in its metabolism. These volatiles are biosynthesized and released from different plant organs (leaves, flowers, fruits, but also roots); they are mainly monoterpenoids, fatty acid degradation products, volatile phenolic and amino acid derivatives. BVOCs are stored in specialised secretory structures, such as glandular trichomes and resin ducts, and are selectively released when the plant is submitted to exogenous or endogenous stimulations or stresses. Indeed, they are fundamental as signals in plant–plant, plant–insect and plant–predator communications, for instance, to attract pollinators and seed dispersers or induce defence reactions against herbivores and pathogens. BVOCs are also ecologically relevant when they act as precursors of tropospheric phytotoxic compounds and have several biological activities of interest for humans [1, 2].

BVOCs are the chemical constituents of the plant volatilome that in 2011 Maffei et al. defined [2, 3] as "*the complex blend of essential oils (EOs) and volatile organic compounds (VOCs) fed by different biosynthetic pathways and produced by plants, constitutively and/or after induction, as a defense strategy against biotic and abiotic stress.*" The plant volatilome to date includes more than 1700 volatile compounds identified in more than 90 plant families; its profile and fingerprint, when studied with a suitable approach (mainly metabolomics) offers a reliable snapshot of the biochemical activity of the investigated biosystem and greatly contributes to understanding the biosynthesis and ecology of plant BVOCs [1, 2, 4]. Specific *in vivo* analytical strategies and in particular dedicated sample preparation techniques are therefore necessary to prevent: (i) chemical degradation of labile metabolites; and (ii) artefacts or new compounds formation due to enzymatic reaction in particular taking place because of stresses due to plant collection [5]. The analytical approach of election to study *in vivo* the plant volatilome composition mostly consists in the combination between static or dynamic headspace sampling techniques (SHS and DHS) and gas chromatography combined with different detection and spectroscopic techniques (mainly flame ionization detectors (FID) and mass spectrometry (MS) or both of them simultaneously operating in parallel [6].

This chapter provides an overview of the principal sampling strategies and analytical methodologies currently employed for *in vivo* and *in situ* investigations of plant volatile compounds. Particular attention is devoted to critical yet challenging aspects of the analytical workflow. The final section presents a detailed discussion of four representative case studies, highlighting methodological solutions applied to examine diverse scenarios of BVOC emissions and to explore the composition of the plant volatilome.

7.2 Sampling Strategies

As mentioned above, the complexity of plant BVOC emissions very often does not allow a direct analysis of the plant volatilome and makes it crucial the adoption of an *in vivo* sampling system representative of the metabolic situation or of the plant relationships within its ecosystem [7–9].

Therefore, in order to obtain a fingerprint/profile representative of the volatilome of the investigated plant, the sampling system must ideally be simple, fast, sustainable, portable, and sensitive, while assuring the recovery of a large number of compounds with different structures, often at very different relative abundances, most of them in traces [8, 10].

The reliable sampling of plant-emitted volatiles requires minimizing disturbances caused by external conditions that can significantly alter the BVOC profile [1, 11, 12]. For this reason, strategies involving the excision of plant parts (e.g. flowers, leaves, fruits, stems, or roots) and conventional sample preparation techniques, such as solvent extraction or distillation, are typically avoided, since they are destructive, labor-intensive, and prone to generating artefacts. Whenever possible, *in situ* techniques are favoured over laboratory-based protocols, as they capture a volatilome that closely reflects the authentic plant emission [8, 13].

Headspace (HS) sampling is the technique that most effectively satisfies these criteria for *in vivo* studies. By definition, HS corresponds to the vapour phase in equilibrium with the volatile content of the emitting system [1, 14, 15]. Depending on the operational mode, HS can be performed dynamically (DHS)

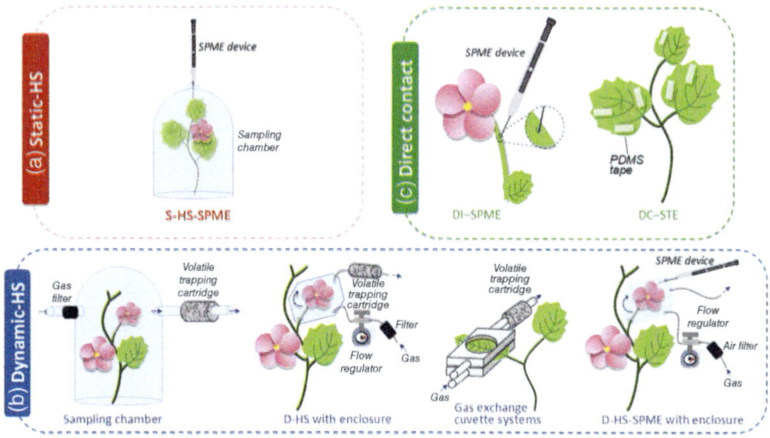

Figure 7.1 Schemes of the most used *in vivo* BVOC sampling techniques from living plants: (a) SHS, (b) DHS, and (c) direct contact (DC) sampling. *Source:* With permission of the copyright owner Cagliero et al. [17] / With permission of Elsevier.

or statically (SHS). HS sampling is generally performed in closed environments, where temperature increases and moisture accumulation may alter the micro-atmosphere [1, 8, 13].

An emerging and particularly effective alternative is direct contact extraction (DCE), in which the extraction device is applied directly to the plant tissue surface [8, 13, 16]. This approach improves upon HS sampling by facilitating the recovery of semi-volatile and polar compounds with high sensitivity.

Figure 7.1 schematically summarizes the main BVOC sampling techniques. Their respective strengths and limitations are discussed in the following sections, and their core features are presented in Table 7.1.

7.2.1 Static Headspace (SHS)

In SHS, the target volatiles of a sample are equilibrated with the surrounding atmosphere and transferred to the vapour phase in function of their partition coefficients and volatility [15]. Static headspace does not require a gas flowing in the sampling chamber, and therefore its composition is "a picture" representing the emitted volatile fraction atmosphere in equilibrium (or not) with the investigated plant when sampling [1, 15, 18]. The introduction over the years of highly efficient automatic systems operating on different approaches have greatly improved SHS precision compared to the original manual transfer with a gas-tight syringe. SHS is a "one-step gas extraction technique" that can successfully be applied to study *in vivo* the volatile fraction emitted by plants because it is

Table 7.1 Main features of the sampling techniques commonly adopted for *in vivo* evaluation of plant volatile emissions.

	Static HS		Dynamic HS		Direct Contact	
Feature	**SHS**	**HS-SPME**	**DHS**	**DHS-SPME**	**DI-SPME**	**DC-STE/SBSE**
Complexity	Low	Low	Medium	Medium	Low to medium	Low to medium
Recovering phases	Not required	PDMS, PDMS/DVB, PDMS/CAR, PDMS/CAR/DVB	PDMS; Carbon-based materials; Tenax	PDMS, PDMS/DVB	PDMS, PDMS/CAR, PDMS/DVB/CAR	PDMS
Enclosure system	Easy to assemble	Easy to assemble	Easy to assemble	Easy to assemble	Not necessary	Not necessary
Analyte concentration factors	No concentration	Medium	High	High	Medium	High
Desorption approach	Not required	Heat desorption	Solvent back-extraction, Heat desorption, Heat desorption (thermal desorption unit- TDS)	Heat desorption	Heat desorption	Heat desorption (TDS)
Analytical platform(s)	GC-MS,GC-ICPMS[c)], GC-ITMS[d)], GC-IMS[e)], GC×GC-MS	GC-MS, GC-ICPMS[c)], GC-ITMS[d)], GC-IMS[e)], GC×GC-MS	GC-MS, (ESI)[a)]MS, GC-FID, GC-DMS[b)], PTR-MS, CO laser-based detector	GC-MS, GC-FID, PTR-MS, GC×GC-MS	GC-MS, GC×GC-MS	GC-MS
Topographical analysis	Not possible	Not possible	Not possible	Not possible	Possible	Possible

a) Electrospray ionization source.
b) Differential mobility spectrometry.
c) Inductively coupled plasma mass spectrometry.
d) Miniature toroidal ion trap mass spectrometry.
e) Ion mobility spectrometry.

simple, flexible and easy to automate; conversely, it does not provide for analytes' enrichment and/or accumulation, sometimes, making it not sensitive enough for the trace compounds that must be sampled/analyzed. The early 1990s marked the introduction of high-concentration capacity headspace (HCC-HS) techniques, designed to surpass the inherent limitations of conventional HS approaches. In these methods, target analytes are progressively concentrated from either gaseous or liquid matrices, generally onto polymeric stationary phases, through mechanisms of sorption (referred to here as *absorption* to prevent confusion) or adsorption. [15, 19]. The most successful HCC-HS technique is undoubtly Headspace Solid Phase Microextraction (HS-SPME) [20], which produces high concentration factors of the volatiles emitted from solid or liquid matrices when used in *in vivo* sampling. As illustrated in Figure 7.1a, the solid-phase microextraction (SPME) device consists of a fiber coated with a thin sorbent and/or adsorbent film, which is placed in contact with the headspace of the plant sample. The efficiency of analyte recovery is governed by their partitioning between the plant and the headspace followed by the equilibrium between the headspace and the fiber coating. Selecting an appropriate coating for *in vivo* applications is therefore essential, as its physicochemical properties determine the recovery mechanism (i.e. absorption or adsorption) [15, 21]. A variety of commercial coatings are available, including mono-, bi-, and tri-material formulations. These range from highly inert phases such as PDMS [22, 23], which operates solely through absorption but exhibits limited performance with polar BVOCs, to mixed coatings (e.g. CAR/PDMS [14, 24–26], PDMS/DVB [27–32], and DVB/CAR/PDMS [12, 33, 34]) that combine absorption and adsorption, thereby extending applicability across different volatility and polarity ranges while enhancing effectiveness and reducing displacement effects.

In most applications, the (part of the) plant is placed in a specially designed chamber that serves to isolate it and to limit interference from the external environment (Figure 7.2a). Among the various options, glass containers (including flasks, vials, bottles, and cage) are the most widely adopted [12, 14, 24–32, 36–40]. A representative example is provided by Zini et al., who employed a silanized glass chamber to sample VOC emissions from the leaves of *Corymbia citriodora* (Hook.) K.D. Hill & L.A.S. Johnson (syn. *Eucalyptus citriodora* Hook). In this study the silanization process rendered the chamber walls more inert, minimizing adsorption of hydroxylated volatiles and thereby avoiding artefactual distortions [27].

Sampling time and temperature are two critical parameters that must carefully be chosen depending on both the physicochemical properties of the investigated target analytes and plant species. In particular, diagnostic results in *in vivo* plant studies are only obtained at temperatures close to those of the plant's environment and with plant–insect interaction compatible with insect survival [24–29, 31–33, 36–40].

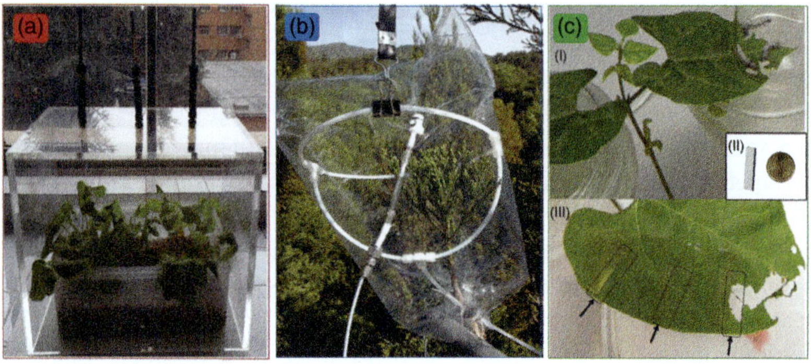

Figure 7.2 Examples of BVOC sampling from living plants: (a) static HS-SPME. *Source:* Adapted from Moreno-Martin et al. [26] / with permission of Elsevier. (b) Teflon bag dynamic enclosure. *Source:* Adapted from Matsunaga et al. [35]. (c) DC-Sorptive Tape Extraction (STE) with PDMS tape. *Source:* Adapted from Boggia et al. [16]. DC-STE sampling strategy. (I) *Spodoptera littoralis* larvae feeding on *Phaseolus lunatus*. (II) tape dimension ($4 \times 15 \times 0.2$ mm, ca. 33 mg). (III) DC-STE tapes placed on the adaxial lamina of the wounded leaf. Arrows indicate the squared translucent tapes. *Source:* Partially reproduced from Boggia et al. [16] / Springer Nature / CC-BY 4.0.

In their study on *Taraxacum* sect. *Ruderalia*, Kirschner, Øllgaard & Štěpánek, Eilers et al. [41] introduced a static passive-sampling system aimed at investigating BVOC emissions from root tissues within the rhizosphere. The system was based on a customized glass mesocosm divided into three compartments (substrate, central, and drain), where PDMS tubes were inserted for sampling and removed afterward. This design enabled the passive accumulation of root-emitted volatiles on PDMS tubes while preventing disturbances from air or water circulation in the root environment.

7.2.2 Dynamic Headspace (DHS)

Dynamic headspace (DHS) was introduced by Wahlroos and Nikkilä in 1966 [42] and almost immediately applied to *in vivo* flower volatile analysis by Herout [43], and nowadays is one of the most popular approaches in this research field [1, 15]. Dynamic headspace (DHS) involves the collection of volatiles carried by a regulated flow of inert gas directed over or through the plant matrix, after which they are concentrated in a trapping device [1]. The trapped analytes are then released from the sorbent phase, mainly via solvent back-extraction or thermal desorption [44], and subsequently subjected to chromatographic analysis, typically GC-FID or GC-MS, most often in on-line mode (Figure 7.1b). DHS is highly flexible since it affords to maximize recovery by both optimizing the trapping systems and (above all) materials to the chemical characteristics of the target analytes and the volume of gas flowing through or over the matrix. Unlike simpler approaches, DHS sampling demands relatively complex apparatus as well as rigorous control and standardization of sampling conditions to achieve reliable *in vivo* results [15]. In practice, the plant or its organ(s) are placed within a specialized enclosure that separates them from ambient air and limits interference from external factors. The selection of the enclosure design is determined by both the physical dimensions of the sample and the specific analytical technology employed. Commonly adopted systems include glass or plastic chambers [44–49] and gas-exchange cuvettes [50–60]. Leaf analysis is more frequently carried out by "sandwich" systems [1]. Figure 7.1b also shows a sketch of the two most common cuvette enclosures; Figure 7.2b shows a system for in field monitoring tree-branch emissions with large and inflatable "bags",

As already mentioned, a reliable *in vivo* DHS analysis requires to manage several variables:

1) The carrier (generally air) must be carefully regulated by a pump or flowmeter to ensure constancy, with a maximum recommended flow rate of 25 L/min. Higher rates may cause condensation, temperature fluctuations, and break-through losses of the target analytes [35]. To mitigate these issues, Balthussen et al. proposed equilibrium sorptive enrichment (ESE) [61], a dynamic configuration in which the carrier gas is recirculated through the system, enabling analytes to equilibrate with a sorbent phase, usually composed of PDMS particles. This design improves the representativeness of the sample, though its application to *in vivo* studies is restricted by the requirement for stable emission rates during the entire collection period [61]. In addition, purification of the carrier gas through charcoal filters is essential to prevent interference from exogenous BVOCs [35, 48, 51, 62–67];
2) The enclosure size and material may affect the natural plant micro-ambient making it mandatory to monitor continuously several parameters during sampling. The system has therefore to be equilibrated for a fixed time (12–48 hours) to stabilise the confined conditions [2, 35, 51, 58, 68–70];
3) Enclosure materials significantly affect the accuracy of *in vivo* BVOC sampling, as they alter microclimatic conditions and potentially contribute interfering compounds. Polyester and glass chambers, for example, may raise the internal temperature by 5.2 and 7.5 °C, respectively, while changes in humidity regulate stomatal conductance and thereby BVOC emission rates. In addition, light availability is reduced since glass and plastic enclosures filter up to 40% and 76% of UVB radiation, respectively. Although materials such as PET, polyacetate, and nylon are attractive for their low cost, portability, and ease of disposal, they can release volatile organics into the sampling atmosphere. By contrast, glass chambers offer higher chemical stability but may adsorb hydroxyl-containing molecules unless properly deactivated, for example by silanization [13, 27, 45, 51, 53, 55, 56, 59, 62, 66, 71–76];

4) Trapping materials are a critical determinant of analyte recovery efficiency and retention mechanism. Commonly used options include PDMS and other inert, thermally resistant polymers [35, 44, 46, 59, 77], which recover analytes predominantly by absorption, or carbon-based sorbents (Carbotrap, Carbograph) and polymeric derivatives (Tenax, Porapak), which function via adsorption [48, 56–58, 60, 62, 63, 65–69, 71, 72, 75, 78, 79]. Both groups exhibit notable benefits as well as limitations [15]. PDMS is valued for its chemical inertness, water resistance, and thermal robustness, but performs poorly with very volatile compounds [61]. Carbon-based adsorbents, on the other hand, efficiently retain light BVOCs, but may lead to irreversible adsorption, analyte displacement, or artefactual transformations, and their water sensitivity often impairs selectivity and recovery. Polymeric adsorbents behave similarly, though the extent of these drawbacks is generally less severe. Cryogenic trapping represents an additional, highly effective strategy, but its application in *in situ* contexts is limited by the need for portable cryosystems and near-immediate analysis to prevent analyte loss during transfer. In some cases, the trapping step can be omitted, with BVOCs sampled via DHS being introduced directly into a PTR-MS detector [45, 51, 54, 55, 74];
5) The BVOCs released from the thermal desorption of the trap must be focussed at the head of the GC column to avoid band broadening and thereby loss of chromatographic efficiency; this is true in particular for the highly volatile components that cannot be re-concentrated by the stationary phase at the head of the GC column because of its minimum operative temperature and/or due to the GC analysis temperature. The re-concentration of the desorbed band can be obtained with a cryogenically cooled GC oven or with a PTV injector or even with cryosystems at the head of the column.

Monitoring volatile emissions under stress factors are usually carried out either by laboratory or greenhouse experiments, mainly because the handling of the stress source, such as insect feeding, mechanical damage and micro/macro nutrient supplements is easier to manage than with field ("*in situ*") experiments [44, 45, 48, 51–53, 55, 56, 59, 62, 72, 73, 75, 77]. On the other hand, although the latter are less invasive and more representative of real-world living systems, outdoor sampling systems are much less flexible and more complex (needing enclosure, gas tanks, pumps and traps).

SPME has also been used in DHS mode for BVOC sampling from living plants [47, 49, 70, 79–81] through the exposition of the SPME fiber into the air stream flowing in the sampling chamber (Figure 7.1b). A significant number of examples includes two simultaneous measurements to produce a complementary and representative matrix profile, namely: (i) the air flow passes through the enclosure/SPME system and is directly channelled to the PTR-MS detector, and

(ii) the SPME fiber is then transferred to the injection port of a GC-MS system to analyze the sampled analytes [49, 70, 79]. Section 7.4 reports a more detailed description of the most commonly employed analytical platforms. The resulting sampling procedure is rather complex and requires a careful optimisation of several variables to obtain a repeatable and representative fiber recovery. The variables to be considered are mainly the coating of the SPME fiber, time and temperature of sampling, percent humidity, and flow rate affording an effective analyte/fiber contact [70, 82].

Sandra et al. developed a fully automatic analytical platform to sample BVOCs emitted from plants by DHS sampling with sorptive trapping of the analytes by PDMS cartridges [44]. Analyses were performed with a single analytical platform that combined on-line DHS sampling, thermodesorption of the trapped analytes from the PDMS cartridges, their subsequent cryotrapping and injection into the GC-MS column. Moreover, this total analysis system included two identical sampling chambers operating alternatively, to allow sequential monitoring and comparison between BVOCs emitted from the plant under different stresses and from the reference.

7.2.3 Direct Contact

An increasingly used approach to study the volatile fraction emitted by a plant *in vivo* is the sampling by direct contact between the investigated matrix and the trapping material (Figure 7.1c). This approach is often carried out by direct immersion of the SPME fiber (DI-SPME) into the living matrix; less frequently, analytes are extracted by absorption with PDMS tapes or twisters directly in contact with the matrix surface (DC-STE) [13, 16].

Investigations of volatile emissions under stress are preferentially conducted in laboratories or greenhouses, as stress sources such as insect herbivory, mechanical wounding, and nutrient supplementation can be applied and controlled more easily than in field experiments [44, 45, 48, 51–53, 55, 56, 59, 62, 72, 73, 75, 77]. Although outdoor sampling better reflects natural conditions and minimizes disturbance, it is technically demanding due to the need for chambers, gas cylinders, pumping systems, and trapping devices.

7.3 Quantification with *in vivo* and *in situ* Analysis: A Complex Task

Quantifying analytes released from solid matrices, particularly from living plants, poses considerable analytical challenges. The volatilome is inherently heterogeneous, analyte release involves complex multiphase equilibria, and

Table 7.2 Main features and limits of quantitative methods for *in vivo* and *in situ* evaluation of plant volatile emissions.

	Methods	Suitable for	Advantages	Limits	Notes	References
Not quant.	Peak areas	Comparison between different extraction conditions, chemometric treatment of the data	The standards of the target analyte(s) are not required	(i) it does not provide the absolute amount of the volatiles in the HS, (ii) it does not consider: matrix effect, analyte discrimination of enrichment rates between analytes, and instrumental fluctuations	An Internal Standard (IS) must be applied when used for comparison and/or chemiometric applications	[5, 74, 83, 84]
Semi-quantitative	Relative % abundance	Target analytes variation of abundance of specific analytes under different conditions, determination of the enantiomeric ratio	The standards of the target analyte(s) are not required	(i) it does not provide the absolute amount of the volatiles in the HS, (ii) it does not consider: matrix effect, different enrichment rates between analytes, and instrumental fluctuations	An IS must be applied when used for comparisonand/or chemiometric applications	[12, 28, 65]
	Normalized relative % abundance	Variation of abundance of specific analytes under different conditions, comparison between samples, determination of the enantiomeric ratio	(i) the standards of the target analytes are not required (ii) it does not depend on instrumental fluctuations (iii) an IS is mandatory	(i) it does not provide the absolute amount of the volatiles in the HS, (ii) it does not consider matrix effect and different enrichment rates between analytes		[13, 75]

Quantitative	Response factor (RF)	Quantification of compounds emitted by the plant for which pre-determined RF are available	(i) it does not need standards of the investigated analyte(s), (ii) it requires only one analysis, (iii) it gives the amount of volatiles in the HS of the plant sample	(i) RF previously calculated or from literature are required, (ii) a frequent and rigorous calibration is mandatory, (iii) it does not consider matrix effect and instrumental fluctuations	RF can be used with FID as GC detector or with PTR-MS (tabulated reaction rates)	[67, 73, 85]
	External calibration without IS normalization	Quantification of compounds emitted by the plant for which standards are available	It provides the amount of volatiles emitted by the plant	(i) the standard(s) of the investigated compound(s) and several calibration analyses are required (ii) it does not consider matrix effect and instrumental fluctuations	It can also be carried out with an authentic standard different (but similar in structure) from the investigated compound	[29, 60, 86]
	External calibration with IS normalization	Quantification of compounds emitted by the plant for which standards are available	(i) it provides the amount of volatiles emitted by the plant, (ii) it is independent on instrumental fluctuations	(i) the standard(s) of the investigated compound(s) and several calibration analyses are required (ii) it does not consider the matrix effect	It can also be carried out with an authentic standard different (but similar in structure) from the investigated compound	[16, 26, 75]
	Internal calibration	Quantification of compounds emitted by the plant with the standard of a similar compound	(i) it provides the amount of volatiles emitted by the plant, (ii) it requires only one standard and one analysis	The standard must be similar in structure and properties to the investigated compound		[69]
	Multiple headspace extraction (MHE)	Quantification of compounds in the plant taking into account the matrix effect	(i) the matrix effect is considered	(i) not applicable to monitoring quick variation of time profiles, (ii) not applicable to DHS, (iii) calibration analysis are necessary		[14]

QUANT, quantitative.

plant emissions are often unstable over time. Depending on the scope of the investigation, assessments of *in vivo* volatile emissions may be qualitative, semi-quantitative, or fully quantitative (Table 7.2).

True quantitative analyses require strict control of sampling parameters, such as temperature, time, and enclosure volume, especially when microextraction methods are applied, as these only recover a small fraction of the HS analytes. Importantly, rigorous standardization permits operation under non-equilibrium conditions, overcoming one of the major obstacles in *in vivo* BVOC analysis that is the difficulty of reaching equilibrium in systems where compounds differ greatly in both abundance and release kinetics [87].

Ideally, absolute quantification is achieved through calibration curves using authentic standards, but these are often unavailable, making *in vivo* quantification difficult. As a result, studies frequently focus on changes in the relative abundance of diagnostic compounds under environmental or stress-related conditions. This has led to the widespread adoption of semi-quantitative methods that normalize abundance values using internal standards. Internal standardization is essential for eliminating variability in analyte responses caused by factors unrelated to the biological process.

Nonetheless, many studies omit this step and mistakenly regard chromatographic peak areas or percentages as quantitative values [5, 8, 12, 13, 22, 23, 25, 27, 28, 30–33, 36–40, 46, 49, 62, 78, 81, 83, 88]. Such practices undermine reliability, since robust quantification with HCC techniques requires reproducible abundance data that account for matrix effects and differing analyte partitioning. Calibration is generally performed using pure standards or chemical analogues [24, 29, 41, 45, 47, 50, 52, 57, 58, 63, 67–70, 78, 79, 82, 87, 89–91]. External calibration, however, is less common because reproducing the dynamic conditions of living systems in blank samples is exceedingly difficult [16, 24, 51, 53, 54, 56, 71, 89].

Internal standards, when used, are usually added directly to the sample [26, 41, 60, 63, 67, 75, 77, 82, 87], but issues such as uneven distribution and surface interactions can limit reproducibility [13, 92]. Vereen et al. [22] reported precisely this problem when adding IS to a Tedlar bag for DHS sampling of foliage, where non-uniform distribution impaired results. To overcome such limitations, Schieberle and Grosch [93] proposed the Stable Isotope Dilution Assay (SIDA), which uses isotopically labelled analogues (^{2}H or ^{13}C) as IS for GC-MS. Similarly, Moreno-Martin et al. applied the in-fiber technique of Wang et al. [94], in which isotope-labelled standards are loaded directly onto the SPME fiber, ensuring comparable sorption and analytical behavior to the target analytes while avoiding spiking. Their isotopic nature also facilitates MS discrimination [26].

Another valuable option is multiple HS extraction (MHE), which involves repeated extractions (at least three) from the same sample to construct an area-decay curve that allows estimation of the total analyte amount [13]. This method has been successfully applied by Deng et al. [14] to quantify BVOCs from *Pelargonium × hortorum* leaves although it is difficult to apply to living plants where the emission and damage-induced release are continuous.

7.4 Overview of Analytical Platforms

Two principal strategies are available for the recovery of analytes following DHS or HCC sampling: (i) solvent back-extraction, typically employing carbon disulfide, methylene chloride, hexane, cyclohexane, or dichloromethane [35, 46, 63, 67, 75, 77, 78], and (ii) thermal desorption [16, 24, 31, 36, 58–60, 65, 66, 70, 78, 81, 83, 89]. Organic solvents are versatile and selective for BVOC recovery; however, their use in automated or on-line systems requires complex instrumentation, longer extraction times, and relatively large solvent volumes, which reduce enrichment factors and also raise environmental concerns. Thermal desorption is generally preferred, as it is solventless and preserves enrichment factors obtained during sampling, although it can induce thermal degradation of labile BVOCs and promote artefact formation [44].

Gas chromatography, together with related multidimensional configurations (MDGC, GC×GC), remains the analytical platform of choice for BVOC analysis, particularly when combined with mass spectrometry or flame ionization detection (GC-MS/FID) [6, 15]. These detectors are complementary: MS enables unambiguous compound identification, while FID is well suited for quantification and for estimating the total organic carbon emitted by plants [47–49, 53, 57, 58, 68, 69, 71, 78]. The most common stationary phases are polydimethylsiloxane (PDMS) and polyethylene glycol (PEG), although other columns (e.g. OV17, OV1701, SPB-50, or ionic liquid-based phases) are used to resolve specific compound groups [95]. Chiral analysis with cyclodextrin-based phases has been applied by Yassaa et al. [78, 79] to separate monoterpene enantiomers and by Chen et al. [63] in defining enantiomeric profiles of *Arabidopsis thaliana* terpenoids. In GC×GC, combinations of polar and nonpolar columns are applied, sometimes in reverse mode, to extend retention and improve separation of highly volatile BVOCs, thereby minimizing wrap-around effects (e.g. PEG-20 M/OV-1701, OV-1701/HP-5). Risticevic et al. [5] achieved enhanced selectivity of apple metabolites using a short ionic liquid column in GC×GC-TOFMS.

Proton-transfer-reaction mass spectrometry (PTR-MS) provides a complementary alternative to GC-MS for real-time monitoring of BVOCs [10, 19, 45, 49, 51, 54–56, 59, 70, 72, 73]. PTR ionization involves a non-dissociative proton transfer from protonated water (H_3O^+) to volatiles with proton affinities higher than water, followed by mass spectrometric detection. Coupled to quadrupole or time-of-flight analyzers, PTR-MS enables rapid, on-line measurement of trace BVOCs with diverse structures [51, 56, 72]. This makes it particularly useful for monitoring dynamic BVOC emission patterns under abiotic or biotic stress. PTR-MS is rapid and direct, while GC-MS remains necessary for definitive compound identification [27, 66, 67, 70, 88, 89]. Bouvier-Brown et al. [70, 80] demonstrated the complementarity of these methods, reporting good agreement in terpene quantification across live vegetation studies (see Section 7.5).

Another promising method is direct-infusion high-resolution MS with a secondary electrospray ionization source (SESI) coupled with an Orbitrap-MS. Barrios-Collado et al. [74] successfully used this approach for real-time monitoring of BVOCs emitted by *Begonia semperflorens*, identifying more than 1200 compounds and confirming key components via HR MS/MS, in agreement with GC-MS data.

Data elaboration constitutes a third critical step. In *in vivo* analysis, complex chemical datasets only gain biological meaning when statistically validated and correlated to physiological information. Detailed discussions of such data treatment strategies are provided by van Dam and Poppy [96].

Finally, portable instrumentation represents a rapidly advancing field for *in situ* BVOC studies. Portable systems minimize sample disturbance and enable direct field analysis under ambient conditions [62]. However, relatively few studies have applied them, as current miniaturized MS detectors still face performance limitations compared with benchtop instruments. Recent advances, including ion mobility spectrometry, and miniature toroidal ion traps (ITMS), have improved field capabilities [97]. For example, McCartney et al. [62] employed a portable GC-IMS system for real-time diagnostics of *Citrus* spp. infected with *Candidatus Liberibacter asiaticus*. Barreira et al. [49] applied a portable low-thermal-mass GC coupled with ITMS to measure forest floor BVOCs, obtaining results comparable to PTR-MS and conventional GC-MS. Similarly, Bouvier-Brown et al. [70, 80] compared PTR-MS, portable GC-FID, and SPME-GC-MS for forest emissions, reporting good consistency among techniques and no significant SPME losses. Reyes-Garcés et al. [98] combined needle-trap devices (NTD) and SPME with portable GC-IMS for diurnal monitoring of pine emissions, demonstrating stability and sensitivity comparable to benchtop systems, albeit requiring adjustments for dynamic range. Wong et al. [84] further demonstrated the potential of SPME coupled to portable low thermal mass GC-ITMS for on-site screening of emissions from diverse plant families, though gas tank requirements limited portability.

7.5 Applications

The literature reports a number of applications of *in vivo* and *in situ* sampling in spite of the difficulties standardizing it. An exhaustive review of this field has been published by Cagliero et al. in 2021 [17]. This section reports a mention to some very recent *in vivo* and *in situ* studies of BVOC emissions appeared in the literature and a detailed description of four applications representative of different topics within the *in vivo* and *in situ* sampling that include (i) the importance of absorption sampling and the effect of the herbivory attack on the plant reaction, (ii) the role of GC×GC in plant/insect interaction, (iii) the complementarity of *in situ* multiple analysis methods to monitor the BVOC emission from plants, and (iv) the development of a portable system for *in situ* BVOCs emission analysis from plants.

Liu et al. [99] examined the influence of increasing nitrogen (*N*) deposition on BVOC emissions in a subtropical forest in southern China, and assessed their potential contribution to secondary atmospheric pollution under both current and projected *N* deposition scenarios. Leaves from *Eucalyptus urophylla* S.T. Blake were enclosed in a leaf cuvette (2 × 3 cm) of a portable photosynthesis measurement system and flushed with purified air. The outflow stream was split into two channels, one directed to an infrared gas analyzer and the other to Tenax-TA stainless steel tubes for BVOC trapping. The trapped volatiles were analyzed by GC-MS after thermal desorption.

Ma et al. [100] evaluated warming-induced changes in isoprenoid emissions from seedlings of *Schima superba* Gardner & Champ. and *Cunninghamia lanceolata* (Lamb.) Hook. using custom-made Teflon cuvettes (40 × 60 cm) under three treatments: control (CK), infrared radiators (IR), and heating wires (HW). Following one, two, and four months of exposure, enclosures were filled with purified air and gas samples (1.0 L) were collected into Tedlar bags. After three-stage preconcentration, samples were analyzed by GC-MS. Isoprenoid emissions increased significantly under both warming conditions, with *S. superba* showing increases of 16.3% (IR) and 72.5% (HW), and *C. lanceolata* showing increases of 2.47- and 5.96-fold, respectively.

Männistö et al. [101] quantified the effect of severe drought and subsequent recovery on BVOC emissions from *Sphagnum* mosses using mesocosms originating from wet and drier boreal fens and bogs. Samples were enclosed in glass chambers (d: 15 cm, h: 26.8 cm, v: 3.6 L) flushed with purified air. Outflow was passed through Tenax TA/Carbopack B adsorption tubes for 90 minutes, followed by GC-MS analysis after thermal desorption. Results showed that drought altered both the qualitative and quantitative BVOC profiles; total BVOC and isoprene emissions recovered post-drought, whereas sesquiterpene emissions remained low six weeks after rehydration.

Chen et al. [102] investigated BVOC emissions from three aquatic plants (*Phragmites australis*, *Typha angustifolia*, and *Iris pseudacorus*) using DHS-GC-MS. Intact branches and leaves were enclosed in 10 L Teflon bags, flushed with purified air, and volatiles were collected on Tenax TA tubes. Thermal desorption followed by cryoconcentration and GC-MS analysis showed that *P. australis* emitted higher total BVOC levels dominated by isoprene, while emissions from *T. angustifolia* and *I. pseudacorus* contained ~50% oxygenated VOCs.

Yuan et al. [103] explored the impact of ozone (O_3) on growth and monoterpene emissions in *Betula pendula* subsp. *mandshurica* (syn. *B. platyphylla*), *Quercus mongolica*, *Larix gmelinii* var. *olgensis* (syn. *L. olgensis*), and *Pinus tabuliformis*. Needle and leaf cuvettes of a portable photosynthesis system were supplied with purified air, and BVOCs were trapped on Carbograph 5TD/Tenax TA cartridges prior to GC-MS analysis. Results indicated that prolonged O_3 exposure decreased monoterpene emissions in broad-leaved species due to reduced biosynthetic capacity, while in conifers monoterpene emissions increased, associated with changes in leaf anatomical traits.

The first application highlights a key question in volatilomics: how BVOC emissions change dynamically in relation to tissue damage, and how this process connects to the biosynthesis of bioactive metabolites during plant–insect interactions. Specifically, the study investigated the emission profiles of *Phaseolus lunatus* L. (lima bean) leaves subjected to herbivory by larvae of *Spodoptera littoralis* Boisd., as well as to mechanical wounding alone and in combination with larval oral secretions (OS) [16, 105, 106]. For this purpose, the authors employed DC-STE (see Section 2.3) together with GC-MS. This approach is particularly suitable because it is non-destructive, yields high enrichment factors, and makes use of non-adhesive PDMS tapes that can be applied directly to leaf surfaces without causing injury. The sampling procedure is shown in Figure 7.2c, and Figure 7.3 illustrates a representative GC-MS chromatogram of volatiles released from *P. lunatus* leaves damaged by *S. littoralis*, following thermal desorption of PDMS tapes after DC-STE sampling (analytical conditions and compound assignments are reported in the caption to Figure 7.3).

In total, 337 time-course experiments were carried out to examine BVOC emissions under three conditions: herbivory wounding by larvae (HW), mechanical wounding with a pattern wheel (MD), and MD combined with larval oral secretions (OS). PCA proved highly effective in distinguishing control samples from damaged ones and in differentiating among the types of damage. The analysis showed that green leaf volatiles (GLVs), such as (E)-2-hexenal, (Z)-3-hexen-1-ol, 1-octen-3-ol, (Z)-3-hexenyl acetate, and (Z)-3-hexenyl butyrate, were associated with both HW and MD, while monoterpenoids ((E)-β-ocimene), sesquiterpenoids ((E)-nerolidol), and homoterpenes (DMNT,

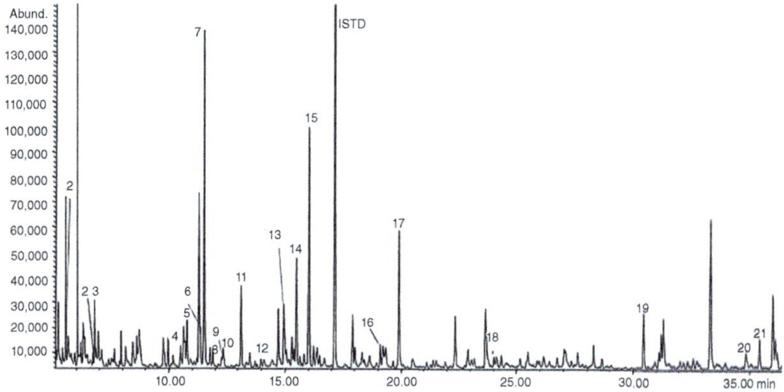

Figure 7.3 Typical GC-MS profile of BVOCs emitted by *Phaseolus lunatus* leaves wounded by *Spodoptera littoralis* obtained after thermal desorption of PDMS tapes after DC-STE sampling Analysis conditions: Instrumentation: automatic MPS-2 combined with a 7890 GC-5975C MSD system (Agilent, Little Falls, DE, USA). Themal desorption and injection: TDU thermal desorption unit with a Gerstel CIS-4 PTV injector. The inlet was operated in the splitless mode. Column: MEGA5MS (30 m × 0.25 mm i.d. × 0.25 μm; MEGA Legnano, Italy). Temperature program: from −30 °C (1 min) to 50 °C at 50 °C/min, then to 165 °C at 3 °C/min, then to 250 °C (5 min) at 25 °C/min. Carrier gas: Helium, flow rate: 1 mL/min. MS operated in EI mode at 70 eV, mass range in full scan mode: 35–350 amu. Peak identification: 1: *n*-hexanal, *2: (E)-2-hexenal, 3: (Z)-3-hexen-1-ol, 4: 1-octen-3-ol*, 5: 6-methyl-5-hepten-2-one, 6: octanal; *7: (Z)-3-hexen-1-ol acetate*, 8: p-cymene, 9: limonene; 10: 2-ethyl-1-hexanol, *11: (E)-β-ocimene, 12: 1-octanol*, 13: linalool, 14: nonanal, *15: (E)-4,8-dimethyl-1,3,7-nonatriene (DMNT), 16: (Z)-3-hexenyl butyrate*, 17: decanal, 18: tridecane, 19: geranylacetone, *20: (E)-nerolidol, 21: (E,E)-4,8,12-trimethyl-trideca-1,3,7,11-tetraene (TMTT)*, ISTD: camphor. Compounds involved with the biotic and abiotic stresses in Italics.

TMTT) were specifically linked to HW. Figure 7.4a,b illustrate the PCA scatterplot and loading plot, highlighting the discriminant variables (see captions for details).

The temporal dimension of BVOC release was also evident: GLVs were emitted shortly after damage, consistent with lipoxygenase activity, while terpenoid emissions occurred later. This emission sequence paralleled the transcriptional upregulation of several genes encoding enzymes involved in VOC biosynthesis, including Farnesyl Pyrophosphate Synthase, Lipoxygenase, Ocimene Synthase, and Terpene Synthase 2. Expression of genes related to (E)-β-ocimene, TMTT, and DMNT was localized to wounded tissues, whereas GLVs and sesquiterpenoids functioned both locally and as systemic signals, inducing gene expression in distant tissues. The study also highlighted that BVOC profiles vary according to the organ sampled, the anatomical characteristics of the tissue (e.g. trichome density), and the size of the wounded area.

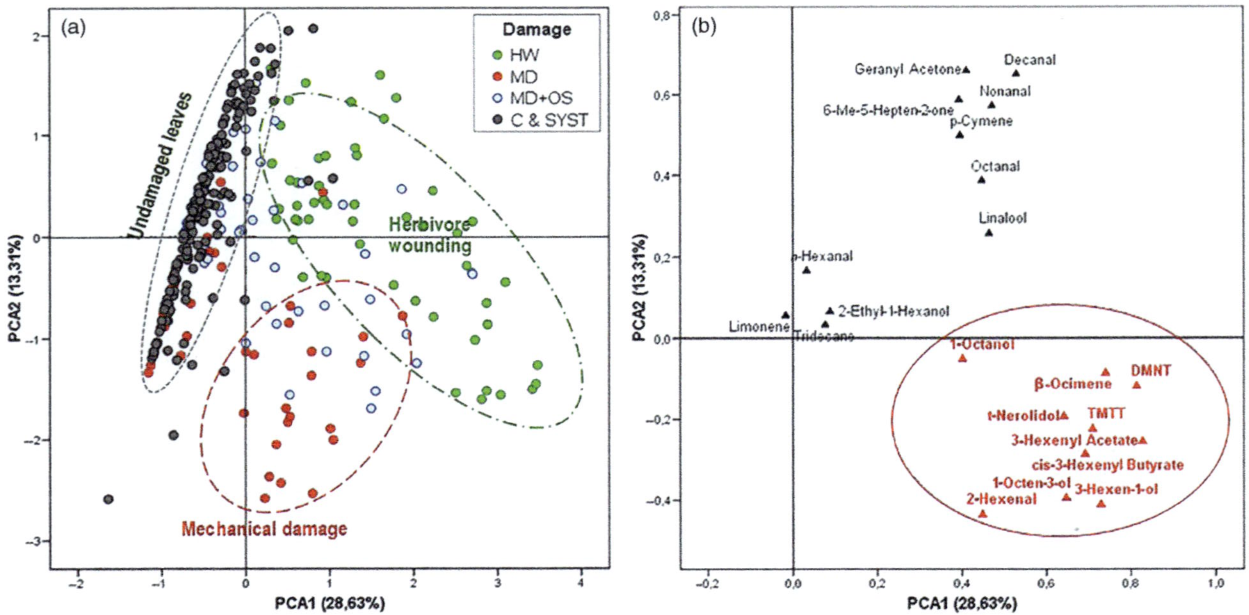

Figure 7.4 PCA of the total dataset of samples (337 samples), discriminating undamaged (controls) from damaged leaves. (a) Scatterplot of the whole set of data, (b) loading plot indicating the discriminant variables. C, control unwounded leaves; H.W., herbivore wounding; M.D., mechanical damage; M.D. + O.S., mechanical damage + oral secretion.

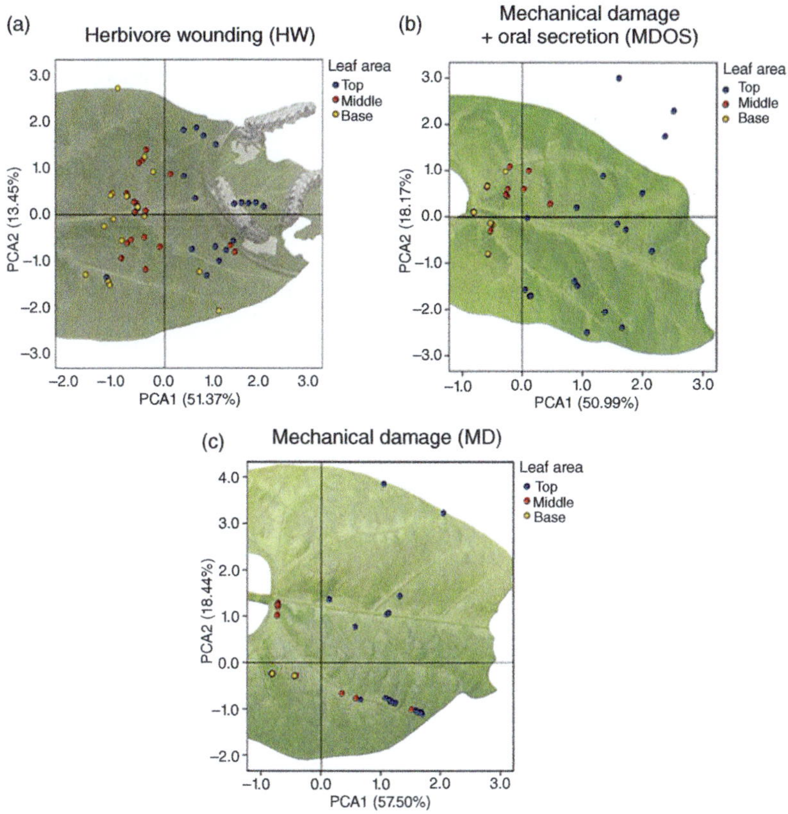

Figure 7.5 PCA score plots showing herbivory-induced plant volatiles (HIPVs) topography; (a) score plot HW leaves (54 samples) indicating the distinction between top samples and other leaf portions, (b) MDOS score plot (52 samples) showing a distribution similar to HW leaves (a), (c) MD score plot (53 samples) with the same topographical distribution. Variables: (*E*)-2-hexenal; (*Z*)-3-hexen-1-ol; 1-octen-3-ol; (*Z*)-3-hexenyl acetate; (*E*)-β-ocimene; DMNT; (*Z*)-3-hexenyl butyrate; (*E*)-nerolidol; TMTT. Top: wounded areas (blue circles), middle: middle of the leaves (red circles), base: base of the leaves (yellow circles). HW, herbivore wounding; MDOS, mechanical damage + oral secretion; MD, mechanical damage. *Source:* Lorenzo Boggia., et al. (2015) / Springer Nature / CC BY 4.0.

DC-STE-GC-MS was also found to be a reliable method for the topographical evaluation of plant responses to biotic and abiotic stresses, by revealing the differential distribution of different classes of emitted BVOCs. Herbivory-induced plant volatiles (HIPVs) topography is clearly illustrated by the PCA score plots: wounded areas (WA) are always well separated from other leaf portions. Figure 7.5 reports the PCA scoreplots of HIPV topography. DC-STE resulted easy, non-distructive and highly effective for *in vivo* reliable and mutiple samplings.

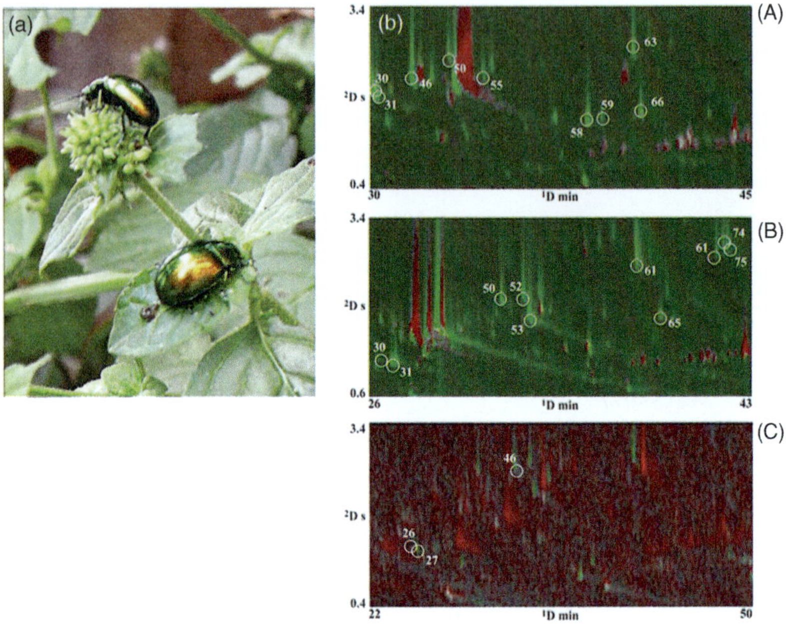

Figure 7.6 (a) *Chrysolina herbacea* feeding on leaves of *Mentha* species, (b) comparative visualization represented as colourized fuzzy ratio of leaf (reference image) versus frass (analyzed image) from three different mint species: (A) *M. spicata*, (B) *M × piperita* and (C) *M. longifolia*. Red colourization: 2D peaks more abundant in the reference image; green colourization: 2D peaks more abundant in the analyzed chromatogram. White circles indicate minutiae features to be investigated as potential bio-transformation and/or degradation products. *Analysis conditions – Sample amounts*: intact leaves (about 50 mg) from living plants of *Mentha* species or frass fluid (4–5 mg) in a 2.0 mL vials. *HS-SPME sampling*: Fiber: Divinylbenzene/Carboxen/Polydimethyl-siloxane (DVB/CAR/PDMS), df 50/30 μm, 2 cm long fiber. Desorption temp.: 250 °C; *GC×GC-QMS instrumentation*: Agilent 6890/5975 MS detector; EI mode at 70 eV. Transfer line: 270 °C. Ion source temp.: 230 °C. Scan range: m/z 35–250, speed: 10,000 amu/s. Modulator: Zoex Corporation KT 2004 loop two-stage thermal modulator; modulation time: 4 seconds. Data elaboration: Agilent – MSD Chem Station ver D.02.00.275 and GC Image software, ver 2.1b0. *GC×GC column set*: 1D SE52 column (95% polydimethylsiloxane, 5% phenyl) (30 m × 0.25 mm ID, 0.25 μm df); 2D OV1701 column (86% polydimethylsiloxane, 7% phenyl, 7% cyanopropyl) (1 m × 0.1 mm ID, 0.10 μm df); loop: 1.0 m from MEGA (Legnano (Milan)-Italy). Inj.: split/splitless; mode: split; split ratio: 1/100; temp.: 280 °C. carrier gas: He; mode: constant flow; flow rate: 1.0 mL/min; temp. progr.: from 45 °C (1 min) to 260 °C (5 min) at 2.5 °C/min. *Source:* With permission of the copyright owner Cordero et al. [106] / with permission of Springer Nature.

The second case study examined how the mint bug, *Chrysolina herbacea*, interacts with three mint species (i.e. *Mentha spicata* L., *M.* × *piperita* L., and *M. longifolia* L.) [106, 107]. Each species has a distinct volatile profile: carvone dominates in *M. spicata*, menthol and menthone in *M.* × *piperita*, and piperitenone oxide in the investigated chemotype of *M. longifolia*. While *C. herbacea* metabolized most compounds from *M. spicata* and *M.* × *piperita* leaves, *M. longifolia* proved toxic and led to insect mortality. The aim was to elucidate how the insect tolerates, detoxifies, or biotransforms terpenoids ingested in large quantities during feeding. Insects were individually fed with fresh leaves from the three species, and their frass was collected for analysis. Both frass and intact leaf volatiles were profiled using automated HS-SPME-GC×GC-QMS, with chromatographic fingerprints subjected to advanced data analysis (details in Figure 7.6). Experimental challenges included the very small frass yield per insect (necessitating pooling from ≥10 insects), differences in sample type (semi-liquid frass versus solid leaves), and the need for reliable SPME performance, addressed by employing in-fiber internal standardization [94, 108]. Peak patterns from GC×GC chromatograms were compared first by visual inspection [109], and subsequently by comprehensive template matching fingerprinting to highlight key chemical differences. Figure 7.6 presents the comparative images, shown as colourized fuzzy ratios of leaf (reference) versus frass (sample) profiles for *M. spicata*, *M.* × *piperita*, and *M. longifolia*.

Figure 7.7 shows the distribution of the main features revealed by the comprehensive template matching fingerprinting for (a) *M. spicata*, (b) *M.* × *piperita* and (c) *M. longifolia*. The results also showed that 1,8-cineole that is present in the leaves of all the *Mentha* species investigated, was oxidized by *C. herbacea* metabolic process as demonstrated by the frass volatile fraction that contained high rates of several hydroxy-1,8-cineoles, including 2α-hydroxy-, 3α-hydroxy-, 3β-hydroxy-, and 9-hydroxy-1,8-cineole.

The next example concerns the complementary use of three analytical platforms (PTR-MS, GC-FID and GC-MS) applied simultaneously in the detection and quantification of BVOCs, mainly terpenes, for a better understanding of both their roles in ecosystem functioning and the chemistry of a forest canopy [70, 80]. The three techniques were combined with the aim to integrate synergically their results. They included the high speed, direct and not destructive PTR-MS technique that was used in parallel to the on-line GC-FID analysis and to the *in situ* DHS-SPME sampling off-line combined to GC-MS analysis to study the emission composition and analyte identification and quantification. The study was carried out on the BVOCs emitted by the branches of three dominant species, *Pinus ponderosa* Douglas ex C.Lawson (pine), *Arctostaphylos* spp. Adans. (manzanita) and *Ceanothus cordulatus* Kellog (whitethorn ceanothus) of a coniferous forest in the Sierra Nevada mountains of California (Blodgett Forest). The BVOCs emitted

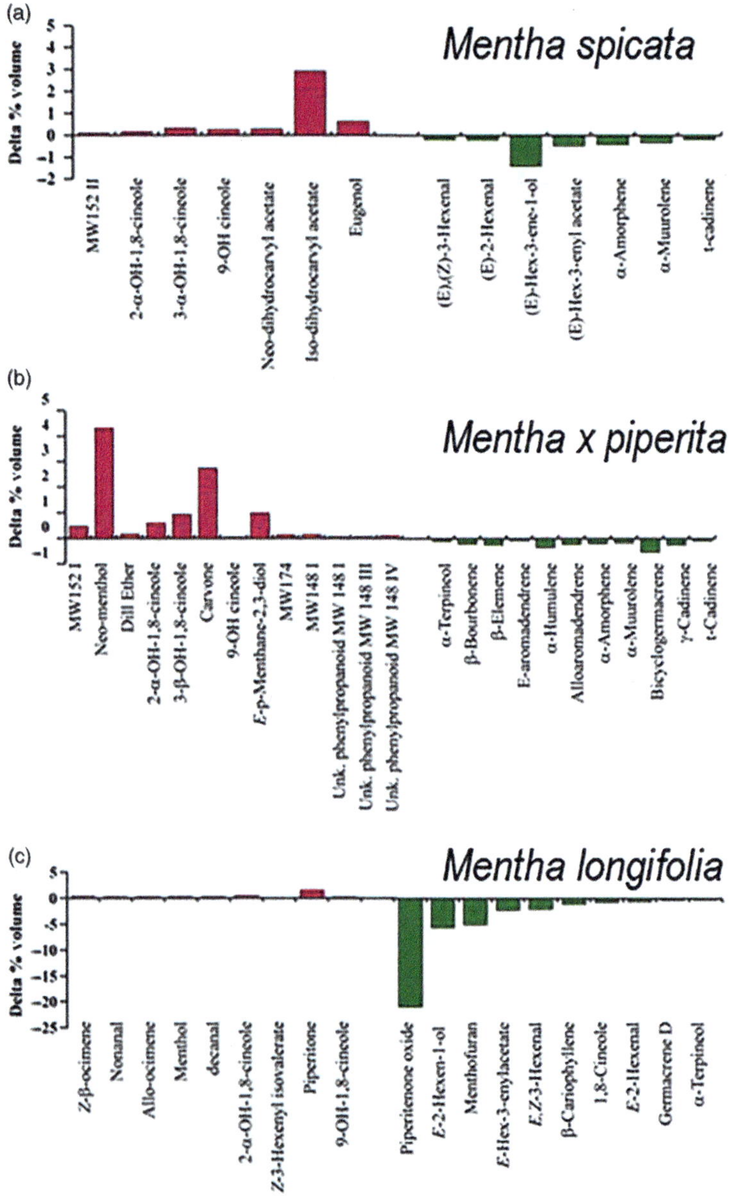

Figure 7.7 Distribution of the main features revealed by the comprehensive template matching fingerprinting for (a) *M. spicata*, (b) *M.* × *piperita* and (c) *M. longifolia*. Variations are represented as the difference between 2D peak volumes percentage (PV%) from frass to leaf. The first group (left violet) collects compounds revealed in frass samples and *not* in leaves, the second (right green) those revealed in the leaf volatiles profile *only*. *Source:* With permission of the copyright owner Cordero et al. [106] / with permission of Springer Nature.

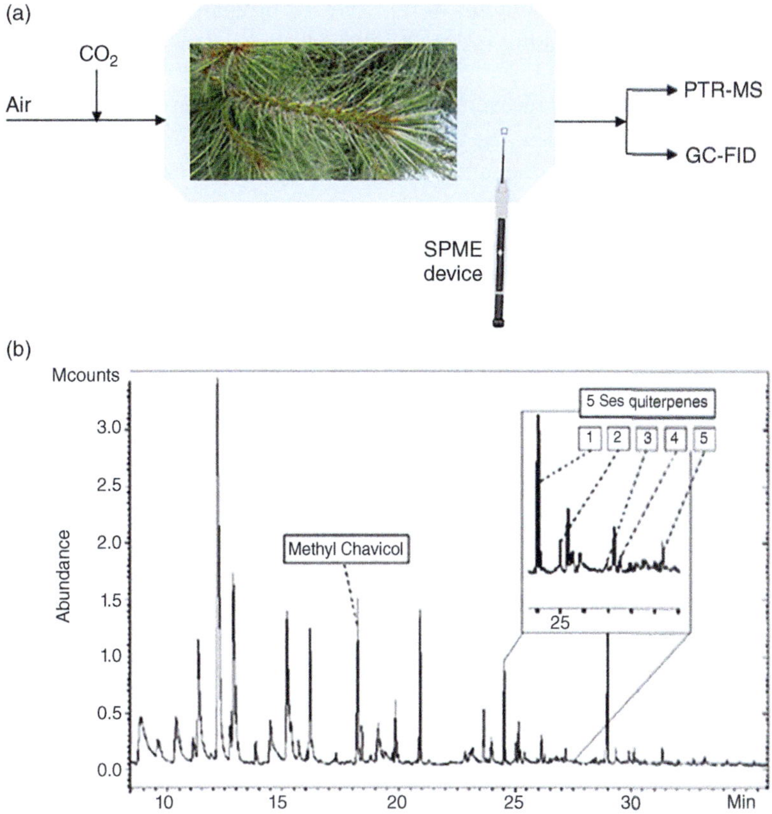

Figure 7.8 (a) Schematic of the branch enclosure setup (*P. ponderosa*) (b) GC-ITMS pattern of the BVOC emission after a 45 minutes exposure sampled by DHS-SPME from to the *P. ponderosa* branch. Analysis conditions: GC-MS system: Varian CP-3800 gas chromatograph with a Saturn 2200 ion trap mass spectrometer. Carrier gas: helium, flow rate 2.2 mL/min; column: dimethylpolysiloxane-5% phenyl; oven temperature: 40 °C (4.25 min)/5 °C/min/160 °C/10 °C/min//220 °C (11.75 min). MS: full scan mode, scan range: 40–650 m/z. Peaks corresponding to methyl chavicol and five sesquiterpenes are indicated. Peak identified: methyl chavicol, 1: α-bergamotene, 2: β-farnesene, 4: α-farnesene. *Source:* With permission of the copyright owner Bouvier-Brown et al. [80] / with permission of Elsevier.

from live branches were recovered using an air flow through a hermetically closed enclosure made of heat-sealed 0.051 mm thick PTFE film fashioned to fit each individual plant branch (Figure 7.8).

The air-flow (zero air, with ambient CO_2), containing the emitted BVOCs at the exit of the enclosure, was split and sent in parallel to both the PTR-MS system for the direct analysis and to the GC-FID after on-line concentration on a Carbopack B/Tenax cartridge and thermal desorption. Simoultaneously, the emitted

BVOCs were sampled *in situ* from the airflow flowing through the enclosure system by DHS-SPME using 65 μm PDMS/DVB fibers installed inside the enclosure. The fiber was then off-line transferred into the GC injector body of the GC-ITMS system where the BVOCs are recovered by heat desorption and on-line transferred to the GC column and analysed. The analytical instrumentation was housed into a temperature-controlled container and connected to the *in situ* chamber sampling ports by tubing about 15 m long.

This method enables the collection and quali-quantitative assessment of monoterpenes, sesquiterpenes, and methyl chavicol with minimal sample transfer, thereby reducing compound losses through tubing. Emissions were monitored from July to September, focusing on nine monoterpenes (β-pinene, Δ-3-carene, α-pinene, limonene, β-phellandrene, camphene, myrcene, α-terpinene, terpinolene, and γ-terpinene), nine sesquiterpenes (α-bergamotene, β-farnesene, α-longipinene, α-cedrene, and α-bisabolene from *Pinus ponderosa*; β-cubebene and β-caryophyllene from *Arctostaphylos* spp.; β-caryophyllene and α-humulene from *Ceanothus cordulatus*; and α-farnesene from all three species), and methyl chavicol from *P. ponderosa*. Results demonstrated that sesquiterpenes and methyl chavicol – previously undetected at this site – make substantial contributions to the overall reactive biogenic emission profile. Because many sesquiterpenes and their oxygenated derivatives are semi-volatile and highly reactive, they are not easily quantified by conventional air-sampling techniques. The findings therefore emphasize the importance of including sesquiterpenes and methyl chavicol in biogenic emission models for atmospheric chemistry and secondary organic aerosol (SOA) formation, as they influence atmospheric oxidative capacity. For well-calibrated compounds, concentration values obtained with the three techniques were consistent within the estimated methodological error.

The fourth application concerns the development of a portable platform combining a NTD and SPME fiber with a portable gas chromatography–ion mobility spectrometry detection system (NTD-GC-IMS) for on-site quantitative analysis of BVOCs (α-pinene and limonene) emitted by a pine branch [98]. The platform consists of a miniaturized GC interface coupled to the handheld IMS analyzer previously modified for SPME applications. Figure 7.9 reports a schematic of the lateral view of the modified GC-IMS system.

The assembly of the analytical platform involved several technical modifications:

1) A newly adapted inlet was designed to enable both SPME and NTD desorption, addressing the incompatibility of the original desorption system with NTD coupling. The NTD was packed with Carboxen (CAR) 1000 (60/80 mesh) and DVB (100/120 mesh), while the SPME fibers were coated with 65 μm PDMS/DVB and 100 μm PDMS.

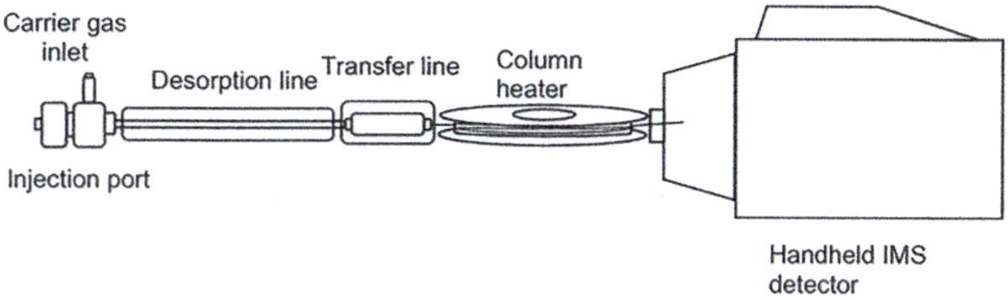

Figure 7.9 Schematic of the lateral view of the modified GC-IMS system. *Source:* With permission of the copyright owner Reyes-Garcés et al. [98] / with permission of Elsevier.

2) The GC unit employed an RTX-5 column (10 m; 0.25 mm I.D.; 0.25 μm film) heated isothermally at 76 °C using a heater derived from a microGC. Nitrogen served as the carrier gas, with flow controlled via a mass flow controller. Separate temperature controllers regulated the desorption line, transfer line, and column.
3) The ion mobility spectrometer (IMS, Sabre 2000) was operated in positive mode with a ^{63}Ni ionization source and water as calibrant/reactant ion. Ionization was achieved by high-energy electrons from the radioactive source colliding with drift gas molecules (air), producing reactant ions that subsequently ionized the analytes. Compounds were then detected and identified based on their mobilities in an electric field [110, 111]. IMS instruments are advantageous due to their compactness, low cost, rapid response, and modest power consumption, though they are limited by narrow dynamic ranges resulting from reactant ion depletion at high analyte concentrations [110].

The NTD-GC-IMS platform was validated for calibration range (ng), linearity, LOD (ng), LOQ (ng), repeatability, and accuracy using α-pinene, limonene, and acetone as model compounds. For *in situ* sampling, SPME and NTD devices were applied to a pine branch enclosed in a glass chamber previously described by Zini et al. [27]. Sampling was performed 20 minutes after enclosure. SPME extractions were carried out first (10 s exposure), followed by NTD sampling with volumes between 1 and 10 mL. Figure 7.10 shows plasmagram amplitude variations obtained from the chamber atmosphere after 10 s SPME (PDMS/DVB fiber) (A) and 1 mL NTD sampling (CAR/DVB) (B), as well as α-pinene concentration profiles at different times of day measured with both techniques ($n = 2$).

7.6 Conclusions and Future Trends

The BVOCs emitted by a plant, in particular when submitted to a biotic or an abiotic stress or when involved in a multitrophic interaction, are important markers

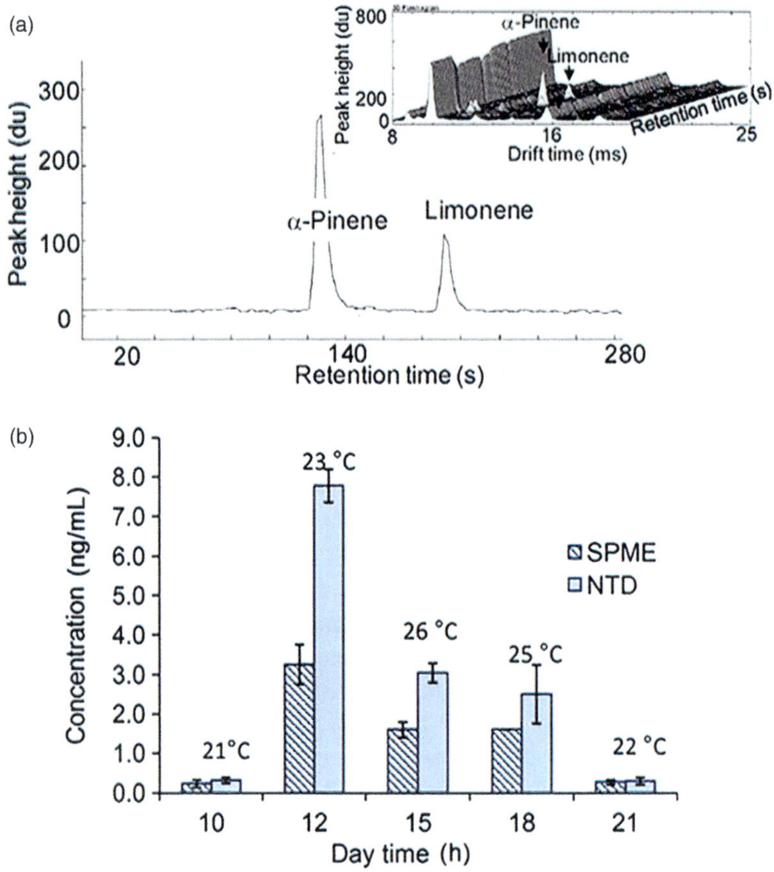

Figure 7.10 (a) Plasmagram amplitude variation from the glass enclosure over the pine branch after sampling of 10 seconds by SPME with a PDMS/DVB fiber (A) and of 1 mL with a CAR/DVB NTD (B). (b) Concentrations of α-pinene at different times along the same day using SPME and NTD ($n = 2$). *Source:* With permission of the copyright owner Reyes-Garcés et al. [98] / with permission of Elsevier.

of the reaction that hexogenous factors can induce. *In vivo* and, when possible, *in situ* analysis is the ideal approach for these studies as it provides meaningful information about both the plant metabolome and, above all, its variation under stress, as well as it reduces to a minimum the possibility of further "external" perturbation to the plant metabolism. This approach makes these analyses a complex task because it ideally requires dedicated analytical platforms involving all three of the main steps in an analytical procedure, i.e. sample preparation, analysis and data handling.

In vivo BVOC collection has found in headspace sampling (both static and dynamic) the approach of choice, that has greatly improved in reliability over the years mainly in quantitative terms also thanks to its ever increasing miniaturization and automation. *In vivo* and *in situ* HS potential has also highly grown in attractiveness thanks to the possibility to run sample preparation and analysis separately thanks to the development of high concentration capacity techniques (HS-SPME, HSSE etc.), in particular with SHS, and of new and very effective trapping material for DHS. BVOCs sampling by direct contact between specific plant parts and absorption/adsorption material is a further *in vivo* and *in situ* approach often complementary to HS providing highly effective analyte recovery and concentration, and very well performing in topographical emission studies.

If sampling allows for a stable storage for sampled BVOCs, analyses are usually performed at a later time over conventional platforms (GC-MS, GC×GC-MS etc.). If not, *in vivo* analyses (mainly vapour phase direct PTR-MS, or FTIR) should be carried out in laboratory (unless specific installations are available, e.g. see Bouvier-Brown et al. [70, 80]), and *in situ* analyses with portable platforms, better if miniaturized. The latter often suffers of lower performance when compared to conventional platforms that probably explains why only relatively low number of examples of *in situ*-analysis with portable platform are available in this field, despite the undoubted advantages of these systems in terms of time saving for experiments and reduction of sample alteration(s). An interesting perspective for field analysis is offered by micro-GC (and micro-GC-MS systems) based on micro-electromechanical systems (MEMS) technology [112]. This technology leads to extreme miniaturisation and great flexibility, which make it worthy of in-depth study and development, as it could provide a great and concrete contribution to overcoming several limitations of the instruments available today.

Data processing and handling, with its new achievements in artificial intelligence and machine learning, is also likely to induce a revolution in *in vivo* and *in situ* emission analysis in the near future, not only as a tool for method development and emission modelling, but also, and above all, in linking the chemistry of BVOCs with the biology associated with their emission.

List of Abbreviations

BVOCs	biogenic volatile organic compounds
CAR	carboxen
DHS	dynamic headspace
DCE	direct contact extraction
DC-SE	direct contact sorptive extraction

DC-STE	direct contact sorptive tape extraction
DC-SBSE	direct contact, stir bar sorptive extraction
DI-SPME	direct immersion SPME
DVB	divinylbenzene
ESE	equilibrium sorptive enrichment
GLVs	green leaf volatiles
HCC	high concentration capacity
HIPVs	herbivore-induced plant volatiles
HR MS	high resolution mass spectrometry
IMS	ion mobility spectrometry
IS	internal standard
IT-MS	miniature toroidal ion trap MS
MEMS	microelectromechanical-system
MHE	multiple headspace extraction
MTBE	methyl tert-butyl ether
MWCNTs	multi-walled carbon nanotubes
NTD	needle trap device
PAR	photosynthetically active radiation
PDMS	polydimethylsiloxane
PCA	principal component analysis
PET	polyethylene terephthalate
PEG	polyethylene glycol
PTFE	polytetrafluoroethylene
PTR	proton transfer reaction
RF	response factor
SHS	static headspace
SBSE	stir bar sorptive extraction
SESI	secondary electrospray ionization source
SIDA	stable isotope dilution assay
SOA	secondary organic aerosol

References

1 Tholl, D., Boland, W., Hansel, A. et al. (2006). Practical approaches to plant volatile analysis. *Plant J.* 45 (4): 540–560. https://doi.org/10.1111/j.1365-313X.2005.02612.x.

2 Maffei, M.E., Gertsch, J., and Appendino, G. (2011). Plant volatiles: Production, function and pharmacology. *Nat. Prod. Rep.* 28 (8): 1359–1380. https://doi.org/10.1039/c1np00021g.

3 Bicchi, C. and Maffei, M. (2012). The Plant Volatilome: Methods of Analysis. *High-Throughput Phenotyping in Plants: Methods and Protocols* (ed. J. Normanly), 289–310. Totowa, NJ: Humana Press https://doi.org/10.1007/978-1-61779-995-2_15.

4 Oliver, S.G., Winson, M.K., Kell, D.B., and Baganz, F. (1998). Systematic functional analysis of the yeast genome. *Trends Biotechnol.* 16 (9): 373–378. https://doi.org/10.1016/S0167-7799(98)01214-1.

5 Risticevic, S., Souza-Silva, E.A., Gionfriddo, E. et al. (2020). Application of *in vivo* solid phase microextraction (SPME) in capturing metabolome of apple (Malus ×domestica Borkh.) fruit. *Sci. Rep.* 10 (1): 6724. https://doi.org/10.1038/s41598-020-63817-8.

6 Cagliero, C., Sgorbini, B., Cordero, C. et al. (2014). Analytical Strategies for Multipurpose Studies of a Plant Volatile Fraction. *Handbook of Chemical and Biological Plant Analytical Methods* (ed. K. Hostettmann, H. Stuppner, A. Marston, and S. Chen), 447–466. Wiley https://doi.org/10.1002/9780470027318.a9919.

7 Ocaña-González, J.A., Fernández-Torres, R., Bello-López, M.Á., and Ramos-Payán, M. (2016). New developments in microextraction techniques in bioanalysis. A review. *Anal. Chim. Acta* 905: 8–23. https://doi.org/10.1016/j.aca.2015.10.041.

8 Risticevic, S., Souza-Silva, E.A., DeEll, J.R. et al. (2016). Capturing Plant Metabolome with Direct-Immersion in Vivo Solid Phase Microextraction of Plant Tissues. *Anal. Chem.* 88 (2): 1266–1274. https://doi.org/10.1021/acs.analchem.5b03684.

9 Mossaddegh, M. and Kazerooni, H. (2019). In Vivo Solid-Phase Microextraction: An Efficient Sample Preparation Method for Plant and Tissue Bioanalysis. *Iran. J. Anal. Chem.* 6 (2): 39–53. https://doi.org/10.30473/ijac.2019.48280.1156.

10 Majchrzak, T., Wojnowski, W., Rutkowska, M., and Wasik, A. (2020). Real-Time Volatilomics: A Novel Approach for Analyzing Biological Samples. *Trends Plant Sci.* 25 (3): 302–312. https://doi.org/10.1016/j.tplants.2019.12.005.

11 Zhu, F., Xu, J., Ke, Y. et al. (2013). Applications of *in vivo* and *in vitro* solid-phase microextraction techniques in plant analysis: A review. *Anal. Chim. Acta* 794: 1–14. https://doi.org/10.1016/j.aca.2013.05.016.

12 Manzo, A., Panseri, S., Vagge, I., and Giorgi, A. (2014). Volatile Fingerprint of Italian Populations of Orchids Using Solid Phase Microextraction and Gas Chromatography Coupled with Mass Spectrometry. *Molecules* 19 (6): 7913–7936. https://doi.org/10.3390/molecules19067913.

13 Kfoury, N., Scott, E., Orians, C., and Robbat, A. Jr. (2017). Direct Contact Sorptive Extraction: A Robust Method for Sampling Plant Volatiles in the

Field. *J. Agric. Food Chem.* 65 (38): 8501–8509. https://doi.org/10.1021/acs.jafc.7b02847.

14 Deng, X., Peng, J., Luo, B. et al. (2004). A direct quantitative analysis method for monitoring biogenic volatile organic compounds released from leaves of Pelargonium hortorum in situ. *Anal. Bioanal. Chem.* 380 (7–8): 950–957. https://doi.org/10.1007/s00216-004-2874-9.

15 Sgorbini, B., Cagliero, C., Cordero, C. et al. (2014). Headspace Sampling and Gas Chromatography of Plants: a Successful Combination to Study the Composition of a Plant Volatile Fraction. *Handbook of Chemical and Biological Plant Analytical Methods* (ed. K. Hostettmann, H. Stuppner, A. Marston, and S. Chen), 245–276. Wiley https://doi.org/10.1002/9780470027318.a9910.

16 L. Boggia, B. Sgorbini, C.M. Bertea et al. (2015). Direct Contact – Sorptive Tape Extraction coupled with Gas Chromatography – Mass Spectrometry to reveal volatile topographical dynamics of lima bean (Phaseolus lunatus L.) upon herbivory by Spodoptera littoralis Boisd.. *BMC Plant Biol.* 15: 102. https://doi.org/10.1186/s12870-015-0487-4.

17 Cagliero, C., Mastellone, G., Marengo, A. et al. (2021). Analytical strategies for in-vivo evaluation of plant volatile emissions - A review. *Anal. Chim. Acta* 1147: 240–258. https://doi.org/10.1016/j.aca.2020.11.029.

18 Huang, S., Chen, G., Ye, N. et al. (2019). Solid-phase microextraction: An appealing alternative for the determination of endogenous substances - A review. *Anal. Chim. Acta* 1077: 67–86. https://doi.org/10.1016/j.aca.2019.05.054.

19 Bicchi, C., Cordero, C., and Rubiolo, P. (2004). A Survey on High-Concentration-Capability Headspace Sampling Techniques in the Analysis of Flavors and Fragrances. *J. Chromatogr. Sci.* 42 (8): 402–409. https://doi.org/10.1093/chromsci/42.8.402.

20 Zhang, Z.Y. and Pawliszyn, J. (1993). Headspace solid-phase microextraction. *Anal. Chem.* 65 (14): 1843–1852. https://doi.org/10.1021/ac00062a008.

21 Souza-Silva, É.A. and Pawliszyn, J. (2017). Recent Advances in Solid-Phase Microextraction for Contaminant Analysis in Food Matrices. *Comprehensive Analytical Chemistry* (ed. E. Ibáñez and A. Cifuentes), 483–517. Elsevier. https://doi.org/10.1016/bs.coac.2017.03.002.

22 Vereen, D.A., McCall, J.P., and Butcher, D.J. (2000). Solid phase microextraction for the determination of volatile organics in the foliage of Fraser fir (Abies fraseri). *Microchem. J.* 65 (3): 269–276. https://doi.org/10.1016/S0026-265X(00)00122-3.

23 Flamini, G. and Cioni, P.L. (2010). Odour gradients and patterns in volatile emission of different plant parts and developing fruits of grapefruit (Citrus paradisi L.). *Food Chem.* 120 (4): 984–992. https://doi.org/10.1016/j.foodchem.2009.11.037.

24 Meija, J., Montes-Bayón, M., Le Duc, D.L. et al. (2002). Simultaneous Monitoring of Volatile Selenium and Sulfur Species from Se Accumulating Plants (Wild Type and Genetically Modified) by GC/MS and GC/ICPMS Using Solid-Phase Microextraction for Sample Introduction. *Anal. Chem.* 74 (22): 5837–5844. https://doi.org/10.1021/ac020285t.

25 Song, G., Xiao, J., Deng, C. et al. (2007). Use of solid-phase microextraction as a sampling technique for the characterization of volatile compounds emitted from Chinese daffodil flowers. *J. Anal. Chem.* 62 (7): 674–679, https://doi.org/10.1134/S1061934807070118.

26 Moreno-Martin, G., Sanz-Landaluze, J., Leon-Gonzalez, M.E., and Madrid, Y. (2019). In-vivo solid phase microextraction for quantitative analysis of volatile organoselenium compounds in plants. *Anal. Chim. Acta* 1081: 72–80. https://doi.org/10.1016/j.aca.2019.06.061.

27 Zini, C.A., Augusto, F., Christensen, E. et al. (2001). Monitoring Biogenic Volatile Compounds Emitted by Eucalyptus citriodora Using SPME. *Anal. Chem.* 73 (19): 4729–4735. https://doi.org/10.1021/ac0103219.

28 Fernandes, F., Pereira, D.M., Guedes de Pinho, P. et al. (2010). Headspace solid-phase microextraction and gas chromatography/ion trap-mass spectrometry applied to a living system: Pieris brassicae fed with kale. *Food Chem.* 119 (4): 1681–1693. https://doi.org/10.1016/j.foodchem.2009.09.046.

29 Danielsson, M., Zhao, T., and Borg-Karlson, A.-K. (2019). Arthropod infestation sites and induced defence can be traced by emission from single spruce needles. *Arth.-Plant Int.* 13 (2): 253–259. https://doi.org/10.1007/s11829-019-09677-0.

30 Stashenko, E.E., Andrés Ordóñez, S., Marín, N.A., and Martínez, J.R. (2009). Determination of the Volatile and Semi-volatile Secondary Metabolites, and Aristolochic Acids in Aristolochia ringens Vahl. *J. Chromatogr. Sci.* 47 (9): 817–821. https://doi.org/10.1093/chromsci/47.9.817.

31 Bar-Akiva, A., Ovadia, R., Rogachev, I. et al. (2010). Metabolic networking in Brunfelsia calycina petals after flower opening. *J. Exp. Bot.* 61 (5): 1393–1403. https://doi.org/10.1093/jxb/erq008.

32 Chen, Y.-J., Cheng, S.-S., and Chang, S.-T. (2010). Monitoring the emission of volatile organic compounds from the leaves of Calocedrus macrolepis var. formosana using solid-phase micro-extraction. *J. Wood Sci.* 56 (2): 140–147. https://doi.org/10.1007/s10086-009-1071-z.

33 Giorgi, A., Manzo, A., Nanayakkara, N.N. et al. (2015). Effect of biotic and abiotic stresses on volatile emission of Achillea collina Becker ex Rchb. *Nat. Prod. Res.* 29 (18): 1695–1702. https://doi.org/10.1080/14786419.2014.997725.

34 Risticevic, S. and Pawliszyn, J. (2013). Solid-Phase Microextraction in Targeted and Nontargeted Analysis: Displacement and Desorption Effects. *Anal. Chem.* 85 (19): 8987–8995. https://doi.org/10.1021/ac4003112.

35 Matsunaga, S.N., Chatani, S., Nakatsuka, S. et al. (2012). Determination and potential importance of diterpene (kaur-16-ene) emitted from dominant coniferous trees in Japan. *Chemosphere* 87 (8): 886–893. https://doi.org/10.1016/j.chemosphere.2012.01.040.

36 Flamini, G., Cioni, P.L., and Morelli, I. (2002). Analysis of the essential oil of the aerial parts of Viola etrusca from Monte Labbro (South Tuscany, Italy) and in vivo analysis of flower volatiles using SPME. *Flavour Fragr. J.* 17 (2): 147–149, https://doi.org/10.1002/ffj.1060.

37 Verdonk, J.C., Ric de Vos, C.H., Verhoeven, H.A. et al. (2003). Regulation of floral scent production in petunia revealed by targeted metabolomics. *Phytochemistry* 62 (6): 997–1008. https://doi.org/10.1016/S0031-9422(02)00707-0.

38 Flamini, G., Cioni, P.L., Morelli, I. et al. (2004). Phytochemical typologies in some populations of Myrtus communis L. on Caprione Promontory (East Liguria, Italy). *Food Chem.* 85 (4): 599–604. https://doi.org/10.1016/j.foodchem.2003.08.005.

39 Flamini, G., Cioni, P.L., and Morelli, I. (2005). Composition of the essential oils and in vivo emission of volatiles of four Lamium species from Italy: L. purpureum, L. hybridum, L. bifidum and L. amplexicaule. *Food Chem.* 91 (1): 63–68. https://doi.org/10.1016/j.foodchem.2004.05.047.

40 Maccioni, S., Baldini, R., Cioni, P.L. et al. (2007). In vivo volatiles emission and essential oils from different organs and pollen of Cistus albidus from Caprione (Eastern Liguria, Italy). *Flavour Fragr. J.* 22 (1): 61–65. https://doi.org/10.1002/ffj.1759.

41 Eilers, E.J., Pauls, G., Rillig, M.C. et al. (2015). Novel Set-Up for Low-Disturbance Sampling of Volatile and Non-volatile Compounds from Plant Roots. *J. Chem. Ecol.* 41 (3): 253–266. https://doi.org/10.1007/s10886-015-0559-9.

42 Wahlroos, O. and Nikkilä, O.E. (1966). Quantitative determination of volatile compounds from food and evaluation of the gas flushing method for their concentration. *Acta Chem. Scand.* 20 (1): 197–205. https://doi.org/10.3891/acta.chem.scand.20-0197.

43 Herout, V. (1967). Gaschromatographie Von Terpenoiden. *Planta Med.* 68–78. https://doi.org/10.1055/s-0028-1100015.

44 Vercammen, J., Pham-Tuan, H., and Sandra, P. (2001). Automated dynamic sampling system for the on-line monitoring of biogenic emissions from living organisms. *J. Chromatogr. A* 930 (1–2): 39–51. https://doi.org/10.1016/S0021-9673(01)01172-4.

45 Beauchamp, J., Wisthaler, A., Hansel, A. et al. (2005). Ozone induced emissions of biogenic VOC from tobacco: relationships between ozone uptake and emission of LOX products. *Plant Cell Environ.* 28 (10): 1334–1343. https://doi.org/10.1111/j.1365-3040.2005.01383.x.

46 Effmert, U., Große, J., Röse, U.S. et al. (2005). Volatile composition, emission pattern, and localization of floral scent emission in Mirabilis jalapa (Nyctaginaceae). *Am. J. Bot.* 92 (1): 2–12. https://doi.org/10.3732/ajb.92.1.2.

47 Baker, B. and Sinnott, M. (2009). Analysis of sesquiterpene emissions by plants using solid phase microextraction. *J. Chromatogr. A* 1216 (48): 8442–8451. https://doi.org/10.1016/j.chroma.2009.10.002.

48 Staudt, M. and Lhoutellier, L. (2011). Monoterpene and sesquiterpene emissions from Quercus coccifera exhibit interacting responses to light and temperature. *Biogeosciences* 8 (9): 2757–2771. https://doi.org/10.5194/bg-8-2757-2011.

49 Barreira, L.M.F., Duporte, G., Parshintsev, J. et al. (2017). Field measurements of biogenic volatile organic compounds in the atmosphere using solid-phase microextraction Arrow. *Boreal Environ. Res.* 22: 393–413. https://doi.org/10.5194/amt-11-881-2018.

50 Geron, C., Guenther, A., Sharkey, T., and Arnts, R.R. (2000). Temporal variability in basal isoprene emission factor. *Tree Physiol.* 20 (12): 799–805. https://doi.org/10.1093/treephys/20.12.799.

51 Holzinger, R., Sandoval-Soto, L., Rottenberger, S. et al. (2000). Emissions of volatile organic compounds from Quercus ilex L. measured by Proton Transfer Reaction Mass Spectrometry under different environmental conditions. *J. Geophys. Res. Atmos.* 105 (D16): 20573–20579, https://doi.org/10.1029/2000JD900296.

52 Pétron, G., Harley, P., Greenberg, J., and Guenther, A. (2001). Seasonal temperature variations influence isoprene emission. *Geophys. Res. Lett.* 28 (9): 1707–1710. https://doi.org/10.1029/2000GL011583.

53 Owen, S.M., Harley, P., Guenther, A., and Hewitt, C.N. (2002). Light dependency of VOC emissions from selected Mediterranean plant species. *Atmos. Environ.* 36 (19): 3147–3159. https://doi.org/10.1016/S1352-2310(02)00235-2.

54 Boamfa, E.I., Steeghs, M.M.L., Cristescu, S.M., and Harren, F.J.M. (2004). Trace gas detection from fermentation processes in apples; an intercomparison study between proton-transfer-reaction mass spectrometry and laser photoacoustics. *Int. J. Mass Spectrom.* 239 (2–3): 193–201. https://doi.org/10.1016/j.ijms.2004.07.017.

55 Graus, M., Schnitzler, J.P., Hansel, A. et al. (2004). Transient Release of Oxygenated Volatile Organic Compounds during Light-Dark Transitions in Grey Poplar Leaves. *Plant Physiol.* 135 (4): 1967–1975. https://doi.org/10.1104/pp.104.043240.

56 Penuelas, J., Filella, I., Stefanescu, C., and Llusia, J. (2005). Caterpillars of Euphydryas aurinia (Lepidoptera: Nymphalidae) feeding on Succisa pratensis leaves induce large foliar emissions of methanol. *New Phytol.* 167 (3): 851–857. https://doi.org/10.1111/j.1469-8137.2005.01459.x.

57 Dindorf, T., Kuhn, U., Ganzeveld, L. et al. (2006). Significant light and temperature dependent monoterpene emissions from European beech (Fagus sylvatica L.) and their potential impact on the European volatile organic compound budget. *J. Geophys. Res.* 111 (D16): https://doi.org/10.1029/2005JD006751.

58 Moukhtar, S., Couret, C., Rouil, L., and Simon, V. (2006). Biogenic Volatile Organic Compounds (BVOCs) emissions from Abies alba in a French forest. *Sci. Total Environ.* 354 (2–3): 232–245. https://doi.org/10.1016/j.scitotenv.2005.01.044.

59 Kreuzwieser, J., Scheerer, U., Kruse, J. et al. (2014). The Venus flytrap attracts insects by the release of volatile organic compounds. *J. Exp. Bot.* 65 (2): 755–766. https://doi.org/10.1093/jxb/ert455.

60 Malik, T.G., Gajbhiye, T., and Pandey, S.K. (2018). Plant specific emission pattern of biogenic volatile organic compounds (BVOCs) from common plant species of Central India. *Environ. Monit. Assess.* 190 (11): 631. https://doi.org/10.1007/s10661-018-7015-6.

61 Baltussen, E., David, F., Sandra, P. et al. (1999). Equilibrium Sorptive Enrichment on Poly(dimethylsiloxane) Particles for Trace Analysis of Volatile Compounds in Gaseous Samples. *Anal. Chem.* 71 (22): 5193–5198. https://doi.org/10.1021/ac990362t.

62 McCartney, M.M., Spitulski, S.L., Pasamontes, A. et al. (2016). Coupling a branch enclosure with differential mobility spectrometry to isolate and measure plant volatiles in contained greenhouse settings. *Talanta* 146: 148–154. https://doi.org/10.1016/j.talanta.2015.08.039.

63 Chen, F., Tholl, D., D'Auria, J.C. et al. (2003). Biosynthesis and Emission of Terpenoid Volatiles from Arabidopsis Flowers. *Plant Cell* 15 (2): 481–494. https://doi.org/10.1105/tpc.007989.

64 Pressley, S. (2004). Monoterpene emissions from a Pacific Northwest Old-Growth Forest and impact on regional biogenic VOC emission estimates. *Atmos. Environ.* 38 (19): 3089–3098. https://doi.org/10.1016/j.atmosenv.2004.01.045.

65 Lin, C.-Y., Chang, T.-C., Chen, Y.-H. et al. (2015). Monitoring the dynamic emission of biogenic volatile organic compounds from Cryptomeria japonica by enclosure measurement. *Atmos. Environ.* 122: 163–170. https://doi.org/10.1016/j.atmosenv.2015.09.046.

66 Vedel-Petersen, I., Schollert, M., Nymand, J., and Rinnan, R. (2015). Volatile organic compound emission profiles of four common arctic plants. *Atmos. Environ.* 120: 117–126. https://doi.org/10.1016/j.atmosenv.2015.08.082.

67 Silva, R., Tinoco, N., Tsukui, A. et al. (2019). Floral Scent and Nectar Sugar Composition of Temnadenia odorifera(Apocynoideae, Apocynaceae). *J. Braz. Chem. Soc.* 30: 388–397. https://doi.org/10.21577/0103-5053.20180188.

68 Helmig, D., Ortega, J., Guenther, A. et al. (2006). Sesquiterpene emissions from loblolly pine and their potential contribution to biogenic aerosol formation in the Southeastern US. *Atmos. Environ.* 40 (22): 4150–4157. https://doi.org/10.1016/j.atmosenv.2006.02.035.

69 Helmig, D., Ortega, J., Duhl, T. et al. (2007). Sesquiterpene Emissions from Pine Trees – Identifications, Emission Rates and Flux Estimates for the Contiguous United States. *Environ. Sci. Technol.* 41 (5): 1545–1553. https://doi.org/10.1021/es0618907.

70 Bouvier-Brown, N.C., Holzinger, R., Palitzsch, K., and Goldstein, A.H. (2009). Large emissions of sesquiterpenes and methyl chavicol quantified from branch enclosure measurements. *Atmos. Environ.* 43 (2): 389–401. https://doi.org/10.1016/j.atmosenv.2008.08.039.

71 Holzke, C., Dindorf, T., Kesselmeier, J. et al. (2006). Terpene emissions from European beech (Fagus sylvatica L.): Pattern and emission behaviour overtwo vegetation periods. *J. Atmos. Chem.* 55 (1): 81–102. https://doi.org/10.1007/s10874-006-9027-9.

72 Crespo, E., Hordijk, C.A., de Graaf, R.M. et al. (2012). On-line detection of root-induced volatiles in Brassica nigra plants infested with Delia radicum L. root fly larvae. *Phytochemistry* 84: 68–77. https://doi.org/10.1016/j.phytochem.2012.08.013.

73 Farré-Armengol, G., Filella, I., Llusia, J. et al. (2015). Enhanced emissions of floral volatiles by Diplotaxis erucoides (L.) in response to folivory and florivory by Pieris brassicae (L.). *Biochem. Syst. Ecol.* 63: 51–58. https://doi.org/10.1016/j.bse.2015.09.022.

74 Barrios-Collado, C., Garcia-Gomez, D., Zenobi, R. et al. (2016). Capturing in Vivo Plant Metabolism by Real-Time Analysis of Low to High Molecular Weight Volatiles. *Anal. Chem.* 88 (4): 2406–2412. https://doi.org/10.1021/acs.analchem.5b04452.

75 Caceres, L.A., Lakshminarayan, S., Yeung, K.K. et al. (2016). Repellent and Attractive Effects of α-, β-, and Dihydro-β-Ionone to Generalist and Specialist Herbivores. *J. Chem. Ecol.* 42 (2): 107–117. https://doi.org/10.1007/s10886-016-0669-z.

76 Stewart-Jones, A. and Poppy, G.M. (2006). Comparison of Glass Vessels and Plastic Bags for Enclosing Living Plant Parts for Headspace Analysis. *J. Chem. Ecol.* 32: 845–864. https://doi.org/10.1007/s10886-006-9039-6.

77 Rose, U.S. and Tumlinson, J.H. (2004). Volatiles released from cotton plants in response to Helicoverpa zea feeding damage on cotton flower buds. *Planta* 218 (5): 824–832. https://doi.org/10.1007/s00425-003-1162-9.

78 Yassaa, N., Meklati, B.Y., and Cecinato, A. (2000). Evaluation of monoterpenic biogenic volatile organic compounds in ambient air around Eucalyptus globulus, Pinus halepensis and Cedrus atlantica trees growing in Algiers city area by chiral and achiral capillary gas chromatography. *Atmos. Environ.* 34 (17): 2809–2816. https://doi.org/10.1016/S1352-2310(99)00436-7.

79 Yassaa, N. and Williams, J. (2007). Enantiomeric monoterpene emissions from natural and damaged Scots pine in a boreal coniferous forest measured using solid-phase microextraction and gas chromatography/mass spectrometry. *J. Chromatogr. A* 1141 (1): 138–144. https://doi.org/10.1016/j.chroma.2006.12.006.

80 Bouvier-Brown, N.C., Holzinger, R., Palitzsch, K., and Goldstein, A.H. (2007). Quantifying sesquiterpene and oxygenated terpene emissions from live vegetation using solid-phase microextraction fibers. *J. Chromatogr. A* 1161 (1–2): 113–120. https://doi.org/10.1016/j.chroma.2007.05.094.

81 Pankow, J.F., Luo, W., Melnychenko, A.N. et al. (2012). Volatilizable Biogenic Organic Compounds (VBOCs) with two dimensional Gas Chromatography-Time of Flight Mass Spectrometry (GC × GC-TOFMS): sampling methods, VBOC complexity, and chromatographic retention data. *Atmos. Meas. Tech.* 5 (2): 345–361. https://doi.org/10.5194/amt-5-345-2012.

82 Souza-Silva, E.A. and Pawliszyn, J. (2012). Optimization of Fiber Coating Structure Enables Direct Immersion Solid Phase Microextraction and High-Throughput Determination of Complex Samples. *Anal. Chem.* 84 (16): 6933–6938. https://doi.org/10.1021/ac301305u.

83 Scott, E.R., Li, X., Kfoury, N. et al. (2019). Interactive effects of drought severity and simulated herbivory on tea (Camellia sinensis) volatile and non-volatile metabolites. *Environ. Exp. Bot.* 157: 283–292. https://doi.org/10.1016/j.envexpbot.2018.10.025.

84 Wong, Y.F., Yan, D., Shellie, R.A. et al. (2019). Rapid Plant Volatiles Screening Using Headspace SPME and Person-Portable Gas Chromatography–Mass Spectrometry. *Chromatographia* 82: 297–305. https://doi.org/10.1007/s10337-018-3605-2.

85 De Saint Laumer, J.Y., Cicchetti, E., Merle, P. et al. (2010). Quantification in Gas Chromatography: Prediction of Flame Ionization Detector Response Factors from Combustion Enthalpies and Molecular Structures. *Anal. Chem.* 82: 6457–6462. https://doi.org/10.1021/ac1006574.

86 Errard, A., Ulrichs, C., Kuhne, S. et al. (2015). Single- versus Multiple-Pest Infestation Affects Differently the Biochemistry of Tomato (Solanum lycopersicum 'Ailsa Craig'). *J. Agric. Food Chem.* 63 (46): 10103–10111. https://doi.org/10.1021/acs.jafc.5b03884.

87 Sgorbini, B., Cagliero, C., Liberto, E. et al. (2019). Strategies for Accurate Quantitation of Volatiles from Foods and Plant-Origin Materials: A Challenging Task. *J. Agric. Food Chem.* 67 (6): 1619–1630. https://doi.org/10.1021/acs.jafc.8b06601.

88 Loi, R.X., Solar, M.C., and Weidenhamer, J.D. (2008). Solid-Phase Microextraction Method For In Vivo Measurement of Allelochemical Uptake. *J. Chem. Ecol.* 34 (1): 70–75. https://doi.org/10.1007/s10886-007-9401-3.

89 Tarvainen, V., Hakola, H., Hellén, H. et al. (2005). Temperature and light dependence of the VOC emissions of Scots pine. *Atmos. Chem. Phys.* 5 (4): 989–998. https://doi.org/10.5194/acp-5-989-2005.

90 Reiche, N., Mothes, F., Fiedler, P., and Borsdorf, H. (2013). A solid-phase microextraction method for the in vivo sampling of MTBE in common reed (Phragmites australis). *Environ. Monit. Assess.* 185 (9): 7133–7144. https://doi.org/10.1007/s10661-013-3089-3.

91 Chen, G., Qiu, J., Liu, Y. et al. (2015). Carbon Nanotubes Act as Contaminant Carriers and Translocate within Plants. *Sci. Rep.* 5: 15682. https://doi.org/10.1038/srep15682.

92 Scott, E.R., Li, X., Wei, J.-P. et al. (2020). Changes in Tea Plant Secondary Metabolite Profiles as a Function of Leafhopper Density and Damage. *Front. Plant Sci.* 11: 636. https://doi.org/10.3389/fpls.2020.00636.

93 Schieberle, P. and Grosch, W. (1987). Quantitative analysis of aroma compounds in wheat and rye bread crusts using a stable isotope dilution assay. *J. Agric. Food Chem.* 35 (2): 252–257. https://doi.org/10.1021/jf00074a021.

94 Wang, Y., OReilly, J., Chen, Y., and Pawliszyn, J. (2005). Equilibrium in-fibre standardisation technique for solid-phase microextraction. *J. Chromatogr. A* 1072 (1): 13–17. https://doi.org/10.1016/j.chroma.2004.12.084.

95 Cagliero, C., Bicchi, C., Cordero, C. et al. (2017). Analysis of essential oils and fragrances with a new generation of highly inert gas chromatographic columns coated with ionic liquids. *J. Chromatogr. A* 1495: 64–75. https://doi.org/10.1016/j.chroma.2017.03.029.

96 van Dam, N.M. and Poppy, G.M. (2008). Why plant volatile analysis needs bioinformatics – detecting signal from noise in increasingly complex profiles. *Plant Biol. (Stuttg.)* 10 (1): 29–37. https://doi.org/10.1055/s-2007-964961.

97 Miller, R.A., Eiceman, G.A., Nazarov, E.G., and King, A.T. (2000). A novel micromachined high-field asymmetric waveform-ion mobility spectrometer. *Sens. Actuator B Chem.* 67 (3): 300–306. https://doi.org/10.1016/S0925-4005(00)00535-9.

98 Reyes-Garcés, N., Gómez-Ríos, G.A., Souza Silva, É.A., and Pawliszyn, J. (2013). Coupling needle trap devices with gas chromatography–ion mobility spectrometry detection as a simple approach for on-site quantitative analysis. *J. Chromatogr. A* 1300: 193–198. https://doi.org/10.1016/j.chroma.2013.05.042.

99 Liu, S., Gonga, D., Wanga, Y. et al. (2024). Responses of plant volatile emissions to increasing nitrogen deposition: A pilot study on Eucalyptus urophylla. *Sci. Total Environ.* 952: 175887. https://doi.org/10.1016/j.scitotenv.2024.175887.

100 Ma, F., Tang, X., Zhang, G., and Yi, Z. (2024). Isoprenoid emissions from Schima superba and Cunninghamia lanceolata: Their responses to elevated temperature by two warming facilities. *Sci. Total Environ.* 930: 172669. https://doi.org/10.1016/j.sctotenv.2024.172669.

101 Mannisto, E., Ylannea, H., Kokkonen, N. et al. (2024). Impact of severe drought on biogenic volatile organic compounds emissions from Sphagnum mosses in boreal peatlands. *Sci. Total Environ.* 951: 175738. https://doi.org/10.1016/j.scitotenv.2024.175738.

102 Chen, W., Wang, L., Wu, J. et al. (2024). Emission Pattern of Biogenic Volatile Organic Compounds from Wetland Vegetation. *Atmos.* 15: 651. https://doi.org/10.3390/atmos15060651.

103 Yuan, X., Du, Y., Agathokleous, E. et al. (2024). Differential responses and mechanisms of monoterpene emissions from broad-leaved and coniferous species under elevated ozone scenarios. *Sci. Total Environ.* 951: 175291. https://doi.org/10.1016/j.scitotenv.2024.175291.

104 Sisalli, S., Adao, A., Lebel, M. et al. (2006). Sorptive Tape Extraction — A Novel Sampling Method for the in vivo Study of Skin. *LC-GC Eur.* 19: 33–39. https://www.chromatographyonline.com/view/sorptive-tape-extraction-novel-sampling-method-vivo-study-skin (last accessed Feb. 2026)

105 Bicchi, C., Cordero, C., Liberto, E. et al. (2007). Sorptive tape extraction in the analysis of the volatile fraction emitted from biological solid matrices. *J. Chromatogr. A* 1148 (2): 137–144. https://doi.org/10.1016/j.chroma.2007.03.007.

106 Cordero, C., Zebelo, S.A., Gnavi, G. et al. (2012). HS-SPME-GC×GC-qMS volatile metabolite profiling of Chrysolina herbacea frass and Mentha spp. leaves. *Anal. Bioanal. Chem.* 402 (5): 1941–1952. https://doi.org/10.1007/s00216-011-5600-4.

107 Pizzolante, G., Cordero, C., Tredici, S.M. et al. (2017). Cultivable gut bacteria provide a pathway for adaptation of Chrysolina herbacea to Mentha aquatica volatiles. *BMC Plant Biol.* 17: 30. https://doi.org/10.1186/s12870-017-0986-6.

108 Setkova, L., Risticevic, S., Linton, C.M. et al. (2007). Solid-phase microextraction–gas chromatography–time-of-flight mass spectrometry utilized for the evaluation of the new-generation super elastic fiber assemblies. *Anal. Chim. Acta* 581: 221–231. https://doi.org/10.1016/j.aca.2006.08.022.

109 Cordero, C., Bicchi, C., and Rubiolo, P. (2008). Group-Type and Fingerprint Analysis of Roasted Food Matrices (Coffee and Hazelnut Samples) by Comprehensive Two-Dimensional Gas Chromatography. *J. Agric. Food Chem.* 56: 7655–7666. https://doi.org/10.1021/jf801001z.

110 Borsdorf, H. and Eiceman, G.A. (2006). Ion Mobility Spectrometry: Principles and Applications. *Appl. Spectrosc. Rev.* 41 (4): 323–375. https://doi.org/10.1080/05704920600663469.

111 Hill, H.H., Siems, W.F., and St. Louis, R.H. (1990). Ion Mobility Spectrometry. *Anal. Chem.* 62 (23): 1201A. https://doi.org/10.1021/ac00222a001.

112 Cagliero, C., Galli, S., Galli, M. et al. (2016). Conventional and enantioselective gas chromatography with microfabricated planar columns for analysis of real-world samples of plant volatile fraction. *J. Chromatogr. A* 1429: 329–339. https://doi.org/10.1016/j.chroma.2015.12.037.

8.1

Specialized Metabolites in Biological Systems (Part I): Focus on Derivatized Metabolites in Human Studies

Jacopo Troisi[1,2,3] *and Martina Lombardi*[1,2,3]

[1]*Department of Medicine, Surgery and Dentistry "Scuola Medica Salernitana", University of Salerno, Baronissi, Italy*
[2]*Theoreo srl, Montecorvino Pugliano, Italy*
[3]*European Institute of Metabolomics (EIM) Foundation, Baronissi, Italy*

8.1.1 Specialized Metabolites and Their Importance in Various Biological Processes

8.1.1.1 Introduction

Specialized metabolites (SMs), also known as secondary metabolites, are small organic molecules synthesized by organisms, primarily plants, fungi, and bacteria, which are not directly involved in growth, development, or reproduction but play crucial roles in interactions with the environment, defense mechanisms, and communication with other organisms. Although SMs are often associated with bacterial or plant studies, animals also possess numerous molecules with pivotal functions. Traditional feeding investigations show that while some come from the diet, many are synthesized within the animals themselves.

Over the past two decades, numerous biosynthetic pathways traditionally attributed to animals have been traced back to symbiotic bacteria, which form close associations with animals for chemical defense [1]. This discovery has led some researchers to hypothesize that most, if not all, animal secondary metabolites originate from microbes. However, this notion is contradicted by the fact that humans synthesize secondary metabolites, including steroids, prostaglandins, lipids, melanins, neurotransmitters, G protein-coupled receptor ligands, and other related compounds, whose biosynthetic processes are now well-established. The apparent familiarity and imprecision of the underlying definitions might make these compounds seem less "secondary." In contrast, the genetic basis for important compounds unique to nonhuman animals has been less understood, with notable exceptions being a few economically significant insect pheromones [2, 3].

Recent advancements in sequencing technologies and bioinformatics, coupled with thorough genetic and biochemical analyses, have revolutionized our understanding of metabolism in nonhuman animals. The origins of myriad secondary metabolites from diverse animal species are beginning to be elucidated. The full scope and significance of these advances are not immediately evident, as the existing literature predominantly focuses on specific groups, such as insects or marine natural products. By adopting a more global perspective, emerging trends can be identified that enhance our understanding of the unique mechanisms through which animals produce complex secondary metabolites [4].

8.1.1.2 Role in Defense Mechanisms

SMs are critical components in the defense mechanisms of animals, providing a multifaceted shield against a variety of biotic and abiotic threats. These compounds encompass a broad spectrum of chemical entities, such as alkaloids, terpenoids, and peptides, each serving unique and essential functions in the survival strategies of their producers.

Alkaloids, for example, are nitrogen-containing compounds that have been extensively studied for their potent toxic properties. In poison dart frogs (*Dendrobatidae*), these alkaloids are sequestered from their diet and concentrated in their skin, rendering the frogs highly toxic to potential predators. The alkaloids act on the nervous systems of these predators, causing paralysis or death, thus providing an effective deterrent [5]. This remarkable chemical defense has not only fascinated scientists but has also contributed significantly to our understanding of predator–prey interactions in ecological systems.

Antimicrobial peptides (AMPs) represent another class of SMs that play a vital role in animal defense. These small, cationic peptides are produced by a variety of organisms, including amphibians, insects, and mammals. In amphibians, AMPs are secreted onto the skin and serve as a frontline defense against microbial infections. They exhibit broad-spectrum antimicrobial activity, targeting bacteria, fungi, and viruses by disrupting their cell membranes [6]. The effectiveness of AMPs in protecting amphibians from pathogenic microorganisms underscores their importance in maintaining the health and survival of these species in their natural habitats.

Terpenoids, another diverse group of SMs, are particularly prominent in marine organisms such as sponges. These compounds have been shown to possess strong antifouling and antimicrobial properties, which help marine sponges deter the colonization of their surfaces by other organisms and inhibit the growth of microbial pathogens. By producing these terpenoids, marine sponges can secure their space and resources in the highly competitive marine environment [7]. This chemical defense mechanism not only ensures the sponges' survival but also plays a crucial role in the dynamics of marine ecosystems. The biosynthesis and regulation of these defensive metabolites involve complex genetic and biochemical pathways, reflecting the evolutionary adaptations that have evolved to enhance an organism's survival and fitness. For instance, the genes involved in the biosynthesis of alkaloids, AMPs, and terpenoids are tightly regulated to ensure their production is efficient and effective in response to environmental threats. Understanding these pathways has provided significant insights into the molecular mechanisms underlying chemical defense in animals. Moreover, the study of SMs extends beyond ecological and evolutionary biology. These compounds have immense potential applications in biotechnology, medicine, and agriculture. For instance, the potent antimicrobial properties of AMPs have spurred research into their potential as novel antibiotics in the face of increasing antibiotic resistance. Similarly, the unique chemical structures of alkaloids and terpenoids offer promising leads for the development of new pharmaceuticals and biocides [8].

8.1.1.3 Role in Signaling and Communication

SMs also play a crucial role in signaling and communication among animals, mediating interactions within and between species. These chemical signals, often referred to as pheromones or semiochemicals, are essential for a variety of behaviors, including mating, territoriality, alarm responses, and social organization. Pheromones, for instance, are highly specific chemical signals that can trigger immediate and profound behavioral or physiological responses in conspecifics. In insects, pheromones are pivotal for reproductive success; for example, female moths release sex pheromones to attract males over long distances. The specificity and potency of these signals ensure that males can accurately locate potential mates, thereby enhancing reproductive efficiency [9]. Alarm pheromones are another class of SMs used for communication, particularly in social insects such as ants, bees, and termites. When an individual is threatened or injured, it releases these compounds to alert colony members to the presence of danger. This chemical cue prompts a coordinated defensive response, including aggressive behavior toward the threat and the mobilization of workers to protect the colony [10]. Such sophisticated chemical communication systems are vital for the survival and success of social insect colonies, highlighting the importance of SMs in facilitating complex social behaviors.

In vertebrates, SMs also play significant roles in social and reproductive behaviors. For example, pheromones are involved in the regulation of reproductive cycles in mammals. Female mice release pheromones in their urine that can accelerate or delay the estrous cycle of other females, a phenomenon known as the Whitten effect. This synchronization of reproductive cycles within a population can increase the chances of successful mating and reproduction [11]. Similarly, male pheromones can induce estrus in females, ensuring that mating occurs when the chances of successful fertilization are highest [12].

SMs also facilitate communication between different species, often mediating mutualistic or antagonistic interactions. For instance, certain plants release volatile organic compounds (VOCs) in response to herbivory, attracting predators or parasitoids of the herbivores. These VOCs act as indirect defense mechanisms, recruiting natural enemies to reduce herbivore pressure [13]. This interspecific chemical communication underscores the complexity and ecological significance of SMs in mediating interactions across trophic levels. In addition, symbiosis mechanisms between plants and fungal species also appear to be mediated by volatile metabolites [14]. The biosynthesis of these signaling compounds is often highly regulated and context-dependent, ensuring that the right signals are produced at the appropriate time. This regulation involves complex biochemical pathways that respond to environmental cues, physiological states, and social contexts. The ability to produce and respond to these chemical signals is a result of evolutionary pressures that have shaped the communication strategies of animals over time. Research into the molecular mechanisms underlying pheromone biosynthesis and detection has provided valuable insights into animal behavior and evolution. For example, studies on the genetic and enzymatic pathways involved in pheromone production in insects have revealed how small changes in these pathways can lead to the evolution of new communication signals and behaviors [15]. Understanding these processes not only sheds light on the biology of communication but also has practical applications in pest management, where synthetic pheromones are used to disrupt mating patterns of agricultural pests [16].

8.1.1.4 Role Adaptation to Environmental Stress

SMs also play a crucial role in the adaptation of animals to environmental stress, enabling them to survive and thrive under challenging conditions. These compounds, which include osmolytes and antioxidants, help organisms mitigate the effects of various stressors such as temperature fluctuations, osmotic changes, and oxidative stress.

As the term suggests, osmolytes, such as betaine, taurine, and trehalose, help animals cope with osmotic stress. These small organic molecules stabilize proteins and cellular structures, thereby protecting cells from the detrimental effects of dehydration and hyperosmolarity. In marine organisms, osmolytes play a vital role in osmoregulation, allowing these animals to survive in environments with varying salinities. For example, the marine invertebrate *Echinocardium cordatum* accumulates high levels of free amino acids and other osmolytes to balance the osmotic pressure of its internal fluids with that of the surrounding seawater [17]. This ability to regulate osmolyte levels is essential for maintaining cellular integrity and function in fluctuating saline conditions.

Antioxidants are SMs that protect animals from oxidative stress by neutralizing reactive oxygen species (ROS). These are by-products of normal cellular metabolism but can increase dramatically under environmental stress, such as exposure to pollutants, UV radiation, or extreme temperatures. Accumulation of ROS can lead to oxidative damage to proteins, lipids, and DNA, potentially causing cellular dysfunction and death. Antioxidants, such as superoxide dismutase, catalase, and glutathione, mitigate this damage by scavenging ROS and repairing oxidative damage. Studies have demonstrated that antioxidant defenses are crucial for the survival of many organisms under oxidative stress conditions. For instance, *Antarctic fish* have evolved enhanced antioxidant enzyme activities to cope with the high levels of ROS generated in their cold, oxygen-rich environments [18]. In addition to these well-known mechanisms, SMs also contribute to the adaptation to environmental stress through more specific and diverse pathways. For example, certain plants and insects produce protective compounds such as polyphenols and flavonoids, which have been shown to enhance resistance to UV radiation and other environmental stressors. These compounds absorb UV light and act as antioxidants, protecting cells from damage [19]. In another example, some amphibians produce glycopeptides in their skin that prevent dehydration and allow them to survive in arid environments by reducing water loss through evaporation. The biosynthesis and regulation of these SMs are tightly controlled and often induced by specific environmental cues. This regulation involves complex genetic and biochemical pathways that enable organisms to respond dynamically to environmental changes. Understanding these adaptive mechanisms provides valuable insights into the resilience and plasticity of animal species in the face of environmental challenges. These researches have broad implications for ecology, evolution, and biotechnology. For instance, understanding how organisms produce and utilize these compounds can inform conservation strategies for species threatened by climate change and habitat loss. Additionally, insights into the molecular mechanisms of stress adaptation can lead to the development of new biotechnological applications, such as stress-resistant crops and novel pharmaceuticals for managing oxidative stress-related disease in humans [20].

8.1.1.5 Significance in Pharmaceutical and Industrial Applications

SMs are of paramount pharmacological relevance due to their diverse and potent bioactivities, which have significantly contributed to drug discovery and development. These naturally occurring compounds, produced by various organisms, exhibit a wide range of therapeutic properties, including antimicrobial, anticancer, anti-inflammatory, and analgesic effects. The study and utilization of these metabolites have led to the development of numerous pharmaceuticals that address critical medical needs and improve human health. One of the most significant contributions of SMs to pharmacology is in the field of antimicrobial therapy. The discovery of antibiotics from microbial metabolites revolutionized medicine in the 20th century, saving countless lives and reducing mortality from bacterial infections. For instance, penicillin, produced by the mold *Penicillium notatum*, was the first antibiotic discovered and remains a cornerstone of antimicrobial therapy. Its discovery by Alexander Fleming in 1928 marked the beginning of the antibiotic era, leading to the development of other vital antibiotics such as streptomycin, tetracycline, and erythromycin, which were derived from soil bacteria [21]. These compounds have been critical in treating infectious diseases and continue to be essential in modern medicine.

SMs also play a crucial role in cancer treatment. Numerous anticancer drugs have been derived from natural products, showcasing the pharmacological potential of these compounds. For example, paclitaxel (Taxol), a diterpenoid isolated from the Pacific yew tree (*Taxus brevifolia*), has been widely used in the treatment of various cancers, including ovarian, breast, and lung cancers. Paclitaxel stabilizes microtubules, preventing cell division and leading to apoptosis in cancer cells [22]. Similarly, camptothecin, derived from the Chinese tree *Camptotheca acuminata*, and its derivatives, such as irinotecan and topotecan, inhibit topoisomerase I, an enzyme critical for DNA replication, and are used in treating colorectal, ovarian, and small-cell lung cancers [23]. These examples highlight the importance of natural metabolites in developing effective cancer therapies. Moreover, several marine-derived compounds have shown potent anticancer activities. For example, trabectedin (Yondelis), derived from the sea squirt *Ecteinascidia turbinata*, is used to treat soft tissue sarcoma and ovarian cancer. Trabectedin works by binding to the DNA minor groove, causing DNA damage and disrupting the cell cycle, which leads to apoptosis in cancer cells [24]. Such marine natural products offer unique structural frameworks and mechanisms of action that are often distinct from those of terrestrial sources, providing new avenues for cancer therapy.

Anti-inflammatory properties of SMs have also been harnessed for therapeutic purposes. For instance, curcumin, a polyphenol derived from the turmeric plant (*Curcuma longa*), exhibits potent anti-inflammatory and antioxidant activities. Curcumin modulates various molecular targets, including transcription factors, enzymes, and cytokines, involved in inflammation and has been investigated for its potential to treat inflammatory diseases, such as arthritis, inflammatory bowel disease, and neurodegenerative disorders [25]. The pharmacological relevance of curcumin and similar compounds underscores the therapeutic potential of natural metabolites in managing chronic inflammatory conditions.

Furthermore, SMs contribute to cardiovascular health through their bioactive properties. Omega-3 fatty acids, primarily eicosapentaenoic acid (EPA) and docosahexaenoic acid (DHA), derived from fish oil, have been shown to reduce the risk of cardiovascular diseases. These fatty acids exhibit anti-inflammatory, antithrombotic, and antiarrhythmic effects, contributing to improved cardiovascular outcomes [26]. Their incorporation into the diet and therapeutic formulations highlights the importance of natural metabolites in promoting heart health.

Recently, the discovery of conotoxins, a group of peptides produced by marine cone snails, has led to significant advancements in pain management. Conotoxins are potent inhibitors of voltage-gated ion channels and have been developed into analgesic drugs, such as ziconotide (Prialt), which is used to treat severe chronic pain in patients who are unresponsive to traditional painkillers [27]. These peptides offer a novel mechanism of action, providing pain relief without the addictive properties associated with opioid analgesics.

Another prominent example is the use of amphibian skin peptides, which have demonstrated strong antimicrobial properties. These peptides, such as magainins from the African clawed frog (*Xenopus laevis*), are effective against a broad spectrum of bacteria, fungi, and viruses. Their unique mode of action, involving disruption of microbial cell membranes, makes them promising candidates for new antibiotics, particularly in the face of rising antibiotic resistance [28]. The pharmaceutical industry is actively researching these peptides to develop new antimicrobial therapies that can overcome the limitations of current drugs.

In the agricultural sector, SMs are utilized as biopesticides and growth enhancers. Insect pheromones, for instance, are employed in integrated pest management (IPM) programs to control pest populations in a more environmentally friendly manner. These pheromones are used in traps or as mating disruptors to reduce the need for synthetic chemical pesticides, which can have harmful ecological effects. The use of pheromones in IPM has been particularly successful in controlling pests like the codling moth (*Cydia pomonella*) in apple orchards, significantly reducing crop damage and pesticide use [16]. Furthermore, the discovery of bioactive compounds from marine organisms has spurred the development of new materials and bio-based products. For instance, chitosan, derived from the exoskeletons of crustaceans, is used in water purification, wound dressings, and biodegradable films due to its biocompatibility and antimicrobial properties. The unique properties of chitosan make it a versatile material for various industrial and biomedical applications [29].

8.1.1.6 Significance in Understanding Biological Interactions

Studies on SMs are also very useful in understanding biological interactions, revealing the intricate chemical dialogues that govern ecological relationships and evolutionary processes. These metabolites, which include pheromones, allelochemicals, and signaling molecules, mediate interactions within and between species, influencing behaviors such as predation, competition, symbiosis, and reproduction. By elucidating these chemical communications, researchers gain insights into the fundamental mechanisms that shape biodiversity and ecosystem dynamics.

Pheromones are a well-known class of SMs that play crucial roles in intra-species communication, particularly in mating behaviors. For instance, the study of sex pheromones in insects, such as the silk moth (*Bombyx mori*), has provided deep insights into the molecular mechanisms of olfaction and mate attraction. Female silk moths release a specific pheromone, bombykol, which is detected by highly sensitive receptors on the male's antennae, triggering a cascade of behavioral responses leading to successful mating [30]. Understanding these interactions not only sheds light on the evolutionary strategies of reproduction but also offers practical applications in pest control. Symbiotic relationships also heavily rely on SMs for communication and coordination. In marine environments, the mutualistic relationship between certain species of fish and bioluminescent bacteria is a fascinating example. These bacteria, such as *Vibrio fischeri*, produce light through a chemical reaction facilitated by the enzyme luciferase. The light emitted by these bacteria – housed within the light organs of fish like the Hawaiian bobtail squid (*Euprymna scolopes*) – provides camouflage and communication advantages for the host [31]. Studying these interactions helps elucidate the coevolution of symbiotic partners and the molecular basis of mutualism.

SMs also play pivotal roles in predator–prey dynamics. Many animals produce toxins as a defense mechanism against predators. For instance, the potent neurotoxin tetrodotoxin (TTX), widely distributed among species such as pufferfish (*Tetraodontidae*), newts, and certain marine invertebrates, exemplifies a highly effective chemical deterrent. Rather than simply acting as a toxicant, TTX is emblematic of an intricate evolutionary dynamic: its molecular target – voltage-gated sodium channels (Nav) – has driven selective pressures in both toxin-bearing species and their predators. Resistance to TTX has evolved in several predators via mutations in Nav channels that reduce toxin binding affinity, highlighting the coevolutionary arms race that characterizes chemical defenses in nature [32]. This example underscores how SMs can influence ecological relationships and shape the genetic architecture of interacting species. Understanding these chemical defenses enhances our knowledge of ecological interactions and evolutionary pressures.

The role of SMs in microbial interactions is another area of significant research interest. Microbes produce a plethora of secondary metabolites that mediate interactions with their environment and other organisms. For example, antibiotics produced by soil bacteria like *Streptomyces* play a crucial role in competitive interactions within the microbial community. The discovery of these antibiotics, such as streptomycin and tetracycline, has not only revolutionized medicine but also provided insights into microbial ecology and evolution [33]. Investigating microbial metabolites helps unravel the complex web of microbial interactions and their impacts on ecosystem health and human welfare.

8.1.1.7 Disease Modeling and Drug Discovery

SMs also play an important role in disease modeling and drug discovery, offering unique molecular structures and mechanisms of action that can be leveraged to develop new therapeutic agents and understand disease pathophysiology.

In disease modeling, SMs are used to simulate disease states *in vitro* and *in vivo*, providing a platform for studying disease progression and testing potential therapeutics. For instance, bacterial endotoxins, such as lipopolysaccharides (LPS), are commonly used to induce inflammatory responses in cell cultures and animal models. These models are instrumental in studying the inflammatory pathways involved in diseases such as sepsis, arthritis, and inflammatory bowel disease. By mimicking the disease environment, researchers can investigate the effects of new drugs and identify key molecular targets for intervention [34].

Similarly, marine toxins, such as okadaic acid, are utilized in neurodegenerative disease models to study the mechanisms of diseases like Alzheimer's and Parkinson's. Okadaic acid inhibits protein phosphatases, leading to the hyperphosphorylation of tau proteins, a hallmark of Alzheimer's disease. This model allows researchers to investigate the molecular pathways involved in tau pathology and screen for compounds that can prevent or reverse these changes [35]. Such models are crucial for understanding the complex molecular interactions underlying neurodegenerative diseases and for discovering therapeutic strategies.

In drug discovery, as described in the previous paragraph, SMs have been a rich source of novel drug leads due to their diverse chemical structures and potent biological activities. Natural products have historically been the foundation for many successful drugs, providing unique scaffolds that are difficult to replicate synthetically. Currently, the advancement of high-throughput screening (HTS) and combinatorial chemistry has enabled the systematic evaluation of natural product libraries for drug discovery. These technologies allow for the rapid identification of bioactive compounds from complex natural extracts, facilitating the discovery of new drugs with novel mechanisms of action. For example, the HTS of marine sponge extracts led to the discovery of eribulin, a synthetic analog of the natural product halichondrin B, which is now used as a chemotherapeutic agent for treating metastatic breast cancer [36].

In addition to providing new drug leads, SMs also serve as valuable tools for chemical biology and target identification. By studying the interactions between these compounds and their biological targets, researchers can gain insights into the molecular pathways involved in disease and identify new therapeutic targets. For instance, the study of natural inhibitors of histone deacetylases (HDACs), such as the fungal metabolite trichostatin A, has advanced our understanding of epigenetic regulation and led to the development of HDAC inhibitors as cancer therapies [37].

8.1.1.8 Environmental Toxicology

SMs have also been very important in environmental toxicology, a field dedicated to understanding the effects of chemical substances on living organisms and ecosystems. These natural compounds, produced by various organisms, can significantly impact environmental health and biological interactions. Studying these metabolites helps researchers assess ecological risks, understand the mechanisms of toxicity, and develop strategies for environmental protection and remediation.

One of the key aspects of environmental toxicology is the assessment of natural toxins produced by plants, fungi, and marine organisms. These toxins can have profound effects on ecosystems and human health. For example, algal blooms caused by certain species of cyanobacteria and dinoflagellates produce potent toxins like microcystins and saxitoxins. These toxins can contaminate water bodies, leading to harmful algal blooms (HABs) that pose serious risks to aquatic life and human health through drinking water and seafood consumption. Understanding the environmental triggers and toxicological effects of these algal toxins is essential for managing and mitigating HABs [38].

Fungal metabolites, such as mycotoxins, also play a significant role in environmental toxicology. Mycotoxins like aflatoxins and ochratoxins, produced by *Aspergillus* and *Penicillium* species, respectively, contaminate crops and pose severe health risks to humans and animals. These toxins can cause acute poisoning, liver damage, and even cancer upon chronic exposure [39]. Monitoring and managing mycotoxin contamination in the environment and food supply are critical for protecting public health.

Marine organisms produce some of the most potent natural toxins known, impacting both marine ecosystems and human health. A representative case is that of TTX, which has gained increasing attention also as an emerging contaminant in marine ecosystems. Rather than focusing solely on its toxicodynamics, current environmental toxicology emphasizes the monitoring of TTX in edible marine species, due to its occurrence in bivalves and gastropods in temperate regions beyond its traditional tropical distribution. These findings raise concerns about bioaccumulation, climate-related range expansion, and public health implications, particularly in coastal communities with high seafood consumption. The development of sensitive analytical tools and early warning systems is crucial to mitigate the risks associated with TTX contamination in marine food chains [40]. The study of these marine toxins not only helps in understanding their ecological roles but also aids in developing detection methods and management strategies to prevent poisoning incidents.

Environmental toxicology also involves the study of bioaccumulation and biomagnification of natural toxins through food webs. Compounds like methylmercury, produced by microbial methylation of inorganic mercury in aquatic environments, accumulate in the tissues of fish and other aquatic organisms. As these organisms are consumed by higher trophic levels, including humans, the concentration of methylmercury increases, leading to severe neurological and developmental effects [41]. Research on the pathways and effects of bioaccumulative toxins is vital for assessing ecological and human health risks and for establishing regulatory guidelines. Additionally, the interaction of natural toxins with anthropogenic pollutants is an emerging area of interest in environmental toxicology. For example, the combined effects of natural algal toxins and synthetic chemicals like pesticides and pharmaceuticals can lead to synergistic or antagonistic interactions, complicating risk assessments [42]. Understanding these interactions is crucial for developing comprehensive environmental management strategies that account for the complex mixtures of natural and synthetic chemicals present in ecosystems.

8.1.1.9 Biomarker Discovery and Diagnostic Applications

The field of biomarker discovery and diagnostic applications also involved the SMs, providing crucial insights into disease states, physiological conditions, and environmental exposures. These metabolites, which include small molecules produced by metabolic processes, serve as indicators that can be measured to diagnose diseases, monitor disease progression, and assess treatment efficacy.

8.1.1.9.1 Biomarker Discovery

Biomarkers are biological molecules that signal normal or abnormal processes occurring in the body, and their discovery is fundamental for early diagnosis, prognosis, and the personalization of medical treatments. SMs are particularly useful as biomarkers due to their specific roles in metabolic pathways and their alteration in response to disease states. For instance, metabolic profiling has identified numerous cancer biomarkers, where altered levels of specific metabolites indicate the presence and progression of malignancies. One notable example is the oncometabolite 2-Hydroxyglutarate (2-HG), which accumulates in certain types of cancer, such as gliomas and acute myeloid leukemia, due to mutations in the isocitrate dehydrogenase (IDH) genes [43]. Elevated levels of 2-HG serve as a diagnostic biomarker and can also inform therapeutic strategies targeting the mutant IDH enzyme.

In cardiovascular diseases, metabolites such as trimethylamine *N*-oxide (TMAO) have emerged as significant biomarkers. TMAO, a product of intestinal microbiota metabolism of dietary choline, phosphatidylcholine, and carnitine, has been linked to an increased risk of cardiovascular events. Elevated plasma TMAO levels are associated with enhanced atherosclerosis and thrombosis, providing a predictive biomarker for cardiovascular disease risk and guiding dietary and therapeutic interventions [44].

8.1.1.9.2 Diagnostic Application

The use of SMs in diagnostic applications extends beyond cancer and cardiovascular diseases to a wide range of conditions. In the context of infectious diseases, metabolites can serve as indicators of pathogen presence and activity. For example, the detection of specific VOCs in the breath can indicate bacterial infections. Tuberculosis, caused by *Mycobacterium tuberculosis*, can be diagnosed by identifying VOCs such as methyl nicotinate in the breath of infected individuals, providing a noninvasive diagnostic tool [45].

SMs are also valuable in the diagnosis and monitoring of neurodegenerative diseases. In Alzheimer's disease, the cerebrospinal fluid (CSF) levels of amyloid-beta peptides and tau proteins are established biomarkers for diagnosis and disease progression. Additionally, metabolic changes, such as alterations in lipid metabolism and oxidative stress markers, are being explored as potential biomarkers for early detection and monitoring the efficacy of therapeutic interventions [46].

8.1.1.9.3 Environmental and Exposure Biomarkers

Beyond disease diagnosis, SMs serve as biomarkers of environmental exposures, providing critical information on the impact of environmental toxins and pollutants on human health. For instance, metabolites of polycyclic aromatic hydrocarbons (PAHs), such as 1-hydroxypyrene, are used as biomarkers of exposure to these carcinogenic compounds commonly found in polluted air, water, and soil. Monitoring these metabolites in urine samples allows for the assessment of exposure levels and the associated health risks [47].

Additionally, biomarkers of oxidative stress, such as 8-iso-prostaglandin F2α (8-iso-PGF2α), reflect the body's response to environmental stressors, including air pollution and radiation. Elevated levels of these biomarkers indicate increased oxidative damage, which is linked to various chronic diseases, including cancer, cardiovascular diseases, and respiratory disorders. Measuring these biomarkers helps in understanding the health impacts of environmental exposures and in developing strategies to mitigate their effects [48].

8.1.2 Biosynthetic Pathways in Animals for Specialized Metabolite Production

8.1.2.1 Shikimate-like Pathways

Shikimic acid plays a crucial role in the synthesis of aromatic amino acids and is central to secondary metabolism. While shikimic acid metabolism is typically absent in animals that obtain aromatic amino acids through their diet [49], recent discoveries have highlighted its presence in certain animal biosynthetic pathways.

The enzyme dehydroquinate synthase (DHQS) catalyzes an early step in shikimate biosynthesis, leading to the production of important compounds such as mycosporine-like amino acids (MAAs) and related sunscreens. These compounds, first characterized in cyanobacteria, have biosynthetic pathways that are also found throughout the animal kingdom, including in vertebrates like fish, but are absent in mammals [49–52].

Three major types of MAA-like biosynthetic pathways have been described, each synthesizing natural sunscreens such as MAAs and related compounds like gadusol and deoxygadusol. These pathways utilize an enzyme that converts sedoheptulose 7-phosphate (SH7P) into a cyclitol through DHQS-like enzymes. One pathway in bacteria synthesizes demethyl 4-deoxygadusol (DDG) via a DDG synthase (DDGS), followed by the action of a methyltransferase to produce deoxygadusol. Nonribosomal peptide synthetase (NRPS)-like and ATP-grasp enzymes then add amino acids to form the final MAA product [53]. In cnidarians, such as corals, a unique fusion of the methyltransferase with the DDGS enzyme is observed [52].

A distinct variation of this pathway exists in fish and other vertebrates, where a different strategy is employed. Instead of directly cyclizing SH7P, this pathway first synthesizes 2-epi-5-epi-valiolone (EEV) through the EEV synthase (EEVS) enzyme. The EEVS enzyme is paired with a fused methyltransferase-oxidase enzyme, which produces gadusol from EEV. The function of this pathway has been strongly supported by evidence from heterologous expression and knockdown experiments in fish [51].

Understanding the biosynthesis of MAAs and related compounds in animals is complicated by their dual origin – both *de novo* synthesis and dietary intake. *De novo* synthesis can occur either through symbiosis with bacteria [54] or via animal-encoded genes [49, 51]. It is believed that animals acquired their

MAA-like biosynthetic machinery through horizontal gene transfer (HGT) from various microbes [51]. The ability to synthesize gadusol appears to have been lost in the ancestral mammal [50]. The widespread presence of these compounds highlights their significant biological role in providing protection against UV damage and other UV-related functions.

8.1.2.2 Terpenes

Isoprenoids or terpenes are natural products ubiquitous across all life forms and include well-known compounds such as steroids. These compounds play a multitude of biological roles, including chemical defense, pheromones, and hormones, among others [55]. Isoprenoids are prevalent in most animals. In bacteria and plants, the isoprene precursors dimethylallyl pyrophosphate (DMAPP) and isopentenyl pyrophosphate (IPP) are synthesized through either the mevalonate or deoxyxylulose phosphate pathways. However, in animals, these precursors are derived exclusively from the mevalonate pathway [56]. The initial steps leading to mevalonate production are almost identical across all animals, with the notable exception that some insect enzymes exhibit relaxed substrate selectivity, producing homo isoprenoids crucial for insect hormones [2]. Subsequently, isoprenyl diphosphate synthases (IDSs) condense DMAPP and IPP, forming methyl-branched oligomers of five-carbon units, such as geranyl pyrophosphate (GPP, 10 carbons), farnesyl pyrophosphate (FPP, 15 carbons), and beyond.

In plants and microbes, terpene synthases/cyclases (TPSs) share structural folds with IDSs, suggesting an evolutionary lineage from ancestral IDS genes [57]. Guided by this idea, a 2016 analysis of a beetle transcriptome identified IDS homologs [58]. Nine transcripts encoding IDS-like sequences were expressed in *E. Coli* and functionally characterized. One of these enzymes, PsTPS1, was found to convert (*Z,E*)-FPP primarily into bicyclic sesquiterpenes. Other identified enzymes were typical IDSs, producing linear isoprenoids, while the substrate specificity for (*Z,E*)-FPP was notable since most TPSs utilize (*E,E*)-FPP. Transcriptional and knockdown studies provided additional evidence confirming the identification of these enzymes.

The discovery of PsTPS1 revealed that other IDS-like TPS enzymes are encoded in insect genomes [58]. These enzymes are more closely related to insect IDSs than to TPSs from other organisms, indicating convergent evolution. Recent characterizations of insect TPSs from a stink bug and a beetle species demonstrated that these enzymes produce various monocyclic, bisabolene-like isoprenoids from (*E,E*)-FPP [59, 60]. The robustness of this experimental evidence is bolstered by heterologous expression and knockdown studies. Evolutionarily, a fascinating revelation is that TPSs have independently evolved from IDSs in multiple insect lineages [59].

8.1.2.3 Polyketide/Fatty Acid Metabolism

Animals produce a wide array of lipid-derived natural products. Traditionally, specialized animal lipid metabolism was considered predominantly degradative. Essential fatty acids like arachidonic acid are modified oxidatively to produce prostaglandins. Recent studies, however, have unveiled the capability of animals to synthesize complex lipid-like compounds, known as polyketides, typically associated with bacterial metabolism. Polyketide products include crucial pharmaceuticals like erythromycin and amphotericin. The first animal polyketide synthase (PKS) was identified more than 20 years ago, and recent advancements in characterization and sequence analysis have revealed extensive PKS families within animal genomes. Insects and other organisms often derive pheromones from simple fatty acids. The underlying proteins for these biosynthetic steps have been identified. For instance, moth pheromones are composed of specific mixtures of acids altered by redox and chain-shortening reactions [61]. A more complex example is the nematode hormones, ascarosides, which are formed through a combination of degradative and biosynthetic steps, producing numerous derivatives [62–66].

Animals synthesize fatty acids *de novo* using type I fatty acid synthase (FAS) enzymes, which are multi-domain polypeptides. These enzymes condense fatty acids from malonyl-CoA and similar units, with each elongation step adding a two-carbon unit. The acyltransferase (AT) domain selects the CoA substrate, and the ketosynthase (KS) domain facilitates carbon–carbon bond formation, resulting in β-ketoesters. The substrates and intermediates are tethered to an acyl carrier protein (ACP), while other domains – ketoreductase (KR), dehydratase (DH), and enoylreductase (ER) – reduce the ketoester to an alkyl ester.

Type I PKSs in animals share a similar domain structure with type I FAS. The first animal PKS, Pks1, was discovered in the sea urchin *Strongylocentrotus purpuratus*. Its knockdown led to a loss of pigmentation, suggesting its role in synthesizing the pigment echinochrome [67]. This finding was later corroborated by creating a stable knockout, which confirmed the role of Pks1 throughout the life cycle [68]. Subsequent studies have identified a family of animal PKSs, revealing an animal-specific PKS family distinct from bacterial PKSs [69, 70].

An intriguing case is carminic acid, a red dye produced by beetles. While the PKS responsible remains unidentified, the glucosyltransferase involved was recently characterized, indicating an insect origin and supporting the notion that animals can biosynthesize aromatic polyketides [71].

OlPKS in zebrafish, characterized through genetic means, is another example of an animal PKS with unknown products. Mutations in the OlPKS KS domain impair otolith formation, indicating its role in calcium accumulation [72], paralleling the role of Pks2 in sea urchins.

A recent genome-wide association study identified a PKS in budgerigars linked to feather coloration. The MuPKS enzyme in budgerigars produces a yellow pigment absent in blue variants. This study was the first comprehensive evidence linking PKS activity to secondary metabolism in animals [70]. Similar PKSs have been found in other bird species, suggesting a common mechanism for coloration. Partial characterization of a PKS from mallard ducks hints at a broader distribution of these enzymes [73].

These studies have uncovered numerous PKSs across various animal phyla, indicating their widespread presence. The conserved domain order at the protein *N*-terminus contrasts with the variable *C*-terminus, which includes different termination domains and even NRPS domains. Phylogenetic analyses show that animal PKSs form a distinct clade, separate from bacterial and fungal PKSs, suggesting an origin from gene duplication and divergence rather than HGT.

8.1.3 The Role of Derivatization in Enhancing Metabolite Analysis

8.1.3.1 Introduction

Derivatization is a fundamental technique in analytical chemistry, employed to enhance the detectability, stability, and separability of metabolites. This process involves chemically altering a compound to produce a derivative that is more amenable to analytical methods. By transforming metabolites into derivatives with improved physical and chemical properties, derivatization significantly broadens the analytical capabilities of techniques such as gas chromatography (GC), liquid chromatography (LC), and mass spectrometry (MS).

One of the primary roles of derivatization is to increase the volatility and thermal stability of metabolites, which is particularly crucial for GC or multidimensional GC analysis. Nonvolatile or thermally labile compounds can be converted into more volatile and stable derivatives, facilitating their detection and quantification. For example, the derivatization of amino acids and organic acids with agents like *N,O*-bis(trimethylsilyl)trifluoroacetamide (BSTFA) or methyl chloroformate can improve their volatility and thermal stability, making them suitable for GC analysis [74].

In addition to enhancing volatility and stability, derivatization can improve the ionization efficiency of metabolites in MS. By modifying the chemical structure of a metabolite, derivatization can introduce functional groups that enhance ionization, thereby increasing the sensitivity and specificity of detection. For instance, the derivatization of metabolites with reagents such as dansyl chloride or trimethylsilyl groups can significantly enhance their ionization efficiency in electrospray ionization (ESI) or atmospheric pressure chemical ionization (APCI)MS, leading to more accurate and sensitive quantification [75].

Derivatization also plays a crucial role in enhancing the selectivity and specificity of chromatographic separation. By introducing specific functional groups, derivatization can alter the interaction of metabolites with the stationary phase in LC or GC, improving their separation. This is particularly important in the analysis of complex biological matrices, where the presence of numerous compounds can complicate the separation and identification of target metabolites. For example, the derivatization of carbohydrates with reagents like 1-phenyl-3-methyl-5-pyrazolone (PMP) or o-phthalaldehyde (OPA) can improve their chromatographic separation and detection [76, 77].

Moreover, derivatization can stabilize labile metabolites, preventing their degradation during analysis. This is essential for accurate quantification, as unstable metabolites may degrade or react during the analytical process, leading to erroneous results. For instance, the derivatization of labile aldehydes with reagents like 2,4-dinitrophenylhydrazine (DNPH) can form stable hydrazone derivatives, enabling reliable detection and quantification [78].

8.1.3.2 Techniques for Derivatization of Metabolites in GC Analysis

8.1.3.2.1 Silylation

Silylation stands out as a prominent derivatization technique in metabolomics analysis, particularly when employing gas chromatography-mass spectrometry (GC-MS). This process involves the substitution of active hydrogen atoms in functional groups, such as hydroxyl, carboxyl, thiol, and amine groups, with silyl groups, typically using reagents like *N,O*-bis(trimethylsilyl)trifluoroacetamide (BSTFA) or trimethylsilyl chloride (TMSCl). The primary goal of silylation is to enhance the volatility, thermal stability, and detectability of metabolites, which are critical parameters for effective GC-MS analysis.

The silylation process significantly enhances the analytical performance of GC-MS by converting polar metabolites, which typically have low volatility and thermal instability, into more volatile and thermally stable silyl derivatives. This transformation is particularly advantageous for metabolites containing hydroxyl and carboxyl groups, such as sugars, amino acids, organic acids, and steroids, making them amenable to GC-MS analysis [79, 80], as schematically illustrated in Figure 8.1.1.

Table 8.1.1 shows in descending order the reactivity of the main functional groups with respect to the silylation reaction.

One of the most profound impacts of silylation is observed in the analysis of complex biological samples, where the technique helps in achieving better chromatographic separation and improved peak shapes. By reducing the polarity of

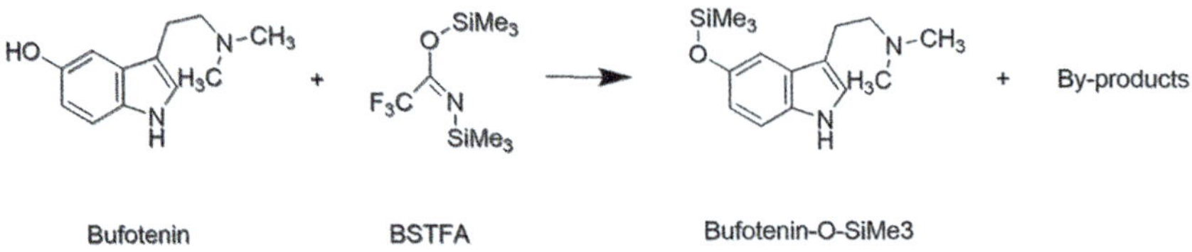

Figure 8.1.1 Schematic representation of the silylation process on bufotenin.

Table 8.1.1 Functional groups ranked by relative ease of silylation.

	Compound	Functional group
1	Alcohol	OH
2	Phenol	OH
3	Thiophenol	SH
4	Aliphatic acid	COOH
5	Aromatic acid	COOH
6	Primary amine	NH_2
7	Thiol	SH
8	Amide	$CONH_2$
9	N-TMS amide	$CONH\text{-}Si(Me)_3$
10	Secondary amide	NH
11	Indole	NH

metabolites, silylation minimizes interactions with the stationary phase of the GC column, leading to sharper peaks and more reliable quantification [79].

In addition to enhancing chromatographic properties, silylation also improves the ionization efficiency of metabolites in the mass spectrometer. The introduction of silyl groups increases the molecular weight and the volatility of the derivatives, leading to more stable and intense signals. This improvement is crucial for detecting low-abundance metabolites in complex biological matrices, as it enhances both the peak intensity and the interpretability of fragmentation patterns in mass spectrometry [81]. Figure 8.1.2 presents an illustrative comparison between the electron ionization (EI) mass spectra of underivatized and silylated bufotenin, highlighting the gain in spectral clarity and the emergence of derivatization-specific fragments.

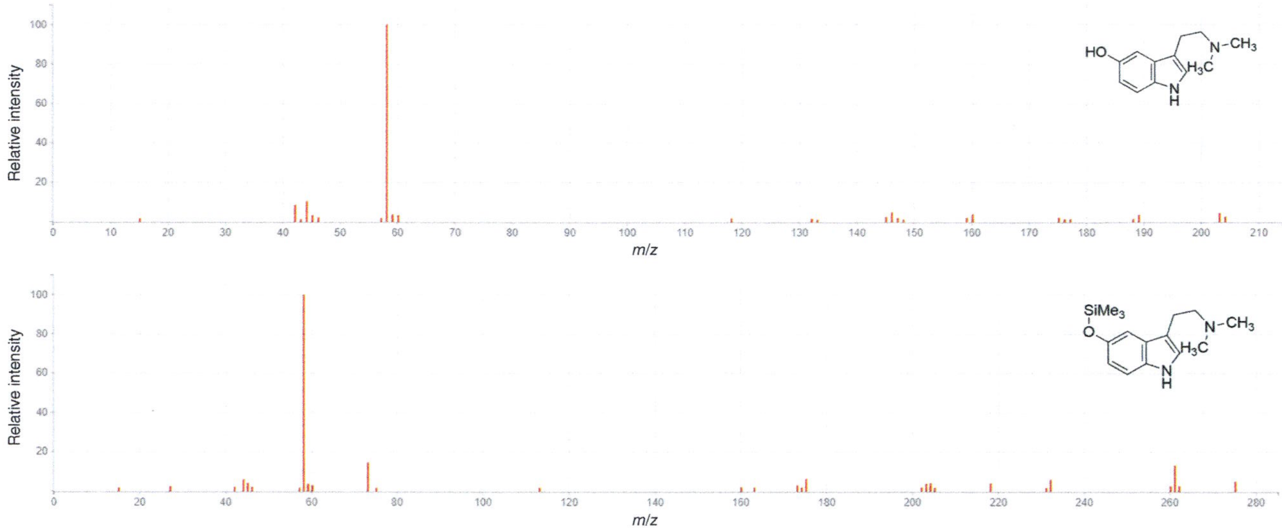

Figure 8.1.2 Electron ionization (EI) mass spectra of bufotenin (top: underivatized; bottom: silylated) obtained from the Human Metabolome Database (HMDB). The spectra were combined and annotated by the authors for comparative purposes. *Note:* The *m/z* axes are scaled differently between the two spectra for visualization purposes, but both represent EI spectra acquired under standardized conditions.

Notably, while the *m/z* scales differ slightly between the spectra for graphical clarity, both were acquired under standardized conditions and are derived from validated reference entries in the Human Metabolome Database (HMDB) [82].

The figure also includes chemical structures for direct visualization of the functional groups affected by silylation, supporting the structural basis for improved ionization efficiency.

A critical preliminary step in the silylation process, especially in metabolomics studies, is the methoximation of carbonyl groups. Carbonyl-containing compounds, such as aldehydes and ketones, are highly reactive and prone to undergoing condensation reactions, which can lead to the formation of unwanted by-products and degradation during analysis. Methoximation involves reacting these carbonyl groups with methoxyamine hydrochloride to form methoxime derivatives, thereby stabilizing them and preventing such reactions, as depicted in Figure 8.1.3 [83].

This preliminary methoximation step is crucial for the subsequent silylation process, as it ensures that all reactive carbonyl groups are protected, allowing for more efficient and complete silylation. By stabilizing these groups, methoximation prevents the formation of side products and enhances the overall accuracy and reliability of metabolite quantification.

The application of silylation in metabolomics has been extensively documented, particularly in the context of large-scale metabolic profiling. For instance, the silylation of primary metabolites in plasma, urine, and tissue extracts allows for comprehensive metabolic profiling, facilitating the identification and quantification of a wide range of metabolites involved in various biochemical pathways and human disease diagnosis [84–89]. This capability is invaluable in understanding metabolic alterations associated with diseases, drug responses, and environmental exposures.

Moreover, the reproducibility and robustness of silylation make it an essential technique in high-throughput metabolomics studies. Automated silylation protocols have been developed to handle large sample sets efficiently, ensuring consistent derivatization and reliable data generation across different experimental batches [90].

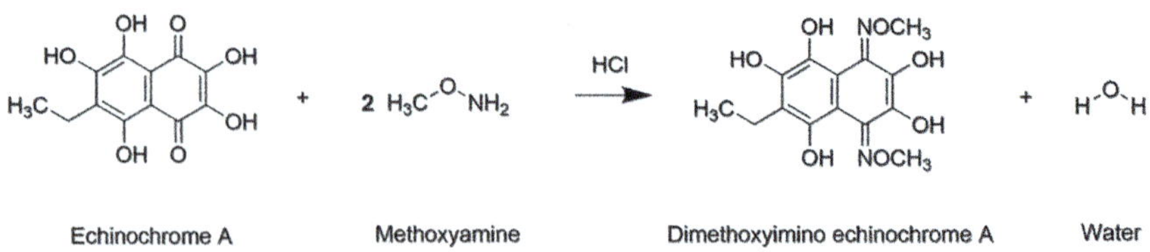

Figure 8.1.3 Methoximation reaction of a carbonyl groups of echinochrome with methoxyamine.

8.1.3.2.2 Acylation

Acylation is a pivotal derivatization technique in metabolomics analysis, especially when employing GC-MS. This process involves the introduction of an acyl group into a compound, typically targeting hydroxyl, amine, and thiol groups to form esters, amides, and thioesters, respectively. Common acylating reagents include acetic anhydride, trifluoroacetic anhydride (TFAA), and heptafluorobutyric anhydride (HFBA). The primary purpose of acylation is to enhance the volatility, thermal stability, and detectability of metabolites, which are essential for effective GC-MS analysis.

The acylation process significantly improves the analytical performance of GC-MS by converting polar metabolites into more volatile and thermally stable acyl derivatives. This transformation is particularly advantageous for metabolites with hydroxyl and amine groups, such as sugars, amino acids, and alcohols, making them more amenable to GC-MS analysis [91], as shown in Figure 8.1.4.

Acylation is extensively utilized in large-scale metabolic profiling. For instance, the acylation of primary metabolites in plasma, urine, and tissue extracts allows for comprehensive metabolic profiling, facilitating the identification and quantification of a wide range of metabolites involved in various biochemical pathways [91]. This capability is invaluable in understanding metabolic alterations associated with diseases, drug responses, and environmental exposures.

8.1.3.2.3 Methylation

Methylation is another highly effective derivatization technique in metabolomics analysis. This method involves the introduction of a methyl group into a compound, typically targeting functional groups such as hydroxyl, carboxyl, amine, and thiol groups, to form methyl esters, methyl ethers, and methylamines. Common methylating reagents include diazomethane, methyl iodide, and trimethylsilyldiazomethane. As for silylation and acylation, the primary goal of methylation is to enhance the volatility, thermal stability, and detectability of metabolites, which are essential for efficient GC-MS analysis, as demonstrated by the reaction scheme in Figure 8.1.5.

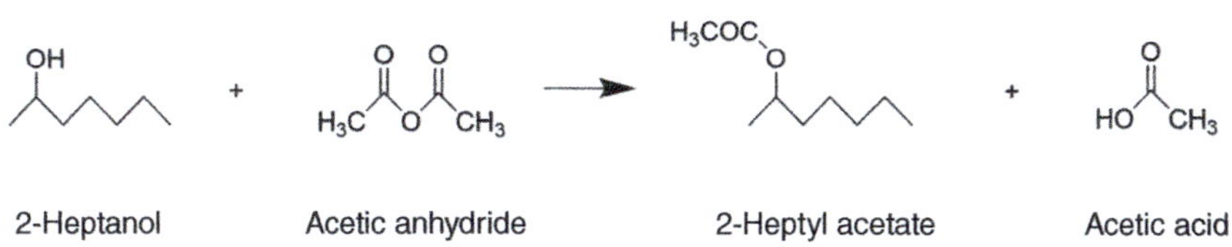

Figure 8.1.4 Schematic representation of the acylation process of 2-Heptanol with acetic anhydride.

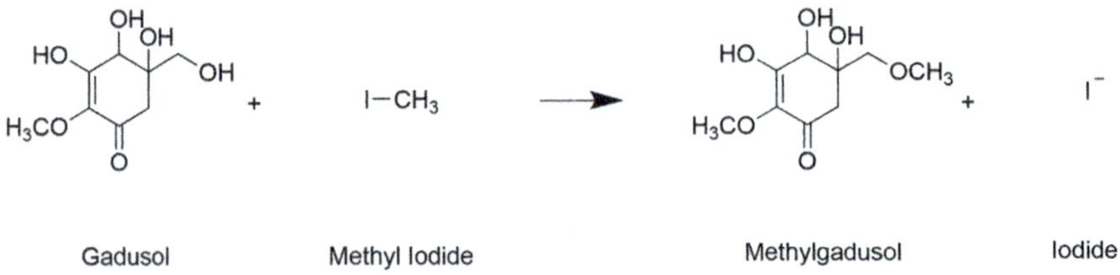

Figure 8.1.5 Schematic representation of the methylation process of gadusol using methyl iodide.

Methylation derivatization has been successfully applied in the analysis of a wide range of secondary metabolites, particularly those with polar functional groups that hinder GC-MS detection. For instance, methyl chloroformate derivatization has enabled the efficient quantification of amino and organic acids, improving detection limits to the low picomole range and allowing absolute quantitation of over 60 metabolites relevant to secondary metabolism [92]. In microbial metabolomics, this method has proven particularly useful for profiling intracellular and extracellular metabolites from fungi and bacteria, including those involved in central and secondary carbon metabolism [93]. In plant systems, derivatization strategies involving methylation have improved the detection of terpenoids, phenolic acids, and flavonoids, significantly expanding the coverage of metabolite classes detected via GC-MS [94]. Such applications underline the versatility and analytical value of methylation derivatization in characterizing complex secondary metabolomes across diverse biological systems.

8.1.3.2.4 Alkylation

This process involves the introduction of an alkyl group into a compound, typically targeting functional groups such as hydroxyl, carboxyl, thiol, and amine groups to form ethers, esters, thioethers, and amines. Common alkylating reagents include methyl iodide (MeI), ethyl iodide (EtI), and various alkyl halides. The primary aim of alkylation is to enhance the volatility, thermal stability, and detectability of metabolites, which are crucial for efficient GC-MS analysis, as illustrated in Figure 8.1.6.

Alternative alkylating agents beyond MCF have also shown strong applicability in secondary metabolite analysis. For example, trialkyloxonium tetrafluoroborates (e.g., triethyloxonium tetrafluoroborate) have been successfully used to derivatize organic acids, phenols, and even degradation products of complex molecules, offering high efficiency in both aqueous and organic media without the hazards associated with diazomethane. These reagents provide a simple one-step protocol and are compatible with both GC-MS and LC-MS platforms [95]. Fluoroalkyl chloroformates, such as trifluoroethyl and heptafluorobutyl chloroformates, have

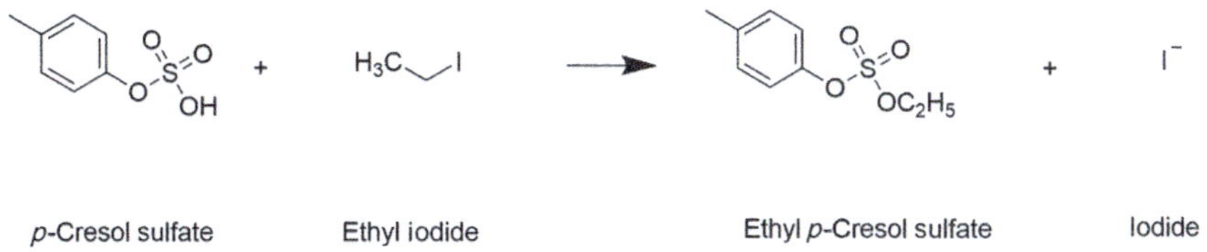

Figure 8.1.6 Schematic representation of the alkylation process of *p*-Cresol sulphate using ethyl iodide.

also been employed to derivatize hydroxyl-containing secondary metabolites like sterols and tocopherols in complex biological matrices. These reagents enable rapid and complete conversion under anhydrous conditions, improving peak shape and sensitivity in GC-MS analysis [96]. Furthermore, in metabolomic profiling of fungi under metal stress, alkylation has proven especially useful for targeting amino acids and organic acids, complementing silylation for broader metabolome coverage [97].

8.1.3.2.5 Esterification

Esterification is another pivotal derivatization technique in GC-MS-based metabolomics analysis. This process involves the conversion of carboxylic acids to esters by reacting them with alcohols in the presence of an acid catalyst. Common esterifying agents include methanol, ethanol, and various alkyl alcohols, often catalyzed by reagents like sulfuric acid, hydrochloric acid, or acidic ion-exchange resins. The primary objective of esterification is to enhance the volatility, thermal stability, and detectability of metabolites, which are essential for effective GC-MS analysis, as represented in Figure 8.1.7.

Esterification has been successfully applied in several metabolomic studies to improve the detection of low-volatility secondary metabolites. For example, in atmospheric chemistry studies, esterification using boron trifluoride (BF_3) in methanol has been effectively used to derivatize low-molecular-weight dicarboxylic acids for GC-MS analysis, achieving low detection limits and high reproducibility even in complex environmental matrices [98]. In natural product

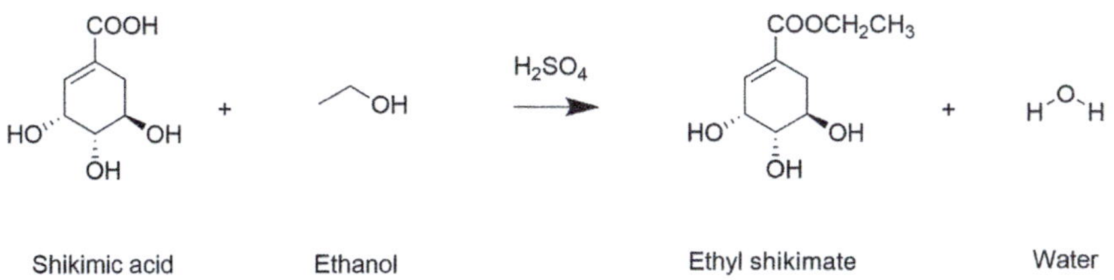

Figure 8.1.7 Schematic representation of the esterification process of shikimic acid using ethanol.

research, esterification has enabled improved chromatographic behavior and identification of carboxylic acid-containing compounds in plant extracts, such as hydroxycinnamic acids and flavonoid conjugates, by enhancing their volatility and thermal stability [94]. Additionally, novel esterification approaches, such as the use of 2,4,6-Trichlorobenzoyl chloride (Yamaguchi reagent), have been employed for regioselective derivatization of hydroxycarboxylic acids, allowing chiral analysis on achiral GC columns and facilitating the stereochemical resolution of complex metabolite mixtures [99]. These examples highlight the versatility of esterification in enhancing the analytical performance of GC-MS workflows in secondary metabolite profiling.

Fatty Acid Profiling by Transesterification and FAMEs Analysis in GC-MS Transesterification is a widely adopted derivatization strategy for the GC-MS analysis of fatty acids, particularly those embedded within complex lipid matrices such as triglycerides, phospholipids, and glycolipids. This process typically involves converting fatty acyl moieties into fatty acid methyl esters (FAMEs) through reaction with methanol in the presence of either acidic or basic catalysts (Figure 8.1.8). The key motivation for this transformation is the inherently low volatility and thermal instability of native fatty acids and lipid-bound acyl chains, which significantly hinders their direct analysis via GC-MS. By forming FAMEs, the analytes gain sufficient volatility, improve chromatographic peak shape, and allow for accurate structural elucidation and quantification, especially when dealing with positional or geometric isomers of unsaturated fatty acids [100].

In a metabolomics context, transesterification plays a crucial role in lipidomics and the broader profiling of energy-related and signaling-related metabolic pathways. FAME analysis enables high-throughput monitoring of dietary lipids, membrane composition, and microbial community structures. Compared to other derivatization techniques like silylation or direct methylation, transesterification offers greater specificity for esterified fatty acids and higher reproducibility when optimized correctly. Moreover, the procedure is compatible with stable isotope labeling and internal standardization strategies, enhancing both qualitative and quantitative robustness [101].

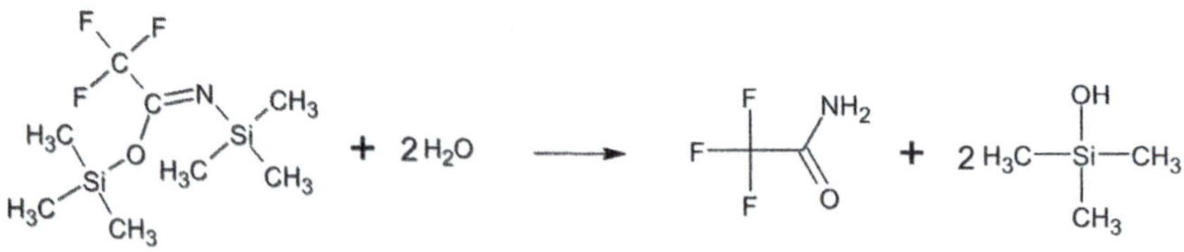

Figure 8.1.8 Schematic representation of a generic transesterification process of a triglyceride.

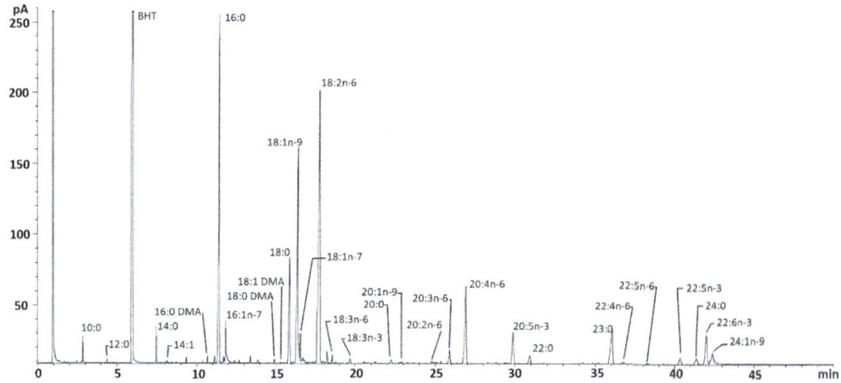

Figure 8.1.9 Representative GC-FID chromatogram of fatty acid methyl esters (FAMEs) derived from human plasma following conventional acid-catalyzed transmethylation. Peaks corresponding to saturated, monounsaturated, and polyunsaturated fatty acids are labeled according to chain length and degree of unsaturation (e.g., 18:1n-9, 22:6n-3). BHT (butylated hydroxytoluene) was used as an antioxidant, and DMA (dimethylacetal) derivatives are observed for aldehydes. The internal standard (23:0 methyl ester) is also indicated. *This figure was obtained by modifying the chromatogram originally published in. Source:* Masood et al. [102] / with permission of Elsevier.

In practice, FAME profiles are typically visualized via gas chromatographic separation followed by flame ionization detection (GC-FID) or MS, producing high-resolution chromatograms where individual peaks correspond to specific FAMEs. Figure 8.1.9 shows a representative GC profile of FAMEs obtained from a biological extract, illustrating the diversity of saturated, monounsaturated, and polyunsaturated fatty acids resolved over a standard chromatographic run. The precise retention times and peak identities depend on the column type and temperature program, but such profiles are routinely used for compositional comparisons, quality control, and quantitative lipidomics workflows.

Multiple protocols exist for FAME preparation, each tailored to specific sample types and analytical goals. Base-catalyzed methods using sodium methoxide or potassium hydroxide in methanol are fast and efficient for transesterifying neutral lipids but can lead to saponification in high-water matrices [103]. Acid-catalyzed methods using methanolic HCl or BF_3 are more universally applicable, including to free fatty acids, but often require longer reaction times and tighter control of reaction conditions [104]. Modern alternatives include microwave-assisted reactions for rapid conversion of complex lipids in under a minute [105], and microfluidic devices that enable automated, low-volume derivatization workflows with enhanced reproducibility [106].

8.1.3.3 Comparative Overview of Derivatization Strategies in GC-MS Metabolomics

As summarized in Table 8.1.2, each derivatization strategy offers specific advantages depending on the chemical nature of the metabolites and the goals of the analytical workflow. Notably, a dedicated derivatization procedure widely used for fatty acid profiling is transesterification, which converts fatty acids into methyl esters to improve volatility, thermal stability, and separation efficiency in GC-MS.

This technique is extensively applied in nutritional, clinical, and environmental studies, particularly for the analysis of total lipid composition in plasma, food, and tissue samples. While transesterification is a form of esterification, its specialized use in lipidomics and its robust standardization justify its inclusion as a distinct category.

8.1.3.3.1 Considerations and Challenges

While derivatization techniques offer numerous benefits for metabolite analysis, such as improved sensitivity, selectivity, and chromatographic resolution, they also introduce complexities and potential artifacts into the analytical workflow. Optimization of derivatization conditions, compatibility with sample matrices, and minimization of side reactions are essential considerations for achieving accurate and reproducible results in metabolomics studies.

Optimization Process The optimization of derivatization conditions is critical to ensure complete and efficient reaction with target metabolites. Parameters such as reagent concentration, reaction time, temperature, and pH must be meticulously controlled. Inadequate derivatization can lead to incomplete reactions, resulting in poor sensitivity and inaccurate quantification. Over-derivatization, on the other hand, can cause excessive reagent consumption, increased costs, and the formation of unwanted by-products that complicate the analysis [79].

Another challenge is ensuring the compatibility of derivatization reagents with various sample matrices. Biological samples such as plasma, urine, and tissue extracts contain a complex mixture of compounds that can interact with derivatizing agents, potentially leading to matrix effects. These interactions can affect the efficiency of derivatization and the stability of the derivatives, thereby impacting the accuracy and reproducibility of the analysis [107]. Careful selection and testing of derivatization reagents are necessary to minimize matrix effects and achieve reliable results.

Several proposals of strategies for derivatization optimization were reported. One of the more interesting were presented by Kumirska et al. [108] whom presented a comprehensive approach to optimizing derivatization reactions for GC-MS analysis of 24 pharmaceuticals spanning six drug classes. Derivatization

Table 8.1.2 Summary of derivatization strategies used in GC-MS metabolomics.

Derivatization type	Target functional groups	Common reagents	Analytical benefits	Typical applications
Silylation	–OH, –COOH, –NH, –SH	BSTFA, MSTFA, TMSI	Increases volatility, thermal stability, enhances MS fragmentation	Broad-spectrum profiling (amino acids, sugars, steroids)
Methoximation	Carbonyl (C=O)	Methoxyamine hydrochloride	Prevents enolization, stabilizes carbonyl compounds	Carbohydrate and keto acid analysis
Acylation	–OH, $-NH_2$, –SH	Acetic anhydride, TFAA, PFPA	Enhances volatility and detector response, increases lipophilicity	Steroids, amines, phenols
Methylation	–COOH, –OH, $-NH_2$	Methyl chloroformate, diazomethane	Converts to methyl esters/ethers; improves volatility and detectability	SCFAs, bile acids, phenolic acids
Alkylation	–OH, –SH, $-PO_4$	Iodomethane, MCF, ethyl chloroformate	Reduces polarity, improves GC peak shape and resolution	Catecholamines, thiols, organic acids
Esterification	–COOH	Ethanol, propanol + HCl or H_2SO_4	Enhances volatility; improves retention and quantification	Amino acids, benzoic acid, aromatic acids
Transesterification	Fatty acid carboxyl groups (–COOH)	Methanol + HCl, BF_3, or KOH (acid/base catalysis)	Converts fatty acids to FAMEs; improves volatility and GC performance	Fatty acid profiling in plasma, serum, food, oils, and tissues

Each row describes a chemical modification method, the targeted functional groups, common reagents, analytical benefits, and representative metabolite classes or sample types. The addition of transesterification for FAMEs production reflects its key role in fatty acid profiling workflows.

was primarily achieved using BSTFA combined with 1% TMCS, tested under various reaction conditions to enhance the derivatization yield and detectability of analytes in complex environmental matrices, including surface and wastewater samples. This study highlighted that optimization is crucial, as derivatization not only increases volatility and stability for GC analysis but also reduces interaction with stationary phases, thus improving chromatographic resolution and sensitivity. This optimization strategy involved a chemometric approach utilizing principal component analysis (PCA) and cluster analysis (CA), which enabled the researchers to interpret complex data sets and identify optimal derivatization conditions efficiently. The study examined three derivatizing reagents: BSTFA/TMCS, *N*-methyl-*N*-(trimethylsilyl)trifluoroacetamide (MSTFA), and 1-(trimethylsilyl)imidazole (TMSI), across a matrix of temperature and reaction time conditions. Each reagent offered distinct reactivity, selectivity, and by-product profiles. Results from PCA and CA analysis identified BSTFA/TMCS at 60 °C for 30 minutes in a solvent mixture of pyridine and ethyl acetate as optimal, yielding high relative response factors (RRFs) across a broad range of pharmaceutical compounds. The researchers used a stepwise approach to optimize both the derivatization reagent and the reaction conditions. Initial experiments (A–I) tested the three derivatizing reagents in a 2:1:1 mixture of pyridine, ethyl acetate, and reagent at 60, 90, and 120 °C for 30 minutes. BSTFA/TMCS emerged as the superior reagent, as TMSI led to incomplete derivatization for several compounds, while MSTFA showed declining efficiency at higher temperatures. In particular, the derivatization efficiency with TMSI was suboptimal for compounds such as salicylic acid and paracetamol, as incomplete reactions persisted across multiple temperature settings. Diethylstilbestrol (DES) and pindolol, which contain multiple active sites, showed significantly higher derivatization yields with BSTFA/TMCS than with TMSI or MSTFA, reaffirming the reagent's suitability for multifunctional analytes.

Following this, further optimization (experiments J–O) focused on varying the reaction time and temperature, specifically with BSTFA/TMCS in the preferred solvent mixture. The study compared the use of pyridine and ethyl acetate, both separately and combined, as pyridine, a Lewis base, activates acid groups and minimizes side reactions, while ethyl acetate improves dissolution [109]. The use of the pyridine-ethyl acetate combination at 60 °C for 30 minutes maintained high RRFs without prolonged reaction times, which could increase degradation risks. Furthermore, PCA and CA confirmed that BSTFA/TMCS in pyridine-ethyl acetate achieved optimal derivatization across temperature and time variations, avoiding excessive heat that could compromise analyte integrity. Experiments using prolonged times or temperatures showed diminished efficiency, indicating that BSTFA/TMCS at 60 °C with a 30-minute reaction time offered the best balance between sensitivity and compound stability. For validation, the optimized derivatization method was applied to solid-phase extraction SPE-GC-MS analysis of water samples from wastewater treatment plants and surface waters in Poland, detecting pharmaceuticals across a range of ng/L to μg/L. The method was rigorously validated for precision, accuracy, linearity, and method detection limits (MDLs) [108].

As proposed by the authors, derivatization conditions should indeed be optimized to ensure complete and reproducible analyte conversion for GC-MS analysis. A general strategy to achieve this could follow these steps:

1) **Preliminary selection of derivatization reagent:** Choose a derivatization reagent based on the specific functional groups in the target compounds. Silylation reagents such as BSTFA with TMCS are highly effective for compounds with acidic or polar functional groups (–OH, $-NH_2$, –COOH). Evaluate alternative reagents, such as MSTFA or TMSI, comparing effectiveness based on derivatization yield and product stability, especially for multifunctional or sterically hindered compounds.
2) **Experimental matrix for reaction conditions:** Create a matrix of reaction conditions (e.g., temperature, time, solvent composition) to evaluate the effects on derivatization efficiency. This can be implemented through Design of Experiment (DOE) approaches to systematically vary and monitor each factor's effect.
3) **Chemometric analysis for data interpretation:** Use PCA and CA or other chemometric tools to visualize and interpret experimental data, identifying patterns that indicate optimal conditions. This step simplifies understanding the relationship between different variables and facilitates the selection of the most suitable reaction conditions.
4) **Validation and application to real samples:** Validate the optimized derivatization conditions through precision, accuracy, and sensitivity tests on real samples, such as environmental water matrices. This validation ensures the robustness of the method in detecting trace analytes accurately in complex matrices.

This methodical optimization ensures the derivatization process is tailored for high sensitivity, reproducibility, and specificity, enhancing the reliability of GC-MS analyses in environmental and metabolomics research [108–110].

Matrix Compatibility The silylation reaction is the most delicate derivatization procedure in terms of compatibility with the matrix under investigation. Indeed, in addition to organic compounds with active hydrogens, certain inorganic compounds, such as H_2O, H_2O_2, and strong inorganic acids, can also react with silylating reagents. Some salts of these acids may undergo silylation as well. The

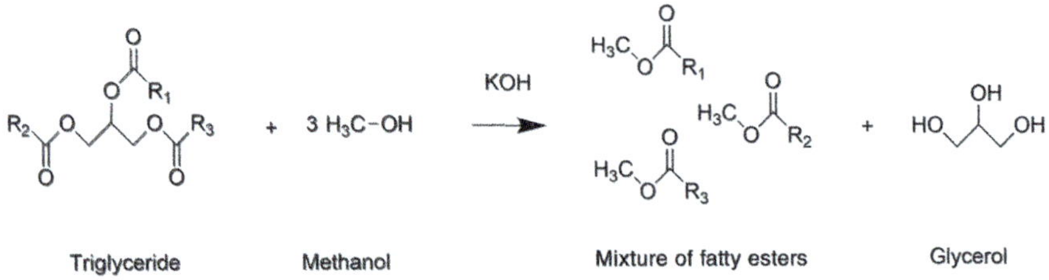

Figure 8.1.10 Schematic representation of the BSTFA inactivation by water.

reactivity of silylating reagents with water requires that water levels in the analyte matrix or solution be minimized, as water reacts, for example, with BSTFA, leading to reagent inactivation, as depicted in Figure 8.1.10, inactivating the derivatizing reagent.

In many solvents used for derivatization, the reaction between water and silylating agents produces trimethylsilanol, which often separates into a distinct layer from the main solvent. This layer formation complicates the sampling process in GC-MS, as it interferes with obtaining a uniform sample of the derivatized compounds. Additionally, excessive water in the sample can inhibit the silylation of other compounds, so silylation is generally not recommended for samples with a water content exceeding approximately 10%. For this reason, the water elimination step is generally a mandatory phase of this process. This can be achieved using methods such as freeze-drying, nitrogen evaporation, SpeedVac systems, and similar techniques.

The silylation reaction is typically conducted in a solvent free of active hydrogens. Common solvents used in silylation include dimethylformamide (DMF), pyridine, and acetonitrile, whose primary function is to dissolve both the analyte and the silylating reagents. The by-product produced in silylation (see Figure 8.1.1) can be acidic, basic, or neutral, depending on the reagent. For example, TMCS yields HCl as a by-product, HMDS produces NH_3, BSTFA forms N-TMS-trifluoroacetamide, and TMSI generates imidazole. When an acidic by-product is formed, it may interfere with the silylation reaction, making the use of an acid-accepting solvent advantageous. Solvents such as pyridine, triethylamine, and, to a lesser extent, DMF, can serve this dual purpose of acting as both solvent and acid acceptor.

Solvent mixtures are often used to improve both solubility and silylation efficiency. For instance, formamide with pyridine can neutralize acidic by-products by generating CO and an ammonium salt. Adding basic compounds can further enhance the silylation reaction, while certain compounds may act as catalysts to increase reaction efficiency.

Side Reactions Minimizing side reactions and artifacts is another critical consideration in derivatization. Side reactions can occur when derivatizing agents react with unintended functional groups, leading to the formation of by-products that complicate the chromatographic profile and mass spectra. These artifacts can interfere with the detection and quantification of target metabolites, reducing the overall quality of the data [111]. To minimize side reactions, it is essential to optimize the derivatization protocol and use selective derivatizing agents that preferentially react with the desired functional groups.

Reproducibility Reproducibility is a significant challenge in derivatization, especially in high-throughput metabolomics studies. Variations in derivatization efficiency can arise from differences in reagent quality, batch-to-batch variability, and slight deviations in reaction conditions. To address these issues, standard operating procedures (SOPs) must be established, and rigorous quality control measures should be implemented.

Automation Automation of derivatization processes can greatly improve reproducibility and efficiency in high-throughput metabolomics studies. Automated systems can precisely control reaction conditions, such as temperature, time, and reagent volumes, reducing human error and variability. This consistency ensures that large sample sets are processed uniformly, leading to more reliable and reproducible data. Automated liquid handling systems and robotic platforms can be programmed to perform derivatization reactions, streamline the workflow, and increase throughput [90]. Automation also enables better integration with subsequent analytical steps, such as GC-MS analysis, further enhancing the overall efficiency of metabolomics studies.

8.1.4 Future Perspectives and Challenges

Reflecting on the evolutionary journey of animals, which dates back over 650 million years, reveals intriguing insights into their biosynthetic capabilities. Unlike bacteria, which are thought to experience frequent HGT and thus possess a diverse array of biosynthetic genes, animals are generally considered to be less exposed to HGT. This suggests that animals might rely on alternative mechanisms for chemical diversification [112, 113].

Recent research highlights instances of relatively recent HGT from microbes to animals. However, convergent evolution appears to be a more prevalent mechanism in animals. Often, the duplication and divergence of ancestral animal genes enable the production of secondary metabolites. This phenomenon is particularly fascinating, as animals sometimes synthesize compounds similar or identical to those found in their diets through convergent evolutionary processes [114, 115]. Given that significant biological activity typically arises from the final product of a biochemical pathway, this trend might imply a form of natural retrobiosynthesis, offering a diverse pool of enzymes with varying properties for biotechnological applications [116].

SMs in animals often serve internal functions, such as UV protection, pigmentation, hormone regulation, and intraspecies communication through pheromones. In contrast, metabolites produced by symbiotic bacteria are predominantly aimed at interactions with other species, such as chemical defense mechanisms [1]. While this trend is notable, it is not absolute. For instance, some insects have evolved cyanide-based defensive systems encoded within their genomes [114]. This pattern may not necessarily reflect a natural phenomenon but could result from a biased research focus to date.

While animals sometimes produce compounds found in other life forms, many unique compounds are exclusive to animals. Examples include intricate and unusual compounds like sponge alkaloids [117], for which the biosynthetic pathways remain largely undiscovered. This challenge is twofold: predicting potential biochemical reactions is difficult, and bioinformatics methods have been limited. Although bioinformatics is continually advancing as part of mainstream biology, predicting unprecedented bond formations remains challenging. Progress in this area requires creativity and robust experimental models, such as the study of animal terpene synthases. Once a pathway is identified, recombinant expression and characterization can be particularly demanding. An example is the biosynthesis of nemamide, where functional analysis has proven difficult despite significant initial findings. Fortunately, methodological advancements are rapidly improving, presenting a promising outlook. MS-based metabolomics has emerged as one of the most promising advancements for studying complex biosynthetic pathways, especially when combined with chromatographic techniques. GC-MS and its multidimensional evolution, in particular, offer superior metabolite separation and minimize ion suppression due to its gas-phase ionization, though it often requires chemical derivatization of metabolic species prior to analysis. Advancements in derivatization methodologies could further enhance these techniques, accelerating research into unique animal metabolites and their biosynthetic pathways.

References

1 Morita, M. and Schmidt, E.W. (2018). Parallel lives of symbionts and hosts: chemical mutualism in marine animals. *Nat. Prod. Rep.* 35 (4): 357–378.

2 Bellés, X., Martín, D., and Piulachs, M.-D. (2005). The mevalonate pathway and the synthesis of juvenile hormone in insects. *Annu. Rev. Entomol.* 50 (1): 181–199.

3 Lee, E., Schooley, D.A., Hall, M.S., and Judy, K.J. (1978). Juvenile hormone biosynthesis: homomevalonate and mevalonate synthesis by insect corpus allatum enzymes. *J. Chem. Soc. Chem. Commun.* 7: 290.

4 Torres, J.P. and Schmidt, E.W. (2019). The biosynthetic diversity of the animal world. *J. Biol. Chem.* 294 (46): 17684–17692.

5 Daly, J.W., Spande, T.F., and Garraffo, H.M. (2005). Alkaloids from amphibian skin: a tabulation of over eight-hundred compounds. *J. Nat. Prod.* 68 (10): 1556–1575.

6 Rollins-Smith, L.A. (2009). The role of amphibian antimicrobial peptides in protection of amphibians from pathogens linked to global amphibian declines. *Biochim. Biophys. Acta Biomembr.* 1788 (8): 1593–1599.

7 Paul, V.J., Arthur, K.E., Ritson-Williams, R. et al. (2007). Chemical defenses: from compounds to communities. *Biol. Bull.* 213 (3): 226–251.

8 Gomes, N., Lefranc, F., Kijjoa, A., and Kiss, R. (2015). Can some marine-derived fungal metabolites become actual anticancer agents? *Mar. Drugs* 13 (6): 3950–3991.

9 Wyatt, T.D. (2014). *Pheromones and Animal Behavior: Chemical Signals and Signatures*. Cambridge University Press.

10 Blum, M.S. (1996). Semiochemical parsimony in the arthropoda. *Annu. Rev. Entomol.* 41 (1): 353–374.

11 Whitten, W.K. (1958). Modification of the oestrous cycle of the mouse by external stimuli associated with the male. *J. Endocrinol.* 17 (3): 307–313.

12 Vandenbergh, J.G. (1969). Male odor accelerates female sexual maturation in mice. *Endocrinology* 84 (3): 658–660.

13 Heil, M. (2008). Indirect defence via tritrophic interactions. *New Phytol.* 178 (1): 41–61.

14 Dini, I., Marra, R., Cavallo, P. et al. (2021). Trichoderma strains and metabolites selectively increase the production of volatile organic compounds (VOCs) in olive trees. *Metabolites* 11 (4): 213.

15 Hart, T., Lopes, L.E., Frank, D.D., and Kronauer, D.J.C. (2024). Pheromone representation in the ant antennal lobe changes with age. *Curr. Biol.* 34 (14): 3233–3240.e4.

16 Witzgall, P., Kirsch, P., and Cork, A. (2010). Sex pheromones and their impact on pest management. *J. Chem. Ecol.* 36 (1): 80–100.

17 Yancey, P.H. (2005). Organic osmolytes as compatible, metabolic and counteracting cytoprotectants in high osmolarity and other stresses. *J. Exp. Biol.* 208 (15): 2819–2830.

18 Ghosh, R., Lokman, P.M., Lamare, M.D. et al. (2013). Changes in physiological responses of an Antarctic fish, the emerald rock cod (*Trematomus bernacchii*), following exposure to polybrominated diphenyl ethers (PBDEs). *Aquat. Toxicol.* 128–129: 91–100.

19 Treutter, D. (2005). Significance of flavonoids in plant resistance and enhancement of their biosynthesis. *Plant Biol.* 7 (6): 581–591.

20 Mittler, R. (2002). Oxidative stress, antioxidants and stress tolerance. *Trends Plant Sci.* 7 (9): 405–410.

21 Katz, L. and Baltz, R.H. (2016). Natural product discovery: past, present, and future. *J. Ind. Microbiol. Biotechnol.* 43 (2–3): 155–176.

22 Cragg, G.M. and Newman, D.J. (2005). Plants as a source of anti-cancer agents. *J. Ethnopharmacol.* 100 (1–2): 72–79.

23 Wall, M.E. (1998). Camptothecin and taxol: discovery to clinic. *Med. Res. Rev.* 18 (5): 299–314.

24 Carter, N.J. and Keam, S.J. (2007). Trabectedin: a review of its use in the management of soft tissue sarcoma and ovarian cancer. *Drugs* 67 (15): 2257–2276.

25 Gupta, S.C., Patchva, S., and Aggarwal, B.B. (2013). Therapeutic roles of curcumin: lessons learned from clinical trials. *AAPS J.* 15 (1): 195–218.

26 Calder, P.C. (2017). Omega-3 fatty acids and inflammatory processes: from molecules to man. *Biochem. Soc. Trans.* 45 (5): 1105–1115.

27 Terlau, H. and Olivera, B.M. (2004). *Conus* venoms: a rich source of novel ion channel-targeted peptides. *Physiol. Rev.* 84 (1): 41–68.

28 Zasloff, M. (2002). Antimicrobial peptides of multicellular organisms. *Nature* 415 (6870): 389–395.

29 Rinaudo, M. (2006). Chitin and chitosan: properties and applications. *Prog. Polym. Sci.* 31 (7): 603–632.

30 Kaissling, K.-E. (2009). Olfactory perireceptor and receptor events in moths: a kinetic model revised. *J. Comp. Physiol. A* 195 (10): 895–922.

31 Visick, K.L. and McFall-Ngai, M.J. (2000). An exclusive contract: specificity in the *Vibrio fischeri-Euprymna scolopes* partnership. *J. Bacteriol.* 182 (7): 1779–1787.

32 Chau, R., Kalaitzis, J.A., and Neilan, B.A. (2011). On the origins and biosynthesis of tetrodotoxin. *Aquat. Toxicol.* 104 (1–2): 61–72.

33 Hopwood, D.A. (2007). Therapeutic treasures from the deep. *Nat. Chem. Biol.* 3 (8): 457–458.

34 O'Neill, L.A.J., Golenbock, D., and Bowie, A.G. (2013). The history of toll-like receptors — redefining innate immunity. *Nat. Rev. Immunol.* 13 (6): 453–460.

35 Gao, Y.-L., Wang, N., Sun, F.-R. et al. (2018). Tau in neurodegenerative disease. *Ann. Transl. Med.* 6 (10): 175–175.

36 Towle, M.J., Salvato, K.A., Budrow, J. et al. (2001). in vitro and in vivo anti-cancer activities of synthetic macrocyclic ketone analogues of halichondrin B. *Cancer Res.* 61 (3): 1013–1021.

37 Vigushin, D.M., Ali, S., Pace, P.E. et al. (2001). Trichostatin A is a histone deacetylase inhibitor with potent antitumor activity against breast cancer in vivo. *Clin. Cancer Res.* 7 (4): 971–976.

38 Carmichael, W.W. (2001). Health effects of toxin-producing cyanobacteria: "the CyanoHABs". *Hum. Ecol. Risk Assess. Int. J.* 7 (5): 1393–1407.

39 Bennett, J.W. and Klich, M. (2003). Mycotoxins. *Clin. Microbiol. Rev.* 16 (3): 497–516.

40 Noguchi, T. (2008). Tetrodotoxin – distribution and accumulation in aquatic organisms, and cases of human intoxication. *Mar. Drugs* 6 (2): 220–242.

41 Clarkson, T.W. and Magos, L. (2006). The toxicology of mercury and its chemical compounds. *Crit. Rev. Toxicol.* 36 (8): 609–662.

42 Schäfer, R.B., Von Der Ohe, P.C., Rasmussen, J. et al. (2012). Thresholds for the effects of pesticides on invertebrate communities and leaf breakdown in stream ecosystems. *Environ. Sci. Technol.* 46 (9): 5134–5142.

43 Dang, L., White, D.W., Gross, S. et al. (2009). Cancer-associated IDH1 mutations produce 2-hydroxyglutarate. *Nature* 462 (7274): 739–744.

44 Tang, W.H.W., Wang, Z., Levison, B.S. et al. (2013). Intestinal microbial metabolism of phosphatidylcholine and cardiovascular risk. *N. Engl. J. Med.* 368 (17): 1575–1584.

45 Phillips, M., Basa-Dalay, V., Bothamley, G. et al. (2010). Breath biomarkers of active pulmonary tuberculosis. *Tuberculosis* 90 (2): 145–151.

46 Henriksen, K., O'Bryant, S.E., Hampel, H. et al. (2014). The future of blood-based biomarkers for Alzheimer's disease. *Alzheimers Dement.* 10 (1): 115–131.

47 Singh, R., Tuček, M., Maxa, K. et al. (1995). A rapid and simple method for the analysis of 1-hydroxypyrene glucuronide: a potential biomarker for poly cyclic aromatic hydrocarbon exposure. *Carcinogenesis* 16 (12): 2909–2915.

48 Dalle-Donne, I., Rossi, R., Colombo, R. et al. (2006). Biomarkers of oxidative damage in human disease. *Clin. Chem.* 52 (4): 601–623.

49 Starcevic, A., Akthar, S., Dunlap, W.C. et al. (2008). Enzymes of the shikimic acid pathway encoded in the genome of a basal metazoan, *Nematostella vectensis*, have microbial origins. *Proc. Natl. Acad. Sci. USA* 105 (7): 2533–2537.

50 Osborn, A.R., Kean, K.M., Karplus, P.A., and Mahmud, T. (2017). The sedoheptulose 7-phosphate cyclases and their emerging roles in biology and ecology. *Nat. Prod. Rep.* 34 (8): 945–956.

51 Osborn, A.R., Almabruk, K.H., Holzwarth, G. et al. (2015). De novo synthesis of a sunscreen compound in vertebrates. *elife* 4: e05919.

52 Starcevic, A., Dunlap, W.C., Cullum, J. et al. (2010). Gene expression in the scleractinian *Acropora microphthalma* exposed to high solar irradiance reveals elements of photoprotection and coral bleaching. *PLoS One* 5 (11): e13975.

53 Balskus, E.P. and Walsh, C.T. (2010). The genetic and molecular basis for sunscreen biosynthesis in cyanobacteria. *Science* 329 (5999): 1653–1656.

54 Donia, M.S., Fricke, W.F., Partensky, F. et al. (2011). Complex microbiome underlying secondary and primary metabolism in the tunicate-*Prochloron* symbiosis. *Proc. Natl. Acad. Sci. USA* 108 (51).

55 Gershenzon, J. and Dudareva, N. (2007). The function of terpene natural products in the natural world. *Nat. Chem. Biol.* 3 (7): 408–414.

56 Walsh, C.T. (2017). *Natural Product Biosynthesis: Chemical Logic and Enzymatic Machinery*. Royal Society of Chemistry.

57 Christianson, D.W. (2017). Structural and chemical biology of terpenoid cyclases. *Chem. Rev.* 117 (17): 11570–11648.

58 Beran, F., Rahfeld, P., Luck, K. et al. (2016). Novel family of terpene synthases evolved from *trans* -isoprenyl diphosphate synthases in a flea beetle. *Proc. Natl. Acad. Sci. USA* 113 (11): 2922–2927.

59 Lancaster, J., Lehner, B., Khrimian, A. et al. (2019). An IDS-type sesquiterpene synthase produces the pheromone precursor (Z)-α-bisabolene in *Nezara viridula*. *J. Chem. Ecol.* 45 (2): 187–197.

60 Lancaster, J., Khrimian, A., Young, S. et al. (2018). De novo formation of an aggregation pheromone precursor by an isoprenyl diphosphate synthase-related terpene synthase in the harlequin bug. *Proc. Natl. Acad. Sci. USA* 115 (37): E8634–E8641.

61 Matsumoto, S. (2010). Molecular mechanisms underlying sex pheromone production in moths. *Biosci. Biotechnol. Biochem.* 74 (2): 223–231.

62 Artyukhin, A.B., Zhang, Y.K., Akagi, A.E. et al. (2018). Metabolomic "dark matter" dependent on peroxisomal β-oxidation in *Caenorhabditis elegans*. *J. Am. Chem. Soc.* 140 (8): 2841–2852.

63 Panda, O., Akagi, A.E., Artyukhin, A.B. et al. (2017). Biosynthesis of modular ascarosides in *C. elegans*. *Angew. Chem. Int. Ed.* 56 (17): 4729–4733.

64 Von Reuss, S.H., Bose, N., Srinivasan, J. et al. (2012). Comparative metabolomics reveals biogenesis of ascarosides, a modular library of small-molecule signals in *C. elegans*. *J. Am. Chem. Soc.* 134 (3): 1817–1824.

65 Von Reuss, S.H. and Schroeder, F.C. (2015). Combinatorial chemistry in nematodes: modular assembly of primary metabolism-derived building blocks. *Nat. Prod. Rep.* 32 (7): 994–1006.

66 Zhou, Y., Wang, Y., Zhang, X. et al. (2018). Biosynthetic tailoring of existing ascaroside pheromones alters their biological function in *C. elegans*. *eLife* 7: e33286.

67 Calestani, C., Rast, J.P., and Davidson, E.H. (2003). Isolation of pigment cell specific genes in the sea urchin embryo by differential macroarray screening. *Development* 130 (19): 4587–4596.

68 Liu, D., Awazu, A., Sakuma, T. et al. (2019). Establishment of knockout adult sea urchins by using a CRISPR-Cas9 system. *Develop. Growth Differ.* 61 (6): 378–388.

69 Castoe, T.A., Stephens, T., Noonan, B.P., and Calestani, C. (2007). A novel group of type I polyketide synthases (PKS) in animals and the complex phylogenomics of PKSs. *Gene* 392 (1–2): 47–58.

70 Cooke, T.F., Fischer, C.R., Wu, P. et al. (2017). Genetic mapping and biochemical basis of yellow feather pigmentation in budgerigars. *Cell* 171 (2): 427–439.e21.

71 Kannangara, R., Siukstaite, L., Borch-Jensen, J. et al. (2017). Characterization of a membrane-bound C-glucosyltransferase responsible for carminic acid biosynthesis in *Dactylopius coccus* costa. *Nat. Commun.* 8 (1): 1987.

72 Thiessen, K.D., Grzegorski, S.J., Chin, Y. et al. (2019). Zebrafish otolith biomineralization requires polyketide synthase. *Mech. Dev.* 157: 1–9.

73 Sabatini, M., Comba, S., Altabe, S. et al. (2018). Biochemical characterization of the minimal domains of an iterative eukaryotic polyketide synthase. *FEBS J.* 285 (23): 4494–4511.

74 Moldoveanu, S.C. and David, V. (2019). Derivatization methods in GC and GC/MS. In: *Gas Chromatography - Derivatization, Sample Preparation, Application* (ed. P. Kusch), 1–33. IntechOpen.

75 Han, X. and Gross, R.W. (2005). Shotgun lipidomics: multidimensional MS analysis of cellular lipidomes. *Expert Rev. Proteomics* 2 (2): 253–264.

76 Altmann, F. (1992). Determination of amino sugars and amino acids in glycoconjugates using precolumn derivatization with o-phthalaldehyde. *Anal. Biochem.* 204 (1): 215–219.

77 Wang, W., Wang, Y., Chen, F., and Zheng, F. (2021). Comparison of determination of sugar-PMP derivatives by two different stationary phases and two HPLC detectors: C18 vs. amide columns and DAD vs. ELSD. *J. Food Compos. Anal.* 96: 103715.

78 Sun, Y., Tang, H., and Wang, Y. (2021). Progress and challenges in quantifying carbonyl-metabolomic phenomes with LC-MS/MS. *Molecules* 26 (20): 6147.

79 Halket, J.M. and Zaikin, V.G. (2003). Derivatization in mass spectrometry—1. Silylation. *Eur. J. Mass Spectrom.* 9 (1): 1–21.

80 Knapp, D.R. (1979). *Handbook of Analytical Derivatization Reactions*. New York: Wiley.

81 Fiehn, O. (2008). Extending the breadth of metabolite profiling by gas chromatography coupled to mass spectrometry. *TrAC Trends Anal. Chem.* 27 (3): 261–269.

82 Wishart, D.S., Feunang, Y.D., Marcu, A. et al. (2018). HMDB 4.0: the human metabolome database for 2018. *Nucleic Acids Res.* 46 (D1): D608–D617.

83 Roessner, U., Wagner, C., Kopka, J. et al. (2000). Simultaneous analysis of metabolites in potato tuber by gas chromatography–mass spectrometry. *Plant J.* 23 (1): 131–142.

84 Troisi, J., Tafuro, M., Lombardi, M. et al. (2022). A metabolomics-based screening proposal for colorectal cancer. *Metabolites* 12 (2): 110.

85 Troisi, J., Colucci, A., Cavallo, P. et al. (2021). A serum metabolomic signature for the detection and grading of bladder cancer. *Appl. Sci.* 11 (6): 2835.

86 Troisi, J., Marciano, F., Scala, G. et al. (2020). Salivary and urinary metabolome in pediatric obesity and metabolic syndrome. In: *Obesity and Diabetes* (ed. Joel Faintuch and Salomão Faintuch), 249–263. Springer.

87 Troisi, J., Landolfi, A., Vitale, C. et al. (2019). A metabolomic signature of treated and drug-naïve patients with Parkinson's disease: a pilot study. *Metabolomics* 15 (6): 90.

88 Troisi, J., Sarno, L., Landolfi, A. et al. (2018). Metabolomic signature of endometrial cancer. *J. Proteome Res.* 17 (2): 804–812.

89 Troisi, J., Sarno, L., Martinelli, P. et al. (2017). A metabolomics-based approach for non-invasive diagnosis of chromosomal anomalies. *Metabolomics* 13 (11): 140.

90 Dettmer, K., Aronov, P.A., and Hammock, B.D. (2007). Mass spectrometry-based metabolomics. *Mass Spectrom. Rev.* 26 (1): 51–78.

91 Zaikin, V.G. and Halket, J.M. (2003). Derivatization in mass spectrometry—2. Acylation. *Eur. J. Mass Spectrom.* 9 (5): 421–434.

92 Kvitvang, H.F.N., Kristiansen, K.A., Lien, S.K., and Bruheim, P. (2014). Quantitative analysis of amino and organic acids by methyl chloroformate derivatization and GC-MS/MS analysis. In: *Methods in Molecular Biology* (ed. Daniel Raftery), 137–145. New York: Springer New York.

93 Smart, K.F., Aggio, R.B.M., Van Houtte, J.R., and Villas-Bôas, S.G. (2010). Analytical platform for metabolome analysis of microbial cells using methyl chloroformate derivatization followed by gas chromatography–mass spectrometry. *Nat. Protoc.* 5 (10): 1709–1729.

94 Da'i, M., Huda, R., and Suhendi, A. (2019). Profile analysis of metabolite fractions from chamber bitter herb (*Phyllanthusniruri L.*) ethanol extract using gas chromatography (GC-MS) method with derivatization. *Test Eng. & Manag.* 81: 3609–3619.

95 Pagliano, E. (2020). Versatile derivatization for GC-MS and LC-MS: alkylation with trialkyloxonium tetrafluoroborates for inorganic anions, chemical warfare agent degradation products, organic acids, and proteomic analysis. *Anal. Bioanal. Chem.* 412 (9): 1963–1971.

96 Řimnáčová, L., Hušek, P., and Šimek, P. (2014). A new method for immediate derivatization of hydroxyl groups by fluoroalkyl chloroformates and its application for the determination of sterols and tocopherols in human serum and amniotic fluid by gas chromatography–mass spectrometry. *J. Chromatogr. A* 1339: 154–167.

97 Jamali, S.A.M., Azizan, K.A., Baharum, S.N., and Taib, M. (2022). Effects of derivatization on the metabolite profiling of the cadmium-tolerant mangrove fungus trichoderma atroviride using GC-MS analysis. *Squalen Bull. Mar. Fish. Postharvest Biotechnol.* 17 (3): 146–154.

98 Pietrogrande, M.C., Bacco, D., and Mercuriali, M. (2010). GC-MS analysis of low-molecular-weight dicarboxylic acids in atmospheric aerosol: comparison between silylation and esterification derivatization procedures. *Anal. Bioanal. Chem.* 396 (2): 877–885.

99 Park, J. and Han, S. (2021). Chiral derivatization of hydroxycarboxylic acids using 2,4,6-trichlorobenzoyl chloride as a highly efficient regioselective esterification reagent for gas chromatography-mass spectrometry. *Anal. Bioanal. Chem.* 396: 877–885.

100 Vodrážka, P., Řimnáčová, L., Berková, P. et al. (2025). A new, validated GC-PICI-MS method for the quantification of 32 lipid fatty acids via base-catalyzed transmethylation and the isotope-coded derivatization of internal standards. *Metabolites* 15 (2): 104.

101 Ecker, J., Scherer, M., Schmitz, G., and Liebisch, G. (2012). A rapid GC–MS method for quantification of positional and geometric isomers of fatty acid methyl esters. *J. Chromatogr. B* 897: 98–104.

102 Masood, A., Stark, K.D., and Salem, N. (2005). A simplified and efficient method for the analysis of fatty acid methyl esters suitable for large clinical studies. *J. Lipid Res.* 46 (10): 2299–2305.

103 Březina, A., Dořmanová, A., Kukučka, P. et al. (2022). Application of transesterification derivatization reaction for GC-MS analysis. *Czech Chem. Soc. Symp. Ser.* 20 (5): 319–322.

104 Seppänen-Laakso, T., Nygren, H., and Rischer, H. (2017). UPLC-ELSD analysis of algal lipid classes and derivatization of bound and free fatty acids and sterols for GC-MS methods. In: *Methods in Molecular Biology* (ed. Kristian Spilling), 223–232. New York, NY: Springer New York.

105 Dasgupta, A., Banerjee, P., and Malik, S. (1992). Use of microwave irradiation for rapid transesterification of lipids and accelerated synthesis of fatty acyl pyrrolidides for analysis by gas chromatography-mass spectrometry:

study of fatty acid profiles of olive oil, evening primrose oil, fish oils and phospholipids from mango pulp. *Chem. Phys. Lipids* 62 (3): 281–291.

106 Duong, C.T. and Roper, M.G. (2012). A microfluidic device for the automated derivatization of free fatty acids to fatty acidmethylesters. *Analyst* 137 (4): 840–846.

107 Villas-Bôas, S.G., Smart, K.F., Sivakumaran, S., and Lane, G.A. (2011). Alkylation or silylation for analysis of amino and non-amino organic acids by GC-MS? *Metabolites* 1 (1): 3–20.

108 Kumirska, J., Plenis, A., Łukaszewicz, P. et al. (2013). Chemometric optimization of derivatization reactions prior to gas chromatography–mass spectrometry analysis. *J. Chromatogr. A* 1296: 164–178.

109 Togola, A. and Budzinski, H. (2008). Multi-residue analysis of pharmaceutical compounds in aqueous samples. *J. Chromatogr. A* 1177 (1): 150–158.

110 Halket, J.M., Waterman, D., Przyborowska, A.M. et al. (2005). Chemical derivatization and mass spectral libraries in metabolic profiling by GC/MS and LC/MS/MS. *J. Exp. Bot.* 56 (410): 219–243.

111 Kind, T., Wohlgemuth, G., Lee, D.Y. et al. (2009). FiehnLib: mass spectral and retention index libraries for metabolomics based on quadrupole and time-of-flight gas chromatography/mass spectrometry. *Anal. Chem.* 81 (24): 10038–10048.

112 Dunning Hotopp, J.C. (2018). Grafting or pruning in the animal tree: lateral gene transfer and gene loss? *BMC Genomics* 19 (1): 470.

113 McDonald, B.R. and Currie, C.R. (2017). Lateral gene transfer dynamics in the ancient bacterial genus *Streptomyces*. *MBio* 8 (3): e00644–e00617.

114 Beran, F., Köllner, T.G., Gershenzon, J., and Tholl, D. (2019). Chemical convergence between plants and insects: biosynthetic origins and functions of common secondary metabolites. *New Phytol.* 223 (1): 52–67.

115 Kubanek, J., Faulkner, D.J., and Andersen, R.J. (2000). Geographic variation and tissue distribution of endogenous terpenoids in the Northeastern Pacific Ocean dorid nudibranch *Cadlina luteomartinata*: implications for the regulation of de novo biosynthesis. *J. Chem. Ecol.* 26 (2): 377–389.

116 Bachmann, B.O. (2010). Biosynthesis: is it time to go retro? *Nat. Chem. Biol.* 6 (6): 390–393.

117 Carroll, A.R., Copp, B.R., Davis, R.A. et al. (2019). Marine natural products. *Nat. Prod. Rep.* 36 (1): 122–173.

8.2

Specialized Metabolites in Biological Systems (Part II): Focus on Volatile Metabolites in Human and Microbial Studies

Allan Polidoro[1], *Riccardo Di Stefano*[2], *Virginia Galliani*[2], *and Flavio A. Franchina*[2]

[1] *Department of Environmental and Prevention Sciences, University of Ferrara, Ferrara, Italy*
[2] *Department of Chemical, Pharmaceutical, and Agricultural Sciences, University of Ferrara, Ferrara, Italy*

8.2.1 Introduction to Volatilomics

The human volatilome, comprising the complete set of volatile organic compounds (VOCs) released by the body, provides a chemically rich, dynamically responsive, and noninvasive window into physiology, pathology, and environmental exposure. VOCs are produced by endogenous metabolic pathways, microbial activity, or exogenous inputs, including diet, medication, pollutants, and lifestyle factors. These compounds, despite their low concentrations, carry valuable information about biological processes occurring across tissues, organs, and microbiomes.

In recent years, volatilomics has been positioned as a potential tool for early diagnosis, disease monitoring, and phenotyping, often praised for its compatibility with noninvasive sample collection and its integration of host-environment–microbiome interactions. However, much of this promise remains unfulfilled. The field is marked by fragmented methodology, limited mechanistic insight, and a proliferation of exploratory studies with insufficient clinical validation. Volatilomics is not yet a fully established solution or a ready diagnostic tool, requiring conceptual maturity, methodological rigor, and clinical grounding.

Multiple techniques have been exploited for the study of volatile metabolites, including or not a chromatographic pre-separation before detection. These include gas chromatography-mass spectrometry (GC-MS), selected ion flow tube mass spectrometry (SIFT-MS), gas chromatography-ion mobility spectrometry (GC-IMS), proton transfer reaction mass spectrometry (PTR-MS), and secondary electrospray ionization mass spectrometry ((SESI)MS), among others.

A key pillar of volatilomics relies on the use of high-resolution analytical techniques, particularly gas chromatography coupled with mass spectrometry, such as one-dimensional GC (GC-MS) or comprehensive two-dimensional gas chromatography (GC×GC-MS). The coupling of high-resolution mass spectrometry (HR MS) is also exploited to achieve the highest separation and identification capabilities.

These tools are essential for detecting, separating, and structurally elucidating trace-level VOCs in complex biological matrices. Unlike sensor-based approaches, high-resolution methods enable reproducible identification, quantitative accuracy, and robust differentiation between endogenous compounds and contaminants, making them indispensable for translational research and future diagnostic applications.

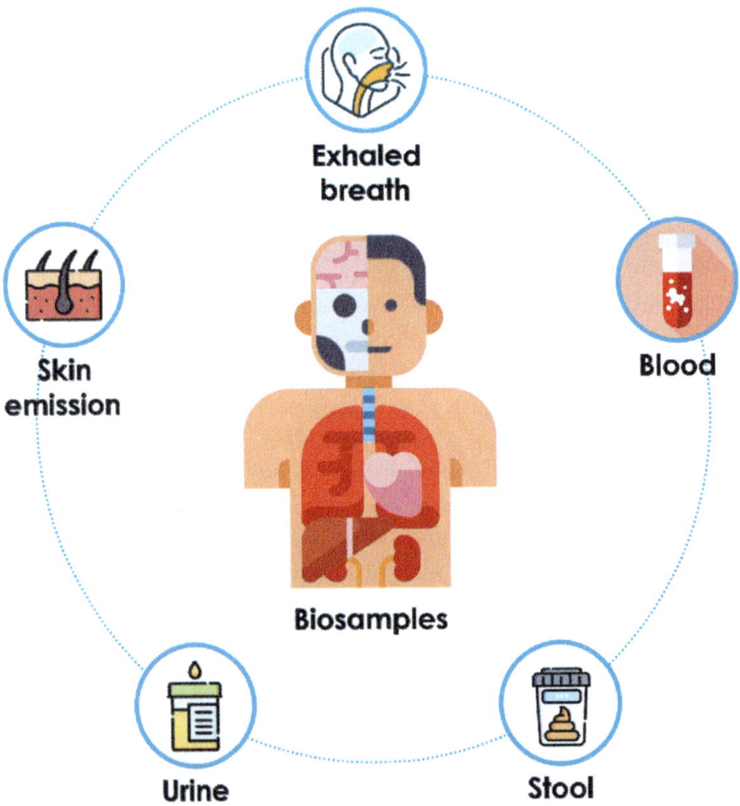

Figure 8.2.1 Biosamples commonly used in volatilomics research. This chapter explores the VOCs emitted from exhaled breath, skin, blood, urine, and stool, highlighting their analytical features, biological origins, and clinical applications.

Therefore, this chapter provides an exploration of human VOCs across five biological matrices: breath, skin, blood, urine, and stool (Figure 8.2.1), examining their biochemical origins, analytical detection methods, and relevance to disease.

8.2.2 Exhaled Breath

8.2.2.1 Biochemistry Considerations

Exhaled breath analysis is an emerging strategy for characterizing the human volatilome and exposome. Although initially applied to monitor targeted pharmacokinetics and exogenous VOC uptake, this approach has attracted attention as a noninvasive alternative to traditional biosampling [1]. The composition of breath VOCs reflects a complex interplay between metabolic activity, pathophysiological conditions, and environmental or lifestyle exposures. More than 1000 VOCs have been detected in human breath at trace levels, with a large fraction being exogenous in origin, stemming from sources such as tobacco smoke, cleaning agents, cosmetics, and food. In contrast, endogenous VOCs are generated through metabolic and inflammatory processes, as well as the activity of the pulmonary and gut microbiota. These VOCs span several chemical classes, including hydrocarbons, oxygenated compounds, and sulfur- and nitrogen-containing substances [2, 3].

Understanding the biochemical pathways underlying VOC formation is essential for advancing our understanding of their physiological roles and potential diagnostic value. Among the main contributors to endogenous VOC production are oxidative stress and cytochrome P450 (CYP) enzymes [4]. Hydrocarbons, for example, are stable end products of lipid peroxidation, typically exhibiting low solubility in blood, allowing for rapid exhalation if not further metabolized. Alkanes, such as ethane and pentane, derived from ω-3 and ω-6 polyunsaturated fatty acids, are well-established markers of oxidative stress [5]. Isoprene, the most abundant VOC in exhaled breath, is believed to be primarily produced via the mevalonate pathway of cholesterol biosynthesis, although a minor portion may also be derived from microbial activity [6, 7].

Another major breath VOC is acetone, primarily produced in the liver through the decarboxylation of acetoacetate, a ketone body formed from excess acetyl-CoA during the metabolism of amino acids. Under metabolic conditions such as fasting, low-carbohydrate dieting, or uncontrolled diabetes mellitus, ketone body levels increase, resulting in elevated acetone concentrations in the breath. However, these levels can also fluctuate significantly in response to exercise, dietary intake, and other physiological factors [6, 8].

In contrast, alcohols such as ethanol and methanol can be introduced exogenously, through food and alcoholic beverages, or produced endogenously through intestinal bacterial fermentation. Their high water solubility facilitates rapid distribution across bodily fluids and tissues, influencing their breath concentration profiles [9, 10].

Aldehydes represent another relevant class of breath VOCs. These molecules are produced through multiple physiological pathways, including the metabolism of endogenous ethanol and methanol by alcohol dehydrogenase (ADH) and CYP2E1, which generate acetaldehyde and formaldehyde, respectively. Lipid peroxidation also contributes, as CYP enzymes reduce hydroperoxides to form aldehydes and ketones [10, 11].

Exogenous sources, especially tobacco smoke, further elevate aldehyde burden. Cigarette smoke contains both saturated (e.g., formaldehyde, acetaldehyde, propionaldehyde, and butyraldehyde) and unsaturated aldehydes (e.g., acrolein and crotonaldehyde). Sugars in the tobacco are considered the primary source of formaldehyde, whereas cellulose yields acetaldehyde. In addition, several tobacco-specific carcinogens, such as 4-(methylnitrosamino)-1-(3-pyridyl)-1-butanone (NNK), undergo CYP-mediated metabolic activation, producing α-hydroxy intermediates that decompose into reactive diazonium ions and aldehydes [4, 12].

Esters, commonly found in natural products such as fats, oils, waxes, and essential oils, are metabolized through enzymatic hydrolysis catalyzed by esterases (e.g., lipases). This process yields corresponding alcohols and acids, which contribute to both endogenous and exogenous VOC pools in breath [13].

Finally, sulfur-containing compounds, such as ethyl mercaptan, dimethyl sulfide, and dimethyl disulfide, are well-known contributors to the characteristic odor of cirrhotic breath [14]. These compounds originate from incomplete methionine metabolism via the transamination pathway, in which mercaptans are oxidized to their corresponding sulfides. Under normal physiological conditions, their concentrations in blood and breath are very low, but impaired liver function disrupts their clearance, leading to systemic accumulation and increased exhalation [2, 6].

8.2.2.2 Diagnostic and Pathophysiological Applications

8.2.2.2.1 Oncological Diseases

Lung Cancer The identification of VOCs in exhaled breath has emerged as a promising alternative for the early detection of lung cancer (LC). Among the most consistently reported markers are aldehydes, including a homologous series from acetaldehyde to octadecanal, as well as branched aldehydes such as 2-methylpropanal, methacrolein, 2-methylbutanal, 3-methylbutanal, and 2-methyl-2-butenal. Except for acetaldehyde, whose levels remain similar between LC patients and healthy controls (HCs), these aldehydes are typically elevated in the breath of LC patients [15, 16]. Notably, formaldehyde levels are significantly higher in LC patients, potentially due to its release by cancer cells during the degradation of tryptophan [15, 17]. Hexanal and heptanal are consistently detected in higher levels across blood, breath, and urine of LC patients, suggesting their direct production from LC cells and systemic distribution [18, 19].

Innovations in sampling and analysis, such as the development of functionalized SPME fibers, have improved VOC detection sensitivity. For instance, Zhang et al. applied an NH2-MON-coated fiber to identify aldehyde levels in breath samples from six LC patients and six HCs, revealing significantly higher levels of hexanal, heptanal, benzaldehyde, octanal, and nonanal in the LC group (Table 8.2.1) [20].

Beyond aldehydes, ketones such as 2-butanone and 2-pentanone have been consistently elevated in LC samples, suggesting a potential involvement of altered ketogenic metabolism in the pathophysiology of LC. Still, acetone, despite being among the most abundant breath VOCs, exhibited inconsistent trends, highlighting the challenge of decoupling disease-related changes from physiological variability or other cofounding factors (e.g., fasting and exercise) [4, 21].

Exogenous compounds such as acetonitrile also feature in LC breath profiles. Acetonitrile, a well-known component of tobacco smoke, is consistently detected at elevated levels in smokers compared to nonsmokers [22, 23]. This compound is slowly metabolized via cytochrome P450 monooxygenase into cyanohydrin, which then decomposes into hydrogen cyanide and formaldehyde. Due to its slow clearance, an expressive portion of acetonitrile is exhaled unchanged. Despite being exogenous, it has been proposed as a potential marker for LC monitoring [4]. Bajtarevic et al. reported higher acetonitrile levels in LC smokers compared to LC ex-smokers [23], whereas Kischkel et al. found increased concentrations in LC ex-smokers compared to healthy smokers [16].

Table 8.2.1 Aldehydes in exhaled breath samples of HCs and LC patients.

Analyte	Conc. in HC (ng/mL)	Conc. in LC patients (ng/mL)
Hexanal	0.41 ± 0.03	5.48 ± 0.31
Heptanal	0.25 ± 0.03	1.59 ± 0.11
Benzaldehyde	0.42 ± 0.04	2.16 ± 0.14
Octanal	0.13 ± 0.02	2.73 ± 0.16
Nonanal	1.63 ± 0.19	11.76 ± 0.89

Source: Zhang et al. [20] / with permission of American Chemical Society.

Table 8.2.2 Significant changes ($p < 0.05$) of VOCs in exhaled breath after surgical resection of lung cancers.

	Peak intensity		
	Preoperative	**Postoperative**	***P*-value**
Acetaldehyde	137	170	0.024
2-Hydroxyacetaldehyde	782	680	0.025
Isoprene	6062	3838	0.017
Pentanal	112	76	0.003
Butyric acid	298	188	0.012
Toluene	3345	1091	0.034
2, 5-Dimethylfuran	19	0	<0.001
Cyclohexanone	150	38	<0.001
Hexanal	26	0	0.018
Heptanal	39	23	0.001
Acetophenone	148	86	0.044
Propylcyclohexane	254	67	<0.001
Octanal	58	27	0.001
Nonanal	31	0	<0.001
Decanal	57	10	<0.001
2, 2-Dimethyldecane	158	47	<0.001

Source: Adapted from Wang et al. [24].

Several studies have investigated the diagnostic potential of breath VOC profiles using both targeted and non-targeted approaches. For instance, Wang et al. utilized high-pressure photoionization time-of-flight mass spectrometry ((HPPI)TOFMS) to analyze exhaled breath from 84 LC patients before and after tumor resection. Sixteen VOCs exhibited significant postoperative decreases, except for acetaldehyde (Table 8.2.2). Validation in a larger cohort, comprising 157 LC patients versus 368 HCs, confirmed the observed trends [24]. These VOC alterations may reflect cancer-induced alterations of cytochrome P450 oxidoreductase (POR) activity, which plays an important role in fatty acid metabolism, including decarboxylation, desaturation, and peroxidation [25]. Given that dysregulated fatty acid metabolism is common in cancer cells, the resulting VOC profiles might serve as biomarkers for malignancy [26, 27].

Building on these findings, Zou et al. developed a noninvasive screening tool for LC by integrating TD-GC-MS with a gradient-boosted decision tree (GBDT) classifier. In a cohort of 236 subjects (60 LC patients and 176 HCs), the model achieved an average accuracy of 85% and an AUC of 0.89. Feature importance analyses identified 22 discriminative peaks, including known LC-associated VOCs such as 2,5-dimethylfuran, styrene, and octanal [28].

In a similar approach, Sakumura et al. applied TD-GC-MS coupled with a support vector machine (SVM) algorithm to distinguish LC patients (n = 107) from HCs (n = 29). From a total of 68 detected VOCs, an optimized panel of five compounds (hydrogen cyanide, methanol, acetonitrile, isoprene, and 1-propanol) (Figure 8.2.2) achieved 89% accuracy, with 92.5% sensitivity, and 75.9% specificity. Notably, the proximity of samples to the SVM decision boundary correlated with the disease stage, with early-stage LC samples clustering closer to the boundary than advanced-stage ones [29].

Liver Cancer The diagnostic potential of exhaled VOCs has also been demonstrated for hepatocellular carcinoma (HCC), offering a promising noninvasive strategy for evaluating hepatic metabolic dysfunction and detecting early-stage tumor development [30–32].

Ferrandino et al. evaluated the feasibility of using exhaled limonene as a biomarker for hepatic metabolic capacity in patients with cirrhosis and HCC. Using TD-GC-MS, the authors analyzed breath samples from 32 cirrhotic patients, 12 patients with cirrhosis-associated HCC, and 40 HCs. Limonene concentrations were elevated in both patient groups compared to the control group. Moreover, breath limonene levels correlated with standard liver function markers, including albumin, bilirubin, and prothrombin time, reflecting impaired hepatic clearance likely associated with compromised cytochrome P450 activity, particularly CYP2C9 and CYP2C19 [33].

Expanding the clinical scope, Miller-Atkins et al. analyzed breath samples from 296 individuals, including patients with cirrhosis, HCC, colorectal cancer, liver metastases, pulmonary hypertension, and HCs. A random forest machine learning model trained on 22 VOCs achieved an overall classification accuracy of 85%, outperforming traditional biomarkers such as α-fetoprotein, especially for HCC detection. Among the most influential VOCs identified were acetaldehyde and acetone, suggesting contributions from both altered tumor metabolism and hepatic dysfunction [30].

Similarly, Sukaram et al. explored the diagnostic potential of VOCs for HCC using GC-MS coupled with an SVM classifier. In a cohort of 97 HCC patients and 111 controls (comprising healthy subjects and cirrhotic patients), a panel of 6 VOCs (acetone, 1,4-pentadiene, methylene chloride, benzene, phenol, and allyl methyl sulfide) distinguished HCC patients from controls with 79.6% accuracy. Furthermore, the samples' distance from the SVM decision boundary increased progressively with advancing HCC stage, suggesting potential for monitoring disease progression. Limonene alone achieved 62.8% sensitivity for detecting

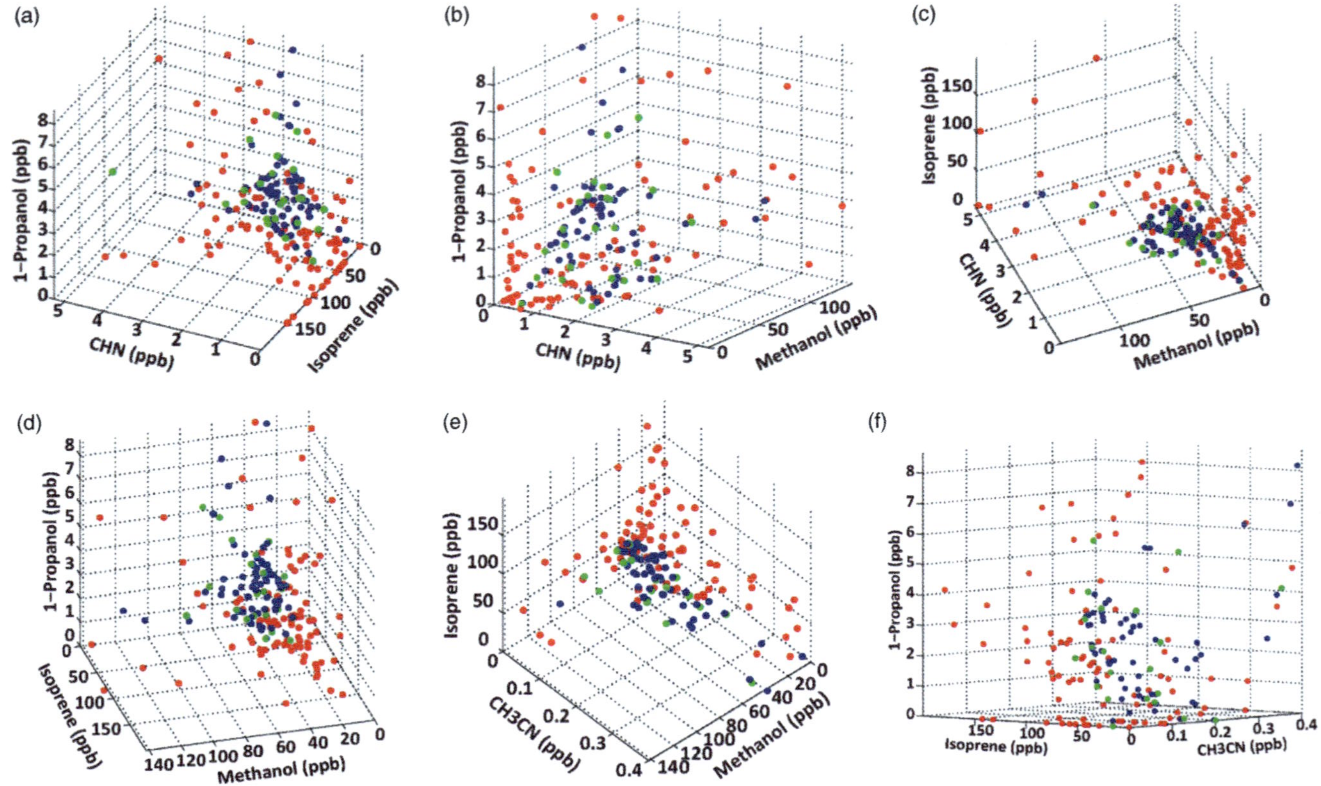

Figure 8.2.2 3D plot of the VOCs combinations: (a) CHN, isoprene, 1-propanol; (b) CHN, methanol, 1-propanol; (c) CHN, methanol, isoprene; (d) isoprene, methanol, 1-propanol; (e) CH_3CN, methanol, isoprene; and (f) isoprene, CH_3CN, 1-propanol. The red and green circles represent LC patients and HC, respectively, and the blue circles indicate the oversampling data. *Source:* Sakumura et al. [29] / MDPI / CC BY 4.0.

early-stage HCC, outperforming α-fetoprotein. Posttreatment changes in acetone concentrations correlated with therapeutic response [31].

Building on these findings, Sukaram et al. further investigated breath biomarkers for early HCC detection using thermal desorption-gas chromatography-field-asymmetric ion mobility spectrometry (TD-GC-FAIMS) in a new cohort comprising 87 early-stage HCC patients, 90 cirrhotic patients, and 72 individuals infected with hepatitis B virus (HBV). Among nine pre-selected VOCs, five differed significantly across groups, with acetone dimer emerging as the most discriminative marker. Logistic regression analysis confirmed a strong correlation between acetone dimer and early-stage HCC, even after adjusting for liver function tests and AFP levels. Similarly, acetone dimer differentiated HCC from HBV-infected patients with 88.5% sensitivity, 90.3% specificity, and an AUROC = 0.922 [34].

Despite these encouraging results, important limitations remain in the field. Sample sizes are usually modest, and external validation is limited. Moreover, most identified VOCs lack well-characterized biochemical pathways, which hinders mechanistic interpretation and limits clinical translation. As such, future research should prioritize the mechanistic elucidation, longitudinal validation, and integration of VOC data with conventional biomarkers and imaging tools to enhance early detection strategies for HCC.

Breast Cancer Breast cancer (BC) is the most commonly diagnosed malignancy among women worldwide, and the analysis of exhaled VOCs has emerged as a promising, noninvasive strategy for its detection and early diagnosis. VOCs in breath may reflect systemic oxidative stress, lipid peroxidation, and tumor-associated metabolic reprogramming, making them attractive candidates for biomarker discovery.

In a targeted study, Li et al. investigated four straight-chain aldehydes (hexanal, heptanal, octanal, and nonanal) in end-tidal breath from 22 BC patients, 17 women with benign breast tumors (BBT), and 24 HCs, using GC-MS. Concentrations of all four aldehydes increased stepwise from HC to BBT and from BBT to BC (Figure 8.2.3) [35]. These findings support a role for aldehyde-related lipid peroxidation in BC pathogenesis and breath-based diagnostics.

Moving from targeted to non-targeted breathomics, Wang et al. analyzed exhaled breath samples from 39 BC patients, 25 individuals with cyclomastopathy, 21 with mammary gland fibroma, and 45 HCs using SPME-GC-MS combined with multivariate statistical analyses. A total of 434, 406, and 408 metabolites were consistently detected in at least half of the samples when comparing the BC and HC groups, the cyclomastopathy group, and the fibroma group, respectively. PCA and OPLS-DA demonstrated clear separation between BC and each comparison group. Three VOCs (2,5,6-trimethyloctane, 1,4-dimethoxy-2,3-butanediol, and cyclohexanone) were consistently elevated in BC breath, highlighting their

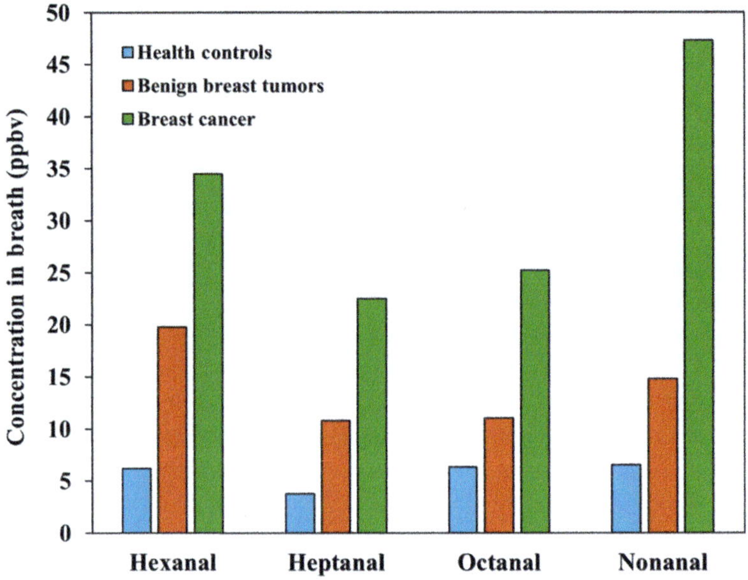

Figure 8.2.3 Stepwise increase in exhaled aldehyde concentrations across healthy controls (HC), individuals with benign breast tumors (BBT), and breast cancer patients (BC). *Source:* Li et al. [35] / with permission of Elsevier.

diagnostic potential [36]. However, the study's relatively small sample size highlights the need for cautious interpretation and further replication.

In a larger cohort, Zhang et al. evaluated breath samples from 203 individuals, 71 BC patients (including ductal carcinoma *in situ*, lymph-node-negative, and lymph-node-positive cases), 78 HCs, and 54 patients with gastric cancer, using SPME-GC-MS. Multivariate models showed good discriminatory performance not only between BC and HC but also between BC and gastric cancer. A seven-compound panel comprising 1,2-propanediol, cyclopentanone, ethylene carbonate, 3-methoxy-1,2-propanediol, 3-methylpyridine, phenol, and tetramethylsilane detected ductal carcinoma *in situ* with 80.8% sensitivity and 100% specificity, and distinguished overall BC from controls with an AUC of 0.919 (93.6% sensitivity, 71.6% specificity) [37]. These results demonstrate the potential for breath VOCs to distinguish between tumor subtypes and stages, though further studies are needed to confirm generalizability across populations and analytical platforms.

These studies demonstrate the promise of breath-based VOC profiling for BC detection, especially when using multivariate models that integrate both known and novel chemical features. Nevertheless, the limited sample sizes, potential confounding factors (e.g., diet, hormonal status), and lack of standardized methodologies remain significant obstacles to clinical translation.

8.2.2.2.2 Respiratory Diseases

Asthma Exhaled VOC analysis has emerged as a promising tool in asthma research, enabling both disease detection and the characterization of inflammatory phenotypes. Unlike conventional biomarkers derived from blood or sputum, exhaled VOCs capture a dynamic interplay between airway inflammation, oxidative stress, and microbiome-host interactions, offering a functional snapshot of pulmonary pathophysiology.

In a study involving 336 women of childbearing age, Sola-Martínez et al. investigated whether exhaled breath TD-GC-MS analysis could discriminate between asthmatic women with coexisting atopic diseases and non-asthmatic controls. Predictive models consistently identified acetone, 2-ethyl-1-hexanol, and a tetrahydroisoquinoline derivative as the most predictive compounds. The same workflow also distinguished asthmatic patients from subjects with atopic conditions alone, highlighting carbon disulfide, decane, and 2-ethyl-1-hexanol as key discriminating VOCs [38]. These findings support the notion that VOC analysis may disentangle overlapping allergic phenotypes, a common challenge in clinical practice.

Focusing on occupational asthma (OA), Heiro et al. analyzed breath samples from 40 affected workers and 45 healthy colleagues using the ReCIVA® device and non-targeted TD-GC-HR MS. Among 536 detected VOCs, 76 were differentially abundant between OA and controls. Logistic regression and random forest models yielded AUC values of 0.94 and 0.90, respectively. Several predictive compounds, including 1-hexadecanol, 2,3-butanediol, phenol, and xylene, are potentially linked to oxidative stress pathways involving NADH and reactive oxygen species (ROS) production. Within irritant-induced asthma (IIA), a distinct phenotype often associated with chemical exposure in the workplace, further stratification distinguished between "probable" and "possible" cases, with 3-methylhexane, methylcyclohexane, and isoprene enriched in the former subgroup [39].

Beyond binary classification, VOC profiling has also shown promise in resolving inflammatory phenotypes of asthma. Schleich et al. evaluated 521 adults across discovery and replication cohorts using GC-MS and GC×GC-HR TOFMS. Seven candidate VOCs were initially identified to differentiate eosinophilic, neutrophilic, and paucigranulocytic subtypes. In the replication cohort, hexane and 2-hexanone effectively distinguished eosinophilic asthma, while nonanal and 1-propanol were elevated in neutrophilic asthma. A model combining hexane, 1-propanol, and nonanal classified neutrophilic asthma with 76% accuracy and an AUC of 0.73, which increased to 97% accuracy (AUC = 0.97) in patients with highly neutrophilic signatures [40]. This stratification capacity is particularly valuable given the heterogeneous nature of asthma and the importance of targeted therapies.

Chronic Obstructive Pulmonary Disease Chronic obstructive pulmonary disease (COPD) is marked by persistent airflow limitation accompanied by systemic inflammation and oxidative stress, both of which are reflected in the composition of exhaled VOCs. Recent studies have suggested that breathomics may aid not only in diagnosis but also in phenotypic differentiation within and beyond COPD.

Using needle-trap devices coupled to GC-MS, Monedeiro et al. profiled breath from 56 participants (16 with LC, 12 with COPD, 8 with asthma, and 20 HCs). Among 29 targeted VOCs, a random forest model based on 12 selected compounds classified samples with 85.7% accuracy. Additionally, multinomial logistic regression using nine VOCs, including 2-propanol, (*E*)-ocimene, and limonene, achieved an average classification accuracy of 95.5%. Notably, hydrocarbons and aldehydes were elevated in both LC and COPD samples, consistent with oxidative stress signatures. Terpineol, benzonitrile, and tridecane emerged as key distinguishing VOCs [41]. However, given the small cohort and high number of predictive features, these findings should be interpreted cautiously and validated in larger, independent populations.

Pizzini et al. applied PTR-TOFMS to a broader cohort of 137 subjects, including 47 with acute exacerbation of COPD (AECOPD), 37 with stable disease, and 53 HCs. Univariate and multivariate screens identified 12 VOCs that differed significantly across groups. Four hydrocarbons, namely isoprene, 2-methylbutane, 2-methylpropane, and 2,4-dimethylpentane, were selectively elevated during AECOPD episodes, whereas acetone and acetonitrile were associated with stable COPD. Dimethyl sulfide and dimethyl disulfide, by contrast, were associated with COPD regardless of clinical state. A random forest classifier built on these markers achieved an AUC of 0.970, with 78% sensitivity and 91% specificity for distinguishing COPD patients from HCs [42].

Although these studies underscore the diagnostic potential of breath VOCs in COPD, the recurrence of similar oxidative-stress-related markers across asthma and lung cancer highlights the need for more specific, mechanistically anchored biomarkers. Further research should aim to disentangle shared inflammatory pathways from disease-specific signatures.

Cystic Fibrosis Cystic fibrosis (CF) is a monogenic disorder characterized by chronic airway inflammation, microbial colonization, and disrupted metabolism, all of which contribute to the generation of a distinct profile of exhaled volatile organic compounds (VOCs). Caused by mutations in the CFTR gene, CF alters electrolyte transport and primarily affects secretory organs such as lungs, pancreas, and reproductive system. In the airways, impaired chloride secretion and enhanced sodium absorption result in dehydrated, viscous mucus, which facilitates microbial colonization, persistent inflammation, and oxidative stress.

These processes generate a distinct profile of exhaled VOCs, offering a potential noninvasive window into disease activity. Pulmonary exacerbations (PEx) accelerate the decline of lung function, reduce quality of life, and shorten life expectancy, making early detection a clinical priority [43, 44].

In a year-long study, Horck et al. collected exhaled breath samples from 49 children with CF every 2 months, analyzing them by GC-TOFMS. Random forest models trained on these data achieved 79% sensitivity, 78% specificity, and an AUC of 0.88 when the sample was obtained within 7 days of PEx onset. However, model performance declined substantially for earlier timepoints, underscoring the need for frequent monitoring and suggesting that these VOC changes reflect ongoing rather than preclinical pathophysiology. Among the discriminant features, compounds such as tetradecanal, camphene, and pentadecane were consistently associated with an increased risk of exacerbation [45].

In a complementary study, Woollam et al. profiled breath from 18 CF patients at baseline and during PEx. Among 125 detected compounds, 74 were selected as candidates for statistical modelling. The alkane 3,7-dimethyldecane was positively correlated with both FEV1% predicted and changes in FEV1% relative to baseline and was significantly downregulated during PEx. Additional differentially abundant VOCs included durene, 2,4,4-trimethyl-1,3-pentanediol 1-isobutyrate, and 5-methyltridecane. Notably, patients not receiving highly effective modulator therapy (HEMT) exhibited elevated levels of octanal, nonanal, and 2,2,8-trimethyldecane (Figure 8.2.4), possibly reflecting persistent oxidative stress or microbial dysbiosis in the absence of effective molecular correction [46].

These studies highlight the promise of exhaled VOC profiling for identifying PEx and monitoring disease status in CF. Nevertheless, despite encouraging diagnostic metrics, the clinical utility of these approaches remains limited by small sample sizes, heterogeneous methodologies, and a lack of longitudinal validation across diverse patient populations. Future research should focus on standardizing sampling protocols, multicenter validation of predictive models, and integrating breathomics into routine care pathways for CF management.

COVID-19 Recent advances have highlighted the application of volatilomics to the noninvasive diagnosis and monitoring of COVID-19. Barberis et al. conducted a non-targeted metabolomics study using GC×GC-TOFMS to analyze exhaled breath condensate (EBC) from COVID-19 patients, healthy volunteers, and individuals with acute or acute-on-chronic cardiopulmonary edema (CPE). Their results identified a unique metabolic signature in EBC associated with SARS-CoV-2 infection, characterized by the differential abundance of 26 small molecules, notably including several monoglycerides of fatty acids. Among the most discriminative markers, monomyristin and monolaurin were found at lower levels in COVID-19 patients than in controls, a pattern consistent with

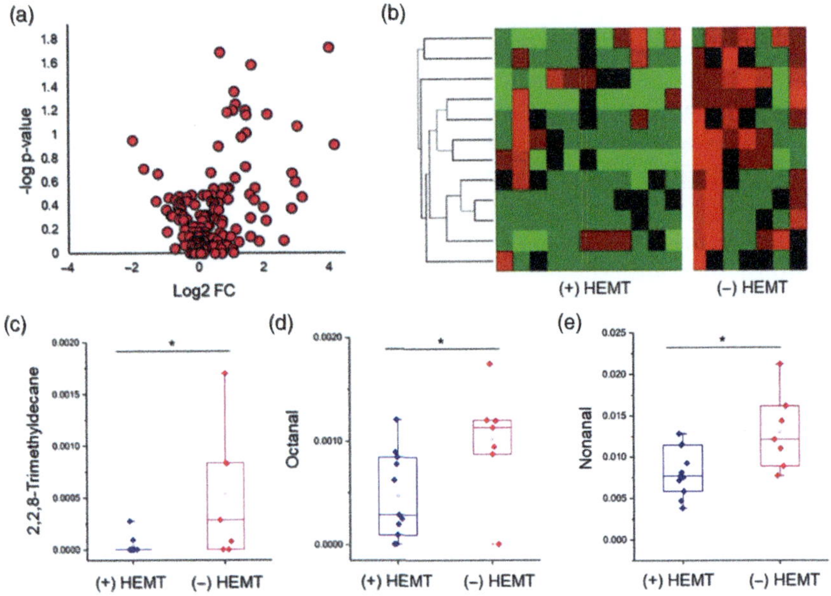

Figure 8.2.4 (a) Volcano plot interpolating statistical significance and a function of fold change for comparing subjects by use of HEMT. (b) Heatmap of the VOCs identified with $p < 0.10$ and (c)–(e) box/whisker plots of the top VOCs with the highest power to discriminate subjects using HEMT. *Source:* Woollam et al. [46] / IOP Publishing / CC BY 4.0.

their known antiviral properties. Multivariate and machine learning analyses enabled robust discrimination between COVID-19, CPE, and healthy subjects (AUC up to 0.98), underscoring the potential of breath condensate metabolomics for distinguishing COVID-19 not only from healthy individuals but also from patients with other respiratory conditions. Importantly, no persistent changes in EBC metabolome were observed in subjects who had recovered from mild COVID-19, suggesting these metabolic alterations are associated with the acute phase of infection [47].

8.2.2.2.3 Liver Cirrhosis

Given the liver's central role in xenobiotic metabolism and systemic VOC clearance, alterations in exhaled breath composition have emerged as promising noninvasive indicators of hepatic dysfunction in liver cirrhosis. In a non-targeted GC-MS survey involving 88 individuals, Ferrandino et al. identified 29 VOCs that significantly differentiated cirrhotic patients from healthy controls, with limonene emerging as the most distinctive marker [48]. Limonene is primarily metabolized by hepatic cytochrome P450 enzymes CYP2C9 and CYP2C19, both of which are functionally impaired in individuals with cirrhosis, leading to its systemic accumulation.

Building on this evidence, a longitudinal study (n = 60) by the same group investigated limonene pharmacokinetics by administering 100 mg of limonene to overnight-fasted patients with cirrhosis and controls (Figure 8.2.5). Peak breath concentrations were observed in both groups 20 minutes after administration, followed by a return to baseline within 2 hours. However, cirrhotic patients exhibited a markedly slower decline, with the most discriminatory time point occurring at 60 minutes (Figure 8.2.5a), where an AUC of 0.91 was achieved (Figure 8.2.5b). Breathing limonene in cirrhotic patients also correlated with the Model for End-Stage Liver Disease (MELD) score, fibrosis markers, and hypertension [49].

Although these findings are promising, several limitations remain. The influence of dietary intake, gut microbiota-derived VOCs, and comorbidities on limonene levels has not been fully disentangled. Additionally, the practical feasibility of administering exogenous VOCs for diagnostic purposes in routine clinical workflows warrants further evaluation. Nonetheless, these studies provide compelling proof of concept that VOC kinetics, particularly for compounds with well-characterized hepatic metabolism, may reflect pathophysiological changes in liver function and complement existing diagnostic strategies.

8.2.3 Skin Emissions

8.2.3.1 Biochemistry Considerations

Skin emissions provide a noninvasive and chemically rich interface between the body and the external environment, reflecting both local physiological activity and systemic metabolic states. As a sampling matrix, skin has attracted growing interest not only for diagnostic applications but also for its relevance in chemical ecology and behavioral signaling. The human skin volatilome includes over 600 VOCs across diverse chemical classes, including hydrocarbons, ketones, aldehydes, alcohols, esters, and organic acids [2]. However, the interpretation of skin VOC profiles requires careful consideration of both endogenous and exogenous sources, including pollutants, food, fragrances, and personal care products. To minimize confounding, environmental controls (e.g., ambient air) are often collected in parallel with skin samples [50].

The chemical complexity of skin VOCs arises from multiple converging sources. Eccrine glands, distributed across regions like the forehead, palms, and soles, produce mostly odorless sweat, rich in water, glycoproteins, and electrolytes. Sebaceous glands, abundant on the scalp, chest, and back, secrete lipids such

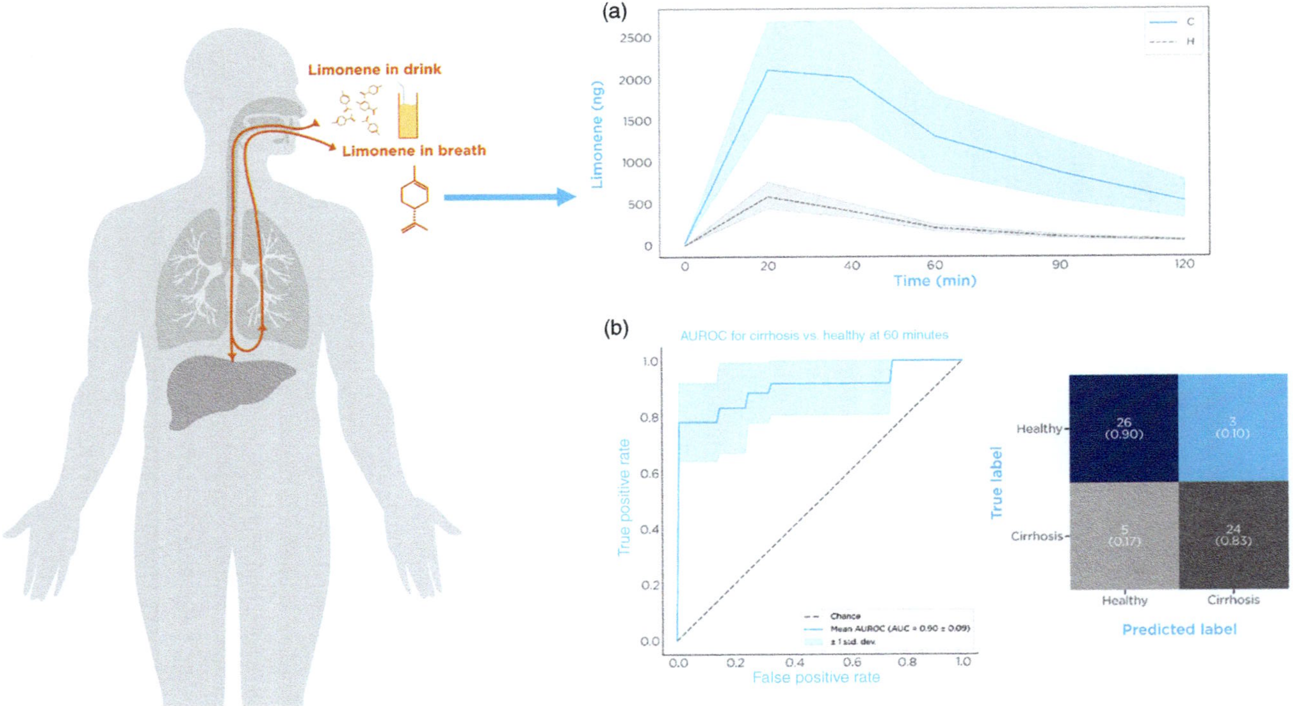

Figure 8.2.5 Pharmacokinetics of orally administered limonene in cirrhotic patients versus healthy controls. (a) Limonene levels peaked at 20 minutes in both groups but declined more slowly in the cirrhosis group. (b) ROC curve and confusion matrix at 60 minutes show strong discrimination (AUC = 0.91). *Source:* Ferrandino et al. [49] / MDPI / CC BY 4.0.

as long-chain fatty acids and cholesterol. Apocrine glands, located primarily in the axillary and genital regions, release proteins, lipids, and steroids in response to adrenergic stimulation and are the primary source of underarm odorants [51]. Resident microbial transform glandular secretions into smaller odorous compounds (e.g., short-chain acids, alcohols, and aldehydes). In addition to these local mechanisms, VOCs can also diffuse from systemic circulation to the skin surface, especially in metabolic or genetic disorders. External factors, such as UV exposure, may also induce oxidative degradation of sebum lipids, further altering the volatilome [50, 52].

Due to the heterogeneous distribution of glands and microbiota, VOC emissions vary across different body regions. For instance, the feet typically emit malodorous carboxylic acids, such as isovaleric acid, while hand emissions are characterized by aldehydes and ketones, including nonanal, decanal, and 6-methyl-5-hepten-2-one. In the axillary and periareolar regions, apocrine output, combined with microbial metabolism, yields a diverse array of alkanes and C_6–C_{11} fatty acids. These regions are frequently studied in the context of maternal–infant chemical communication, with VOCs such as 3-methyl-2-butenal, benzaldehyde, and octanal being consistently detected [53–57].

Because numerous intrinsic and extrinsic variables influence skin VOCs, rigorous standardization of sampling protocols is crucial. Pre-sampling guidelines often include avoiding spicy foods, fragrances, or alcohol-based hygiene products, and sometimes involve light exercise to stimulate secretion [51, 58]. Methodological variability can also bias findings, as highlighted by Gallagher et al., who compared SPME and solvent extraction for skin sampling. While using SPME and solvent extraction, 58 and 49 compounds were identified, respectively; only 15 compounds were common to both methods [51]. Dormont et al. compared four techniques (solvent extraction, HS-SPME, contact SPME, and DHS) and found that the methods captured the major known skin odorants, with aldehydes (mainly nonanal and decanal) predominating the volatilomic profiles. Remarkably, contact SPME required only 3 minutes to yield results comparable to those obtained from a 45-minute HS-SPME collection, demonstrating the feasibility of rapid sampling [59].

Finally, emerging evidence suggests that skin VOCs may encode information about transient physiological and psychological states. Martin et al. employed PDMS sampling and GC-MS to analyze forehead VOCs from 15 volunteers during resting and stress-inducing tasks. Benzoic acid, *n*-decanoic acid, xylene, and 3-carene were identified as differentiators between calm and stressed states, possibly reflecting shifts in microbial activity or oxidative metabolism [60].

Taken together, these findings point to the untapped potential of skin volatilomics. Nevertheless, realizing this potential will require not only technological refinement but also a deeper understanding of how internal metabolism, skin physiology, and environmental exposure intersect to shape the observed VOC profiles.

8.2.3.2 Diagnostic and Pathophysiological Applications

Although less established than breath analysis, skin volatilomics is gaining traction as a promising, noninvasive strategy for disease detection and physiological monitoring. The anatomical accessibility of the skin and the composite origin of its VOCs, from glandular secretions, microbial metabolism, systemic diffusion, and environmental deposition, make it a chemically rich but methodologically challenging biospecimen. Importantly, many disease-related processes, including inflammation, oxidative stress, and dysregulated lipid metabolism, leave distinct biochemical signatures that may be detectable at the skin surface.

8.2.3.2.1 Infectious Diseases

Infectious disease applications include a study by Vishinkin et al., who designed a polymeric pouch system to collect skin-emitted VOCs for tuberculosis (TB) diagnosis. Using GC-MS in cohorts from South Africa and India ($n = 636$), they identified four TB-associated VOC alterations in the South-African cohort (acetic acid, 2-ethyl-1-hexanol, hexyl butyrate, and toluene) and three in the Indian cohort (toluene, octanoic acid, and ethyl cyclopropane), with toluene being consistently elevated at both sites (Figure 8.2.6). Although toluene is typically associated with environmental pollution, an altered hepatic degradation pathway, possibly linked to *Mycobacterium tuberculosis* infection, may explain its elevated levels in TB patients [58]. These findings highlight the potential of skin VOCs in high-burden, low-resource settings for disease screening contexts.

Among recent advances in the use of skin VOCs for infectious disease diagnosis, Pretorius et al. explored cutaneous volatile and semi-volatile organic compounds as markers of malaria infection. The study employed wearable PDMS wristbands for noninvasive sampling of the epidermal chemical profile from malaria-positive and malaria-negative individuals in South Africa, followed by analysis with TD-GC×GC-TOFMS. Feature selection by machine learning identified a variety of compounds, including mid- to long-chain fatty acids, unsaturated alcohols, aldehydes, ketones, and alkylbenzenes, several of which have been previously reported as potential markers of *Plasmodium* infection. Notably, (*E*)-2-octen-1-ol was detected at high abundance in malaria-positive individuals, although it was also found in a few malaria-negative participants, potentially indicating latent infection. The use of wearable epidermal samplers presents a promising alternative to blood-based diagnostics, with the potential to detect both active and latent malaria cases, and is applicable in resource-limited settings. The study also highlights the challenges of biomarker specificity, as many volatiles are influenced by comorbidities and other infections, underlining the need for careful interpretation and larger validation cohorts [61].

Expanding the application of cutaneous volatilomics to viral infections, Tungkijanansin et al. applied GC×GC-MS with artificial modulation to profile sweat

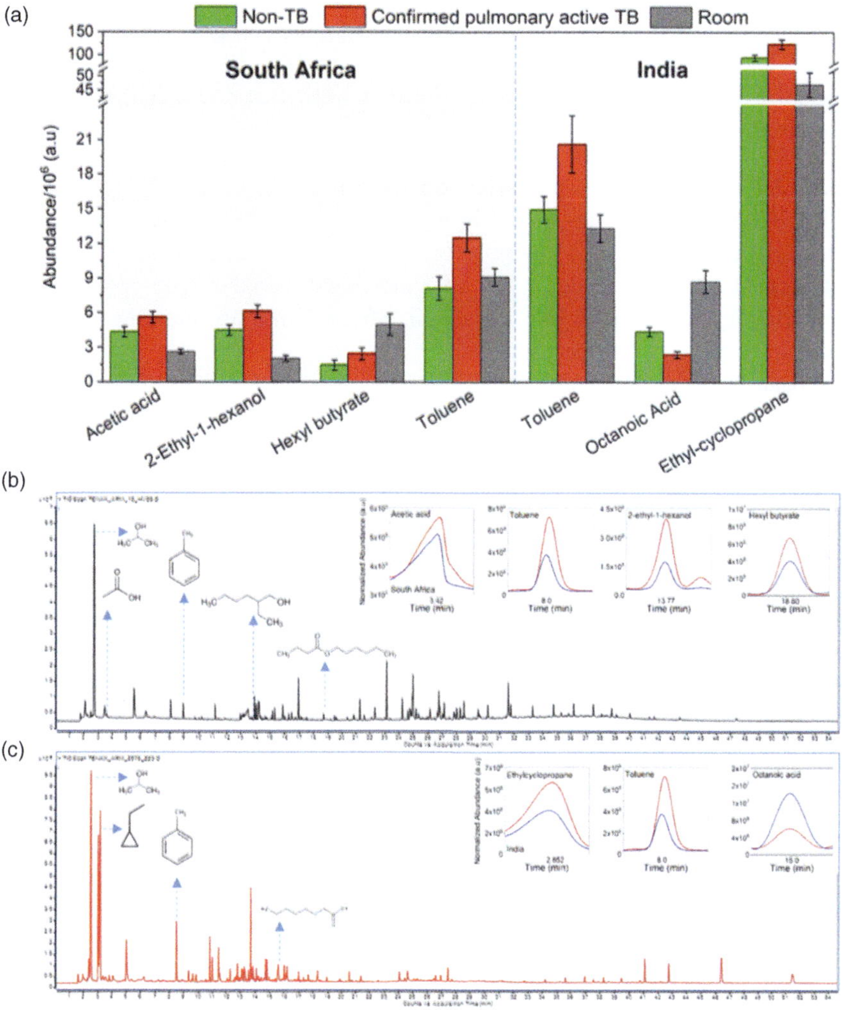

Figure 8.2.6 GC-MS analysis of VOCs in breath samples from confirmed pulmonary TB patients, non-TB individuals, and room air across clinical sites in South Africa and India. (a) Relative abundances of key discriminatory VOCs, including toluene, acetic acid, 2-ethyl-1-hexanol, ethyl-cyclopropane, hexyl butyrate, and octanoic acid. (b, c) Representative chromatograms from both cohorts highlight statistically significant differences. *Source:* Vishinkin et al. [58] / John Wiley & Sons / CC BY 4.0.

VOCs in individuals with and without COVID-19. Using axillary swab samples, the authors identified 14 discriminatory metabolites, including *p*-cymene, linalool, and 2,6,11-trimethyldodecane, capable of distinguishing COVID-19-positive and negative cases with classification accuracies, sensitivities, and specificities above 93% [62].

8.2.3.2.2 Skin VOCs and Host-vector Interactions

Skin VOCs play a fundamental role in mediating host-vector interactions, particularly in the context of mosquito behavior and disease transmission. Several studies have demonstrated that inter-individual variability in skin-emitted volatiles influences mosquito attraction, offering insights into the chemical ecology of vector-borne diseases.

Logan et al. explored differential human attractiveness to *Aedes aegypti* by linking behavioral responses and electrophysiological activity to skin-derived volatiles. Using gas chromatography coupled with electroantennographic (GC-EAG) recordings from the antennae of *Aedes aegypti*, the authors detected 33 physiologically active compounds, including aldehydes (e.g., octanal, nonanal, and decanal), ketones (e.g., 6-methyl-5-hepten-2-one, geranylacetone), and aromatic compounds (e.g., benzaldehyde and naphthalene). Interestingly, individuals less attractive to mosquitoes emitted higher levels of certain aldehydes and ketones. In behavioral experiments, five compounds (namely 6-methyl-5-hepten-2-one, octanal, nonanal, decanal, and geranylacetone) caused a significant reduction in the upwind flight of *Aedes aegypti* to human hands. Thus, the unattractiveness of individuals may result from a repellent or attractant "masking" mechanism [63].

Verhulst et al. connected this differential attractiveness to genetic determinants. In a study with 48 male volunteers, the authors investigated the relationship between human leukocyte antigen (HLA) genotypes, skin-emitted VOCs, and the differential attractiveness of humans to the malaria vector *Anopheles gambiae*. The study reported that individuals highly attractive to *Anopheles gambiae* emitted higher levels of lactic acid, tetradecanoic acid, and octanal, while those less attractive produced more limonene and 2-phenylethanol. Importantly, genotyping revealed a significant overrepresentation of the HLA-Cw*07 allele among highly attractive individuals, suggesting that host genetics may shape skin VOC composition through modulation of immune function or cutaneous microbiota [64]. These findings underscore the multifactorial nature of mosquito attraction, implicating a complex interplay between metabolic, microbial, and genetic determinants.

8.2.3.2.3 Oncological Diagnostics

Skin-emitted VOCs have garnered interest as noninvasive biomarkers for cancer detection, particularly due to the accessibility of the skin and its potential to reflect both local and systemic biochemical changes. Early investigations into cutaneous volatilomics in oncology have yielded promising, albeit preliminary, findings.

Abaffy et al. profiled the volatiles from 10 patients (5 melanomas and, as controls, 5 non-melanoma lesions, each paired with nearby uninvolved skin) using HS-SPME-GC-MS. Melanoma samples showed higher levels of the saturated fatty acids lauric, myristic, and palmitic, as well as 2-ethylhexyl-4-methoxycinnamate and 1-eicosene, whereas toluene was decreased. No VOC alterations were observed between non-melanoma lesions and their paired healthy skin, aligning with their benign histology. The elevated fatty acids in melanoma tissue are suggested to originate from enhanced *de novo* lipogenesis and oxidative stress, indicating potential for VOC profiling in melanoma detection [65].

Expanding this concept, Monedeiro et al. developed an HS-GC-MS method to profile sweat VOCs from cancer patients (n = 32) and HCs (n = 32). VOC profiles were analyzed via PCA, hierarchical clustering, and random forest classification. A panel of VOCs, including aldehydes, alcohols, and terpenoids, successfully differentiated between controls and cancer patients, with 2-ethyl-1-hexanol, hexanal, and octanal yielding 100% classification accuracy. Additional models discriminated between tumor types and stages, with n-aldehydes correlating with the staging of adenocarcinoma [66].

8.2.3.2.4 Wounds and Other Conditions

Skin VOC analysis is also being investigated for its utility in evaluating chronic wounds, where metabolic disruption and microbial colonization alter the local chemical environment. Thomas et al. sampled volatiles from arterial leg ulcers and adjacent healthy skin in five male patients using PDMS patches and analyzed them by GC-MS. PCA and PLS-DA revealed clear separation between the ulcer, wound margin, and control sites. Site-specific markers included several alcohols, hydrocarbons, and esters, indicating that local metabolic and microbial processes create a distinct volatilomic signature that could aid bedside evaluation of chronic wounds [67].

Beyond wound diagnostics, skin VOCs have also been linked to broader metabolic dysfunction. For instance, acetone emitted from the skin has been correlated with circulating β-hydroxybutyrate. This association indicates that skin VOCs may serve as noninvasive proxies for ketone metabolism, offering potential applications in the monitoring of glycemic control, ketosis, or metabolic disorders such as diabetes [50, 68].

8.2.4 Blood

8.2.4.1 Biochemistry Considerations

As the body's principal transport medium, blood serves as a systemic integrator of biochemical activity, carrying VOCs released from various organs and tissues. Unlike excretory fluids such as breath or urine, which reflect VOCs after substantial physicochemical filtering, blood offers a more direct and temporally immediate snapshot of endogenous metabolic processes. Its chemical composition captures compounds shortly after production, often before enzymatic degradation or exhalation. Despite these advantages, blood has been relatively neglected in volatilomic research, primarily due to its invasive sampling requirements, higher operational costs, and the need for trained personnel [1, 50].

Nonetheless, the diagnostic potential of blood-based VOC analysis is considerable. Circulating VOCs exhibit rapid turnover, from minutes to hours, via pulmonary exhalation, renal excretion, or metabolic conversion [50, 69, 70]. Hepatic cytochrome P450 enzymes oxidize many VOCs into reactive intermediates, which are then conjugated, often via the mercapturic acid pathway, and excreted [71]. For example, isoprene is metabolized by CYP2E1 into mono- and diepoxides, such as 3,4-epoxy-3-methyl-1-butene, which are subsequently converted by epoxide hydrolase into vicinal diols [72, 73]. Similarly, benzene is metabolized via CYP enzymes into benzene oxide, which is further processed into sulfate and glucuronide conjugates, yielding phenol, catechol, and hydroquinone [74]. These pathways illustrate how blood VOCs reflect both the origin and metabolic fate of volatile species.

Despite its physiological centrality, blood remains a relatively undercharacterized matrix in volatilomics. According to Drabińska et al., only 379 VOCs have been identified in blood, representing less than 15% of the known human volatilome. The most frequently reported classes are halogenated compounds (19%), often originating from environmental exposures such as solvents and anesthetics, followed by aliphatic hydrocarbons (15%), nitrogen-containing compounds (13%), ketones (11%), alcohols (8%), and ethers (8%). Remarkably, only about 6% of known human VOCs are shared between blood and breath, likely due to volatility-dependent partitioning, enzymatic transformation during pulmonary passage, or selective excretion mechanisms [2, 50].

A more targeted profiling by Mochalski et al. employed HS-SPME-GC-MS to analyze the blood of healthy volunteers. Among 90 detected VOCs, 62 were positively identified and quantified. Hydrocarbons (19) and ketones (9) constituted the majority, with additional contributions from heterocycles (7), sulfur compounds (6), aromatics (7), terpenes (7), and esters (3). Only two aldehydes, propanal and 2-propenal, were detected, possibly due to their high reactivity and short half-life in circulation. A subset of 10 compounds, including acetone, isoprene, dimethyl sulfide, and limonene, was consistently present across all samples, suggesting a core set of background VOCs in the blood of healthy individuals. Another six compounds appeared in over 80% of samples, reinforcing the existence of a reproducible basal volatilomic fingerprint [75].

Altogether, these findings underscore both the diagnostic promise and current knowledge gaps in blood volatilomics. Greater methodological standardization, mechanistic insight into VOC metabolism, and large-scale population studies are essential to elevate blood VOC profiling from an experimental tool to a clinically applicable modality, a topic explored in the following section.

8.2.4.2 Diagnostic and Pathophysiological Applications

8.2.4.2.1 Established Applications of Blood VOC Analysis

While blood is a cornerstone of modern diagnostics for nonvolatile biomarkers, the clinical use of VOCs in blood remains limited to a few well-established applications. Among the most routine is blood alcohol testing, typically performed via HS-GC. This approach minimizes matrix effects by avoiding the direct injection of whole blood, enabling the accurate quantification of low-boiling-point alcohols. Standard clinical panels include methanol, ethanol, isopropanol, 1-propanol, acetaldehyde, and acetone, while forensic investigations may also target diethyl ether, butane, ethyl acetate, hexane, toluene, xylene, and halogenated hydrocarbons, especially in postmortem toxicology where exogenous exposure is suspected [2, 76, 77].

Another routinely analyzed blood-borne VOC is ammonia, a key marker of nitrogen metabolism. Elevated blood ammonia levels are indicative of impaired hepatic detoxification and are clinically associated with hyperammonemia and hepatic encephalopathy. Quantification may be achieved either through direct measurement of gaseous ammonia released from plasma or by quantifying plasma ammonium ions using enzymatic or spectrometric methods [78–81]. These measurements, though technically distinct from broader volatilomic profiling, demonstrate that volatile species in blood can provide actionable clinical information under appropriate analytical frameworks.

8.2.4.2.2 Oncological Diagnostics

Blood VOC profiling has increasingly drawn attention as a potential tool for oncological diagnostics particularly due to its capacity to capture systemic metabolic alterations associated with tumorigenesis. In one study, Wang et al. used HS-SPME-GC-MS to analyze serum samples from 16 colorectal cancer (CRC) patients and 20 HCs. A total of 341 VOC features were detected and filtered using PCA and PLS-DA. Compared to controls, CRC samples exhibited decreased levels of phenyl methylcarbamate, ethylhexanol, and 6-*tert*-butyl-2,2,9,9-tetramethyl-3,5-decadien-7-yne, alongside increased levels of 1,1,4,4-tetramethyl-2,5-dimethylene-cyclohexane. The PLS-DA model demonstrated good discriminatory performance, supporting the feasibility of using blood VOCs for cancer detection [82].

Pushing the boundaries of analytical sensitivity, Einoch-Amor et al. demonstrated that even a single circulating tumor cell can alter the VOC composition of blood. Using an artificially intelligent nanoarray with cross-reactive chemiresistive sensors, the authors analyzed the headspace of blood, plasma, and culture media spiked with either 1 or 100,000 A549 lung cancer cells. The system achieved >90% accuracy, >85% sensitivity, and >95% specificity in distinguishing spiked samples from controls. GC-MS analysis identified 51 discriminative VOCs, with 2-ethyl-1-hexanol and 2-methyl-2-propanol among the most prominent. Notably, some VOCs varied in opposite directions between matrices, indicating complex interactions between tumor cells and hematopoietic elements [83].

These findings highlight the sensitivity of blood VOCs as metabolic surrogates for oncological processes, even at low tumor burdens. However, they also underscore the need for deeper mechanistic insight and methodological standardization before such approaches can be translated into robust clinical tools.

8.2.4.2.3 Pathogen Detection and Infectious Disease Diagnostics

Beyond oncology, blood VOC profiling has shown growing potential in the realm of infectious disease diagnostics, particularly for detecting bloodstream infections (bacteremia). Upon inoculation with common pathogens such as *Escherichia coli*, *Staphylococcus aureus*, *Neisseria meningitidis*, or *Pseudomonas aeruginosa*, blood produces distinct volatilomic signatures, driven by microbial metabolism and host-pathogen interactions [84–86].

For instance, Rees et al. analyzed VOCs emitted by *Klebsiella pneumoniae* cultured in human blood using HS-SPME-GC×GC-TOFMS. Representative chromatograms are displayed in Figure 8.2.7. Subtraction of sterile blood chromatograms from *K. pneumoniae* culture chromatograms (Figure 8.2.7c) resulted in peaks that were either uniquely present in cultures or more abundant in cultures relative to blood. Thirty-three compounds were significantly elevated in infected samples, including 22 VOCs absent from sterile blood. Nine of these had not been previously reported for *K. pneumoniae*, and four were novel across all bacterial species. Discriminant metabolites included 2,3-butanedione, 3-hydroxy-2-butanone, methyl nitrite, and pyrrole. Random forest analysis and PCA identified a 27-compound signature capable of distinguishing infected from control samples [70].

These findings demonstrate that blood VOC analysis can capture pathogen-specific metabolic fingerprints, offering a promising foundation for rapid, culture-independent diagnostics. However, translating such findings into clinical practice will require addressing challenges related to reproducibility, inter-individual variability, and matrix complexity.

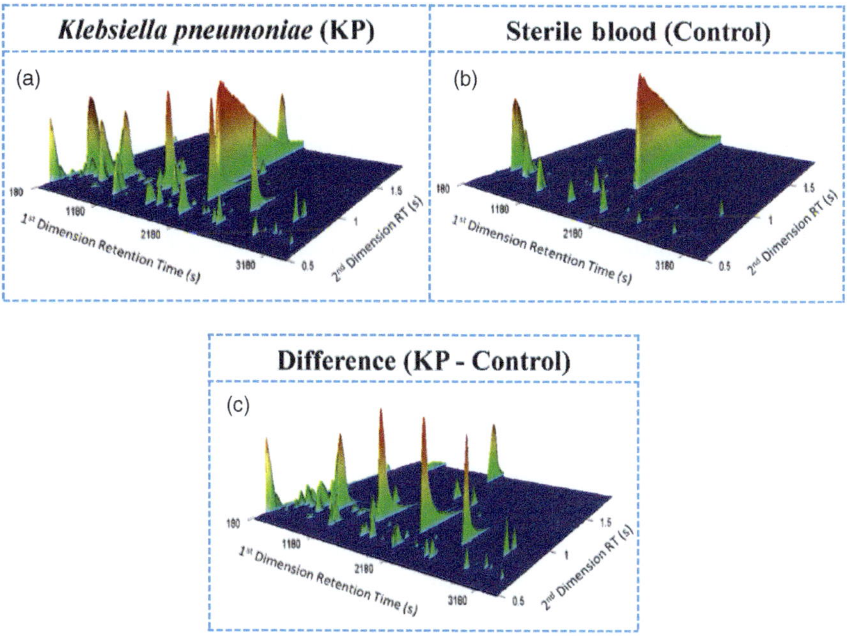

Figure 8.2.7 Representative GC×GC-TOFMS chromatograms of headspace volatiles sampled from *K. pneumoniae* cultures and sterile blood (control). Culture volatiles are presented in (a), while the control is shown in (b). Subtraction of sterile blood chromatograms from *K. pneumoniae* culture chromatograms results in (c). *Source:* Rees et al. [70] / with permission of IOP Publishing.

8.2.5 Urine

8.2.5.1 Biochemistry Considerations

Urine is a metabolite-rich excretory fluid that serves as a highly informative matrix for VOC analysis. Its noninvasive collection, high chemical stability, and ability to concentrate soluble volatiles and their conjugates make it particularly well-suited for diagnostic and exposomic applications. Unlike breath, which reflects transient fluctuations, or blood, which requires invasive sampling, urine integrates metabolic activity over longer timeframes, providing a broader snapshot of physiological and pathological states.

Urinary volatilomics has been explored using GC-MS, typically following pre-concentration steps such as SPME, purge-and-trap (P&T), or TD. The urinary VOC profile encompasses a broad array of chemical classes, including acids, alcohols, ketones, aldehydes, amines, heterocycles (nitrogen- and oxygen-containing), sulfur compounds, and hydrocarbons. Among the most consistently detected compounds are acetone, 2-butanone, 2-pentanone, 2-heptanone, 3-hexanone, 4-heptanone, 2,5-dimethylfuran, 2-ethyl-5-methylfuran, and toluene [50, 87–92].

The biochemical origins of urinary VOCs are equally diverse. For example, 4-heptanone has been linked to the oxidative metabolism of the plasticizer 2-ethylhexanoic acid, while acetone can result from the nonenzymatic decarboxylation of acetoacetate, a ketone body elevated in diabetic states [88]. These examples highlight the dual endogenous and exogenous nature of the urinary volatilome, reinforcing its diagnostic potential not only for monitoring metabolic disorders but also for evaluating environmental and toxicological exposures.

8.2.5.2 Diagnostic and Pathophysiological Applications

8.2.5.2.1 Oncological Diagnostics

Urinary VOC analysis has shown considerable promise as a noninvasive diagnostic tool for various malignancies, offering a complementary approach to tissue-based or blood-based methods. Cancer-related metabolic shifts, including oxidative stress, lipid peroxidation, and altered microbiome activity, contribute to distinctive volatilomic signatures in urine.

Prostate and colorectal cancers are among the most thoroughly investigated. In prostate cancer, Lima et al. used SPME-GC-MS to reveal significantly lower levels of hexanal, 4-methylhexan-3-one, and methylglyoxal in prostate cancer patients, alongside elevated levels of phenylpropionaldehyde, 2,5-dimethylbenzaldehyde, ethylbenzene, and methylbenzoate [93]. For CRC, Mozdiak et al. employed FAIMS and GC-IMS in a cohort of 163 subjects to discriminate CRC from adenomas and other gastrointestinal disorders using machine learning models [94]. Similarly, Arasaradnam et al. analyzed urine from 83 CRC patients and 50 controls, achieving 88% sensitivity but lower specificity (60%) via FAIMS-based VOC profiling [95].

Breast cancer has been explored in both human and animal studies. In a mouse model, HS-SPME-GC-QTOFMS distinguished between localized and metastatic breast tumors using VOC panels that included (*Z*)-5-octen-1-ol and 3,3,6-trimethylbicyclo[3.1.0]hexan-2-one [96]. Human urine samples analyzed by SPME-GC-MS showed increased levels of 4-carene, 3-heptanone, 1,2,4-trimethylbenzene, methoxythiophene, and phenol in cancer patients, while dimethyl disulfide was reduced [97]. Another investigation identified 14 VOCs associated with invasive ductal carcinoma, of which 5 (2-ethyl-1-hexanol, isolongifolenone, furan, dodecanoic acid, and 2-methoxyphenol) were validated in an external cohort [98].

Other cancers have also been studied. A SPME-GC-MS approach in head and neck cancer identified 28 altered VOCs, including 2,6-dimethyl-7-octen-2-ol, 1-butanol, *p*-xylene, and 4-methyl-2-heptanone, demonstrating high diagnostic performance [99]. For non-Hodgkin lymphoma (NHL), Mesquita et al. applied HS-SPME-GC-MS to analyze urine from patients before and during chemotherapy. Twenty-eight metabolites, including 1-octanol, 1-hydroxyanthraquinone, and 7,9-di-*tert*-butyl-1-oxaspiro-4,5-deca-6,9-diene-2,8-dione, discriminated cases from controls. PLS-DA models separated groups and identified 43 key molecular features. Longitudinal analysis revealed two metabolites (7-exo-ethenyl-bicyclo[4.2.0]oct-1-ene and 1-dodecyn-4-ol) whose levels changed throughout treatment, with the latter showing higher abundance in patients who exhibited clinical improvement [100].

In bladder cancer, Pinto et al. reported a 17-VOC signature characterized by increased levels of certain alkanes and aromatics, as well as decreased levels of aldehydes and monoterpenes [101]. Urinary biomarkers have also been explored for renal cell carcinoma, using SPME-GC-MS, with 2-oxopropanal and 2,5,8-trimethyl-1,2,3,4-tetrahydronaphthalene-1-ol identified as potential markers [102].

Lastly, environmental exposures may modulate urinary VOC profiles. A study comparing children and adults exposed to secondhand smoke reported significantly higher urinary levels of 17 VOCs in children, along with increased biomarkers of oxidative stress such as 8-hydroxy-2′-deoxyguanosine and malonaldehyde [103]. This underscores the sensitivity of the urinary volatilome to both endogenous disease processes and external toxicological exposures.

8.2.5.2.2 Nephrotic Diseases

Urinary VOC profiling has emerged as a promising tool for investigating kidney-related diseases, particularly those involving glomerular damage. The kidney's central role in filtration and excretion makes urine an ideal matrix for detecting metabolic changes and volatile signatures associated with nephrotic syndromes.

Wang et al. investigated urinary VOCs as potential noninvasive biomarkers for idiopathic membranous nephropathy (iMN). Using HS-SPME-GC-MS, they analyzed urine samples from 63 biopsy-confirmed iMN patients and 15 HCs. Six VOCs (carbamic acid monoammonium salt, 2-pentanone, 2,4-dimethyl-pentanal, hydrogen azide, thiourea, and 4-heptanone) were significantly elevated in iMN patients. These compounds are associated with oxidative stress, lipid peroxidation, and β-oxidation, suggesting underlying mitochondrial dysfunction [104].

Ligor et al. employed an HS-SPME-GC×GC-TOFMS approach to study glomerular diseases in 27 patients and 20 controls. From a total of 282 detected VOCs, 13 showed significant group differences, including elevated levels of cyclohexanone, methyl hexadecanoate, 9-hexadecen-1-ol, 6,10-dimethyl-5,9-undecadien-2-one, and 2-pentanone, which are likely linked to oxidative stress and altered fatty acid metabolism [105].

Liu et al. focused on minimal change nephrotic syndrome (MCNS), analyzing urine samples from 38 patients and 15 HCs by HS-SPME-GC-MS. Six VOCs were significantly altered in MCNS patients: *trans*-2,2-dimethyl-4-decene, pyrrole, carbamic acid monoammonium salt, 1-butyne-3,3-dimethyl, diisopropylamine, and 4-heptanone. Notably, levels of 4-heptanone and pyrrole were reduced in MCNS patients, suggesting impaired renal excretion. The altered profile reflects changes in ketone body metabolism, nitrogenous compound handling, and olefin degradation, consistent with tubular and glomerular dysfunction [106].

These studies demonstrate that urinary VOCs can capture subtle metabolic alterations associated with glomerular diseases. While the findings are preliminary and require replication in larger, standardized cohorts, they highlight the potential of volatilomics as a noninvasive diagnostic and monitoring tool in nephrology.

8.2.6 Stool

8.2.6.1 Biochemistry Considerations

Fecal VOC analysis offers a chemically rich and biologically informative window into host-microbiota interactions, dietary composition, and gut health. Stool is a complex biological matrix composed of microbial cells, shed intestinal epithelium, and undigested dietary components, making it especially suitable for biomarker discovery in gastrointestinal and systemic disorders [50].

Fecal VOCs originate from at least four sources: diet, gut microbiota, host metabolism, and environmental contaminants. For instance, limonene, commonly found in citrus peel, can also be synthesized by *Bacillus pallidus*, blurring distinctions between exogenous and endogenous sources [107]. After most dietary nutrients are absorbed in the small intestine, unabsorbed components are fermented by colonic bacteria into short- and medium-chain fatty acids. These fatty acids can be absorbed, excreted unchanged, converted into alcohols, or esterified. Similarly, the fermentation of branched-chain amino acids produces branched fatty acids that may be esterified or excreted. The gut microbiome is critical in generating and metabolizing VOCs. For example, particular bacterial species (*Clostridium* spp., *Fusobacterium varium*, *Klebsiella pneumoniae*, and *Enterobacter aerogenes*) convert tyrosine into phenol, while benzaldehyde is produced from phenylalanine via phenylpyruvate, a reaction mediated by aromatic amino acid aminotransferase and transaminases in Firmicutes and *Lactobacilli* [107–110]. The oral microbiome may further influence the gut, as Olsen and Yamazaki demonstrated that chronic periodontitis involving *Porphyromonas gingivalis* induces dysbiosis that disrupts gut microbial balance [111].

Fatty acids such as acetic, propanoic, and butanoic acids are central metabolites with pyruvate as a common precursor. Over 50 alcohols have been reported, including ethanol, likely arising from bacterial fermentation. Esters, often the most abundant VOC class, are biosynthesized via esterification of microbial fatty acids [2, 112]. Aromatic compounds such as benzenoids, furans, and nitrogen-containing heterocycles reflect amino acid catabolism. While alkylfurans are typically associated with fungal metabolism, they may also derive from lipid oxidation [2, 10].

Ketones and aldehydes are also prominent in feces. Acetone and 2-butanone, commonly detected, arise from fatty acid and carbohydrate metabolism. Methyl ketones such as 2,3-butanedione are produced by bacteria and fungi and can influence microbial dynamics. A complete homologous series of aldehydes, ranging from acetaldehyde to octadecanal, has been identified in feces. Although some aldehydes might originate from dietary sources, they are unlikely to survive digestion unchanged [2, 113].

Hydrocarbons such as methane and isoprene are also present. Methane may form via microbial reduction of carbon dioxide, conversion of acetic acid, or oxidation of unsaturated lipids [113–115]. Ethers like 2-ethoxyethanol and 1,3-dimethoxybenzene are often detected in fecal HS, but they usually originate from exogenous sources [2, 116].

Some nitrogenous VOCs (e.g., methylpyrazine, pyridine, and pyrrole) are found in coffee but may degrade in gastric acid, suggesting that fecal detection often reflects microbial synthesis. Ammonia primarily results from microbial activity.

A diverse range of sulfur compounds has been reported in feces, including methanethiol and dimethyl sulfide; methanethiol is partly produced from methionine by *Clostridia* and may also arise via the methylation of hydrogen sulfide by mucosal thiol S-methyltransferase. Dimethyldisulfide and dimethyltrisulfide are also frequently detected [2].

8.2.6.2 Diagnostic and Pathophysiological Applications

8.2.6.2.1 Colorectal Cancer (CRC)

Volatilomic profiling of stool has shown promising diagnostic value for colorectal cancer (CRC), leveraging noninvasive access to metabolically active gut environments. Through predictive modeling, GC-MS investigations of fecal HS have successfully distinguished CRC and advanced adenomas from HCs. For example, Bond et al. identified propan-2-ol as a promising single biomarker (83% sensitivity, 71% specificity), while a two-VOC model combining 3-methylbutanoic acid with propan-2-ol improved diagnostic performance to 88% sensitivity and 85% specificity, highlighting the value of multivariate VOC signatures for noninvasive CRC screening [117].

8.2.6.2.2 Inflammatory Bowel Disease (IBD) and Pouchitis

Fecal volatilomics has also advanced the understanding of inflammatory bowel disease (IBD), including ulcerative colitis (UC), Crohn's disease (CD), and pouchitis. The metabolic alterations observed in IBD are largely driven by inflammation-induced shifts in microbial composition and function, which directly influence VOC profiles [118, 119].

In a longitudinal GC-MS study, Walton et al. analyzed fecal VOCs from patients with active and inactive CD and UC, as well as healthy controls. They targeted 13 compounds previously associated with gut inflammation. At baseline, eight VOCs, particularly short-chain fatty acid esters and 1-propanol, were elevated in IBD patients. After clinical treatment, CD patients exhibited a more pronounced decline in VOC levels compared to UC patients, supporting the potential of volatilomics for monitoring therapeutic response [120].

Expanding on these findings, Ahmed et al. employed HS-SPME-GC-MS and multivariate modeling in a large cohort comprising 117 CD, 100 UC, and 109 control subjects. They found elevated levels of heptanal, 1-octen-3-ol, and 6-methyl-2-heptanone in active CD, with reduced levels of methanethiol, 3-methylphenol, and several esterified fatty acids. Notably, the VOC profile successfully distinguished active from inactive CD, small-bowel from colonic CD, and CD from UC, demonstrating the potential of fecal VOCs for subtyping within IBD [118].

Bosch et al. evaluated the diagnostic performance of fecal VOC analysis in 280 adults with IBD, including CD and UC, and 227 healthy controls using GC-IMS. VOC profiles discriminated IBD patients from controls with high accuracy (AUCs ~0.95–0.96), both during active disease and remission. However, VOC signatures poorly differentiated CD from UC (AUC = 0.55) and showed limited value in distinguishing active from latent disease states [121].

In a study by Hassani et al., the diagnostic potential of both urinary and fecal VOCs was investigated in treatment-naive pediatric IBD. The study enrolled 10 newly diagnosed IBD patients (5 CD, 5 UC) and 10 age- and sex-matched symptomatic controls from an endoscopy-confirmed cohort. Samples were analyzed using GC-IMS, with machine learning models trained to discriminate groups based on VOC signatures. Fecal VOCs achieved an AUC of 0.73, while urinary VOCs yielded a comparable AUC of 0.78, suggesting both matrices hold similar diagnostic accuracy [122].

De Preter et al. reported that patients with IBD and pouchitis showed significant reductions in several medium-chain fatty acids, specifically butanoic, pentanoic, hexanoic, heptanoic, octanoic, and nonanoic acids, while acetic and propanoic acids remained stable. Additionally, microbial fermentation byproducts such as *p*-cresol and 3-methylindole were diminished in active CD, possibly reflecting a dysbiotic shift [118, 120].

Across studies, consistent VOC trends have emerged: increased phenol and propan-1-ol in CD, decreased butanoic acid in UC, and lower medium-chain fatty acids overall. However, IBD monitoring remains challenging due to intraindividual variability and overlapping metabolic signatures.

8.2.6.2.3 Irritable Bowel Syndrome (IBS)

Volatilomic profiling of irritable bowel syndrome (IBS) remains relatively underexplored, particularly when compared to inflammatory bowel conditions. Nevertheless, emerging evidence suggests that VOCs may hold diagnostic potential, especially for distinguishing IBS subtypes. Ahmed et al. found that esters from short-chain fatty acids and cyclohexanecarboxylic acid were elevated in diarrhea-predominant irritable bowel syndrome (IBS-D), distinguishing IBS-D from controls with 82% sensitivity and 78% specificity. Esters and specific organic acids were more abundant in IBS than in IBD, suggesting enhanced fermentation pathways in the former (Figure 8.2.8) [123].

8.2.7 *In vitro* and *ex vivo* Models

In vitro and *ex vivo* models offer a robust platform for examining cellular metabolism under controlled conditions, enabling the isolation of biochemical pathways and cellular responses that might otherwise be obscured in biofluid analyses. These systems are particularly valuable for elucidating disease mechanisms, evaluating drug interactions, and assessing the impact of environmental factors on cellular processes. In this context, the application of GC-MS to analyze VOCs released by cells as metabolic byproducts represents an important advancement in this field, as these volatile signatures serve as dynamic indicators of cellular states, providing real-time insights into metabolic shifts, oxidative stress, and toxicological effects. Advanced techniques, such as GC×GC-TOFMS, further enhance the detection and quantification of low-abundance VOCs through increased resolution and sensitivity [124–127].

The enhanced analytical capabilities of these techniques have enabled the identification of VOCs and the tracing of their origins, metabolic pathways, and biological significance, as exemplified by the findings of Acevedo et al. The authors developed a SPME-GC-MS approach to profile VOCs released by cultured human skin fibroblasts at varying stages of senescence. Using a PDMS-DVB fiber, the authors identified 12 VOCs in the culture medium, 9 of which overlapped with compounds previously reported in human skin emissions. They demonstrated that senescence alters the VOC profile, with changes in compounds such as styrene, benzaldehyde, β-pinene, benzyl alcohol, and acetophenone. Notably, β-pinene was suggested to have endogenous origins

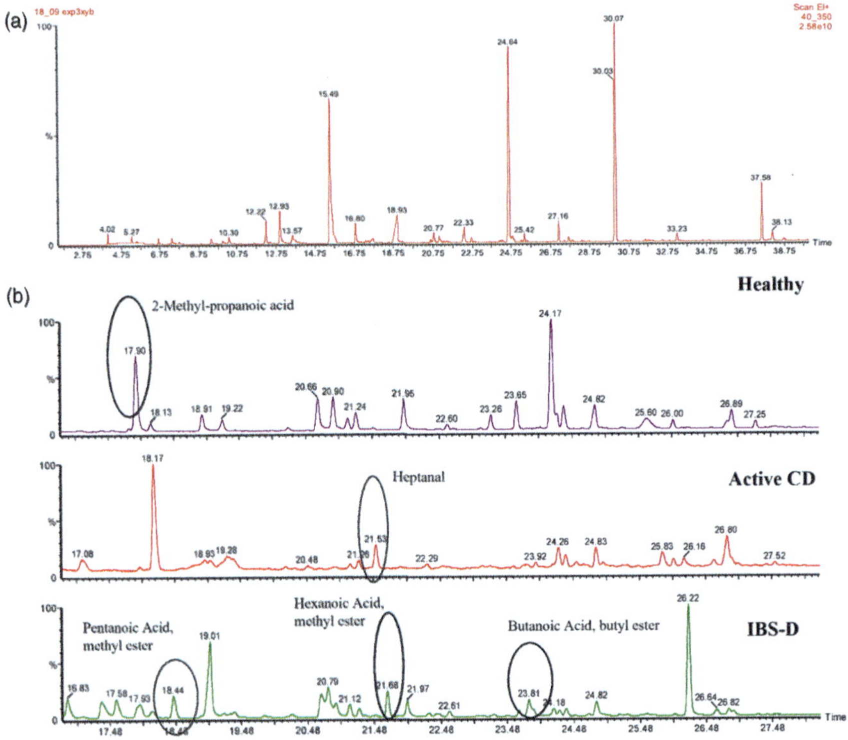

Figure 8.2.8 (a) Representative chromatogram of fecal headspace gas from a healthy volunteer. (b) Comparative chromatograms from Crohn's disease (CD), diarrhea-predominant irritable bowel syndrome (IBS-D), and healthy controls, highlighting differences in VOC peaks. *Source:* Ahmed et al. [123] / PLOS / CC BY 4.0.

and possible receptor-mediated bioactivity. Multivariate chemometric analyses (PCA and PLS) revealed that VOC patterns were distinct between pre-senescent and senescent cell cultures. Bioinformatic mapping associated these volatiles with metabolic pathways linked to glycolysis, which was found to increase with cellular aging. Correlations between VOC fluxes, lactate production, and established senescence markers (SA-β-Gal) further supported the connection between metabolic reprogramming and volatile output [128].

8.2.7.1 Diagnostic and Pathophysiological Applications

Beyond metabolic characterization, VOC profiling has emerged as a valuable approach for investigating disease states and pathogenic processes. Cellular responses to inflammation, oxidative stress, and infection lead to distinctive volatilomic signatures that can serve as biomarkers for disease detection and progression. While bacteria and fungi produce unique volatile fingerprints originating directly from their metabolic pathways, viruses alter host-cell metabolism, indirectly shaping the VOC profile. Similarly, tumor cells exhibit specific volatile signatures reflecting metabolic reprogramming and adaptation to the tumor microenvironment. The analysis of VOCs generated in response to inflammatory stimuli, oxidative stress, pathogenic infections, and oncogenic transformation holds considerable potential for both noninvasive diagnostic applications and elucidating disease-specific biochemical mechanisms.

8.2.7.1.1 Cellular Responses to Inflammation and Oxidative Stress

Inflammation and oxidative stress are markers of a wide range of chronic diseases. *In vitro* cellular models provide a controlled environment to elucidate the metabolic consequences of these stressors and the biochemical pathways underlying disease-associated metabolic shifts.

Zanella et al. applied this approach by investigating the volatilomic responses of lung epithelial (A549) cells subjected to oxidative stress (induced by H_2O_2) and inflammation (simulated by co-culture with sputum from asthmatic patients). Oxidative stress significantly elevated VOCs associated with lipid peroxidation and fatty acid metabolism, including 2-ethyl-1-hexanol and cyclohexanol. Inflammatory co-cultures yielded dose-dependent cell proliferation and unique VOC profiles, characterized by elevated aldehydes (hexanal, nonanal) and hydrocarbons (3-ethylhexane, dodecane), linking these metabolites to immune activation and fatty acid degradation pathways. Importantly, the VOC patterns differed significantly between oxidative stress and inflammatory models, with the inflammatory model more closely mimicking the metabolic signatures observed in asthma patients *in vivo* [127].

Extending this concept to viral infections, Purcaro et al. evaluated whether volatile metabolites could discriminate between virally infected and uninfected respiratory cell cultures. Using SPME-GC×GC-TOFMS combined with machine learning algorithms (random forest, linear SVM, PLS-DA), the authors analyzed headspace VOCs from human cells (HEp-2) infected with respiratory syncytial virus (RSV) and murine (MLE-Kd) cells infected with influenza A virus (IAV). The approach differentiated infected from uninfected cells (AUROC: 0.78–0.84), highlighting hydrocarbons, aldehydes, aromatics, alcohols, and ketones as discriminatory metabolites. Notably, an elevated abundance of hydrocarbons was detected consistently across RSV- and IAV-infected cultures [129].

Both studies demonstrate that diverse cellular insults trigger characteristic VOC signatures, especially those associated with lipid peroxidation and oxidative stress. This convergence highlights the potential of VOC analysis as a noninvasive tool for disease detection and mechanistic exploration.

8.2.7.1.2 Bacterial Volatilomics

The characterization of VOCs produced by pathogenic bacteria has emerged as a promising strategy for the development of noninvasive diagnostic methods in infectious diseases. Experimental models, particularly those employing *in vitro* bacterial cultures and *in vivo* animal infections, are important for elucidating the biochemical origins, diversity, and diagnostic relevance of microbial VOCs.

An example of this approach is provided by Mellors et al., who investigated volatile biomarkers of *Mycobacterium tuberculosis* using both *in vitro* culture and an *in vivo* nonhuman primate model. In their study, headspace volatiles from *M. tuberculosis* cultures were characterized using GC×GC-TOFMS and compared to the volatile profiles found in the exhaled breath of cynomolgus macaques experimentally infected with the same strain of *M. tuberculosis*. Of the 76 discriminatory features identified in culture, 37 were also detectable in the breath of macaques, enabling discrimination between infected and uninfected animals with an AUC of 87%. Conversely, a set of 38 volatile molecules selected directly from breath samples yielded even higher diagnostic accuracy (AUC 98%). Notably, two compounds, 4-(1,1-dimethylpropyl)phenol and 4-ethyl-2,2,6,6-tetramethylheptane, were consistently detected both *in vitro* and *in vivo*, underscoring their potential as translational biomarkers [130].

Another translational approach was employed by Zhu et al., who investigated whether VOC fingerprints established from *in vitro* bacterial cultures could be used to diagnose acute lung infections in mice. Using (SESI)MS, the authors compared the volatile profiles of *Pseudomonas aeruginosa* (two strains) and *Staphylococcus aureus* in culture with the breathprints of mice experimentally infected with the same pathogens. The breath analysis enabled clear discrimination between infected and uninfected animals, as well as between infections caused by different species and even by different *P. aeruginosa* strains. However, direct comparison revealed that only 25–34% of the VOC peaks detected *in vitro* were present in the *in vivo* breathprints, underscoring a substantial divergence between bacterial volatiles produced in culture and those detectable during active infection in the host [126].

Bean et al. contributed to the expansion of bacterial volatilomics by applying HS-SPME-GC×GC-TOFMS to analyze volatile metabolites produced by *Pseudomonas aeruginosa* PA14 *in vitro*. The authors identified 56 compounds present at least twofold higher than in medium blanks. The detected metabolites encompassed alcohols, ketones, aldehydes, acids, esters, aromatic, and heteroaromatic compounds. Notably, 28 of these volatiles had not been previously reported for *P. aeruginosa*, nearly doubling the known repertoire for this species, which highlights the resolving power of GC×GC for detecting low-abundance volatiles and underscores the importance of comprehensive volatilome characterization in developing robust diagnostic fingerprints for clinically relevant pathogens [125].

However, despite significant advances in identifying VOC biomarkers using experimental models, translating these findings to *in vivo* and clinical contexts remains challenging due to the complex interplay of host and environmental factors. In this regard, high-resolution analytical techniques such as GC×GC-MS are crucial, as they enable the comprehensive detection and characterization of low-abundance and previously unrecognized VOCs, thus supporting the discovery of more specific and robust diagnostic markers.

8.2.7.1.3 Tumor Metabolism

Metabolic reprogramming is an important feature of cancer cells, enabling them to sustain rapid proliferation, evade apoptosis, and adapt to the changing tumor microenvironment. These alterations extend beyond intracellular metabolites and are reflected in the VOCs released by tumor cells. *In vitro* models of cancer metabolism provide an opportunity to understand these biochemical pathways, allowing for the identification of specific VOC signatures associated with oncogenic processes.

Recent studies have demonstrated that breast cancer cells display distinct volatilomic signatures under various microenvironmental conditions and across different molecular subtypes. For instance, Taware et al. investigated how hypoxia, a mark of the tumor microenvironment, affects the extracellular volatilome of breast cancer cell lines with differing metastatic potential. Using HS-SPME-GC-MS and multivariate statistics, the authors identified 20 VOCs significantly altered by hypoxic exposure. Notably, aggressive (MDA-MB-231) cells exhibited a unique volatilomic profile driven by hypoxia, distinct from that of nonaggressive (MCF-7) cells, characterized by higher levels of formic acid and a reprogramming of metabolic pathways related to nicotinate and nicotinamide metabolism under hypoxic conditions [131].

Similarly, Silva et al. performed a comprehensive analysis of VOCs released by several human breast cancer cell lines and compared them to those released by normal mammary epithelial cells. Using HS-SPME-GC-MS, they established a volatile metabolomic signature that distinguishes between cancerous and normal cells, as well as between different types of cancer. Their results revealed six VOCs (2-pentanone, 2-heptanone, 3-methyl-3-buten-1-ol, ethyl acetate, ethyl propanoate, and 2-methylbutanoate) detected exclusively in cancer cell lines. In contrast, higher levels of 2-ethyl-1-hexanol and cyclohexanol were observed across all breast cancer cells compared to normal cells. Multivariate statistical analyses (PCA, PLS-DA, LDA) further demonstrated that VOC profiles could classify and discriminate between normal and cancerous cell lines according to molecular subtype [132].

Gastric cancer remains one of the most lethal malignancies globally, often diagnosed at an advanced stage due to the lack of reliable, noninvasive early detection tools. In this context, Mochalski et al. performed an *ex vivo* analysis of gastric tissues from 94 gastric cancer patients across 2 independent cohorts (Latvia and Ukraine) using HS-SPME-GC-MS. A total of 19 VOCs were found to differ significantly between cancerous and noncancerous tissues. Among them, 7 compounds (namely hexanal, nonanal, cyclohexanone, 2-nonanone, pyrrole, pyridine, and phenol) were elevated in cancerous tissues, while 11 VOCs, such as ethyl acetate, acetoin, 2,3-butanedione, and 3-methyl-1-butanol, were more abundant in noncancerous samples. Notably, pyridine emerged as the most promising biomarker, exhibiting a 2.2- to 6.3-fold increase in cancerous tissues across both study populations. Pyridine is hypothesized to originate from the degradation of nicotinamide adenine dinucleotide phosphate ($NADP^+$), nicotinamide adenine dinucleotide (NAD^+), and NADPH. The elevated catabolism of these coenzymes within cancer cells likely accounts for the observed pyridine elevation, supporting its potential as a biomarker for gastric cancer.

8.2.8 Current Limitations and Future Directions in Volatilomics

Volatilomics continues to captivate researchers with the promise of noninvasive diagnostics based on VOCs. Across biofluids such as exhaled breath, skin, blood, urine, and stool, VOCs reflect complex physiological, microbial, metabolic, and environmental interactions in near real time. Despite growing interest and increasingly refined analytical technologies, the field remains hindered by conceptual ambiguity, methodological fragmentation, and limited clinical translation.

In breath analysis, the most widely explored matrix, significant progress has been made in VOC detection; however, the mechanistic understanding of their origin, kinetics, and modulation remains shallow. Most studies rely on associations rather than causality and overlook the profound influence of breathing patterns, airway microbiota, contamination, and compound-specific transport dynamics. Without standardization in sampling, preconcentration, and statistical modeling, breath-based diagnostics risk being attractive but unreliable.

Skin volatilomics, although less studied, offers unique advantages such as passive collection, low participant burden, and relevance to local and systemic physiology. However, it remains underpowered by small studies, heterogeneous protocols, and the challenge of distinguishing endogenous compounds from environmental contamination. In its current state, skin VOC analysis serves more as a complementary tool than a standalone diagnostic approach.

Blood, paradoxically, remains underutilized in volatilomics despite being a central circulatory medium. Its invasiveness partly explains this, but more fundamentally, the complex interactions of VOCs with plasma proteins, lipids, and enzymatic pathways are still poorly understood. Many reported biomarkers may reflect contamination or non-biological processes. Advancing blood volatilomics will require a mechanistic mapping of VOC metabolism and a stronger connection between biochemistry and clinical relevance.

Urine is one of the most promising matrices, due to its ease of collection and biochemical stability. Numerous studies have demonstrated its potential for detecting cancer, endocrine disruption, and renal disease. Nevertheless, the field also suffers from a lack of biochemical clarity: the metabolic origins and excretion routes of many VOCs remain speculative. Moreover, inconsistent sample handling and extraction protocols complicate cross-study comparisons and slow biomarker validation.

Stool, while offering a chemically rich and microbiota-informed profile, remains one of the most unstandardized and heterogeneous matrices. Fecal VOCs exhibit significant variations within and between individuals, and the mechanisms linking microbiome activity, diet, inflammation, and host metabolism to specific volatile signatures remain poorly understood. Most studies are exploratory, with limited cohort sizes, inconsistent methods, and little external validation.

The continued advancement of volatilomics also depends on the adoption and optimization of high-resolution analytical platforms, particularly GC-MS, GC×GC-TOFMS, and related hyphenated techniques. These methods are essential for detecting and structurally elucidating trace-level compounds within complex biological matrices. Unlike lower-resolution or sensor-based approaches, high-resolution GC-MS provides the necessary selectivity, sensitivity, and mass accuracy to distinguish endogenous VOCs from background noise, artifacts, and contaminants. It also enables robust compound identification through spectral matching and retention index alignment, which are critical for reproducibility and interlaboratory validation. As the field advances toward clinical translation, coupling high-resolution detection with standardized workflows and annotated spectral libraries will be crucial to achieving diagnostic reliability and ultimately securing regulatory approval.

The technology for VOC detection is already mature; what is needed now is a shift in research culture. Volatilomics holds significant promise, particularly in contexts where other biofluids are impractical or where early detection is crucial. However, realizing this potential depends on addressing several persistent structural weaknesses in the field. Across all matrices, studies often suffer from poor reproducibility, underpowered designs, insufficient attention to confounders, and overreliance on statistical models lacking mechanistic support. Furthermore, analytical workflows remain largely unstandardized. Too frequently, efforts are focused on identifying statistically significant VOCs without validating their diagnostic robustness or clarifying their biological relevance. This has resulted in a proliferation of candidate VOC "biomarkers" that rarely withstand independent verification. To overcome these limitations, future studies must prioritize harmonization, embrace multicenter validation, invest in pathway-oriented mechanistic research, and treat VOCs not as standalone diagnostic tools but as one component within a broader integrative framework that includes host genetics, microbiome profiling, and clinical phenotyping.

In short, the promise of volatilomics will not be fulfilled by broader compound lists or more complex models, but by asking better questions, designing better studies, and establishing stronger biological foundations. If that transition is made, VOCs may finally move from a seductive promise to a reliable tool in the diagnostic arsenal.

References

1 Beauchamp, J., Davis, C., and Pleil, J. (2020). *Breathborne Biomarkers and the Human Volatilome*. Elsevier.

2 Drabińska, N., Flynn, C., Ratcliffe, N. et al. (2021). A literature survey of all volatiles from healthy human breath and bodily fluids: the human volatilome. *J. Breath Res.* 15 (3): 034001.

3 Beccaria, M., Stefanuto, P.-H., Purcaro, G., and Franchina, F.A. (2022). The diagnostic purpose of odorant patterns for clinical applications using GC×GC. *Compr. Anal. Chem.* 96: 367–409.

4 Hakim, M., Broza, Y.Y., Barash, O. et al. (2012). Volatile organic compounds of lung cancer and possible biochemical pathways. *Chem. Rev.* 112 (11): 5949–5966.

5 Frank Kneepkens, C.M., Lepage, G., and Roy, C.C. (1994). The potential of the hydrocarbon breath test as a measure of lipid peroxidation. *Free Radic. Biol. Med.* 17 (2): 127–160.

6 Miekisch, W., Schubert, J.K., and Noeldge-Schomburg, G.F.E. (2004). Diagnostic potential of breath analysis—focus on volatile organic compounds. *Clin. Chim. Acta* 347 (1–2): 25–39.

7 Kuzma, J., Nemecek-Marshall, M., Pollock, W.H., and Fall, R. (1995). Bacteria produce the volatile hydrocarbon isoprene. *Curr. Microbiol.* 30 (2): 97–103.

8 King, J., Unterkofler, K., Teschl, G. et al. (2011). A mathematical model for breath gas analysis of volatile organic compounds with special emphasis on acetone. *J. Math. Biol.* 63 (5): 959–999.

9 Cope, K., Risby, T., and Diehl, A.M. (2000). Increased gastrointestinal ethanol production in obese mice: implications for fatty liver disease pathogenesis. *Gastroenterology* 119 (5): 1340–1347.

10 Ratcliffe, N., Wieczorek, T., Drabińska, N. et al. (2020). A mechanistic study and review of volatile products from peroxidation of unsaturated fatty acids:

an aid to understanding the origins of volatile organic compounds from the human body. *J. Breath Res.* 14 (3): 034001.

11 Vaz, A.D.N. and Coon, M.J. (1987). Hydrocarbon formation in the reductive cleavage of hydroperoxides by cytochrome P-450. *Proc. Natl. Acad. Sci. USA* 84: 1172–1176.

12 Branton, P.J., McAdam, K.G., Winter, D.B. et al. (2011). Reduction of aldehydes and hydrogen cyanide yields in mainstream cigarette smoke using an amine functionalised ion exchange resin. *Chem. Cent. J.* 5 (1): 15.

13 de Lacy Costello, B., Amann, A., Al-Kateb, H. et al. (2014). A review of the volatiles from the healthy human body. *J. Breath Res.* 8 (1): 014001.

14 Chen, S., Zieve, L., and Mahadevan, V. (1970). Mercaptans and dimethyl sulfide in the breath of patients with cirrhosis of the liver: effect of feeding methionine. *J. Lab. Clin. Med.* 75 (4): 628–635.

15 Fuchs, P., Loeseken, C., Schubert, J.K., and Miekisch, W. (2010). Breath gas aldehydes as biomarkers of lung cancer. *Int. J. Cancer* 126: 2663–2670.

16 Kischkel, S., Miekisch, W., Sawacki, A. et al. (2010). Breath biomarkers for lung cancer detection and assessment of smoking related effects — confounding variables, influence of normalization and statistical algorithms. *Clin. Chim. Acta* 411 (21–22): 1637–1644.

17 Byrne, G.I., Lehmann, L.K., Kirschbaum, J.G. et al. (1986). Induction of tryptophan degradation *in vitro* and *in vivo*: a γ-interferon-stimulated activity. *J. Interf. Res.* 6 (4): 389–396.

18 Lili, L., Xu, H., Song, D. et al. (2010). Analysis of volatile aldehyde biomarkers in human blood by derivatization and dispersive liquid–liquid microextraction based on solidification of floating organic droplet method by high performance liquid chromatography. *J. Chromatogr. A* 1217 (16): 2365–2370.

19 Deng, C. (2004). Investigation of volatile biomarkers in lung cancer blood using solid-phase microextraction and capillary gas chromatography? mass spectrometry. *J. Chromatogr. B* 808 (2): 269–277.

20 Zhang, F., Li, P., Lu, Y. et al. (2024). Advancing lung cancer diagnosis through NH_2-MON-SPME-GC-MS/MS: enhanced sensitivity in aldehyde biomarker detection from exhaled breath. *Anal. Chem.* 96: 15422–15429.

21 Filipiak, W., Sponring, A., Filipiak, A. et al. (2010). TD-GC-MS analysis of volatile metabolites of human lung cancer and normal cells in vitro. *Cancer Epidemiol. Biomarkers Prev.* 19: 182–195.

22 Greenberg, M. (1999). Toxicological review of acetonitrile. In: *Risk Assessment Division* (ed. A. Katherine and L. David), 25–31. Washington: US Environmental Protection Agency Publishing.

23 Bajtarevic, A., Ager, C., Pienz, M. et al. (2009). Noninvasive detection of lung cancer by analysis of exhaled breath. *BMC Cancer* 9 (1): 348.

24 Wang, P., Huang, Q., Meng, S. et al. (2022). Identification of lung cancer breath biomarkers based on perioperative breathomics testing: a prospective observational study. *EClinicalMedicine.* 47: 101384.

25 Munro, A.W., McLean, K.J., Grant, J.L., and Makris, T.M. (2018). Structure and function of the cytochrome P450 peroxygenase enzymes. *Biochem. Soc. Trans.* 46 (1): 183–196.

26 Koundouros, N. and Poulogiannis, G. (2019). Reprogramming of fatty acid metabolism in cancer. *Br. J. Cancer* 122 (1): 4–22.

27 Hoy, A.J., Nagarajan, S.R., and Butler, L.M. (2021). Tumour fatty acid metabolism in the context of therapy resistance and obesity. *Nat. Rev. Cancer* 21 (12): 753–766.

28 Zou, Y., Wang, Y., Jiang, Z. et al. (2021). Breath profile as composite biomarkers for lung cancer diagnosis. *Lung Cancer* 154: 206–213.

29 Sakumura, Y., Koyama, Y., Tokutake, H. et al. (2017). Diagnosis by volatile organic compounds in exhaled breath from lung cancer patients using support vector machine algorithm. *Sensors (Switzerland)* 17: 287.

30 Miller-Atkins, G., Acevedo-Moreno, L.A., Grove, D. et al. (2020). Breath metabolomics provides an accurate and noninvasive approach for screening cirrhosis, primary, and secondary liver tumors. *Hepatol. Commun.* 4: 1041–1055.

31 Sukaram, T., Tansawat, R., Apiparakoon, T. et al. (2022). Exhaled volatile organic compounds for diagnosis of hepatocellular carcinoma. *Sci. Rep.* 12: 5326.

32 Qin, T., Liu, H., Song, Q. et al. (2010). The screening of volatile markers for hepatocellular carcinoma. *Cancer Epidemiol. Biomarkers Prev.* 19 (9): 2247–2253.

33 Ferrandino, G., Orf, I., Smith, R. et al. (2020). Breath biopsy assessment of liver disease using an exogenous volatile organic compound-toward improved detection of liver impairment. *Clin. Transl. Gastroenterol.* 11: e00239.

34 Sukaram, T., Tansawat, R., Phathong, C. et al. (2024). Volatile organic compounds for diagnosis of early hepatocellular carcinoma in at-risk patients. *Clin. Chim. Acta* 556: 117831.

35 Li, J., Peng, Y., Liu, Y. et al. (2014). Investigation of potential breath biomarkers for the early diagnosis of breast cancer using gas chromatography-mass spectrometry. *Clin. Chim. Acta* 436: 59–67.

36 Wang, C., Sun, B., Guo, L. et al. (2014). Volatile organic metabolites identify patients with breast cancer, cyclomastopathy, and mammary gland fibroma. *Sci. Rep.* 4: 5383.

37 Zhang, Y., Guo, L., Qiu, Z. et al. (2020). Early diagnosis of breast cancer from exhaled breath by gas chromatography-mass spectrometry (GC/MS) analysis: a prospective cohort study. *J. Clin. Lab. Anal.* 34: e23526.

38 Sola-Martínez, R.A., Lozano-Terol, G., Gallego-Jara, J. et al. (2021). Exhaled volatilome analysis as a useful tool to discriminate asthma with other coexisting atopic diseases in women of childbearing age. *Sci. Rep.* 11 (1): 13823.

39 Heiro, H., Hildre, T.T., Craster, A. et al. (2024). Exploring exhaled breath volatile organic compounds in occupational asthma: a pilot cross-sectional study. *J. Breath Res.* 18: 046012.

40 Schleich, F.N., Zanella, D., Stefanuto, P.H. et al. (2019). Exhaled volatile organic compounds are able to discriminate between neutrophilic and eosinophilic asthma. *Am. J. Respir. Crit. Care Med.* 200: 444–453.

41 Monedeiro, F., Monedeiro-Milanowski, M., Ratiu, I.A. et al. (2021). Needle trap device-GC-MS for characterization of lung diseases based on breath VOC profiles. *Molecules* 26: 1789.

42 Pizzini, A., Filipiak, W., Wille, J. et al. (2018). Analysis of volatile organic compounds in the breath of patients with stable or acute exacerbation of chronic obstructive pulmonary disease. *J. Breath Res.* 12: 036002.

43 Goss, C.H. (2019). Acute pulmonary exacerbations in cystic fibrosis. *Semin. Respir. Crit. Care Med.* 40: 792–803.

44 Roesch, E.A., Nichols, D.P., and Chmiel, J.F. (2018). Inflammation in cystic fibrosis: an update. *Pediatr. Pulmonol.* 53 (S3): S30–S50.

45 van Horck, M., Smolinska, A., Wesseling, G. et al. (2021). Exhaled volatile organic compounds detect pulmonary exacerbations early in children with cystic fibrosis: results of a 1 year observational pilot study. *J. Breath Res.* 15: 026012.

46 Woollam, M., Siegel, A.P., Grocki, P. et al. (2022). Preliminary method for profiling volatile organic compounds in breath that correlate with pulmonary function and other clinical traits of subjects diagnosed with cystic fibrosis: a pilot study. *J. Breath Res.* 16: 027103.

47 Barberis, E., Amede, E., Khoso, S. et al. (2021). Metabolomics diagnosis of Covid-19 from exhaled breath condensate. *Meta* 11 (12): 847.

48 Ferrandino, G., De Palo, G., Murgia, A. et al. (2023). Breath biopsy® to identify exhaled volatile organic compounds biomarkers for liver cirrhosis detection. *J. Clin. Transl. Hepatol.* 11: 638–648.

49 Ferrandino, G., Ricciardi, F., Murgia, A. et al. (2023). Exogenous volatile organic compound (EVOC®) breath testing maximizes classification performance for subjects with cirrhosis and reveals signs of portal hypertension. *Biomedicine* 11: 2957.

50 Haick, H. (ed.) (2022). *Volatile Biomarkers for Human Health: From Nature to Artificial Senses*. The Royal Society of Chemistry.

51 Gallagher, M., Wysocki, C.J., Leyden, J.J. et al. (2008). Analyses of volatile organic compounds from human skin. *Br. J. Dermatol.* 159: 780–791.

52 Dormont, L., Bessière, J.M., and Cohuet, A. (2013). Human skin volatiles: a review. *J. Chem. Ecol.* 39 (5): 569–578.

53 Grice, E.A. and Segre, J.A. (2011). The skin microbiome. *Nat. Rev. Microbiol.* 9 (4): 244–253.

54 Penn, D.J., Oberzaucher, E., Grammer, K. et al. (2007). Individual and gender fingerprints in human body odour. *J. R. Soc. Interface* 4 (13): 331–340.

55 Stevens, D., Cornmell, R., Taylor, D. et al. (2015). Spatial variations in the microbial community structure and diversity of the human foot is associated with the production of odorous volatiles. *FEMS Microbiol. Ecol.* 91 (1): 1–11.

56 Curran, A.M., Prada, P.A., and Furton, K.G. (2010). The differentiation of the volatile organic signatures of individuals through SPME-GC/MS of characteristic human scent compounds. *J. Forensic Sci.* 55 (1): 50–57.

57 Nicolaides, N. (1974). Skin lipids: their biochemical uniqueness. *Science (1979)* 186: 19–26.

58 Vishinkin, R., Busool, R., Mansour, E. et al. (2021). Profiles of volatile biomarkers detect tuberculosis from skin. *Adv. Sci.* 8: e2100235.

59 Dormont, L., Bessiere, J.M., McKey, D., and Cohuet, A. (2013). New methods for field collection of human skin volatiles and perspectives for their application in the chemical ecology of human-pathogen-vector interactions. *J. Exp. Biol.* 216: 2783–2788.

60 Martin, H.J., Turner, M.A., Bandelow, S. et al. (2016). Volatile organic compound markers of psychological stress in skin: a pilot study. *J. Breath Res.* 10 (4): 046012.

61 Pretorius, D.T., Rohwer, E.R., and Naudé, Y. (2025). Cutaneous volatile and semi-volatile organic compounds as markers of malaria-infection by wearable samplers and two-dimensional gas chromatography—time-of-flight-mass spectrometry. *Journal of Chromatography Open* 8: 100233.

62 Tungkijanansin, N., Giebelhaus, R.T., Schmidt, S.A. et al. (2024). Identification of coronavirus disease marker compounds in sweat with comprehensive two dimensional gas chromatography using multiloop splitter-based non-cryogenic artificial trapping modulation system. *J. Chromatogr. Open* 5: 100113.

63 Logan, J.G., Birkett, M.A., Clark, S.J. et al. (2008). Identification of human-derived volatile chemicals that interfere with attraction of Aedes aegypti mosquitoes. *J. Chem. Ecol.* 34: 308–322.

64 Verhulst, N.O., Beijleveld, H., Qiu, Y.T. et al. (2013). Relation between HLA genes, human skin volatiles and attractiveness of humans to malaria mosquitoes. *Infect. Genet. Evol.* 18: 87–93.

65 Abaffy, T., Möller, M.G., Riemer, D.D. et al. (2013). Comparative analysis of volatile metabolomics signals from melanoma and benign skin: a pilot study. *Metabolomics* 9 (5): 998–1008.

66 Monedeiro, F., Dos Reis, R.B., Peria, F.M. et al. (2020). Investigation of sweat VOC profiles in assessment of cancer biomarkers using HS-GC-MS. *J. Breath Res.* 14: 026009.

67 Thomas, A.N., Riazanskaia, S., Cheung, W. et al. (2010). Novel noninvasive identification of biomarkers by analytical profiling of chronic wounds using volatile organic compounds. *Wound Repair Regen.* 18: 391–400.

68 Yamane, N., Tsuda, T., Nose, K. et al. (2006). Relationship between skin acetone and blood β-hydroxybutyrate concentrations in diabetes. *Clin. Chim. Acta* 365: 325–329.

69 Smolinska, A., Tedjo, D.I., Blanchet, L. et al. (2018). Volatile metabolites in breath strongly correlate with gut microbiome in CD patients. *Anal. Chim. Acta* 1025: 1–11.

70 Rees, C.A., Smolinska, A., and Hill, J.E. (2016). The volatile metabolome of Klebsiella pneumoniae in human blood. *J. Breath Res.* 10: 027101.

71 de Montellano, P.R.O. (2015). *Cytochrome P450*. Cham: Springer International Publishing.

72 Kushch, I., Arendacká, B., Štolc, S. et al. (2008). Breath isoprene – aspects of normal physiology related to age, gender and cholesterol profile as determined in a proton transfer reaction mass spectrometry study. *Clin. Chem. Lab. Med.* 46 (7): 1011–1018.

73 Watson, W.P., Cottrell, L., Zhang, D., and Golding, B.T. (2001). Metabolism and molecular toxicology of isoprene. *Chem. Biol. Interact.* 135–136: 223–238.

74 Waidyanatha, S., Rothman, N., Li, G. et al. (2004). Rapid determination of six urinary benzene metabolites in occupationally exposed and unexposed subjects. *Anal. Biochem.* 327: 184–199.

75 Mochalski, P., King, J., Klieber, M. et al. (2013). Blood and breath levels of selected volatile organic compounds in healthy volunteers. *Analyst* 138: 2134–2145.

76 Kimberley Molina, D. (2010). A characterization of sources of isopropanol detected on postmortem toxicologic analysis. *J. Forensic Sci.* 55 (4): 998–1002.

77 Ojanperä, I., Pihlainen, K., and Vuori, E. (1998). Identification limits for volatile organic compounds in the blood by purge-and-trap GC-FTIR. *J. Anal. Toxicol.* 22 (4): 290–295.

78 Romero-Gómez, M., Ramos-Guerrero, R., Grande, L. et al. (2004). Intestinal glutaminase activity is increased in liver cirrhosis and correlates with minimal hepatic encephalopathy. *J. Hepatol.* 41 (1): 49–54.

79 Müting, D., Heinze, J., Reikowski, J. et al. (1968). Enzymatic ammonia determinations in the blood and cerebrospinal fluid of healthy persons. *Clin. Chim. Acta* 19 (3): 391–395.

80 Barsotti, R.J. (2001). Measurement of ammonia in blood. *J. Pediatr.* 138 (1): S11–S20.

81 Cooke, R.J. and Jensen, R.L. (1983). Micromethod for determining plasma ammonia nitrogen with use of an ion-selective electrode. *Clin. Chem.* 29 (5): 867–869.

82 Wang, C., Peng, L., Ailing, L. et al. (2014). Blood volatile compounds as biomarkers for colorectal cancer. *Cancer Biol. Ther.* 15 (2): 200–206.

83 Einoch-Amor, R., Broza, Y.Y., and Haick, H. (2021). Detection of single cancer cells in blood with artificially intelligent nanoarray. *ACS Nano* 15: 7744–7755.

84 Umber, B.J., Shin, H., Meinardi, S. et al. (2013). Gas signatures from *Escherichia coli* and *Escherichia coli* -inoculated human whole blood. *Clin. Transl. Med.* 2 (1): 13.

85 Allardyce, R.A., Langford, V.S., Hill, A.L., and Murdoch, D.R. (2006). Detection of volatile metabolites produced by bacterial growth in blood culture media by selected ion flow tube mass spectrometry (SIFT-MS). *J. Microbiol. Methods* 65 (2): 361–365.

86 Chingin, K., Liang, J., Hang, Y. et al. (2015). Rapid recognition of bacteremia in humans using atmospheric pressure chemical ionization mass spectrometry of volatiles emitted by blood cultures. *RSC Adv.* 5 (18): 13952–13957.

87 Janasik, B., Jakubowski, M., and Jałowiecki, P. (2008). Excretion of unchanged volatile organic compounds (toluene, ethylbenzene, xylene and mesitylene) in urine as result of experimental human volunteer exposure. *Int. Arch. Occup. Environ. Health* 81 (4): 443–449.

88 Walker, V. and Mills, G.A. (2001). Urine 4-heptanone: a β-oxidation product of 2-ethylhexanoic acid from plasticisers. *Clin. Chim. Acta* 306: 51–61.

89 O'Lenick, C.R., Pleil, J.D., Stiegel, M.A. et al. (2019). Detection and analysis of endogenous polar volatile organic compounds (PVOCs) in urine for human exposome research. *Biomarkers* 24 (3): 240–248.

90 Zou, X., Lu, Y., Xia, L. et al. (2018). Detection of volatile organic compounds in a drop of urine by ultrasonic nebulization extraction proton transfer reaction mass spectrometry. *Anal. Chem.* 90: 2210–2215.

91 Smith, S., Burden, H., Persad, R. et al. (2008). A comparative study of the analysis of human urine headspace using gas chromatography–mass spectrometry. *J. Breath Res.* 2 (3): 037022.

92 Mills, G.A. and Walker, V. (2001). Headspace solid-phase microextraction profiling of volatile compounds in urine: application to metabolic investigations. *J. Chromatogr. B Biomed. Sci. Appl.* 753 (2): 259–268.

93 Lima, A.R., Pinto, J., Carvalho-Maia, C. et al. (2020). A panel of urinary volatile biomarkers for differential diagnosis of prostate cancer from other urological cancers. *Cancers (Basel)* 12: 1–8.

94 Mozdiak, E., Wicaksono, A.N., Covington, J.A., and Arasaradnam, R.P. (2019). Colorectal cancer and adenoma screening using urinary volatile organic compound (VOC) detection: early results from a single-centre bowel screening population (UK BCSP). *Tech. Coloproctol.* 23: 343–351.

95 Arasaradnam, R.P., McFarlane, M.J., Ryan-Fisher, C. et al. (2014). Detection of colorectal cancer (CRC) by urinary volatile organic compound analysis. *PLoS One* 9 (9): e108750.

96 Woollam, M., Teli, M., Angarita-Rivera, P. et al. (2019). Detection of volatile organic compounds (VOCs) in urine via gas chromatography-mass spectrometry QTOF to differentiate between localized and metastatic models of breast cancer. *Sci. Rep.* 9: 2526.

97 Silva, C.L., Passos, M., and Câmara, J.S. (2012). Solid phase microextraction, mass spectrometry and metabolomic approaches for detection of potential urinary cancer biomarkers—a powerful strategy for breast cancer diagnosis. *Talanta* 89: 360–368.

98 Taunk, K., Taware, R., More, T.H. et al. (2018). A non-invasive approach to explore the discriminatory potential of the urinary volatilome of invasive ductal carcinoma of the breast. *RSC Adv.* 8: 25040–25050.

99 Taware, R., Taunk, K., Pereira, J.A.M. et al. (2017). Investigation of urinary volatomic alterations in head and neck cancer: a non-invasive approach towards diagnosis and prognosis. *Metabolomics* 13: 111.

100 de Sousa Mesquita, A., Zamora-Obando, H.R., Neves dos Santos, F. et al. (2020). Volatile organic compounds analysis optimization and biomarker discovery in urine of Non-Hodgkin lymphoma patients before and during chemotherapy. *Microchem. J.* 159: 105479.

101 Pinto, J., Carapito, Â., Amaro, F. et al. (2021). Discovery of volatile biomarkers for bladder cancer detection and staging through urine metabolomics. *Meta* 11: 199.

102 Monteiro, M., Moreira, N., Pinto, J. et al. (2017). GC-MS metabolomics-based approach for the identification of a potential VOC-biomarker panel in the urine of renal cell carcinoma patients. *J. Cell. Mol. Med.* 21 (9): 2092–2105.

103 Kuang, H., Feng, J., Li, Z. et al. (2022). Volatile organic compounds from second-hand smoke may increase susceptibility of children through oxidative stress damage. *Environ. Res.* 207: 112227.

104 Wang, M., Xie, R., Jia, X., and Liu, R. (2017). Urinary volatile organic compounds as potential biomarkers in idiopathic membranous nephropathy. *Med. Princ. Pract.* 26: 375–380.

105 Ligor, T., Zawadzka, J., Straczyński, G. et al. (2021). Searching for potential markers of glomerulopathy in urine by HS-SPME-GC×GC TOFMS. *Molecules* 26: 1817.

106 Liu, D., Zhao, N., Wang, M. et al. (2018). Urine volatile organic compounds as biomarkers for minimal change type nephrotic syndrome. *Biochem. Biophys. Res. Commun.* 496 (1): 58–63.

107 Marmulla, R. and Harder, J. (2014). Microbial monoterpene transformations–a review. *Front. Microbiol.* 5: 346.

108 Nierop Groot, M.N. and De Bont, J.A.M. (1998). Conversion of phenylalanine to benzaldehyde initiated by an aminotransferase in Lactobacillus plantarum. *Appl. Environ. Microbiol.* 64: 3009–3013.

109 Saito, Y., Sato, T., Nomoto, K., and Tsuji, H. (2018). Identification of phenol- and p-cresol-producing intestinal bacteria by using media supplemented with tyrosine and its metabolites. *FEMS Microbiol. Ecol.* 94 (9): fiy125.

110 Ashkar, F. and Wu, J. (2023). Effects of food factors and processing on protein digestibility and gut microbiota. *J. Agric. Food Chem.* 71 (23): 8685–8698.

111 Olsen, I. and Yamazaki, K. (2019). Can oral bacteria affect the microbiome of the gut? *J. Oral Microbiol.* 11: 1586422.

112 Mortensen, P.B. and Clausen, M.R. (1996). Short-chain fatty acids in the human colon: relation to gastrointestinal health and disease. *Scand. J. Gastroenterol.* 31 (sup216): 132–148.

113 Garner, C.E., Smith, S., Lacy Costello, B. et al. (2007). Volatile organic compounds from feces and their potential for diagnosis of gastrointestinal disease. *FASEB J.* 21: 1675–1688.

114 Deneris, E.S., Stein, R.A., and Mead, J.F. (1985). Acid-catalyzed formation of isoprene from a mevalonate-derived product using a rat liver cytosolic fraction. *J. Biol. Chem.* 260 (3): 1382–1385.

115 Salerno-Kennedy, R. and Cashman, K.D. (2005). Potential applications of breath isoprene as a biomarker in modern medicine: a concise overview. *Wien. Klin. Wochenschr.* 117 (5–6): 180–186.

116 Johnson, W. (2002). Final report on the safety assessment of ethoxyethanol and ethoxyethanol acetate. *Int. J. Toxicol.* 21 (1_suppl): 9–62.

117 Bond, A., Greenwood, R., Lewis, S. et al. (2019). Volatile organic compounds emitted from faeces as a biomarker for colorectal cancer. *Aliment. Pharmacol. Ther.* 49 (8): 1005–1012.

118 Ahmed, I., Greenwood, R., Costello, B. et al. (2016). Investigation of faecal volatile organic metabolites as novel diagnostic biomarkers in inflammatory bowel disease. *Aliment. Pharmacol. Ther.* 43 (5): 596–611.

119 De Preter, V., Machiels, K., Joossens, M. et al. (2015). Faecal metabolite profiling identifies medium-chain fatty acids as discriminating compounds in IBD. *Gut* 64 (3): 447–458.

120 Walton, C., Fowler, D.P., Turner, C. et al. (2013). Analysis of volatile organic compounds of bacterial origin in chronic gastrointestinal diseases. *Inflamm. Bowel Dis.* 19: 2069–2078.

121 Bosch, S., Wintjens, D.S.J., Wicaksono, A. et al. (2020). The faecal scent of inflammatory bowel disease: detection and monitoring based on volatile organic compound analysis. *Dig. Liver Dis.* 52 (7): 745–752.

122 El Hassani, S.E.M., Bosch, S., Lemmen, J.P.M. et al. (2019). Simultaneous assessment of urinary and fecal volatile organic compound analysis in de novo pediatric IBD. *Sensors (Switzerland)* 19: 4496.

123 Ahmed, I., Greenwood, R., Costello, B.d.L. et al. (2013). An investigation of fecal volatile organic metabolites in irritable bowel syndrome. *PLoS One* 8 (3): e58204.

124 Franchina, F.A., Zanella, D., Dejong, T., and Focant, J.-F. (2021). Impact of the adsorbent material on volatile metabolites during in vitro and in vivo bio-sampling. *Talanta* 222: 121569.

125 Bean, H.D., Dimandja, J.M.D., and Hill, J.E. (2012). Bacterial volatile discovery using solid phase microextraction and comprehensive two-dimensional gas chromatography-time-of-flight mass spectrometry. *J. Chromatogr. B Analyt. Technol. Biomed. Life Sci.* 901: 41–46.

126 Zhu, J., Bean, H.D., Wargo, M.J. et al. (2013). Detecting bacterial lung infections: in vivo evaluation of in vitro volatile fingerprints. *J. Breath Res.* 7 (1): 016003.

127 Zanella, D., Henket, M., Schleich, F. et al. (2020). Comparison of the effect of chemically and biologically induced inflammation on the volatile metabolite production of lung epithelial cells by GC×GC-TOFMS. *Analyst* 145 (15): 5148–5157.

128 Acevedo, C.A., Sanchez, E.Y., Reyes, J.G., and Young, M.E. (2010). Volatile profiles of human skin cell cultures in different degrees of senescence. *J. Chromatogr. B* 878 (3–4): 449–455.

129 Purcaro, G., Rees, C.A., Wieland-Alter, W.F. et al. (2018). Volatile fingerprinting of human respiratory viruses from cell culture. *J. Breath Res.* 12 (2): 026015.

130 Mellors, T.R., Nasir, M., Franchina, F.A. et al. (2019). Identification of Mycobacterium tuberculosis using volatile biomarkers in culture and exhaled breath. *J. Breath Res.* 13 (1): 016004.

131 Taware, R., Taunk, K., Kumar, T.V.S. et al. (2020). Extracellular volatilomic alterations induced by hypoxia in breast cancer cells. *Metabolomics* 16 (2): 21.

132 Silva, C.L., Perestrelo, R., Silva, P. et al. (2017). Volatile metabolomic signature of human breast cancer cell lines. *Sci. Rep.* 7: 43969.

9

Pollution and Contaminants: Environmental and Individual Impact

Lourdes Ramos

Department of Instrumental Analysis and Environmental Chemistry, Institute of Organic Chemistry, CSIC, Madrid, Spain

9.1 Introduction

The most recent report from the International Council of Chemical Associations (ICCA) estimates 40,000–60,000 as the total number of industrial chemicals in commerce worldwide [1]. Such an estimation was made using the most updated data on manufactured chemicals listed in the EU, USA, Canada, Japan, and China, and it is smaller than the 100,000+ chemicals reported using other estimation strategies [2, 3]. According to ICCA, a reason for such a discrepancy can be the existence of numerous duplications in the different lists used for this purpose. On the other hand, it should be mentioned that the ICCA list does not include chemicals that are no longer in use. This is an important point to consider when trying to translate this figure to the number of potential pollutants present in the environment. In this latter case, it should be considered that some of the already banned or restricted products may remain in the environment for long periods of time after being released due to their resistance to physical, chemical, and biological degradation. Therefore, whatever the real number of industrial chemicals designed, synthesized, and commercialized for a specific purpose at present, it is evident that the total number of manufactured substances present in the environment is most probably higher than the ICCA threshold of 40,000–60,000 due to the recalcitrant nature of some of the chemicals used in the past. In addition, it should also be considered that some chemicals are produced and spread in the environment in an unintended way as a consequence of different natural or industrial processes. That is the case for numerous, toxic, and sometimes recalcitrant classes of chemicals, such as polyaromatic hydrocarbons (PAHs) and polychlorinated dibenzo-*p*-dioxins and furans (PCDD/Fs).

These properties, together with their proven capacity to bioaccumulate in the organisms' tissues, bioconcentrate through the food webs, and their ability for long-range transportation, made these families of pollutants (among several others) to be regulated under the Stockholm Convention for Persistent Organic Pollutants (POPs) [4], and included in most environmental and human monitoring programs. Finally, other chemicals, like in-use pesticides or human and veterinary pharmaceuticals, are more prone to undergo breakdown and transformation processes once released into the environment or as a consequence of their metabolism. These reactions usually yield more mobile and, frequently, less toxic compounds. Nevertheless, in some cases, their degradation products and metabolites are also toxic for the aquatic and terrestrial ecosystems, as well as for humans [5, 6], making their control in the affected matrices and the evaluation of their environmental behavior and fate similarly advisable [7, 8].

Previous considerations indicate that, at present, there is a plethora of chemicals that can reach (or have already reached) the various compartments of the environment through different ways and that these compounds can affect the ecosystems, including humans, in a manner that is still most often unknown. Although these statements are generally accepted by the scientific community, several recent bibliometric studies have evidenced a consistent bias toward the repetitive assessment of the environmental presence of specific compounds (or classes of compounds), and a lack of studies intended to include broader (and less evaluated) sets of substances [9, 10]. This trend apparently started in the mid-1970s, when the lists of priority substances and exposure guidelines were introduced. However, it looks to remain nowadays [9], demonstrating the influence of such types of prioritization lists and current regulations on the selection of individual chemicals considered for environmental analysis and monitoring programs. To say differently, this biased tendency seems to point to a scientist's preference for the analysis of well-known, prioritized chemicals, for which appropriate pure analytical standards and well-established methods are available [10], over those substances requiring new analytical methodologies or instrumentation not always easily available. In addition, it should be mentioned that the complete characterization of a given analyte (or class of analytes) is not necessary enough to attract scientists' attention. Factors like its anticipated global (versus local) widespread and, probably more importantly, the evidence of its toxicity and (possible) environmental persistence have been pointed out as key aspects driving the selection of the studied analyte [11]. Altogether, it would explain why a significant portion of all environmental measurements published over the years invariably correspond to a relatively small number of well-known substances, such as the 16 priority PAHs, phase-out POPs like some organochlorinated pesticides (OCPs) or polychlorinated biphenyls (PCBs) and, only more recently, other flame retardants (FRs) that are also now regulated, just to mention a few examples.

Apart from these considerations, a key factor with a major influence on the nature and number of the chemicals measured in environmental media has been the available analytical instrumentation. Thereby, although gas chromatography (GC) amenable contaminants have been uninterruptedly analyzed over time, since the mid-1990s, pivotal advances in the liquid chromatography-mass spectrometry (LC-MS) techniques have contributed to expanding analytical scope into polar and ionizable substances, which have been progressively incorporated into the monitoring programs, with special emphasis on water quality [9]. In the 2000s, the increasing implementation in laboratories of high-resolution mass spectrometry (HR MS) instruments for the accurate determination of contaminants contributed to consolidating this trend. Thereby, although some families of gas chromatography-mass spectrometry (GC-MS) amenable substances clearly benefited from these instrumental advances (e.g., specific classes of novel FRs), most newly investigated chemicals corresponded to pollutants already known to exist in commerce (or known-unknowns) but difficult to analyze with available GC-MS-based instruments. This is particularly the case of in-use pesticides, pharmaceuticals and their metabolites (which accounted for 50–60% of the measured chemicals in the period 2000–2015), and poly/perfluoroalkyl substances (PFASs) related compounds, as evidenced by a recent bibliometric study [9] on the major classes of substances determined in environmental media in the last years (Figure 9.1). Interestingly, this study identified various emerging classes of chemicals, in some cases associated with the plastic industry (e.g., plastic additives, bisphenols and alkyl phenols, siloxanes, or benzotriazoles), probably reflecting the increasing concern regarding the environmental behavior and fate of plastic litter and microplastics [3]. In addition, a significant number of globally reported compounds categorized as "not characterized" did not belong to commonly used priority and monitoring pollutant lists. In our opinion, this last finding probably demonstrates the current need to (i) review and broaden these lists, and (ii) increase our knowledge about the possible presence in the media of known-unknown (i.e., suspected) or fully unknown chemicals that can affect the ecosystems and human health.

In line with these considerations, this chapter aims to provide an overview of the current state-of-the-art concerning the use of GC-MS-based techniques for the evaluation of organic contaminants in environmental and human samples. Emphasis will be placed on omic-like approaches, providing a broader and more realistic estimate of the occurrence and potential global impact derived from the presence of these pollutants on ecosystems, including humans.

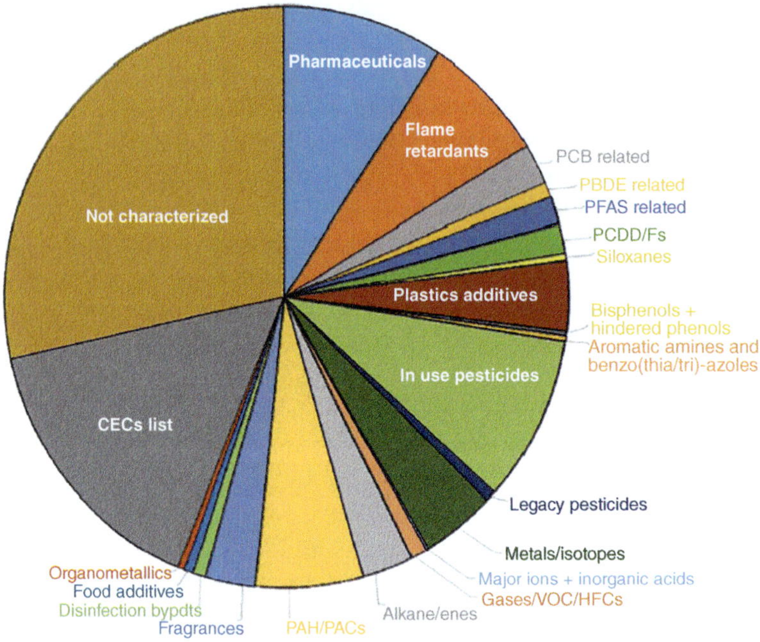

Figure 9.1 Overview of the 22 major classes of chemicals determined in environmental media. The "not characterized" class is a varied group of substances not found in priority and monitoring pollutant lists. *Source:* Muir et al. [9] / with permission of American Chemical Society / CC BY 4.0.

9.2 Target, Suspect, and Non-targeted Screening Analyses of Pollutants: General Considerations and Instrumentation

The concepts of suspect screening analysis (SSA) and non-target screening (NTS) as a complement to the conventional target determination of a suite of well-known (i.e., preselected) pollutants using GC-based techniques are not new. Over the years, target analysis of contaminants has relied on the availability of appropriate pure standards and the application of carefully optimized sample preparation protocols for each analyte or class of analytes targeted [12–14]. These treatments are typically multistep procedures that, due to the low levels at which the analytes of interest are found in environmental and human samples, start with their exhaustive extraction from the matrix in which they are entrapped. Due to the essentially nonselective nature of this process, a number of treatments should follow to isolate the investigated pollutants, firstly from co-extracted non-chemically related matrix components (e.g., lipids from PCDD/Fs), and then from other chemically and structurally related compounds that can interfere with their final instrumental separation-plus-detection (e.g., PCDD/Fs from other POP families). The high selectivity and specificity of these sample treatment protocols facilitated the accurate determination and quantification of the preselected analytes. However, a number of pollutants with physicochemical properties close to those of the targeted analytes have also been tentatively identified in these analyses through close inspection of the mass spectra of "unexpected" analytes eluting within the retention window of the well-known pollutants of interest. In the case of monodimensional GC-based determinations, and despite the great resolution offered by capillary GC columns, such identifications were often hampered by the close elution of structurally related compounds [12, 15, 16], and the frequent use (especially in the past) of low (i.e., unit) resolution MS instruments. In some cases, the higher concentrations at which the target compounds were present in the samples compared to these other new analytes contributed to obscuring their detection and identification. However, different halogenated pollutants (and classes of halogen-containing contaminants) were discovered in this way [12]. In other cases, a truly non-targeted approach was adopted for the analysis of a variety of organic pollutants in contaminated aqueous samples [17] but also more complex matrices, such as sediments [18] and biological tissues [16, 19]. In these studies, the application of a somewhat simplified sample treatment allowed for the retention of other compounds with similar physicochemical properties present in the matrix through the entire sample preparation procedure. This type of *generic* sample treatment protocol is effective in reducing main interferences while providing appropriate sensitivity and reproducibility. However, one should be aware that they intrinsically result in a discernible loss of analytes. The need for a proper balance among the retained analytical features is always an important matter, but it becomes particularly relevant for analyses intended for the discovery of suspected or unknown pollutants requiring an omic-like approach. In this field, novel, robust, and standardized sample treatment procedures minimizing the co-extraction of interferences while preserving broad chemical coverage are still required [20, 21], and the use of strict QC/QA protocols (internal standards to reduce the large uncertainty associated with these determinations, blank and control samples, etc.) is definitely mandatory [22].

In any case, these earlier studies demonstrated the feasibility of the adopted simplified sample preparation methods for their intended purposes, although they also pointed out the convenience of using complementary MS ionization techniques and/or chromatographic separation mechanisms for proper tentative annotation or identification of the detected analyte(s). It is also important to mention that, in some of these pioneer studies carried out in the late 1980s, a meritorious positive identification was achieved by the authors through the synthesis of the annotated analyte(s) [19].

In the 1990s, the implementation of multidimensional gas chromatography (MDGC) contributed to solving coelution problems unsolvable with monodimensional GC. The technique involved two capillary GC columns with different stationary phases serially connected and a valve system that allowed the selective and accurate transfer of specific fractions of the eluent from the first column to the second one for further separation and eventual resolution of coeluting peaks, thus facilitating the analyte identification and the quantitation. In the environmental field, the MDGC technique was initially used for the accurate quantitation of individual toxic congeners in complex mixtures of pollutants containing numerous isomers at widely divergent concentrations. A typical and pioneering application example was the determination of non-ortho PCBs without any previous isolation from other PCBs present in the sample [23]. The improved resolution achieved in these types of target analysis enabled MDGC to be rapidly assayed for the non-oriented qualitative characterization or classification of the organic pollutants present in contaminated environmental samples [24]. These studies demonstrated that the complexity of the applied chromatographic approach varied greatly depending on the evaluated matrix. Thereby, while the analysis of fairly simple matrices (e.g., water) could be completed using relatively simple chromatographic programs and with a single-column run, the accurate characterization of more complex extracts, like those obtained from contaminated sediments, made the second-stage separation mandatory and required a high number of heart-cuts (e.g., up to 25 heart-cuts [24]). The following studies demonstrated the feasibility of the technique for NTS analyses, but also highlighted the complexity of some of the chromatographic methodologies required for such determinations.

The emergence of comprehensive two-dimensional gas chromatography (GC×GC) in the 2000s contributed to facilitating the analysis of complex environmental extracts. GC×GC allowed direct and sequential transfer of the whole eluent from the first column as small fractions collected and compressed in the modulator, and then reinjected in a second short and narrow column for further chromatographic resolution of the analytes within a few seconds under a different separation mechanism [25]. The compression of the chromatographic bands in the modulator resulted in narrower bands, with a consequent improvement in sensitivity compared with equivalent monodimensional approaches. The separate analytes were distributed in a chromatographic plane, the so-called contour-plot, and defined by two retention times, one per column separation (or dimension), which made identification more specific and accurate. It was considered that GC×GC provided at least one order of magnitude more resolved peaks than conventional single-column GC approaches [26]. Besides, the possibility of generating organized chromatograms, in which the structurally related analytes appeared as ordered roof-tile structures, was an added value that facilitated the visual comparison among complex samples, analyte grouping, and the tentative identification of class-related and known-unknown analytes [27, 28]. This unprecedented chromatographic resolution attracted the attention of scientists. In particular, when GC×GC was combined with MS as a detector, the enhanced separation achieved among the analytes present in the extract led to single-component peaks, for which interference-free mass spectra were obtained. This was a distinctive advantage, allowing a detailed fragmentation analysis and a more accurate matching with mass spectral libraries that facilitated the preliminary identification of suspect and unknown contaminants [29]. However, somehow surprisingly, due to its many positive features, GC×GC was not massively adopted for these types of determinations.

Multiple dimension GC-based setups are not only possible on the chromatographic side, as discussed for MDGC and GC×GC. Hyphenation of the selected GC system with any technique providing additional identification capabilities should result in an extra identification criterion (i.e., an additional separation dimension) that may facilitate the unambiguous identification of the analytes of interest. In this sense, GC-based approaches have benefited greatly from the advances made over the past decades in the field of MS. In-depth discussion on the basic and main features of conventional and novel MS instruments can be found in previous chapters of this monography. Therefore, only a brief discussion of the main advantages and limitations of the mass analyzers most frequently used for the analysis of traces of organic pollutants in environmental samples has been included here.

Mass spectrometers are usually designed for specific analytical purposes. Therefore, each instrument has its own advantages and shortcomings or, to say it differently, its own application aim (Table 9.1).

GC-MS-based techniques are feasible for the direct determination of a wide variety of low- and medium-polar, (semi-)volatile chemicals considered in current monitoring programs, but also for many other less volatile and polar pollutants included in these control lists after an appropriate derivatization step [32, 33].

Many of these target determinations have been traditionally done using low-resolution MS (LR MS), initially with single quadrupole (Q) analyzers (combined with both mono- and multidimensional GC instruments) and more recently with ion traps (ITs) and triple quadrupole (QqQ) MS instruments. QMS instruments are easy to handle and maintain, and extremely robust. These features made them be used as a general workhorse in the environmental and control laboratories. QMS instruments can be operated in the full-scan and SIM modes. They show working mass intervals typically in the 2–4000 Da, which makes them appropriate for the analysis of a wide variety of organic microcontaminants in complex environmental and human matrices. However, their m/z accuracy at the unit level and limited sensitivity are sometimes insufficient for the accurate determination of certain classes of pollutants at the low levels at which they are often present in real-life samples, in particular when operated in the

Table 9.1 MS techniques used in combination with GC for the determination of organic micropollutants, their most relevant features, and application field.

Mass analyzer	Acronym	Advantages	Limitations	Environmental application field
Single quadrupole	Q	Easy to operate and maintain Robust and very reproducible Broad dynamic linear range Mass spectral libraries available Low initial investment Small size	Nominal mass accuracy Limited sensitivity, in particular in full-scan mode	Target analysis SSA
Ion trap	IT	MS^2 allowed Improved selectivity and sensitivity compared to QMS Compact design Relatively low initial investment	Nominal mass accuracy Limited ion capacity At least two transitions (two product ions of the precursor ion) and their intensity ratio required Limited mass spectral libraries available	Sensitive target analysis Structure elucidation
Triple quadrupole	QqQ	MS^n allowed Good selectivity and sensitivity Broad dynamic linear range Fragmentation patterns can help in the identification of functional groups Relatively low initial investment	Nominal mass accuracy At least two transitions (two product ions of the precursor ion) and their intensity ratio required Limited mass spectral libraries available	High sensitivity target analysis SSA Structure elucidation

Orbitrap	Orbi	High mass accuracy ($R^a \cong 100{,}000$) High mass range High sensitivity	Limited dynamic linear range Large data set generated Complex data set generated High initial investment Relatively large size instrument Limited mass spectral libraries available	High sensitivity target analysis SSA and NTS Structure elucidation
Magnetic sector	M	High mass accuracy ($R > 10{,}000$) High sensitivity (in SIM[b] mode) High reproducibility	Limited data acquisition rate Limited dynamic linear range High initial investment Large size instrument	High sensitivity target analysis
Time-of-flight	TOF	(i) Low-to-medium or (ii) high mass accuracy ($R < 50{,}000$) Ultra-fast data acquisition rates (i) Virtually unlimited m/z range High sensitivity in (i) full-scan mode or (ii) small mass scan ranges	(i) Medium or (ii) high initial investment Relatively large size instruments	(i) Fast GC and GC×GC target analysis, SSA and NTA (ii) Target analysis and SSA structure elucidation
Ion mobility spectrometry	IMS	Separation based on mass, charge, size, and shape of the analytes Separation of isobars allowed Wide dynamic linear range	Sophisticated instrument Large data set generated High complex data sets Very high initial investment Large size instrument	SSA and NTS Structure elucidation

(Continued)

Table 9.1 (Continued)

Mass analyzer	Acronym	Advantages	Limitations	Environmental application field
Fourier transform ion cyclotron resonances	FT ICRMS	Ultra-high mass accuracy (*R* up to 1,000,000)	Limited data acquisition rate Limited scan rate Ultra-high vacuum needed Sophisticated instrument Large and complex data set generated Very high initial investment Large size instrument	SSA and NTA Structure elucidation

a) Resolution.
b) Selected ion monitoring.
Source: Partially adapted from Refs. [30, 31].

full-scan mode. Depending on the investigated contaminants, these limitations can be partially circumvented by using ionization strategies [34] different from the conventional electron ionization (EI) at 70 eV [35].

IT mass spectrometers are versatile instruments also with unit resolution, but they allow us to perform tandem mass spectrometry (MS/MS), which is a distinctive advantage resulting in improved selectivity and sensitivity compared to QMS. With these systems, the analyte identification relies on the simultaneous detection, at the retention time of the precursor ion, of at least two product ions, and in the appropriate maintenance of their corresponding intensity ratio. The possibility of performing tandem MS has been used to confirm chemical structure proposals made from LR MS data [36]. One of the most pressing shortcomings of these instruments is the limited ion capacity of the traps, which can result in space-charge effects that ultimately affect their sensitivity, mass accuracy, and mass resolution. The linear design of QqQ mass spectrometers solved this shortcoming, allowing the storage of a large number of ions and, consequently, of transitions being acquired within the same run with improved specificity and sensitivity [31]. Thereby, although these instruments only achieve m/z accuracy at the unit level, they exhibit broad dynamic linear ranges and sensitivities low enough to permit accurate quantitation of trace organic contaminants in complex extracts, in particular when using the selected reaction monitoring (SRM) mode. When operated in the multiple reaction monitoring (MRM) mode, the latest generation of QqQMSs allows the analysis of hundreds of compounds in a single run with low limits of detection (LODs) due to the reduced background noise [37]. The possibility of performing MS^n experiments, in particular when allowing positive and negative ionization in a single run, is an additional positive feature providing extra structural information and confirmation capabilities in SSA and NTS studies.

Modern time-of-flight mass spectrometers (TOFMS) provide medium to high mass accuracy with a virtually unlimited m/z acquisition range. The TOFMS sensitivity is not affected by the registered m/z range, and those with unit mass accuracy allow ultra-fast data acquisition rates, which makes them ideal instruments to be hyphenated with fast-GC and GC×GC instruments. TOFMS instruments do not allow performing tandem ionization. However, in recent years, several new TOFMSs and hybrid QTOF and ITTOF mass analyzers capable of mass resolution above 10,000 have been introduced in the market to obtain accurate mass at the MS^n level at very high data acquisition rates [38].

HR MSs provide high mass resolutions (in many instances, higher than the standard 10,000), exact molecular mass, and high sensitivity (depending on the instrument, also in full-scan mode). High mass resolution power is needed when the resolution of overlapping interferences relies exclusively on mass spectrometry. Instruments like the magnetic sector (M), Fourier transform ion cyclotron resonance (FT ICRMS), and Orbitrap fulfil these specifications. However, these instruments need long cycle time to acquire high-resolution broadband mass spectra (e.g., up to 2 seconds for a single scan at 1,000,000 resolving power with FT ICRMS [39]), thus limiting their practicality for specific application studies. Furthermore, M and, in particular, FT ICRMS are sophisticated instruments whose high price and operating costs limit their implementation in commercial and control laboratories.

Similarly to FT ICR-analyzers, the Orbitrap analyzer uses a Fourier transform to convert the signal produced by ions oscillating in a trap from the time domain to a frequency domain. However, unlike the FT ICR analyzer, which uses a magnetic field to induce ion oscillations, an electric field is applied, thus avoiding the use of cryogenic refrigerants. In addition, the size of the instrument and its operating cost are lower than for FT ICR, while providing high resolving power and sensitivity, and high mass accuracy, although the latter at the price of slow scan cycle times for broad mass ranges [39, 40], as mentioned earlier. As in the case of modern QTOFMS instruments, new hybrid Q-Orbitrap (QOrbi) instruments, in which the initial quadrupole acts as a mass filter, are versatile and sensitive instruments with fast scan velocities, broad dynamic ranges, and the possibility of performing MS^n. These features, together with their increasing accessibility for laboratories, have led to the use of these hybrid analyzers in many recent environmental SSA and NTS investigations, as will be shown later.

Despite the attractiveness of the idea of incorporation of ion mobility spectrometry (IMS) into current HR MS-based analytical methods as an additional separation dimension [41], or even to expand the dynamic linear range of other MS, the penetration of this technology in the environmental field is still limited [42–45]. Unlike previously discussed analyzers, where separation is based on the mass and charge of the analytes, in IMS, two other parameters affect such a separation: the size and shape of the molecule, which result in an enhanced separation power and the unique capability to differentiate isobars of the same charge by their steric properties. The integration of the IMS dimension on a separation-plus-HR MS technique contributes to the removal of unwanted matrix components and interferences, with the consequent improvement of the method selectivity and sensitivity due to the reduction of the chemical noise. The IMS-derived collision cross-section (CCS) is a physicochemical parameter of ionized molecules that remains stable irrespective of the sample matrix, the chromatographic and mass spectrometric conditions [43], and even for different separation instruments (i.e., for values measured with either GC or LC) [46]. Thereby, CCS values provide additional evidence for analyte identification, complementary to that of mass spectra and retention time, and compounds can be identified with more confidence, either in targeted or non-targeted approaches [45]. On the other hand, and as expected due to the high density of accurate data generated, the application of IMS to the analysis of complex extracts results in large, complex data sets whose processing and interpretation can be extremely demanding.

One should be aware that, in general, a similar consideration applies to the treatment of the large data files generated by the untargeted multidimensional and/or HR MS analysis of complex environmental and human matrices. The use of appropriate statistical and multivariate chemometric tools can help to efficiently extract relevant information from these challenging datasets. Therefore, the selection (and, in some cases, development) of friendly software allowing to export the acquired data under appropriate formats recognizable by other (e.g., state-of-the-art open source) data evaluation and mass spectral exchange platforms is another (or even equally [39]) relevant issue to be considered in this research area. The setting up of appropriate general workflows [22, 33, 42] and suitable and flexible data workstations [47] are also relevant aspects to consider. The development of novel algorithms and computational tools for big-data treatment, and the implementation of artificial intelligence-aided workflows, will be key in these contexts, as already demonstrated for LC-based approaches [48]. These types of advances will also play an essential role in combining the data generated by multidimensional and HR MS analytical platforms with those obtained by other research disciplines, as required, for instance, for the prioritization of the detected analytes and risk evaluation [49].

A final issue concerning the analysis of previously undescribed chemicals is the way in which the identification confidence in the proposed new chemical candidate is graded and transmitted to other researchers and to the general audience. This is actually a major challenge with direct implications on the recommendation and application, if required, of specific actions to be taken by the competent authorities with regard to the environmental monitoring, production, and control of the newly detected analytes. Concern about this issue is not exclusive to the environmental field. Several systems have been proposed to score the confidence in the identification of small molecules during the last years, including the metabolomics standard initiative [50], further redefined for LC-HR MS/MS data in the search for new metabolites [51], and later expanded to contain additional score levels relevant to LC-HR MS/MS screening of pollutants [52]. In contrast, the slowest developments in GC-HR MS-based techniques compared to their equivalent LC-HR MS counterparts have made the application of GC-HR MS to NTS studies somewhat more limited and relatively less developed [53, 54]. As a side effect, in many instances, a similar score scale to those described for LC-based evaluations has been assumed in GC-based analyses. This is particularly the case of the classification system proposed by Schymanski et al. [52], which has been the most generally accepted and applied by now. This system establishes the following levels of confidence:

- Confirmed identification with two independent properties against an authentic standard analyzed under similar conditions (Level 1).
- Probable identification with all evidence indicating that only one structure is possible, but with no authentic standard available for confirmation (Level 2: 2.a: library spectrum match; 2.b: other diagnostic evidence).
- Tentative identification/putative compound class: multiple structures are possible or insufficient evidence to rule out other structures; only the analyte class is clear (Level 3).
- Unknown compound of interest: unequivocal molecular formula is provided (Level 4) or only exact mass is available (Level 5), which allows evaluation of the presence of the substance in other samples, although its identity remains unknown.

Based on this scoring system, other classification scales have been proposed in recent years that incorporate specific conventions and tools to facilitate the identification and classification of analytes belonging to numerous classes, such as the perfluoroalkyl and polyfluoroalkyl substances [55], or the use of less conventional MS-based detectors, such as IMS [56], but again in combination with LC. To our knowledge, only one annotation scoring framework for GC-HR MS data related to organic pollutants has been reported so far [22]. Although based on Schymanski's schema, this new annotation system combines qualitative and some quantitative scoring metrics in an attempt to decrease the number of false positives. It also includes a blank-feature filtering step, and chemical-series and class-based annotations missing in the Schymanski schema. The feasibility of the developed open-source software for NTS was illustrated through the evaluation of datasets corresponding to the analysis of spiked analytes in plasma (112 compounds) and validated outdoor and indoor air samples (70 native analytes and 11 isotopically labelled chemicals). Results showed a false positive rate of 12% for chemical identifications assigned as Level 2. This annotation system represents an improvement on previous frameworks for GC data. However, further work is still required to implement additional identification criteria not yet considered, e.g., the second dimension retention indexes (RIs) in GC×GC [57].

9.3 Application of GC-MS-based Technique to the Environmental Analysis of Pollutants

Apart from the intrinsic difficulty associated with the determination of individual compounds at the residue level in the presence of structurally related (and frequently interfering) analytes, the analysis of micropollutants in environmental samples is additionally complicated by the concurrent presence in the final purified extract of other matrix components that can hamper the chromatographic separation of the analytes of interest and/or their ionization process. These problems become more pressing when the sample preparation procedure is simplified to preserve as many compounds as possible, i.e., in SSA and NTS studies, the former in many instances is considered as a particular case of the latter. Generally speaking, and as can be inferred from the previous discussion, two main strategies have typically been adopted to avoid (or at least to minimize) these difficulties: either increasing the separation power on the chromatographic front and/or enhancing the resolution and identification capability of the MS detector. This last goal can also be achieved by subsequent confirmation of the detected analytes using a different MS instrument providing higher resolution and/or structural confirmation capabilities [36].

As previously indicated, in principle, the instrumentation used for separation-plus-detection of the investigated organic pollutants in the analyzed extracts is determined by the specific analysis aim. Nevertheless, a variety of approaches have been used over time for essentially similar studies, but also quite similar methodologies have been used for completely different purposes. This variety of instrumental approaches has probably been determined by the evolution of the instrumentation accessible to laboratories over the years and by the advances achieved in the field of MS during the last decades. A general overview of this evolution and the different GC-MS-based strategies used for the analysis of organic pollutants in environmental and human samples is shown in Table 9.2.

QMS instruments combined with single-column GC have been (and are) successfully used for the target determination of individual components belonging to a variety of classes of organic pollutants in widely divergent environmental, food, and human matrices. Depending on the nature of the evaluated contaminant(s), different ionization strategies (e.g., positive or negative chemical ionization, P/NCI, instead of the more conventional EI) have been applied to maximize the specificity and sensitivity of the instrumental determination [34, 35, 58, 108]. Irrespective of the selected ionization procedure and assuming an optimal chromatographic separation, the limited resolution of this type of MS analyzer makes the accuracy of the determination essentially rely on the effectiveness of the sample preparation procedure applied to isolate the analytes (or class of analytes) of interest from other co-extracted matrix components and structurally-related interferences. In addition, m/z data acquisition is typically performed in the SIM mode to reach the required sensitivity. Despite these limitations, it should be mentioned that, in earlier times, GC-QMS was also used for SSA and NTS of organic contaminants in complex contaminated matrices. In these studies, the limited resolution offered by the use of single-column GC was circumvented by previous fractionation of the analytes in the extract under investigation according to their polarity, thus reducing

Table 9.2 Selected representative application studies involving different GC-MS-based systems for the determination of organic pollutants in environmental and human samples.

GC system	Detector	Matrix	Analytes	Year	References
Target analysis					
GC	QMS	Human milk	PCBs	1997	[58]
GC	QMS	Soils	PAHs	2000	[35]
GC	QMS/ENCI QMS	Wild bird eggs	OCP, PCBs, PBDEs, and DP[a)]	2017	[34]
GC	QMS	Synthetic seawater	5 volatile PFASs	2025	[59]
GC	ITMS	Drinking water	Chlorophenols	1996	[60]
GC	ITMS	Biological tissues	OCP, PCBs and PBDEs[b)]	2009	[61]
GC	ITMS	Minute biotic samples	PCBs	2013	[62]
GC	QqQMS	Surface, ground, and wastewater samples	50 priority organic pollutants from the European Water Policy	2007	[63]
GC	QqQMS	Sediments	97 legacy and emerging hydrophobic and polar pollutants	2016	[64]
GC	QqQMS	Human serum	22 legacy POPs	2019	[65]
MDGC	QMS	Biotic reference materials and sediment	Chiral OCPs and PCBs	2002	[66]
MDGC	QMS	Apple juice	24 pesticides	2009	[67]
GC×GC	QMS	Grape fruit extract	92 pesticides	2007	[68]
GC×GC	QqQMS	Coriander	Diazinon and permethrin	2008	[69]
GC×GC	TOFMS	Human serum and milk	Legacy POPs (OCPs, PCBs, PBDEs and PBBs[c)])	2004	[70]

Table 9.2 (Continued)

GC system	Detector	Matrix	Analytes	Year	References
GC×GC	TOFMS	River water	97 pesticides, pharmaceuticals and related organic pollutants	2010	[71]
GC×GC	TOFMS	Water from an industrial site	9 volatile PFAS	2025	[72]
GC×GC	HR TOFMS	Sediments	47 legacy COPs, PCBs, and PBDEs	2019	[73]
GC	QTOFMS	Household dust	59 toxic organic compounds	2018	[74]
GC	QTOFMS	Air around metallurgical plants	187 PAHs, including 38 Cl/Br-PAHs	2022	[75]
GC	OrbiMS	Tomato, leek, orange	54 pesticides	2016	[54]
GC	OrbiMS	Fresh-food commodities	31 pesticides	2020	[76]
GC	(APPI)OrbiMS	River water	12 neutral PFASs	2020	[77]
GC	OrbiMS	Air $PM_{2.5}$	PAE[d] and DEHA[e]	2024	[78]
GC	(APCI)FT ICRMS	Food samples	Cl-/Br-FRs	2019	[79]
Suspect screening					
GC	QqQMS	Sediments	Halogenated POPs, pesticides, and other emerging pollutants	2020	[80]
GC	QqQMS	Liver tissue	209 pesticides and POPs	2021	[81]
GC	QqQMS	Plant protection products	Undeclared ingredients (263 pesticides)	2023	[37]
MGGC	QMS	Peach, orange, and pineapple juice	46 pesticides	2012	[82]

(Continued)

Table 9.2 (Continued)

GC system	Detector	Matrix	Analytes	Year	References
GC×GC	QMS	Drinking water	170 DBPs[f)]	2016	[83]
GC×GC	TOFMS	Crops	235 pesticides	2011	[84]
GC×GC	TOFMS	Naturally contaminated sediments	PAHs	2012	[85]
GC×GC	QTOFMS	Wastewater treatment plant influents and effluents	291 organic pollutants	2023	[86]
GC×GC	(EI, NCI) HR TOFMS; QqQMS	Sediment and flight ash CRMs	Cl-/Br-containing pollutants	2018	[87]
GC	QTOFMS	Wastewater treatment plant influents and effluents	Organic pollutants	2019	[88]
GC	QTOFMS	Sediments	LCMs[g)]	2022	[89]
GC	QTOFMS	Air from fluorochemical industries	Neutral PFASs	2025	[90]
GC	(ECNI)OrbiMS	Salmon from different origins	SCCP[h)], and MCCP[i)]	2018	[91]
GC	OrbiMS	Sea cucumbers	289 industrial and agricultural pollutants	2022	[92]
GC	IMS-HR MS	Crude oil	>100 isomeric and isobaric PAHs	2021	[93]
GC	QqQ-FT ICRMS	Great Lakes fish and Arctic beluga whales	Dechlorane 602 and dechlorination products	2012	[94]
Non-target screening					
GC	QMS	Wastewater from tire plant	41 organic pollutants +4 unknown	1976	[17]
GC	(EI, CI, NCI)QMS	Sediments and fish	New series of bis(trifluoro-methyl)-substituted PCBs	1985	[19]

Table 9.2 (Continued)

GC system	Detector	Matrix	Analytes	Year	References
GC×GC	TOFMS	Tuna muscle	Anthropogenic and naturally produced brominated compounds	2011	[36]
GC×GC	TOFMS	Contaminated soils	PAHs	2018	[95]
GC×GC	TOFMS	Gas phase in ambient air	Potential emerging pollutants	2022	[96]
GC×GC	(EI, ENCI) HR TOFMS	Dust from an electronic facility	Halogenated pollutants	2015	[97]
GC×GC	HR TOFMS	Sewage sludge	99 organic pollutants	2017	[98]
GC× GC	HR TOFMS	Cat hair	Legacy POPs and OPFRs[j)]	2018	[99]
GC	QTOFMS	Household dust	76 toxic organic compounds	2018	[74]
GC	QTOFMS	Harbor porpoise, guillemot, and white-tailed sea eagle	Organic contaminants	2022	[100]
GC	TOFMS/Orbi MS	Honeybees	13/15 PAHs, phthalates, synthetic musks, and veterinary drugs	2018	[101]
GC	OrbiMS	Fresh-food commodities	86 pesticides	2020	[76]
GC	OrbiMS	Silicone personal passive samplers	123 environmentally relevant standards (pesticides, FR, organophosphate esters, and PAHs)	2021	[53]
GC	(EI, PCI, NCI) OrbiMS	Incineration samples	141 PFAS	2023	[102]

(Continued)

Table 9.2 (Continued)

GC system	Detector	Matrix	Analytes	Year	References
GC	(API)cIMS	Indoor dust	New PFAS: chlorofuoro phthalimides and chlorofluoro *n*-alkanes	2022	[103]
GC	(API)cIMS	Indoor dust	Fluorotelomer ethoxylates	2023	[104]
GC	(APCI)IMS-HR MS	Marine bivalves	157 halogenated pollutants	2023	[105]
GC	QqQ-FR ICR HR MS	Contaminated soil	PAHs and halo-PAHs	2014	[106]
GC	(APPI)FR-ICR HR MS	Atmospheric particulate matter	PAHs and alkyl-PAHs	2021	[107]

a) Dechlorane-plus.
b) Polybromodiphenyl ethers.
c) Polybromodiphenyls.
d) Phthalate esters.
e) bis(2-Ethylhexyl)adipate.
f) Disinfection by-products.
g) Liquid cristal monomers.
h) Short-chain chlorinated paraffins.
i) Medium-chain chlorinated paraffins.
j) Organophosphorous flame retardants.

chromatographic overlapping. Although two GC-QMS runs were required per analyzed sample, the strategy provided really satisfactory results, as illustrated for the analysis of wastewater from a tire plant. In this study, the authors managed to identify 41 aromatic contaminants associated with tire production and informed on the presence of four extra unknown chemicals [17]. An essentially similar approach led the same authors to identify up to 77 aromatic and chlorinated POPs, and two not previously reported chloro-fluoro-derivatives in watersheds collected from contaminated dumping sites in or near the Niagara River. The newly identified contaminants, dichloro(trifluoromethyl)benzophenone and dichloro(trifluoromethyl)-α,α-difluorodiphenylmethane, were considered possible byproducts of the industrial production of benzyl chloride and (trifluoromethyl)chlorobenzene, respectively [18]. In a subsequent study, Hites' group also identified a new series of bis(trifluoromethyl)-substituted PCBs in sediments and fishes collected in the same area [19]. Unlike previous studies from the group, where EI was exclusively used for MS analysis, in this latter investigation, the authors also used methane negative and positive CI to obtain as much structural information concerning the analytes as possible. Several isomers belonging to various series of homologues were identified, and one of the most abundant isomers was confirmed through the synthesis and instrumental characterization of the corresponding pure standard. This standard was also used to provide semiquantitative results on the presence of these novel chemicals in the samples. On the basis of the obtained values, the authors concluded that these newly identified pollutants should partition into sediments and bioaccumulate in biota even more effectively than the equivalent PCB congeners.

As previously mentioned, only a limited number of ions can be stored in an IT. However, the possibility of removing all ions other than the parent ions selected for the target compounds from the detector reduces the chemical noise. This allows the ionization time to be extended in order to store as many parent ions as possible, which ultimately results in a significant reduction of the LODs. In an earlier study, Turnes et al. [60] reported a 125 to 800-fold reduction in the LODs calculated for mono- to penta-chlorophenols using GC-ITMS, and demonstrated the potential of the approach for the routine determination of these pollutants at parts-per-trillion levels in drinking water when combined with solid phase extraction (SPE) for analyte preconcentration. With such a low LOD, the application scope of the technique for the target determination of trace organic pollutants was rapidly expanded to the analysis of more complex matrices. Reported studies ranged from the determination of selected chlorobenzenes in wastewater [109], s-triazines and their metabolites in the atmosphere [110], priority PCBs in serum of a breast cancer patient population [111], or toxic OCPs [112] and PCBs [113] in foodstuffs. Alternatively, the possibility of taking advantage of the enhanced sensitivity and selectivity provided by the technique to scale down the sample size to a few milligrams, thus minimizing reagent consumption and significantly increasing sample throughput, was also demonstrated [62, 114]. Despite the analytical performance of the reported GC-ITMS methods, the requirement of pre-selecting the parent ions to be isolated in the trap limited the applicability of this technique for SSA and NTS studies, and makes it essentially used for the accurate determination of a more or less numerous set of preselected microcontaminants [61].

The linear configuration of QqQMS instruments avoided the limitation of former IT and allowed both increasing the number of organic pollutants simultaneously analyzed in target studies and performing SSA. Concerning the former topic, the potential of GC-QqQMS instruments for sensitive and selective determination of organic pollutants in multiclass analyses has been widely illustrated through a number of investigations that demonstrated their suitability for routine monitoring and control analyses. Typical application studies included, for example, the simultaneous and accurate determination of 57 priority substances from the European Water Framework Directive (WFD) in liquid and particulate matter effluents from wastewater treatment plants after minimum sample treatment [115], 57 legacy POPs (including COPs, PCBs and PAHs) in human claustrum and milk [116], 56 PCBs and pesticides and their degradation products in hair and urine [117], 97 legacy (PCBs and PAHs), pesticides and emerging organic pollutants (i.e., fragrances, UV filters, repellents, endocrine disruptors, biocides, and organophosphorus flame retardants) in sediments [64], or 100 legacy POPs (i.e., OCPs, PCBs, and PAHs) in fine dust ($PM_{2.5}$) [118], among many others. A large majority of the reported studies involving GC-QqQMS for environmental applications focused exclusively on GC-amenable contaminants. However, a few of them included an extra derivatization step prior to the instrumental analysis of the purified extracts to improve the signal and sensitivity of some target polar compounds [64].

EI was the most frequently used ionization source in methods involving GC-QqQMS for instrumental determination of organic pollutants, and the one used in all previously mentioned studies. However, the atmospheric pressure chemical ionization (APCI) can be a valuable alternative to EI for some specific determinations. The APCI source is a soft ionization technique that operates at atmospheric pressure. Therefore, it results in abundant molecular ion formation with minimal fragmentation, in contrast to the extensive fragmentation caused by EI. In an exhaustive characterization study involving 85 legacy halogenated POPs, Lee et al. [119] found that both ionization sources showed similar satisfactory linear ranges and accuracies for all tested analytes. Similar LODs were also obtained for mono- to tetra-CBs and mono- to di-BDEs with both sources. However, for highly substituted congeners, APCI provided LODs 2- to 50-fold lower than those determined using EI. A similar trend was observed for OCPs, and although similar LODs were obtained for some of the investigated pesticides, in general, the APCI provided LODs 2 to 10-fold lower than EI. The lowest LODs obtained with APCI represent a particularly advantageous feature when a multiresidue determination of these pollutants should be carried out and the sample size is limited, as it is usually the case for human samples (e.g., in this and other studies, ca. 200 μL of human serum were employed for the analysis [119, 120]).

As evidenced by previously mentioned studies, technical advances in the MS instruments allow the last generation of QqQMS instruments to acquire a large number of transitions within the same run. The optimized MRM methods are now able to analyze hundreds of preselected compounds in a single run [37, 64, 81]. Several approaches and selection criteria have been applied to set the list of suspected analytes considered when developing these methodologies, with the main study goal and the nature of the investigated matrix usually being the major constraints [37, 64, 80, 121]. To our knowledge, GC-QqQMS has not been used in pure NTS studies yet. Nevertheless, the technique has successfully contributed to providing insight and confirming the chemical structures proposed for novel compounds detected using other LR MS-based techniques [36].

The enhanced separation power provided by MDGC and GC×GC opens new application opportunities for target, SSA, and NTS of organic pollutants in complex environmental samples with LR MS instruments, in particular for fast acquisition rate QMS, TOFMS, and QqQMS instruments.

In general, application studies involving MDGC and GC×GC for the analysis of contaminants followed trends similar to those observed for single-column GC-MS methods over time, firstly focusing on target analysis and then evolving to SSA and NTS. However, the enhanced chromatographic resolution provided by these techniques allowed the analysis of more complicated mixtures and facilitated determinations that are hard to perform with single-column GC systems. The use of MDGC-QMS for the evaluation of the selective enantiomeric enrichment of chiral OCPs and PCBs during the uptake and environmental biotransformation of these pollutants [66, 122], or the fast determination of pesticides (up to 24 in an 8 min/run) after minimal sample treatment [67] can be mentioned as representative examples. The development of faster QMS allowed an increase in the number of contaminants determined in a single run and, in particular, contributed to avoiding false positives detected with single-column GC due to interference from matrix components, as demonstrated by del Castillo et al. during the screening and quantitation of 46 pesticides in fruit juices [82].

The expanded peak capacity provided by GC×GC allowed increasing the number of organic pollutants monitored in a single run compared to MDGC-based determinations. Thereby, with the appropriate column combination, GC×GC made possible the simultaneous analysis of various classes of legacy POPs commonly interfering with the determination of each other when using single-column GC [123]. When adding up the identification capabilities derived from its hyphenation with TOFMS, the technique was demonstrated to allow the simultaneous and unambiguous identification and determination of toxic OCPs, PCBs, and PBDEs in, for example, human serum and milk [70]. More importantly, the approach proposed in that study maintained the standards of quality required for these types of determinations according to the official methodology, which required the separate analysis of the several classes of investigated pollutants with GC-(M)HR MS [70]. As a whole, the developed methodology represented a valuable, cost- and time-efficient analytical alternative to conventional procedures involved in these monitoring and epidemiological studies. Similar conclusions were drawn when applying GC×GC-TOFMS to the multi-target and SSA of other legacy pollutants, such as bromine-containing compounds [36],

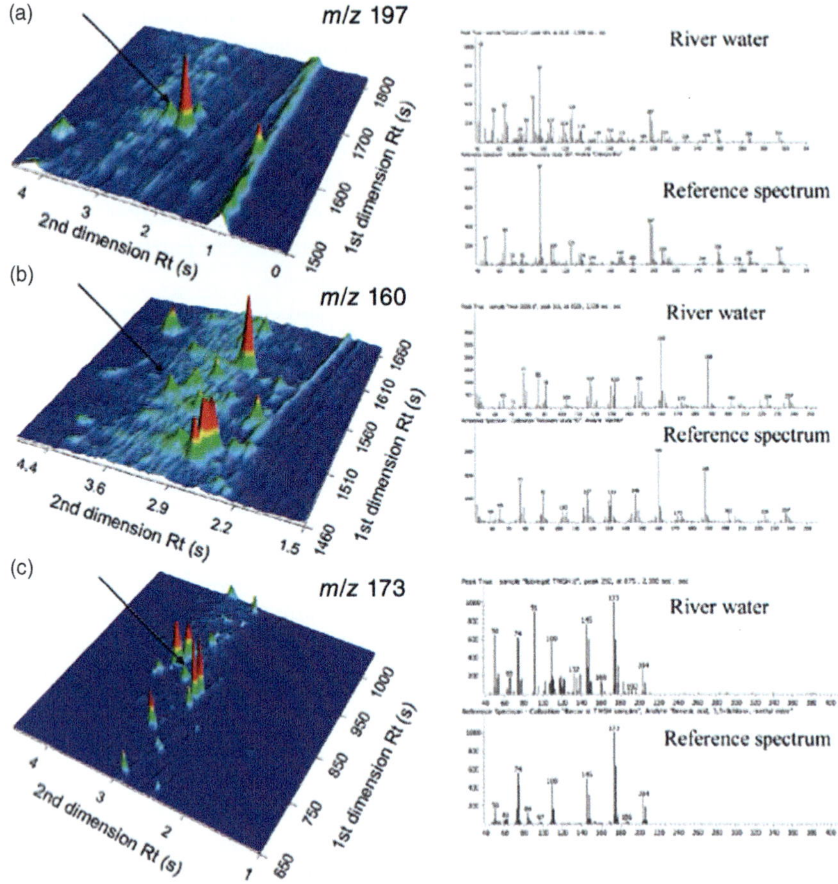

Figure 9.2 Example of the separation achieved with GC×GC-TOFMS (contour plots at left) for (a) clorpyrifos, (b) alachlor, and (c) 3,5-dichlorobenzoic acid in naturally contaminated river water, and comparison between the sample spectrum and the reference (at right). *Source:* Matamoros et al. [71] / with permission of American Chemical Society.

PAHs [85], pesticides and pharmaceuticals [71] or, more recently, volatile PFASs [72] in naturally contaminated environmental samples. A typical example of the chromatographic resolution and accuracy of the library-based detection achieved in these studies is shown in Figure 9.2 for the determination of a variety of organic legacy and emerging pollutants at the part-per-trillion level in river water [71].

The development of faster QMS and QqQMS instruments allowed the hyphenation of these rapid scanning mass spectrometers with GC×GC over a decade ago [68, 69]. However, these determinations frequently involved either reduced mass scanning ranges [83, 87] or complementary instrumental [69] or identification [68] strategies. In any case, these instruments have essentially been used for target or SSA, while TOFMS remains the most widely used GC×GC-MS platform for NTS.

GC×GC-TOFMS has been used for a wide variety of NTS evaluations, covering from the well-known research fields associated with oil, agricultural, and industrial contamination [72, 124], to other emerging areas of concern, such as those related to the lixiviation of toxic compounds from wastes [125] or from virgin (and, increasingly, recycled) polymeric materials [126], among others. For obvious reasons, in this type of analysis, the number of analytes detected and, consequently, the complexity of the contour plots, depend largely on the nature of the investigated matrix, but also on the sample treatment protocol adopted to obtain the extract submitted to instrumental analysis. Thereby, although some earlier studies eventually included a rough fractionation of the analytes of interest before GC×GC-TOFMS analysis [36], in most instances, rather generic sample preparation procedures have been applied [127] in an attempt to preserve as many analytes as possible. The latter approach yields very complex chromatograms in which the manual (i.e., peak-by-peak) evaluation of the detected compounds is simply unfeasible. In this context, the application of well-defined data processing protocols, including appropriate classification tools and script functions for automatic data filtration and analyte identification, becomes mandatory [95, 128–130]. The efficiency of such approaches for fast and reliable data treatment and for the discovery of non-previously described pollutants – including unknown-knowns (e.g., novel polybrominated hexahydroxanthene derivatives in tuna [36]), but also fully unknown contaminants – is already accepted. The identification of several novel perchlorinated dichloro-diphenyl-trichloroethane (DDT-) derivatives in dolphins [131] and deep-sea fishes [127], a novel mixed chloro/bromo-carbazole in sediments [128], or different potential emerging organic pollutants in the gas phase of ambient air [70], can be mentioned as representative examples. It is also worth mentioning that, despite the efforts carried out in these studies to positively identify the detected analytes, many of them remain as annotated or tentatively identified analytes (Levels 2 and 3), or as unidentified features (Levels 4 and 5) [131–133]. In such cases, reporting the mass spectra of the detected analytes or their incorporation in public repositories becomes particularly advisable so that other researchers can compare these findings with their results [134, 135], and the scientific community can have a realistic idea of the potential use and distribution in the environment of the detected chemicals.

When the goal of the study is to identify features that can contribute to differentiating among various groups of samples, an alternative to these exhaustive (and laborious) characterization studies [130] (Figure 9.3) is the use of the Fisher ratio method [136]. This method allows identifying the components that contribute the most to between-class variation. Then, the generated reduced data

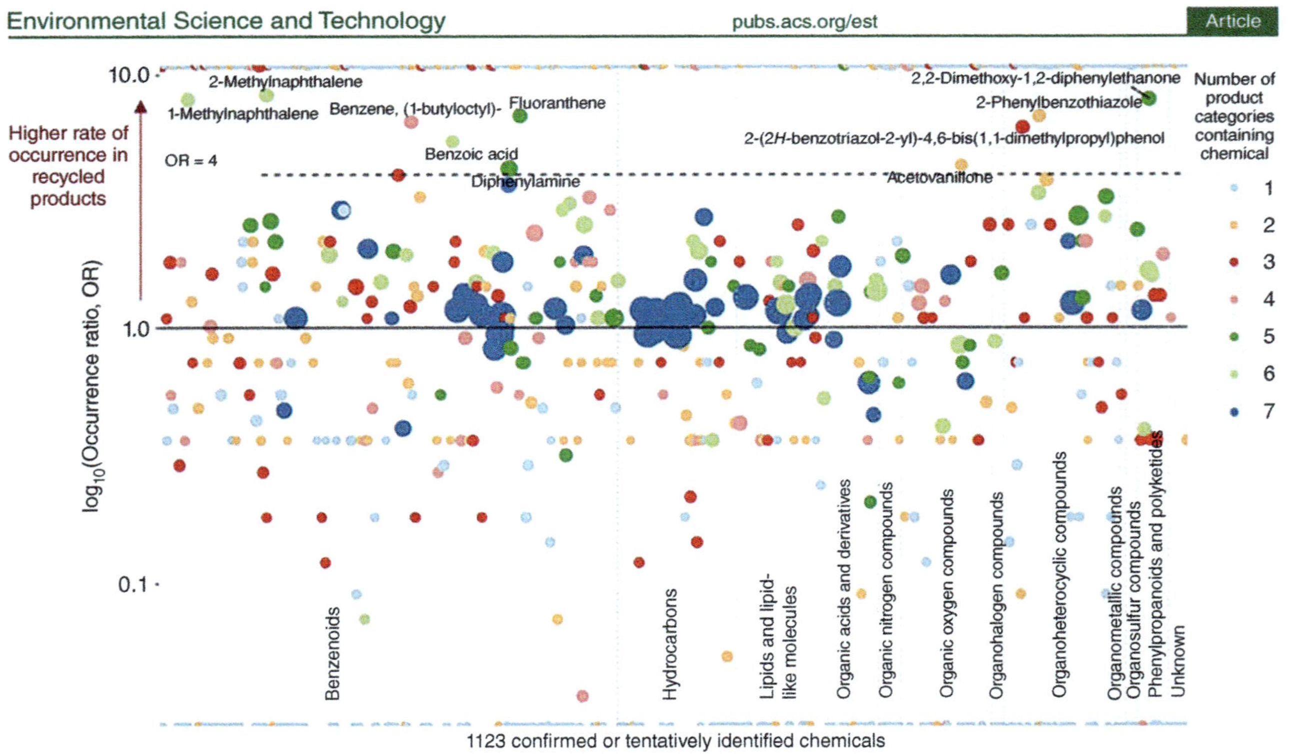

Figure 9.3 Occurrence ratios (OR) for 1123 chemicals detected in virgin and recycled products and materials. OR was defined as the ratio between recycled products with occurrence and virgin products with occurrence. Point sizes indicated the number of individual products with occurrence, while colors referred to the number of product categories with at least one occurrence. *Source:* Lowe et al. [130] / American Chemical Society.

set can be used for multivariate statistics, and the identification efforts can focus on a limited collection of features. The efficiency of the Fisher ratio method as a data analysis technique for NTS has also been illustrated in a number of studies, ranging from the comparison of contaminated soil pre- and post-bioremediation [95], the identification of the contaminants that may be impacting specific organism health [132], to the identification of putative markers for the early diagnosis of lung cancer in human exhaled breath [137].

The recent hyphenation of GC×GC with fast HR TOFMS instruments contributed to both improving the confidence in the analyte assignments and providing new filtering and identification tools due to the accurate mass information provided by the instrument. On the other hand, although in principle these improved features should facilitate data curation, in practice, the large volume of data generated by the technique makes highly advisable the application of well-defined workflows with consistent and clearly defined objectives and pass rate criteria for analyte classification and identification. For this purpose, commercial [98, 138] and home-made [73, 139] software have been applied, both with positive results. These studies have also demonstrated the validity of common low-resolution MS libraries (e.g., NIST) for analyte identification. Nevertheless, the high-resolution mass spectra generated also made possible other approaches, like the Kendrich mass defect plots [97] or the use of *in silico* fragmentation tools [98], such as MetFrag [140], to identify unknown analytes. These tools are actually applicable to ultra-high [106, 141] or high-resolution mass spectrometry data [97, 98, 139]. Thereby, although a good separation is essential to obtain a clean mass spectrum and, in this sense, multidimensional separation techniques represent an advantageous choice, new HR MS-based instruments open new opportunities, not only for target analysis, but also for SSA and NTS studies using single-column GC (Table 9.2).

QTOF and Orbitrap are among the most common high-resolution mass spectrometers used in combination with one-dimensional GC in recent years. Similarly to other HR MS techniques, due to the massive data generated, their effective application requires the implementation of well-defined procedures for spectral deconvolution, library search, and post-data treatment [74, 100]. In a representative example of this fact, Moschet et al. [74] evaluated the presence of organic pollutants in household dust by applying a combination of target, SSA, and NTS approaches. The authors highlighted the complementary nature of GC-QTOFMS and LC-QTOFMS for the comprehensive characterization of this type of complex extract and to ensure a realistic evaluation of the risks to human health associated with this exposure source. The study involved 38 households collected in the areas of Sacramento and Fresno (CA, USA). Regarding GC-QTOFMS determinations, 59 out of the 76 frequently detected chemicals used for target analysis were detected and quantified in the investigated samples. For SSA and NTS analyses, the RIs of all non-targeted features were calculated, and their mass spectra deconvoluted for subsequent comparison with a commercial spectral library. In this part of the data treatment workflow, the authors used the parameters calculated by the MassHunter Unknown Analysis software for the preselection of the features used for further analysis (i.e., NIST library match factor > 60, and a chromatographic peak width between 3 and 15 s). The next data post-treatment step consisted of the careful and manual grouping of features detected in the various investigated samples. Selection was based on the mutual agreement among the assigned candidate names and the mass retention time combinations. The rejection limits for these parameters were set on the basis of the results obtained for target analysis (i.e., mass deviations of ±50 ppm and retention time deviations of ±0.2 min). Then, analytes considered not relevant for the study goal (e.g., long-chain alkanes and their acids, esters, and similar compounds) were excluded from the list of features. In the next step, the RIs calculated for the selected analytes were compared with the NIST library values (experimental or estimated). On the basis of the experience with target compounds, a deviation of ±2% in RI was considered acceptable for the calculated values, while an arbitrary limit of ±10% in RI was applied when the NIST library contained only an estimated RI value. Finally, only compounds with intensities 10 times higher than those of the corresponding peak in the method blank were considered. This approach led to the identification of 78 compounds in multiple samples out of the 3000–5000 features initially detected. Twenty-six of these compounds were discarded due to high presence in the blank or RI deviation above the selected criterion. Of the remaining 48 compounds, 22 were target compounds (PBDE, OPFR, pyrethroids, and phthalates), while 27 were uniquely identified through the proposed NTS workflow. Seventeen of these compounds were positively identified by the analysis of the corresponding pure standard, and 10 remained as tentatively identified with confidence Level 2. Interestingly, 7,9-di-*tert*-butyl-1-oxaspiro-(4,5)deca-6,9-diene-2,8-dione (CAS 82304-66-3), a leachable from plastics, was identified for the first time in dust, demonstrating the growing presence of contaminants associated with this industry in the environment. This point has also been evidenced in other studies. For example, a time-trend guided NTS study by Rebryk et al. [100] on the presence of organic contaminants in marine mammals and birds confirmed the increasing incidence over time of emerging pollutants and replacement chemicals associated with the plastic industry in these species. Some of these chemicals were reported in biota for the first time within the frame of this study, in which between 8 and 49 features remained as unknown compounds, depending on the considered species.

Concerning the various NTS protocols applied by researchers in their investigations, it must be mentioned that authors frequently set variable, but reasonably, rejection values for the several criteria used to identify and filter the detected analytes. There may also be slight differences in the order in which some of the filtration steps are carried out, and home-made libraries are sometimes used [89, 100, 129, 142] in combination with commercial ones to enhance the number of tentatively identified features. In addition, specific pieces of external (sometimes open-source) software can be applied to convert spectral data to other readable formats for subsequent treatment with alternative databases and programs [100, 143]. In any case, and whatever the approach applied, SSA and NTS remain very laborious, time-consuming, and prone to human error studies, while (semi)quantification of the detected analytes remains as an under-development goal [144]. Altogether highlights the already mentioned pressing need for automated workflows allowing (semi-)unsupervised target and screening analysis [145].

The potential of GC-OrbiMS for target and SSA of environmental pollutants was demonstrated by Mol et al. in 2016, immediately after the introduction of the technology [54]. In that pioneer study, the authors set a workflow for target analysis using 54 pesticides of different natures as test compounds. The instrument showed high mass accuracy and consistent ion ratios over a very broad concentration range, and its feasibility for quantitative multi-residue analysis was demonstrated by analyzing different fruits and vegetables. The authors also evaluated the potential of the approach for qualitative screening, i.e., SSA. The satisfactory results obtained made them consider the instrument an attractive alternative to GC-QqQMS for this type of determinations [54]. This statement was later evaluated by other authors who used the same workflow for SSA and NTS of pollutants in honeybees [101]. In this latter study, 15 contaminants, including PAHs, phthalates, synthetic musks, and veterinary drugs, were tentatively identified without any prior hypothesis. The analysis of the samples by GC-(EI)TOFMS resulted in the identification of 13 additional pollutants belonging to the evaluated classes. Although some differences were observed among the contaminants identified with each instrument, the authors concluded that both were equally appropriate for the intended determination. Another illustrative example of the identification power of this technology was reported by Krätschme et al. [91] in 2018. In that study, GC-OrbiMS operated in full-scan acquisition in electron capture negative ion (ECNI) mode, and at a resolution of 60,000 was used to unravel the composition of short-chain chlorinated paraffins (SCCP) and medium-chain chlorinated paraffins (MCCP) in salmon samples. The authors concluded that, under these operational conditions, coelutions with isobaric compounds had only minimal influence on the CPs analysis. This enhanced resolution power, combined with the excellent reproducibility and sensitivity (low pg/μL) of the technique, allowed the identification of small differences in the CP homologue patterns that permitted distinguishing samples from different geographic origins.

Further studies involving GC-OrbiMS contributed to support these initial statements through the target determination of specific pollutants in complex minute samples (e.g., selected plasticizers in air $PM_{2.5}$ particles [78]), the SSA of hundreds of industrial and agricultural contaminants in a single run (as shown, e.g., for sea cucumbers [92]), or through the identification of 86 pesticides in fresh-food commodities [76], 123 environmentally relevant compounds in silicone personal passive samplers [53], and 141 diverse PFAS in incineration samples [102] using carefully designed NTS workflows.

With the growing number of reported experimental CCS databases and the increasing resolving power of commercial IMS techniques, application studies involving GC-IMS-HR MS for the determination of small molecules, such as pollutants, have grown during the last years [43]. The elimination of interfering ions, the reduction of false positives and false negatives, the capability to separate isomers, and the enhanced sensitivity are distinctive features that make the GC-IMS-HR MS platforms particularly suitable for SSA and NTS of pollutants in complex environmental matrices. By now, GC-IMS-HR MS has typically been used in difficult determinations that required additional characterization metrics not available with more conventional GC-HR MS-based instruments, e.g., the identification of more than 100 isomeric and isobaric PAHs in crude oil samples [93]. Another illustrative example of the identification power of GC-(APCI)IMS-HR MS is the NTS of potentially bioaccumulative and bioactive polyhalogenated organic pollutants and natural products in marine bivalves recently reported by Singh et al. [105]. The identification of 157 halogenated features, belonging to 11 classes, was facilitated by constructing a Kendrick mass defect versus CCS plot in which the potentially structurally-related compounds cluster together (Figure 9.4). Emerging PFASs have also been the target of different studies involving a cyclic ion mobility mass spectrometer (cIMS) hyphenated with GC-(APCI). These investigations led to the identification of four new chlorofluoro phthalimides (confirmed at Level 1 by the synthesis of the corresponding pure standard) in indoor dust samples and revealed the presence of chlorofluoro *n*-alkanes as an emerging class of PFASs in this type of matrix [103]. Different fluorotelomer ethoxylates were also identified in indoor dust and industrial wastewater samples using the same instrumentation [104].

The Fourier transform ion cyclotron resonance mass spectrometry (FT ICRMS) offers a more powerful resolution and a higher mass accuracy than the techniques discussed above. These features make it particularly suitable for the simultaneous molecular-level characterization of pollutants, their precursors, and/or transformation products. The potential of FT ICRMS for the comprehensive characterization of natural and anthropogenic DBPs and PFASs [146], or PAHs, including less investigated alkylated- or heteroatomic-isomers [147], has been highlighted in several review papers. However, despite its enormous identification potential, the penetration of this technique in the environmental field is much more limited than that of previously discussed GC-MS-based techniques. Its higher complexity and price could be possible reasons behind this observation.

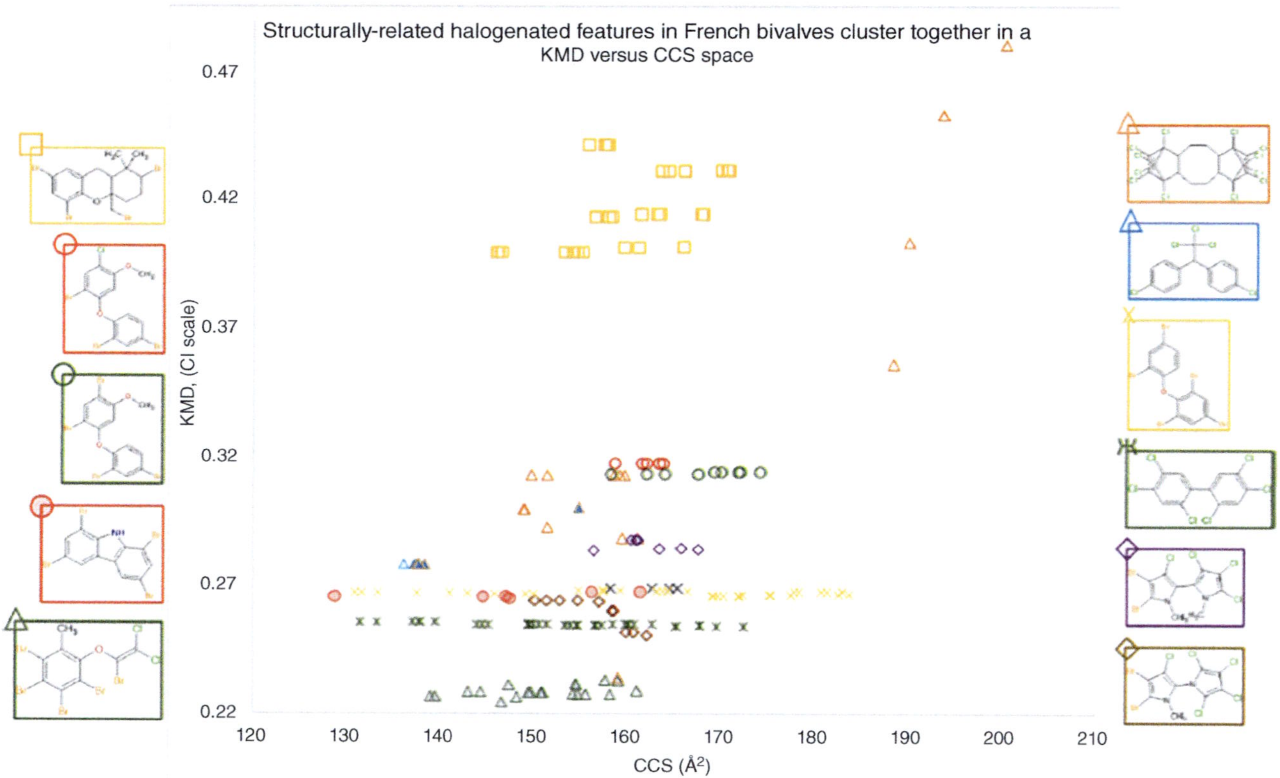

Figure 9.4 Kendrick mass defect versus CCS plot of the natural and exogenous organohalogenated compounds belonging to 11 classes detected in marine bivalves using GC-(APCI)IMS-HR MS and a NTS strategy. *Source:* Singh et al. [105] / with permission of Elsevier / CC BY 4.0.

The feasibility of GC-(APCI)FT ICRMS for target and SSA has been illustrated through the analysis of Cl-/Br-FRs in food commodities, including meat, whole milk, eggs, and cereals [79] or, for example, the SSA of degradation products of FRs in marine organisms [94]. However, the technique is particularly suitable for NTS analysis [141]. In an early application study in this field, Fernando et al. [106] used QqQ-FT ICRMS for the retrospective NTS of halogen-containing pollutants in soils affected by a fire in a plastic recycling facility that involved around 400 tons of polyvinyl chloride (PVC). The authors proved that halogenated-PAHs were among the main released halo-pollutants. The FT ICRMS experiments led to the identification of approximately 150 molecular formulas of halogen-containing organic pollutants. A large number of these features corresponded to Cl- and mixed Cl-/Br-PAHs. Among them, several highly substituted (e.g., $C_{14}HCl_9$ and $C_{16}HCl_9$) and high-molecular-weight (e.g., $C_{28}H_{12}C_{14}$) Cl-PAHs were identified and reported in environmental samples for the first time. However, the separation of isomers could only be achieved by hyphenation with chromatography. The structured chromatograms provided by GC×GC-HR TOFMS, combined with the mass spectrometric data obtained from the FT ICRMS experiments, enable the resolution of individual halo-PAHs. GC×GC-TOFMS provided nominal mass resolution, but was capable of acquiring data at a higher speed, thus providing better peak definition. Therefore, this technique was used to positively confirm and quantify 25 halo-PAH isomers with available pure standards, and to obtain semiquantitative data for several groups of structurally related isomers. A similar approach, but involving either (ESI+)FT ICRMS or (APPI)FT ICRMS combined with GC×GC, was later used for the characterization of polyaromatic compounds (PACs) in sludge [148] and atmospheric particulate matter [107]. In this latter study, high concentrations of toxic dibenzo[a,e]pyrene and 1-methylpyrene were found, pointing out the need to expand the traditional lists of legacy PAHs to improve pollution control and risk assessment accuracy.

9.4 The Exposome Evaluation

Exposome studies aim to unravel the impact of exposure to pollutants on human health to prevent noncommunicable chronic diseases, such as obesity, diabetes, nonalcoholic fatty liver disease, atherosclerosis, and cardiovascular diseases [42]. Different approaches are possible to investigate the connection between "exposure" (i.e., contamination) and "response" (i.e., illness). Bio-based studies focus on evaluating the response of model systems (e.g., model animals, organ tissues, or *in vitro* bioassays) exposed to specific chemical(s) [149, 150]. Meanwhile, technology-based studies focus on the development of new sampling and measurement strategies and techniques [151, 152], which are subsequently applied to the NTS of chemicals in the studied sample(s). Irrespective of the selected approach, water, soil, food, consumer products, or air (and atmospheric particles) are among the most evaluated exposure matrices, whereas human hair, skin, urine, blood, and other bodily fluids are considered relevant response matrices [42, 152]. Previous sections discussed the potential and relative merits of state-of-the-art (GC-)HR MS-based instrumentation for the characterization and quantitation of contaminants in a wide variety of environmental samples (Table 9.2). That discussion also highlighted the relevance of SSA and NTS analyses regarding the identification of unexpected and unknown pollutants and toxicants in a variety of matrices, thus allowing a more realistic evaluation of the so-called "external exposome" [153] or, put differently, the exposure assessment. When applied to human samples, SSA and NTS analyses provide equivalent information regarding the presence of contaminants in the investigated human matrices. However, these determinations actually integrate information concerning the long-term exposure of individuals, their occupational exposure, diet, habits, and other general aspects, such as socioeconomic status, education, climate, and so on, i.e., they inform about the "internal exposome" of the sampled individual.

Environmental analytical researchers have provided accurate data concerning the presence of contaminants in food, environmental, and human samples before the term exposome was coined. These investigations have evidenced that passive sampling devices (PSDs), widely used for the time-integrated and *in situ* preconcentration of chemicals from environmental compartments ([135] and references therein), also have an undebatable value for the assessment of personal exposure to pollutants. Among the several possible PSDs used for human passive sampling, silicone samplers, commonly in a wristband format, are one of the most convenient strategies for population exposure evaluation [154, 155]. When combined with GC-MS-based NTS, the approach can provide a comprehensive chemical assessment for exposure studies, as demonstrated by Reiter et al. [156] or Travis et al. [53]. In this latter study, silicone wristband samplers were combined with GC-(EI)OrbiMS for the non-targeted evaluation (actually, in part SSA) of human exposition to 123 relevant organic contaminants, including pesticides, FRs, organophosphate esters, and PAHs. The optimized NTS workflow allowed the positive identification of 70 analytes and reduced the number of false positives by over 50%, but it also evidenced the relevance of applying a proper sample clean-up in this type of analytical approach [135]. An average of 7 compounds were positively identified in worn wristbands (Level 1), while 19 were labelled as Level 2 analytes. SSA allowed the identification of 36 extra compounds. As a whole, these results demonstrated the potential of the optimized procedure for characterizing and assessing the chemical exposure of the general population.

Regarding human samples, previous investigations have pointed out that, apart from the invasive nature of some of these determinations [157–160], one should be aware that the obtained information can be biased by the nature of the investigated matrix [117, 161, 162], the differentiated bioaccessibility of the various contaminant classes of interest [163, 164], and their physicochemical properties (some are suitable for GC analysis, some are better detected by LC). Despite these limitations, and always referring to GC compatible chemicals, NTS of organic pollutants present in human samples have evidenced that state-of-the-art techniques based on GC-HR MS are suitable for performing this type of determination in an accurate and robust manner, despite the size-limited samples typically used in these analyses. In some instances, these studies involved samples collected from the general population. Whenever possible, easily accessible health-related data, such as body weight or blood pressure, were also collected as response information [152]. In other cases, the investigation focused on groups of population affected by a specific health problem, such as type 2 diabetes [165, 166], hypertension [167], obesity [168, 169], or cardiovascular illnesses [170, 171], among other chronic diseases [172]. In a representative study within this field, Wada et al. [170] evaluated the link between exposure to selected industrial and agricultural contaminants and the cardiometabolic health status of individuals from the Great Region, in central Europe (i.e., the NESCAV study [173]). Thirteen POPs (i.e., 3 PCBs, 1 PBDE, and 9 OCPs) and 20 nonpersistent pesticides were analyzed in hair samples (n = 33) from a subset of this population (941 participants), whose social status and factors, lifestyle, and environmental exposures were known. The authors used a multi-trait Bayesian approach for variable selection to model the complexity of the exposure profiles and cardiometabolic health. The model pointed out strong associations between hexachlorobenzene and trifluralin exposure and adverse cardiometabolic status, and included traits of obesity, dyslipidemia, and hypertension.

Similarly to this exposome study, most of the exposure assessment evaluations involving human samples reported so far have focused on selected well-known, and frequently regulated or phase-out pollutants, as target analytes. The toxicity, persistence, and bioaccumulative nature of these chemicals, combined with the complexity and, in particular, high cost of these investigations, can be possible explanations for this trend. However, it is our opinion that the increasing evidence of human internalization of less investigated contaminants and additives, for example, those related to the plastic industry, probably makes it advisable to expand such evaluations to these and other nonpersistent, but widespread, pollutants [163, 174].

Up to now, only a limited number of exposure assessment studies, including human samples considered the simultaneous monitoring of the metabolic status of the evaluated individuals [160, 175]. Maybe, the still emerging nature of this type of complementary investigations combining knowledge from two disciplines that, until recently, were separate, i.e., environmental analysis and metabolomics, and the need for different sophisticated analytical instruments not always simultaneously available in the research groups, could explain this observation. Yet, the potential of the approach has been demonstrated by various recent investigations. In a pilot study, Valvi et al. [160] measured 18 POPs recognized as endocrine and metabolic disruptors in visceral and subcutaneous abdominal adipose tissue, and liver samples from 11 adolescent patients subjected to bariatric surgery using GC-QOrbiMS. Paired plasma samples from the same patients were subjected to untargeted metabolic profiling using GC-QOrbiMS. The results evidenced that, except for PBDE-154 that exhibited similar average levels across all analyzed tissues, the rest of the investigated POPs showed a preferential accumulation in adipose tissues. High correlations between tissue-specific POPs and metabolic alterations in pathways of amino acid metabolism, lipid and fatty acid metabolism, and carbohydrate metabolism were observed. As an alternative to this separate determination of pollutants and metabolites, González-Dominguez et al. [175] proposed the use of an optimized and validated UHPLC-QqQMS equipped with an Ion Drive Turbo V ion source for the analysis of different fractions of contaminants obtained from urine and plasma samples, as well as for a high-throughput metabolomic fingerprint. The use of the same sample and instrumental protocol for the determination of both pollutants and metabolites should increase the repeatability and robustness of these determinations, already affected by the anticipated interindividual variability [175].

In these integrated studies, either GC- or LC-HR MS-based techniques were used for the evaluation of the contaminants, while the metabolites were always analyzed by LC-based methods. The recent development of several GC-MS-based procedures and workflows for target and untargeted metabolomics analysis [176–178] opens a door to these types of integrated studies to also be developed by laboratories in which only GC-based instrumentation is available.

The relevance of the conclusions of previously discussed studies is undebatable. However, it should be noted that most of them focus on the evaluation of a specific class (or classes) of compounds with similar (or closely related) physicochemical properties. That is, contaminants that can be extracted together from the investigated matrix and that show essentially similar (i.e., appropriate) chromatographic behaviors and responses during MS ionization and detection. This book chapter

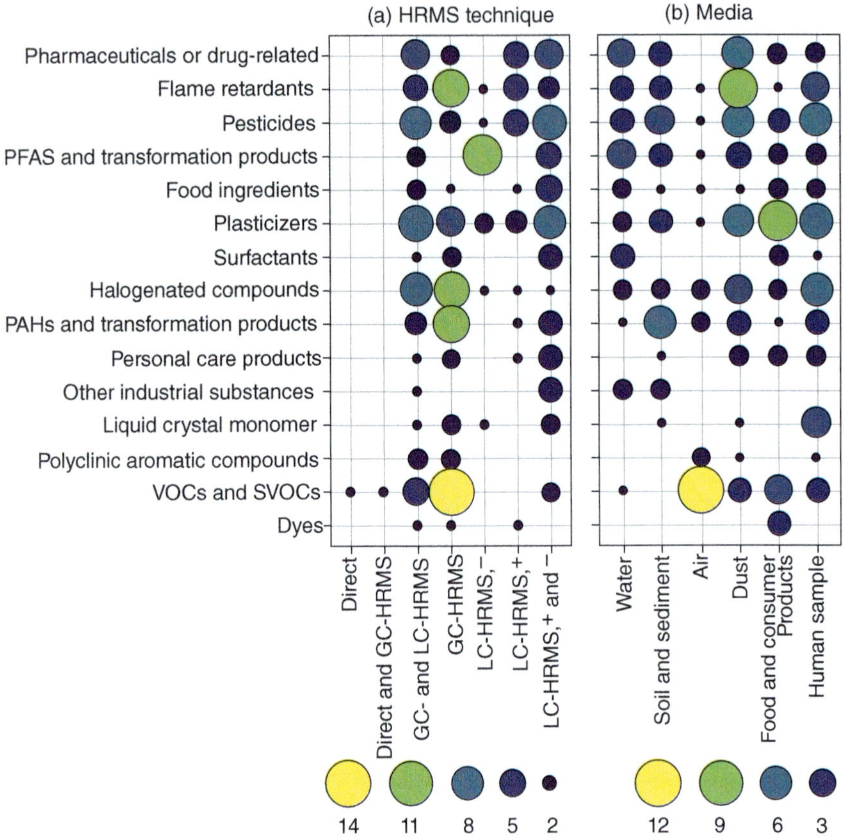

Figure 9.5 Chemical space detected in NTS using HR MS-based techniques and environmental samples. The dot size and color represent the reported frequency of the several chemical classes. *Source:* Manz et al. [179] / Springer Nature / CC BY 4.0.

focuses on GC and, therefore, studies dealing with GC-amenable (i.e., volatile and semi-volatile) contaminants have been reviewed. However, the exposure spectrum also includes polar pollutants that can hardly be recovered with the solvents used to extract the nonpolar or medium-polar chemicals discussed above. These polar contaminants and other nonvolatile chemicals can be better determined by LC-based techniques. Figure 9.5 graphically summarizes the chemical space detected in NTS with HR MS-based techniques and in samples of different natures. The figure evidences the complementary nature of these techniques and makes clear that the complete coverage of the external exposome can be better afforded through parallel analyses of fractions obtained from the same sample on both GC- and LC-HR MS-based techniques [179].

Finally, although not discussed here, it should be noted the relevance of biologically directed non-targeted identification, also named as effects-directed analysis (EDA). The approach consists of the combination of bioassay-guided identification of the fraction of pollutants present in a sample that promotes a particular toxic effect in order to facilitate the subsequent identification of the contaminant(s) or transformation product(s) responsible for the detected effect. The approach is not new, but until recently, its application has been almost restricted to abiotic matrices [180, 181], probably due to the difficulty of the preparation of biotic matrices before biochemical analyses [182]. However, EDA integrates aspects such as bioavailability, bioaccumulation, and possible metabolization, and can assist in revealing mixture effects and, when integrated in SSA and NTS workflows, can help in feature prioritization, thus contributing to high-throughput routine identification of unknown-knowns and unknown pollutants.

9.5 Conclusions

Although the concepts of SSA and NTS in the context of environmental analysis are not really new and these approaches led to the discovery of some new pollutants already in the late 1970s, it has not been until the last decade that this topic has been consolidated as a recognizable and established research area. A possible reason for this observation is that the advances in this field have been strongly dependent on the improvements achieved in the HR MS technology, which explains that the current level of maturity is higher for LC-HR MS-based approaches than for the corresponding GC-HR MS counterpart. In any case, application examples discussed in this chapter demonstrated that, at present, a variety of single-column and multidimensional GC-MS-based procedures can be used for these determinations, which in many instances are combined with the regular target analysis of known toxic pollutants. The latter analyses are essential for environmental and human exposure assessment. Meanwhile, SSA and NTS are relevant for the discovery of not previously described or fully known pollutants, and to obtain a more realistic, i.e., integrated estimation of the global burden of pollutants affecting the ecosystems, including humans.

Research carried out in the field of target analysis of pollutants and the analytical knowledge behind these determinations are essential to determine the best sampling, sample preparation, and instrumental analysis strategy to be applied at the individual level. However, the exposome concept has granted these exposure assessment evaluations a more comprehensive and multidisciplinary dimension, closer to that of other omic sciences, and is making the interaction of analytical and environmental researchers with other disciplines and omics sciences mandatory. This change from the individual to the population level is essential to establish connections between exposures and health outcomes. At the same time, it conveys a number of analytical challenges, thus stimulating new developments and technological advances to facilitate biomonitoring at the population level under an NTS perspective.

References

1 ICCA (2024). Debunking the myths: are there more than 100,000 chemicals in commerce? (accessed 8 February 2024).

2 Wang, Z., Walker, G.W., Muir, D.C.G., and Nagatani-Yoshida, K. (2020). Toward a global understanding of chemical pollution: a first comprehensive analysis of regulatory industrial chemical inventories. *Environ. Sci. Technol.* 54: 2575–2584.

3 Persson, L., Carney-Almroth, B.M., Collins, C.D. et al. (2022). Outside the safe operating space of the planetary boundary for novel entities. *Environ. Sci. Technol.* 56: 1510–1521.

4 UN (2024). Stockholm convention on persistent organic pollutants (accessed 8 February 2024).

5 Diamond, J.M., Latimer, H.A., Munkittrick, K.R. et al. (2011). Prioritizing contaminants of emerging concern for ecological screening assessments. *Environ. Toxicol. Chem.* 30: 2385–2394.

6 Meijer, J., Lamoree, M., Hamers, T. et al. (2020). S71 | CECSCREEN | HBM4EU CECscreen: screening list for chemicals of emerging concern plus metadata and predicted phase 1 metabolites (accessed 10 February 2024).

7 Kiefer, K., Müller, A., Singer, H., and Hollender, J. (2019). New relevant pesticide transformation products in groundwater detected using target and suspect screening for agricultural and urban micropollutants with LC-HRMS. *Water Res.* 165: 114972.

8 Tröger, R., Ren, H., Yin, D. et al. (2021). What's in the water? - target and suspect screening of contaminants of emerging concern in raw water and drinking water from Europe and Asia. *Water Res.* 198: 117099.

9 Muir, D.C.G., Getzinger, G.J., McBride, M., and Ferguson, P.L. (2023). How many chemicals in commerce have been analyzed in environmental media? A 50 year bibliometric analysis. *Environ. Sci. Technol.* 57: 9119–9129.

10 Grandjean, P., Eriksen, M.L., Ellegaard, O., and Wallin, J.A. (2011). The Matthew effect in environmental science publication: a bibliometric analysis of chemical substances in journal articles. *Environ. Health* 10: 96.

11 Hites, R.A. (2022). Mass spectrometric identification of pollutants in the environment: a personal and bibliometric perspective. *J. Am. Soc. Mass Spectrom.* 33: 620–626.

12 de Boer, J. (1999). Capillary gas chromatography for the determination of halogenated micro-contaminants. *J. Chromatogr. A* 843: 179–198.

13 Beyer, A. and Biziuk, M. (2008). Applications of sample preparation techniques in the analysis of pesticides and PCBs in food. *Food Chem.* 108: 669–680.

14 Jia, Q., Liao, G.-Q., Chen, L. et al. (2024). Pesticide residues in animal-derived food: current state and perspectives. *Food Chem.* 438: 137974.

15 Vetter, W., Gall, V., and Skírnisson, K. (2015). Polyhalogenated compounds (PCBs, chlordanes, HCB and BFRs) in four polar bears (*Ursus maritimus*) that swam malnourished from East Greenland to Iceland. *Sci. Total Environ.* 533: 290–296.

16 Weichbrodt, M., Vetter, W., Scholza, E. et al. (1999). Determination of organochlorine levels in Antarctic skua and penguin eggs by application of combined focused open-vessel microwave-assisted extraction, gel-permeation chromatography, adsorption chromatography, and GC/ECD. *Int. J. Environ. Anal. Chem.* 73: 309–328.

17 Jungclaus, G.A., Games, L.M., and Hites, R.A. (1976). Identification of trace organic compounds in tire manufacturing plant wastewaters. *Anal. Chem.* 48: 1894–1896.

18 Elder, V.A., Procter, B.L., and Hites, R.A. (1981). Organic compounds found near dump sites in Niagara Falls, New York. *Environ. Sci. Technol.* 15: 1237–1243.

19 Jaffe, R. and Hites, R.A. (1985). Identification of new, fluorinated biphenyls in the Niagara River-Lake Ontario area. *Environ. Sci. Technol.* 19: 736–740.

20 Gu, Y., Peach, J.T., and Warth, B. (2023). Sample preparation strategies for mass spectrometry analysis in human exposome research: current status and future perspectives. *TRAC Trends Anal. Chem.* 166: 117151.

21 Black, G., Lowe, C., Anumol, T. et al. (2023). Exploring chemical space in non-targeted analysis: a proposed ChemSpace tool. *Anal. Bioanal. Chem.* 415: 35–44.

22 Koelmel, J.P., Xie, H., Price, E.J. et al. (2022). An actionable annotation scoring framework for gas chromatography-high-resolution mass spectrometry. *Exposome* 2: osac007.

23 Kannan, N., Petrick, G., Schultz-Bull, D.E., and Duinker, J.C. (1993). Chromatographic techniques in accurate analysis of chlorobiphenyls. *J. Chromatogr. A* 642: 425–434.

24 Krock, K.A. and Wilkins, C.L. (1996). Qualitative analysis of contaminated environmental extracts by multidimensional gas chromatography with infrared and mass spectral detection (MDGC-IR-MS). *J. Chromatogr. A* 726: 167–178.

25 Ramos, L. (2009). *Comprehensive Two-Dimensional Gas Chromatography*. Elsevier.

26 Phillips, J.B. and Beens, J. (1999). Comprehensive two-dimensional gas chromatography: a hyphenated method with strong coupling between the two dimensions. *J. Chromatogr. A* 856: 331–347.

27 Gaines, R.B., Frysinger, G.S., Hendrick-Smith, M.S., and Stuart, J.D. (1999). Oil spill source identification by comprehensive two-dimensional gas chromatography. *Environ. Sci. Technol.* 33: 2106–2112.

28 Frysinger, G.S. and Gaines, R.B. (2002). Forensic analysis of ignitable liquids in fire debris by comprehensive two-dimensional gas chromatography. *J. Forensic Sci.* 47: 471–482.

29 Frysinger, G.S., Gaines, R.B., and Reddy, C.M. (2002). GC × GC—A new analytical tool for environmental forensics. *Environ. Forensics* 3: 27–34.

30 Pastor, K., Ilić, M., Vujić, D. et al. (2023). Gas chromatography and mass spectrometry: the technique. In: *Emerging Food Authentication Methodologies Using GC/MS* (ed. K. Pastor), 3–31. Springer.

31 Moeder, M. (2014). Gas chromatography-mass spectrometry. In: *Practical Gas Chromatography: A Comprehensive Reference* (ed. K. Dettmer-Wilde and W. Engewald), 303–350. Berlin: Springer.

32 Zaikin, V. and Halket, J.M. (2009). *A Handbook of Derivatives for Mass Spectrometry*. West Sussex: IM Publications.

33 Devers, J., Pattison, D.I., Hansen, A.B., and Christensen, J.H. (2025). Comprehensive two-dimensional gas chromatography as a tool for targeted and non-targeted analysis of contaminants of emerging concern in wastewater. *Talanta* 282: 127032.

34 Roscales, J.L., Vicente, A., Ramos, L., and Jiménez, B. (2017). Miniaturised sample preparation method for the multiresidual determination of regulated organohalogenated pollutants and related compounds in wild bird eggs. *Anal. Bioanal. Chem.* 409: 4905–4913.

35 Ramos, L., Vreuls, J.J., and Brinkman, U.A.T. (2000). Miniaturised pressurised liquid extraction of polycyclic aromatic hydrocarbons from soil and sediment with subsequent large volume injection-gas chromatography. *J. Chromatogr. A* 891: 275–286.

36 Pena-Abaurrea, M., Covaci, A., and Ramos, L. (2011). Comprehensive two-dimensional gas-chromatography time-of-flight mass spectrometry for the identification of organobrominated compounds in bluefin tuna. *J. Chromatogr. A* 1218: 6995–7002.

37 Li, X.H., Li, C.M., Li, S.M. et al. (2023). High-throughput screening and quantitative analysis of hidden ingredients in plant protection products by GC-MS/MS. *Ecotoxicol. Environ. Safety* 266: 115556.

38 Glish, G.L. and Burinsky, D.J. (2008). Hybrid mass spectrometers for tandem mass spectrometry. *J. Am. Soc. Mass Spectrom.* 19: 161–172.

39 Kind, T. and Fiehn, O. (2010). Advances in structure elucidation of small molecules using mass spectrometry. *Bioanal. Rev.* 2: 23–60.

40 Makarov, A., Denisov, E., and Lange, O. (2009). Performance evaluation of a high-field Orbitrap mass analyzer. *J. Am. Soc. Mass Spectrom.* 20: 1391–1396.

41 Metz, T.O., Baker, E.S., Schymanski, E.L. et al. (2017). Integrating ion mobility spectrometry into mass spectrometry-based exposome measurements: what can it add and how far can it go? *Bioanalysis* 9: 91–98.

42 Hyötyläinen, T. (2021). Analytical challenges in human exposome analysis with focus on environmental analysis combined with metabolomics. *J. Sep. Sci.* 44: 1769–1787.

43 Song, X.C., Canellas, E., Dreolin, N. et al. (2023). Application of ion mobility spectrometry and the derived collision cross section in the analysis of environmental organic micropollutants. *Environ. Sci. Technol.* 57: 21485–21502.

44 Hernández-Mesa, M. and Moreno-González, D. (2022). Current role of mass spectrometry in the determination of pesticide residues in food. *Separations* 9: 148.

45 Hernández-Mesa, M., Ropartz, D., García-Campaña, A.M. et al. (2019). Ion mobility spectrometry in food analysis: principles, current applications and future trends. *Molecules* 24: 2706.

46 Izquierdo-Sandoval, D., Fabregat-Safont, D., Lacalle-Bergeron, L. et al. (2022). Benefits of ion mobility separation in GC-APCI-HRMS screening: from the construction of a CCS library to the application to real-world samples. *Anal. Chem.* 95: 9040–9047.

47 Perrault-Uptmor, K.A. and Harynuk, J. (2024). Moving beyond minimum specs to set up your chromatography data workstation. *LCGC Int.* 20: 25–27.

48 Chao, A., Minucci, J.M., Ferland, T.M. et al. (2025). Prioritizing chemical candidates from non-targeted analysis using metadata, spectral similarity, and hazard scoring within INTERPRET NTA. *Anal. Chem.* 97: 15904–15912.

49 Li, P., Zhu, B., Fu, J. et al. (2025). Nontarget screening analysis combined with computational toxicology: a promising solution for identification and risk assessment of environmental pollutants in the big data era. *Environ. Sci. Technol.* 59: 14842–14852.

50 Sumner, L.W., Amberg, A., Barrett, D. et al. (2007). Proposed minimum reporting standards for chemical analysis. *Metabolomics* 3: 211–221.

51 Jeon, J., Kurth, D., and Hollender, J. (2013). Biotransformation pathways of biocides and pharmaceuticals in freshwater crustaceans based on structure elucidation of metabolites using high resolution mass spectrometry. *Chem. Res. Toxicol.* 26: 313–324.

52 Schymanski, E.L., Jeon, J., Gulde, R. et al. (2014). Identifying small molecules via high resolution mass spectrometry: communicating confidence. *Environ. Sci. Technol.* 48: 2097–2098.

53 Travis, S.C., Kordas, K., and Aga, D.S. (2021). Optimized workflow for unknown screening using gas chromatography high-resolution mass spectrometry expands identification of contaminants in silicone personal passive samplers. *Rapid Comm. Mass Spec.* 35: e9048.

54 Mol, H.G.J., Tienstra, M., and Zomer, P. (2016). Evaluation of gas chromatography electron ionization–full scan high resolution Orbitrap mass spectrometry for pesticide residue analysis. *Anal. Chim. Acta* 935: 161–172.

55 Charbonnet, J.A., McDonough, C.A., Xiao, F. et al. (2022). Communicating confidence of per- and polyfluoroalkyl substance identification via high-resolution mass spectrometry. *Environ. Sci. Technol. Letters* 9: 473–481.

56 Celma, A., Sancho, J.V., Schymanski, E.L. et al. (2020). Improving target and suspect screening high-resolution mass spectrometry workflows in environmental analysis by ion mobility separation. *Environ. Sci. Technol.* 54: 15120–15131.

57 Devers, J., Pattison, D.I., and Christensen, J.H. (2022). Second dimension retention indices in "normal" orthogonality comprehensive two-dimensional gas chromatography using single standard injection generated isovolatility curves. *J. Chromatogr. A* 1683: 463548.

58 Ramos, L., Hernández, L.M., and González, M.J. (1997). Variation of PCB congener levels during lactation period and relationship to their molecular structure. *Arch. Environ. Contam. Toxicol.* 33: 97–103.

59 Martínez-Pérez-Cejuela, H., Williams, M.L., McLeod, C., and Gionfriddo, E. (2025). Effective preconcentration of volatile per- and polyfluoroalkyl substances from gas and aqueous phase via solid phase microextraction. *Anal. Chim. Acta* 1345: 343746.

60 Turnes, I., Rodriguez, I., Garcia, C.M., and Cela, R. (1996). Determination of chlorophenols in drinking water with high resolution gas chromatography-tandem mass spectrometry. *J. Chromatogr. A* 743: 283–292.

61 van de Merwe, J.P., Hodge, M., Whittier, J.M., and Lee, S.Y. (2009). Analysing persistent organic pollutants in eggs, blood and tissue of the green sea turtle (*Chelonia mydas*) using gas chromatography with tandem mass spectrometry (GC-MS/MS). *Anal. Bioanal. Chem.* 393: 1719–1731.

62 Pena-Abaurrea, M., García de la Torre, V.S., and Ramos, L. (2013). Ultrasound-assisted extraction followed by disposable pipette purification for the determination of polychlorinated biphenyls in small-size biological tissue samples. *J. Chromatogr. A* 1317: 223–229.

63 Pitarch, E., Medina, C., Portolés, T. et al. (2007). Determination of priority organic micro-pollutants in water by gas chromatography coupled to triple quadrupole mass spectrometry. *Anal. Chim. Acta* 583: 246–258.

64 Pintado-Herrera, M.G., González-Mazo, E., and Lara-Martín, P.A. (2016). In-cell clean-up pressurized liquid extraction and gas chromatography-tandem mass spectrometry determination of hydrophobic persistent and emerging organic pollutants in coastal sediments. *J. Chromatogr. A* 1429: 107–118.

65 Gasull, M., Pumarega, J., Kiviranta, H. et al. (2019). Methodological issues in a prospective study on plasma concentrations of persistent organic pollutants and pancreatic cancer risk within the EPIC cohort. *Environ. Research* 169: 417–433.

66 Wong, C.S., Hoekstra, P.F., Karlsson, H. et al. (2002). Enantiomer fractions of chiral organochlorine pesticides and polychlorinated biphenyls in standard and certified reference materials. *Chemosphere* 49: 1339–1347.

67 Cunha, S.C., Fernandes, J.O., and Oliveira, M.B.P.P. (2009). Fast analysis of multiple pesticide residues in apple juice using dispersive liquid-liquid microextraction and multidimensional gas chromatography-mass spectrometry. *J. Chromatogr. A* 1216: 8835–8844.

68 Mondello, L., Casilli, A., Tranchida, P.Q. et al. (2007). Comprehensive gas chromatography coupled to mass spectrometry for the separation of pesticides in a very complex matrix. *Anal. Bioanal. Chem.* 389: 1755–1763.

69 Poliak, M., Fialkov, A.B., and Amirav, A. (2008). Pulsed flow modulation twodimensional comprehensive gas chromatography-tandem mass spectrometry with supersonic molecular beams. *J. Chromatogr. A* 1210: 108–114.

70 Focant, J.F., Sjödin, A., Turner, W.R., and Patterson, D.G. Jr. (2004). Measurement of selected polybrominated diphenyl ethers, polybrominated and polychlorinated biphenyls, and organochlorine pesticides in human serum and milk using comprehensive two-dimensional gas chromatography isotope dilution time-of-flight mass spectrometry. *Anal. Chem.* 76: 6313–6320.

71 Matamoros, V., Jover, E., and Bayona, J.M. (2010). Part-per-trillion determination of pharmaceuticals, pesticides, and related organic contaminants in river water by solid-phase extraction followed by comprehensive two-dimensional gas chromatography time-of-flight mass spectrometry. *Anal. Chem.* 82: 699–706.

72 Corviseri, M.C., Polidoro, A., De Poli, M. et al. (2025). Targeted determination of volatile fluoroalkyl pollutants and non-targeted screening for environmental monitoring. *Talanta* 292: 127944.

73 Ieda, T., Hashimoto, S., Isobe, T. et al. (2019). Evaluation of a data-processing method for target and non-target screening using comprehensive two-dimensional gas chromatography coupled with high-resolution time-of-flight mass spectrometry for environmental samples. *Talanta* 194: 461–468.

74 Moschet, C., Anumol, T., Lew, B.M. et al. (2018). Household dust as a repository of chemical accumulation: new insights from a comprehensive high-resolution mass spectrometric study. *Environ. Sci. Technol.* 52: 2878–2887.

75 Yang, L.L., Wu, J.J., Zheng, M.H. et al. (2022). Non-target screening of organic pollutants and target analysis of halogenated polycyclic aromatic hydrocarbons in the atmosphere around metallurgical plants by high-resolution GC/Q-TOF-MS. *Environ. Sci. Eur.* 32: 96.

76 Vargas-Pérez, M., Domínguez, I., Egea-González, F.J., and Garrido-Frenich, A. (2020). Application of full scan gas chromatography high resolution mass spectrometry data to quantify targeted-pesticide residues and to screen for additional substances of concern in fresh-food commodities. *J. Chromatogr. A* 1622: 46111.

77 Ayala-Cabrera, J.F., Contreras-Llin, A., Moyano, E., and Santos, F.J. (2020). A novel methodology for the determination of neutral perfluoroalkyl and polyfluoroalkyl substances in water by gas chromatography-atmospheric pressure photoionisation-high resolution mass spectrometry. *Anal. Chim. Acta* 1100: 97–106.

78 Argamino, C.R.A., Sebben, B.G., da Costa, G. et al. (2024). Development and validation of a GC Orbitrap-MS method for the analysis of phthalate esters (PAE) and bis(2-ethylhexyl)adipate (DEHA) in atmospheric particles and its application for screening $PM_{2.5}$ from Curitiba, Brazil. *Anal. Methods* 16: 1579–1592.

79 Zacs, D., Perkons, I., and Bartkevics, V. (2019). Evaluation of analytical performance of gas chromatography coupled with atmospheric pressure chemical ionization Fourier transform ion cyclotron resonance mass spectrometry (GC-APCI-FT-ICR-MS) in the target and non-targeted analysis of brominated and chlorinated flame retardants in food. *Chemosphere* 225: 368–377.

80 Chiaia-Hernández, A.C., Scheringer, M., Müller, A. et al. (2020). Target and suspect screening analysis reveals persistent emerging organic contaminants in soils and sediments. *Sci. Total Environ.* 740: 140181.

81 Schanzer, S., Kroener, E., Wibbelt, G. et al. (2021). Miniaturized multiresidue method for the analysis of pesticides and persistent organic pollutants in non-target wildlife animal liver tissues using GC-MS/MS. *Chemosphere* 279: 130434.

82 del Castillo, M.L.R., Rodriguez-Valenciano, M., de la Peña Moreno, F., and Blanch, G.P. (2012). Evaluation of pesticide residue contents in fruit juice by solidphase microextraction and multidimensional gas chromatography coupled with mass spectrometry. *Talanta* 89: 77–83.

83 Li, C., Wang, D., Li, N. et al. (2016). Identifying unknown by-products in drinking water using comprehensive two dimensional gas chromatography–quadrupole mass spectrometry and in silico toxicity assessment. *Chemosphere* 163: 535–543.

84 Mol, H.G.J., van der Kamp, H., Weg, G.V.D. et al. (2011). Validation of automated library-based qualitative screening of pesticides by comprehensive two-dimensional gas chromatography/time-of flight mass spectrometry. *J. AOAC Int.* 94: 1722–1740.

85 Pena-Abaurrea, M., Ye, F., Blasco, J., and Ramos, L. (2012). Evaluation of comprehensive two-dimensional gas-chromatography time-of-flight mass spectrometry for the analysis of polyaromatic hydrocarbons in sediments. *J. Chromatogr. A* 1256: 222–231.

86 Kilpinen, K., Devers, J., Castro, M. et al. (2023). Catchment area, fate, and environmental risks investigation of micropollutants in Danish wastewater. *Environ. Sci. Pollut. Res.* 30: 121107–121123.

87 Hashimoto, S., Zushi, Y., Takazawa, Y. et al. (2018). Selective and comprehensive analysis of organohalogen compounds by GC × GC-HRToFMS and MS/MS. *Environ. Sci. Pollut. Res.* 25: 7135–7146.

88 Wang, Y.J., Gao, W., Wang, Y.W., and Jiang, G.B. (2019). Suspect screening analysis of the occurrence and removal of micropollutants by GC-QTOF MS during wastewater treatment processes. *J. Hazard. Mat.* 376: 153–159.

89 Su, H.J., Ren, K.F., Li, R.R. et al. (2022). Suspect screening of liquid crystal monomers (LCMs) in sediment using an established database covering 1173 LCMs. *Environ. Sci. Technol.* 56: 8061–8070.

90 Mok, S., Lee, S., Choi, Y. et al. (2023). Target and non-target analyses of neutral per- and polyfluoroalkyl substances from fluorochemical industries using GC-MS/MS and GC-TOF: insights on their environmental fate. *Environ. Int.* 182: 108311.

91 Krätschmer, K., Cojocariu, C., Schächtele, A. et al. (2018). Chlorinated paraffin analysis by gas chromatography Orbitrap high-resolution mass spectrometry: method performance, investigation of possible interferences and analysis of fish samples. *J. Chromatogr. A* 1539: 53–61.

92 Ma, C., Cao, R., Sun, S. et al. (2022). Screening of risky substances in sea cucumbers based on gas chromatography-orbitrap high-resolution mass spectrometry. *Chin. J. Chromatogr.* 40: 944–951.

93 Olanrewaju, C.A., Ramirez, C.E., and Fernandez-Lima, F. (2021). Comprehensive screening of polycyclic aromatic hydrocarbons and similar compounds using GC-APLI-TIMS-TOFMS/GC-EI-MS. *Anal. Chem.* 93: 6080–6087.

94 Shen, L., Jobst, K.J., Helm, P.A. et al. (2012). Identification and determination of the dechlorination products of Dechlorane 602 in Great Lakes fish and Arctic beluga whales by gas chromatography-high resolution mass spectrometry. *Anal. Bioanal. Chem.* 404: 2737–2748.

95 Titaley, I.A., Ogba, M., Chibwe, L. et al. (2018). Automating data analysis for two-dimensional gas chromatography/time-of-flight mass spectrometry non-targeted analysis of comparative samples. *J. Chromatogr. A* 1541: 57–62.

96 Qiao, L., Gao, L.R., Liu, Y. et al. (2022). Recognition and health impacts of organic pollutants with significantly different proportions in the gas phase and size-fractionated particulate phase in ambient air. *Environ. Sci. Technol.* 56: 7153–7162.

97 Ubukata, M., Jobst, K.J., Reiner, E.J. et al. (2015). Non-targeted analysis of electronics waste by comprehensive two-dimensional gas chromatography combined with high-resolution mass spectrometry: using accurate mass information and mass defect analysis to explore the data. *J. Chromatogr. A* 1395: 152–159.

98 Veenaas, C. and Haglund, P. (2017). Methodology for non-target screening of sewage sludge using comprehensive two-dimensional gas chromatography coupled to high-resolution mass spectrometry. *Anal. Bioanal. Chem.* 409: 4867–4883.

99 Brits, M., Gorst-Allman, P., Rohwer, E.R. et al. (2018). Comprehensive two-dimensional gas chromatography coupled to high resolution time-of-flight mass spectrometry for screening of organohalogenated compounds in cat hair. *J. Chromatogr. A* 1536: 151–162.

100 Rebryk, A., Gallampois, C., and Haglund, P. (2022). A time-trend guided non-target screening study of organic contaminants in Baltic Sea harbor porpoise (1988–2019), guillemot (1986–2019), and white-tailed sea eagle (1965–2017) using gas chromatography-high-resolution mass spectrometry. *Sci. Total Environ.* 829: 154620.

101 Gómez-Ramos, M.M., Ucles, S., Ferrer, C. et al. (2018). Exploration of environmental contaminants in honeybees using GC-TOF-MS and GC-Orbitrap-MS. *Sci. Total Environ.* 647: 232–244.

102 Casey, J.S., Jackson, S.R., Ryan, J., and Newton, S.R. (2023). The use of gas chromatography – high resolution mass spectrometry for suspect screening and non-targeted analysis of per- and polyfluoroalkyl substances. *J. Chromatogr. A* 1693: 463884.

103 MacNeil, A., Li, X.L., Amiri, R. et al. (2022). Gas chromatography-(cyclic) ion mobility mass spectrometry: a novel platform for the discovery of unknown per-/polyfluoroalkyl substances. *Anal. Chem.* 94: 11096–11103.

104 Steeves, K.L., Bissram, M.J., Kleywegt, S. et al. (2023). Nontargeted screening reveals fluorotelomer ethoxylates in indoor dust and industrial wastewater. *Environ. Int.* 171: 107634.

105 Singh, R.R., Aminot, Y., Heas-Moisan, K. et al. (2023). Cracked and shucked: GC-APCI-IMS-HRMS facilitates identification of unknown halogenated organic chemicals in French marine bivalves. *Environ. Int.* 178: 108094.

106 Fernando, S., Jobst, K.J., Taguchi, V.Y. et al. (2014). Identification of the halogenated compounds resulting from the 1997 Plastimet Inc. fire in Hamilton, Ontario, using comprehensive two-dimensional gas chromatography and (ultra)high resolution mass spectrometry. *Environ. Sci. Technol.* 48: 10656–10663.

107 Xu, C., Gao, L.R., Zheng, M.H. et al. (2021). Nontarget screening of polycyclic aromatic compounds in atmospheric particulate matter using ultrahigh resolution mass spectrometry and comprehensive two-dimensional gas chromatography. *Environ. Sci. Technol.* 55: 109–119.

108 Ramos, J.J., González, M.J., and Ramos, L. (2009). Comparison of gas-chromatography-based approaches after fast miniaturised sample preparation for the monitoring of selected pesticide classes. *J. Chromatogr. A* 1216: 7307–7313.

109 Borrelli, R., Fiorani, T., and Golfetto, P. (1996). Ultratrace determination of 1,2,4-trichlorobenzene in waste water by purge and trap/gas chromatography coupled to different detectors: flame ionization (FID), electron capture (ECD), and multiple ion detection-mass spectrometry (MID-MS). *J. High Resol. Chromatogr.* 19: 457–461.

110 Sauret, N., Millet, M., Mirabel, P., and Wortham, H. (2000). Analytical method using gas chromatography and ion trap tandem mass spectrometry for the determination of *s*-triazines and their metabolites in the atmosphere. *Environ. Pollut.* 110: 243–252.

111 Charlier, C., Pitance, F., and Plomteux, G. (2003). PCB residues in a breast cancer patient population. *Bull. Environ. Contan. Toxicol.* 71: 887–891.

112 Masci, M., Orban, E., and Nevigato, T. (2014). Organochlorine pesticide residues: an extensive monitoring of Italian fishery and aquaculture. *Chemosphere* 94: 190–198.

113 Gómara, B., Fernández, M.A., González, M.J., and Ramos, L. (2006). Feasibility of gas chromatography-ion trap tandem mass spectrometry for the determination of polychlorinated biphenyls in food. *J. Sep. Sci.* 29: 123–130.

114 Ramos, J.J., González, M.J., and Ramos, L. (2004). Miniaturised sample preparation of fatty foodstuffs for the determination of polychlorinated biphenyls. *J. Sep. Sci.* 27: 595–601.

115 Robles-Molina, J., Gilbert-López, B., García-Reyes, J.F., and Molina-Díaz, A. (2013). Gas chromatography triple quadrupole mass spectrometry method for monitoring multiclass organic pollutants in Spanish sewage treatment plants effluents. *Talanta* 111: 196–205.

116 Luzardo, O.P., Ruiz-Suárez, N., Almeida-González, M. et al. (2013). Multi-residue method for the determination of 57 persistent organic pollutants in human milk and colostrum using a QuEChERS-based extraction procedure. *Anal. Bioanal. Chem.* 405: 9523–9536.

117 Hardy, E.M., Duca, R.C., Salquebre, G., and Appenzeller, B.M.R. (2015). Multi-residue analysis of organic pollutants in hair and urine for matrices comparison. *Forens. Sci. Int.* 249: 6–19.

118 Kim, I., Lee, S., and Kim, S.D. (2019). Determination of toxic organic pollutants in fine particulate matter using selective pressurized liquid extraction and gas chromatography-tandem mass spectrometry. *J. Chromatogr. A* 1590: 39–46.

119 Lee, J.E., Oh, H.B., Im, H. et al. (2020). Multiresidue analysis of 85 persistent organic pollutants in small human serum samples by modified QuEChERS preparation with different ionization sources in mass spectrometry. *J. Chromatogr. A* 1623: 461170.

120 Palát, I., Kukučka, P., Codling, G.P. et al. (2022). Application of 96-well plate SPE method for analysis of persistent organic pollutants in low volume blood serum samples. *Chemosphere* 287: 132300.

121 Williams-Clayson, A.M., Vane, C.H., Jones, M.D. et al. (2023). Characterisation of former manufactured gas plant soils using parent and alkylated polycyclic aromatic hydrocarbons and Rock-Eval(6) pyrolysis. *Environ. Pollut.* 339: 122658.

122 Blanch, G.P., Glausch, A., Schurig, V. et al. (1996). Quantification and determination of enantiomeric ratios of chiral PCB 95, PCB 132, and PCB 149 in shark liver samples (C-coelolepis) from the Atlantic Ocean. *J. High Resol. Chromatogr.* 19: 392–396.

123 Bordajandi, L.R., Ramos, J.J., Sanz, J. et al. (2008). Comprehensive two-dimensional gas chromatography in the screening of persistent organohalogenated pollutants in environmental samples. *J. Chromatogr. A* 1186: 312–324.

124 Vaye, O., Ngumbu, R.S., and Xia, D. (2022). A review of the application of comprehensive two-dimensional gas chromatography MS-based techniques for the analysis of persistent organic pollutants and ultra-trace level of organic pollutants in environmental samples. *Rev. Anal. Chem.* 41: 63–73.

125 Altamirano, J.C., Yin, S., Belova, L. et al. (2024). Exploring the hidden chemical landscape: non-target and suspect screening analysis for investigating solid waste-associated environments. *Environ. Research* 245: 118006.

126 Zanella, D., Romagnoli, M., Malcangi, S. et al. (2023). The contribution of high-resolution GC separations in plastic recycling research. *Anal. Bioanal. Chem.* 415: 2343–2355.

127 Escobar-Arnanz, J., Pena-Abaurrea, M., Santos, F.J., and Ramos, L. (2023). Non-target analysis of organohalogenated contaminants in deep-sea fishes from the Mediterranean Sea by comprehensive two-dimensional gas chromatography–time-of-flight mass spectrometry. *Sci. Total Environ.* 877: 162873.

128 Pena-Abaurrea, M., Jobst, K.J., Ruffolo, R. et al. (2014). Identification of potential novel bioaccumulative and persistent chemicals in sediments from Ontario (Canada) using scripting approaches with GC × GC-TOF MS analysis. *Environ. Sci. Technol.* 48: 9591–9599.

129 Escobar-Arnanz, J., Mekni, S., Blanco, G. et al. (2018). Characterization of organic aromatic compounds in soils affected by an uncontrolled tire landfill fire through the use of comprehensive two-dimensional gas chromatography–time-of-flight mass spectrometry. *J. Chromatogr. A* 1536: 163–175.

130 Lowe, C.N., Phillips, K.A., Favela, K.A. et al. (2021). Chemical characterization of recycled consumer products using suspect screening analysis. *Environ. Sci. Technol.* 55: 11375–11387.

131 Hoh, E., Dodder, N.G., Lehotay, S.J. et al. (2012). Nontargeted comprehensive two-dimensional gas chromatography/time-of-flight mass spectrometry method and software for inventorying persistent and bioaccumulative contaminants in marine environments. *Environ. Sci. Technol.* 46: 8001–8008.

132 Teehan, P., Schall, M.K., Blazer, V.S. et al. (2020). Modified QuEChERS extraction for the analysis of young-of-year smallmouth bass using GC × GC-TOFMS. *Anal. Methods* 12: 3697–3704.

133 Carrero-Carralero, C., Escobar-Arnanz, J., Ros, M. et al. (2019). An untargeted evaluation of the volatile and semi-volatile compounds migrating into food simulants from polypropylene food containers by comprehensive two-dimensional gas chromatography–time-of-flight mass spectrometry. *Talanta* 195: 800–806.

134 Vosough, M., Schmidt, T.C., and Renner, G. (2024). Non-target screening in water analysis: recent trends of data evaluation, quality assurance, and their future perspectives. *Anal. Bioanal. Chem.* 416: 2125–2136.

135 Hollender, J., Schymanski, E.L., Ahrens, L. et al. (2023). NORMAN guidance on suspect and non-target screening in environmental monitoring. *Environ. Sci. Eur.* 35: 75.

136 Pierce, K.M., Hoggard, J.C., Hope, J.L. et al. (2006). Fisher ratio method applied to third-order separation data to identify significant chemical components of metabolite extracts. *Anal. Chem.* 78: 5068–5075.

137 Pesesse, R., Stefanuto, P.H., Schleich, F. et al. (2019). Multimodal chemometric approach for the analysis of human exhaled breath in lung cancer patients by TD-GC × GC-TOFMS. *J. Chromatogr. B* 1114: 146–153.

138 Veenaas, C., Ripszam, M., and Haglund, P. (2020). Analysis of volatile organic compounds in indoor environments using thermal desorption with comprehensive two-dimensional gas chromatography and high-resolution time-of-flight mass spectrometry. *J. Sep. Sci.* 43: 1489–1498.

139 Zushi, Y., Hashimoto, S., Tamada, M. et al. (2014). Retrospective analysis by data processing tools for comprehensive two-dimensional gas chromatography coupled to high resolution time-of-flight mass spectrometry: a challenge for matrix-rich sediment core sample from Tokyo Bay. *J. Chromatogr. A* 1338: 117–126.

140 Ruttkies, C., Schymanski, E.L., Wolf, S. et al. (2016). MetFrag relaunched: incorporating strategies beyond in silico fragmentation. *J. Cheminform.* 8: 1–16.

141 Tang, C., Chen, G., Jiang, B. et al. (2022). High-performance nontarget analysis of halogenated organic compounds in tap water, fly ash, soil and sediment using ultrahigh resolution mass spectrometry and scripting approaches based on Cl/Br-specific search algorithms. *Anal. Chim. Acta* 1204: 339618.

142 Castro, V., Quintana, J.B., López-Vázquez, J. et al. (2021). Development and application of an in-house library and workflow for gas chromatography-electron ionization-accurate-mass/high-resolution mass spectrometry screening of environmental samples. *Anal. Bioanal. Chem.* 414: 6327–6340.

143 Schymanski, E.L., Singer, H.P., Slobodnik, J. et al. (2015). Non-target screening with high-resolution mass spectrometry: critical review using a collaborative trial on water analysis. *Anal. Bioanal. Chem.* 407: 6237–6255.

144 Devers, J., Pattison, D.I., Hansen, A.B., and Christensen, J.H. (2025). New strategies for non-targeted quantification in comprehensive two-dimensional gas chromatography: the potential of reconstructed TIC response factor surfaces. *J. Chromatogr. A* 1747: 465811.

145 Braun, G., Krauss, M., Spahr, S., and Escher, B.I. (2024). Handling of problematic ion chromatograms with the automated target screening (ATS) workflow for unsupervised analysis of high-resolution mass spectrometry data. *Anal. Bioanal. Chem.* 416: 2983–2993.

146 Wu, S.X., Fujii, M., Yang, X., and Fu, Q.L. (2023). Characterization of halogenated organic compounds by the Fourier transform ion cyclotron resonance mass spectrometry: a critical review. *Water Res.* 246: 120694.

147 Peng, B., Dong, Q.L., Li, F.Z. et al. (2023). A systematic review of polycyclic aromatic hydrocarbon derivatives: occurrences, levels, biotransformation, exposure biomarkers, and toxicity. *Environ. Sci. Technol.* 57: 15314–15335.

148 Dittrich, N., Zuber, J., and Rathsack, P. (2018). Screening of organic compounds in complexly contaminated scrubber dust by GC × GC-MS and FT-ICR-MS. *Microchem. J.* 139: 85–93.

149 Smith, M.R., Walker, D.I., Uppal, K. et al. (2019). Benzo[a]pyrene perturbs mitochondrial and amino acid metabolism in lung epithelial cells and has similar correlations with metabolic changes in human serum. *J. Occup. Environ. Med.* 61: S73–S81.

150 Baumer, A., Jäsch, S., Ulrich, N. et al. (2021). Chemical mixtures in human post-mortem tissues assessed by a combination of chemical analysis and in vitro bioassays after extraction with silicone. *Environ. Int.* 157: 106867.

151 Zhang, P., Carlsten, C., Chaleckis, R. et al. (2021). Defining the scope of exposome studies and research needs from a multidisciplinary perspective. *Environ. Sci. Technol. Lett.* 8: 839–852.

152 Weggler, B.A., Gruber, B., and Focant, J.F. (2020). Comprehensive two-dimensional gas-chromatography to study the human exposome: current trends and perspectives. *Curr. Opin. Environ. Sci. Health* 15: 16–25.

153 Wild, C.P. (2012). The exposome: from concept to utility. *Int. J. Epidemiol.* 41: 24–32.

154 Fuentes, Z.C., Schwartz, Y.L., Robuck, A.R., and Walker, D.I. (2022). Operationalizing the exposome using passive silicone samplers. *Curr. Pollut. Rep.* 8: 1–29.

155 Samon, S.M., Hammel, S.C., Stapleton, H.M., and Anderson, K.A. (2022). Silicone wristbands as personal passive sampling devices: current knowledge, recommendations for use, and future directions. *Environ. Int.* 169: 107339.

156 Reiter, E.B., Escher, B.I., Rojo-Nieto, E. et al. (2023). Characterizing the marine mammal exposome by iceberg modeling, linking chemical analysis and in vitro bioassays. *Environ. Int.* 25: 1802–1816.

157 Wang, A., Abrahamsson, D.P., Jiang, T. et al. (2021). Suspect screening, prioritization, and confirmation of environmental chemicals in maternal-newborn pairs from San Francisco. *Environ. Sci. Technol.* 55: 5037–5049.

158 Hu, X., Walker, D.I., Liang, Y. et al. (2021). A scalable workflow to characterize the human exposome. *Nat. Commun.* 12: 5575.

159 Dusza, H.M., Manz, K.E., Pennell, K.D. et al. (2022). Identification of known and novel nonpolar endocrine disruptors in human amniotic fluid. *Environ. Int.* 158: 106904.

160 Valvi, D., Walker, D.I., Inge, T. et al. (2020). Environmental chemical burden in metabolic tissues and systemic biological pathways in adolescent bariatric surgery patients: a pilot untargeted metabolomic approach. *Environ. Int.* 143: 105957.

161 Lv, Y.-Z., Luo, X.-J., Qi, X.-M. et al. (2024). A comprehensive assessment of external exposure to persistent halogenated organic pollutants for residents in an e-waste recycling site, South China. *Environ. Pollut.* 343: 123120.

162 Amann, A., Costello, B., Miekisch, W. et al. (2014). The human volatilome: volatile organic compounds (VOCs) in exhaled breath, skin emanations, urine, feces and saliva. *J. Breath Res.* 8: 034001.

163 Estévez-Danta, A., Ayala-Cabrera, J.F., López-Vázquez, J. et al. (2024). Analytical methodology for unveiling human exposure to (micro)plastic additives. *TRAC Trends Anal. Chem.* 173: 117653.

164 Rodríguez-Navas, C., Rosende, M., and Miró, M. (2017). In-vitro physiologically based extraction of solid materials: do we have reliable analytical methods for bioaccessibility studies of emerging organic contaminants? *TRAC Trends Anal. Chem.* 91: 42–52.

165 Evangelou, E., Ntritsos, G., Chondrogiorgi, M. et al. (2016). Exposure to pesticides and diabetes: a systematic review and meta-analysis. *Environ. Int.* 91: 60–68.

166 Yipei, Y., Zhilin, L., Yuhong, L. et al. (2022). Assessing the risk of diabetes in participants with DDT DDE exposure - a systematic review and meta-analysis. *Environ. Res.* 210: 113018.

167 Hernández-Mariano, J.A., Baltazar-Reyes, M.C., Salazar-Martínez, E., and Cupul-Uicab, L.A. (2022). Exposure to the pesticide DDT and risk of diabetes and hypertension: systematic review and meta-analysis of prospective studies. *Int. J. Hyg. Environ. Health* 239: 113865.

168 Di Ciaula, A. and Portincasa, P. (2019). Diet and contaminants: driving the rise to obesity epidemics? *Curr. Med. Chem.* 26: 3471–3482.

169 Czajka, M., Matysiak-Kucharek, M., Jodłowska-Jędrych, B. et al. (2019). Organophosphorus pesticides can influence the development of obesity and type 2 diabetes with concomitant metabolic changes. *Environ. Res.* 178: 108685.

170 Wada, R., Peng, F.-J., Lin, C.-A. et al. (2024). Hair-derived exposome exploration of cardiometabolic health: piloting a bayesian multitrait variable selection approach. *Environ. Sci. Technol.* 58: 5383–5393.

171 Mariana, M., Castelo-Branco, M., Soares, A.M., and Cairrao, E. (2023). Phthalates' exposure leads to an increasing concern on cardiovascular health. *J. Hazard. Mat.* 457: 131680.

172 Zhang, Y.Z., Gao, Y.R., Liu, Q.S. et al. (2024). Chemical contaminants in blood and their implications in chronic diseases. *J. Hazard. Mat.* 466: 133511.

173 Alkerwi, A., Guillaume, M., Zannad, F. et al. (2010). Nutrition, environment and cardiovascular health (NESCAV): protocol of an inter-regional cross-sectional study. *BMC Public Health* 10: 698.

174 Carli, F., Ciociaro, D., and Gastaldelli, A. (2022). Assessment of exposure to di-(2-ethylhexyl) phthalate (DEHP) metabolites and bisphenol A (BPA) and its importance for the prevention of cardiometabolic diseases. *Metabolites* 12: 167.

175 González-Domínguez, R., Jáuregui, O., Queipo-Ortuño, M.I., and Andrés-Lacueva, C. (2020). Characterization of the human exposome by a comprehensive and quantitative large-scale multianalyte metabolomics platform. *Anal. Chem.* 92: 13767–13775.

176 Robbat, A. Jr., Kfoury, N., Baydakov, E., and Gankin, Y. (2017). Optimizing targeted/untargeted metabolomics by automating gas chromatography/mass spectrometry workflows. *J. Chromatogr. A* 1505: 96–105.

177 Rey-Stolle, F., Dudzik, D., Gonzalez-Riano, C. et al. (2022). Low and high resolution gas chromatography-mass spectrometry for untargeted metabolomics: a tutorial. *Anal. Chim. Acta* 1210: 339043.

178 Wang, Y., Zhou, L., Zhou, Y. et al. (2020). A rapid GC method coupled with quadrupole or time of flight mass spectrometry for metabolomics analysis. *J. Chromatogr. B* 1160: 122355.

179 Manz, K.E., Feerick, A., Braun, J.M. et al. (2023). Non-targeted analysis (NTA) and suspect screening analysis (SSA): a review of examining the chemical exposome. *J. Exposure Sci. Environ. Epidemiol.* 33: 524–536.

180 Ma, Q.C., Liu, Y.N., Yang, X.X. et al. (2022). Effect-directed analysis for revealing aryl hydrocarbon receptor agonists in sediment samples from an electronic waste recycling town in China. *Environ. Pollut.* 308: 119659.

181 Lopez-Herguedas, N., Mijangos, L., Alvarez-Mora, I. et al. (2023). Suspect screening of chemicals in hospital wastewaters using effect-directed analysis approach as prioritization strategy. *Molecules* 28: 1212.

182 Simon, E., Lamoree, M.H., Hamers, T., and de Boer, J. (2015). Challenges in effect-directed analysis with a focus on biological samples. *TRAC Trends Anal. Chem.* 67: 179–191.

10

Markers of Quality, Authenticity, and Safety in Processed Food Samples: The Omics Strategy

Andrea Caratti, Angelica Fina, Fulvia Trapani, Eloisa Bagnulo, Erica Liberto, Carlo Bicchi, and Chiara Cordero

Department of Drug Science and Technology, University of Turin, Turin, Italy

10.1 Introduction

Comprehensive two-dimensional gas chromatography (comprehensive 2D GC or GC×GC) effectively exploits and realizes the concept of a multidimensional (MD) technique as defined by Giddings in 1987 [1]: "*a separation system is considered multidimensional when (i) all components of a sample undergo two or more separation steps, where their movements depend on different variables, and (ii) if two or more components are substantially separated in any single step, they should remain separated until the completion of the separation process.*" This definition clearly established the principle of multidimensionality, at that time primarily applicable to two-dimensional thin-layer chromatography (TLC) and paper chromatography, which were widely utilized by biochemists. Within just 4 years, John Phillips, together with Zi Liu, translated this theoretical framework into practice for gas chromatography (GC) and developed a specialized instrument for this purpose [2]. The combined contributions of the concepts introduced by Giddings and Phillips have laid the groundwork for further MD chromatographic techniques, extending beyond GC to include liquid chromatography (LC) and supercritical fluid chromatography (SFC), such as LC×LC, SFC×SFC, and SFC×LC.

Like all innovative techniques, GC×GC had to progress through three distinct phases to gain full acceptance. The first phase involved the introduction and optimization of its theoretical foundations and operational parameters, spanning approximately 10 years (1991–2000). During this period, the petrochemical industry was the main contributor to the development of GC×GC due to its ongoing need to characterize highly complex samples [3, 4]. The subsequent phase, lasting about another decade (2001–2012), was marked by the dissemination of the technique across a diverse range of scientific disciplines [5–7]. This era also saw the routine implementation of fingerprinting methods in biological research, where a comprehensive analysis of complex mixtures in a single step became crucial, especially in omics fields such as metabolomics (as discussed below) [8]. Concurrently, there was significant advancement in hardware, standardization of instrumentation, optimization of GC×GC coupling with mass spectrometry (MS), and the development of dedicated software [5, 6, 9–13]. Finally, in the current phase, GC×GC has transitioned from academic research settings into routine applications, where ongoing analytical innovation continues to shape new analytical strategies, including the adoption of "dilute-and-shoot" methods.

In this context, research studies utilizing GC×GC as the primary analytical tool should: (i) configure the analytical platform appropriately by integrating all relevant dimensions of analysis; (ii) establish a data processing strategy that optimizes the extraction of meaningful information from raw data; and (iii) employ effective data mining techniques to interpret the results and gain deeper insights into the studied phenomenon. These components of the systematic approach are illustrated in Figure 10.1, showing their connection to the investigative goals in food omics. While a detailed discussion of each key element is beyond the scope of this chapter, readers interested in recent developments in data processing and mining for "omics" applications are encouraged to consult recent reviews [9, 11–17].

GC×GC has evolved into a fully independent yet complementary technique to conventional one-dimensional GC (1D GC), allowing for the analysis of highly complex mixtures that exceed the capabilities of 1D GC, which typically struggles with samples containing more than 250 components.

Following a brief introductory overview of the principles of systems biology and the concept of analytical dimensions, this chapter will examine the role of GC×GC in various disciplines and subdisciplines related to the food sector (food-omics), including food metabolomics, nutrimetabolomics, sensomics, volatilomics, and food safety.

10.1.1 Analytical Chemistry Toolbox to Advance Knowledge at the Boundary Between Chemistry and Biology

The complex nature of biological systems and the numerous levels of interactions underlying biological processes have driven the growth of systems biology, a field focused on modeling this complexity by revealing higher-order network structures and relationships between different conditions, while considering the impact of molecular patterns on multiple targets [18]. In contrast, chemical biology, which has developed alongside systems biology, employs well-established

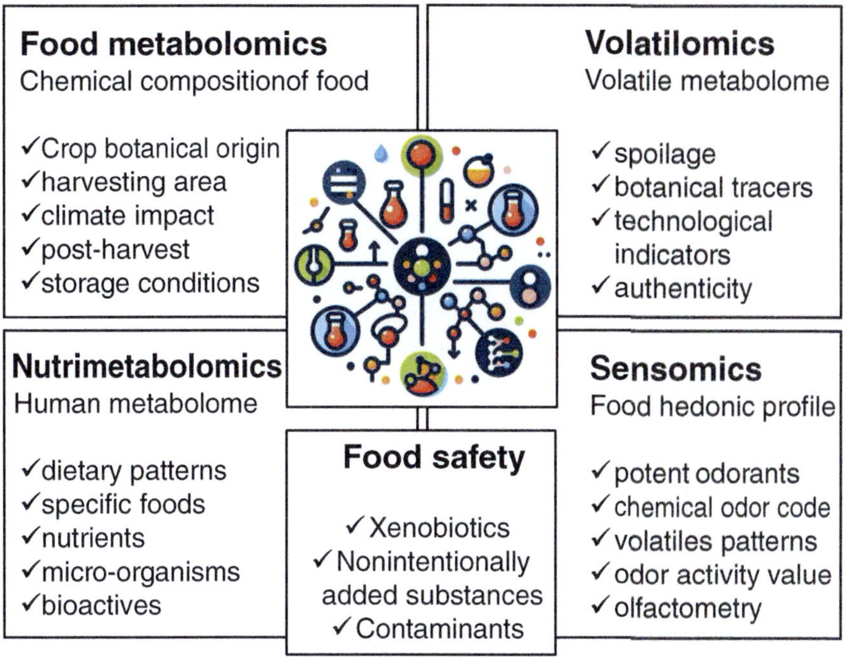

Figure 10.1 Main investigation goals of food-omics (i.e., food metabolomics, food volatilomics, sensomics, nutrimetabolomics, and food safety).

strategies aimed at linking individual compounds to specific targets, following a "one compound/one target" paradigm, often referred to as a reductionist approach in biology [18].

Chemical research has undergone a transition from a reductionist approach to an integrative mindset that is characteristic of systems biology. This has been driven by two factors. Firstly, there is a growing demand for high-quality food that is rich in nutritional value and bioactive components. Secondly, analytical chemistry methods and tools have undergone significant advancements. This shift was anticipated in 1991 by R. Wilson in an editorial for Analytical Chemistry [19], where he noted: "*a persistent research frontier is the analytical chemistry of the mixtures of chemical substances generated by living organisms. The problems of separation, molecular identification, and quantification of these mixtures are enormous. They are the ultimate molecular mishmash. The challenges they offer have demanded, and produced, many new concepts in chemical measurement science.*"

For chromatographers, 1991 was a pivotal year with the introduction of comprehensive two-dimensional separation in the gas phase [2]. Liu and

Phillips demonstrated this by separating a standard mixture using two capillary columns in series: a first-dimension (^{1}D) polyethylene glycol (PEG) column [21 m × 0.25 mm ID × 0.25 μm d_f] and a second-dimension (^{2}D) apolar dimethyl polysiloxane column [1 m × 0.10 mm ID × 0.10 μm d_f], coupled with a thermal modulator. The separation was completed in 2.5 minutes with a modulation period (P_M) of 2 seconds, resulting in a well-ordered retention pattern across the separation space, shaped by the relative volatility and polarity of the analytes.

Since then, despite an initial induction period marked by skepticism from some expert chromatographers who were not fully convinced of the benefits and potential of GC×GC, the technique has gradually been adopted in various application areas, either as an alternative to or a replacement for 1D GC [3–5, 20–26]. The enhanced separation capabilities of GC×GC, which offer greater separation power, efficiency, and resolution, along with band compression in space that boosts method sensitivity, make it an ideal tool for exploring the chemical dimensions of complex samples [27]. This aligns with the integrationist approach, which seeks detailed patterns of bioactive molecules.

To reach a sufficient level of informative detail about sample composition, it is essential to properly combine multiple analytical dimensions. The following section introduces the concept of multidimensional analytical (MDA) platforms and outlines common GC×GC configurations along with their informative dimensions.

10.1.2 Analytical Dimensions

MDA platforms integrate orthogonal measurement principles by combining chromatographic, electromigration, size exclusion, and/or other physicochemical discrimination methods with spectroscopic techniques such as MS, nuclear magnetic resonance (NMR), infrared, and ultraviolet spectroscopy. By gathering multiplex analytical data, these platforms can effectively discriminate, identify, and quantify sample components or homologous classes, even in highly complex samples with potential interferents. The greater the "degree of orthogonality" in the data collected, the more comprehensive the approach to analyzing food samples becomes [11, 18, 21, 28, 29].

While the integrationist approach seeks patterns of chemicals rather than individual components or biomarkers, the analytical process must still provide reliable information about the identities and quantities – whether relative or absolute – of analytes. To achieve these objectives, MS with unit mass resolution or exact mass determination using high-resolution MS (HR MS) is crucial. The molecular characteristics and functions can be elucidated through fragmentation patterns obtained via electron ionization MS ((EI)MS) at standard

70 eV or lower energies [30], by detecting molecular ions or adducts generated through chemical ionization MS ((CI)MS), or by employing multiple fragmentation steps in tandem MS (MS/MS or MS^n). However, due to the high compositional complexity of food samples, which may contain hundreds of isomeric or isobaric components, clear identification and accurate quantification are not feasible without adequate physicochemical separation, which is provided by chromatography (either 1D GC or GC×GC) [11, 31].

Additionally, when the sensory properties of food are of interest or are the primary focus, odor quality assessment via olfactometry becomes a crucial dimension of the analytical system. The section on sensomics offers further insights into the role of online or offline olfactometry when combined with GC×GC platforms [29, 32–34].

The separation power of GC×GC is a crucial feature that allows for the generation of highly detailed 2D chromatographic fingerprints with systematically organized retention patterns, opening up new possibilities for data analysis, such as group-type classification [15, 35, 36], and improving the confident identification of unknown compounds [37]. For instance, the strategic combination of column dimensions, stationary phase chemistries, and modulation dynamics to clarify the retention behavior of alkyl-substituted pyrazines in roasted coffee has been demonstrated by Ryan et al. [38] and Mondello et al. [39]. These studies were among the first to showcase the benefits of GC×GC in group-type analyses of food. The authors used a polar × apolar column combination (^{1}D PEG × ^{2}D 5% phenyl – 95% dimethylpolysiloxane) and achieved unambiguous identification by integrating linear retention indices (RI) in the ^{1}D, MS fragmentation patterns via EI, time-of-flight MS (TOFMS), and 2D retention logic.

Figure 10.2 shows the structured retention pattern for homolog series in butter volatilome. Colored circles highlight hydrocarbons (light gray) and terpenes hydrocarbons (light green), highly retained in the ^{2}D (semi-polar stationary phase OV1701), primary alcohols (red), and short-chain fatty acids (SCFAs) (green), with polar functions and therefore highly retained in the ^{1}D (polar stationary phase – polyethylenglycol PEG) linear saturated aldehydes (light blue), esters, and lactones (cyan).

Many fermented foods or their primary ingredients exhibit distinct profiles of homologous compounds that arise from common biosynthetic pathways, such as SCFAs and primary alcohols in cocoa fermentation [40–43] or butter [44], or are formed through shared reaction mechanisms, like the Maillard reaction in roasted foods [45–47]. Compound classes such as SCFAs, aliphatic alcohols, linear saturated and unsaturated aldehydes, esters, furans, and pyrroles display rational retention patterns, which greatly aid in their identification and differentiation.

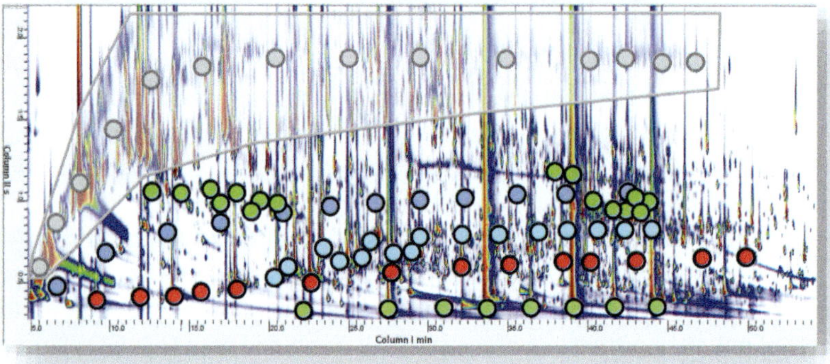

Figure 10.2 Ordered retention patterns for homolog series in butter volatilome. Colored circles highlight hydrocarbons (light gray), primary alcohols (red), short-chain FAs (green), linear saturated aldehydes (light blue), esters and lactones (cyan), and terpenes hydrocarbons (light green).

Beyond the ordered retention patterns, the use of ^{1}D retention indices is essential for distinguishing isomers or isobaric compounds that exhibit similar (EI)MS fragmentation patterns. Most common MS databases, such as those provided by NIST [48] and McLafferty [49], include both experimental and estimated retention indices on common stationary phases, allowing MS search software to apply retention-index constraints. This approach was first introduced by Mondello and coworkers [50, 51], who created one of the earliest interactive databases for flavor and fragrance analysis. More recently, Reichenbach et al. [52] developed an algorithm for calculating RI that does not require calibration with standard *n*-alkanes or fatty acid methyl esters (FAMEs). With this method, putative identifications generated through an automatic and comprehensive, untargeted MS search on the sample components use their retention times as reference points to calculate piecewise linear or logarithmic interpolations of RI, such as Kovats indices (in isothermal analysis) or Van den Dool indices (i.e., linear retention indices). Figure 10.3a shows a plot using a piecewise linear interpolation function of ${}^{1}t_{R}$ versus library RI for 21 analytes in a terpene standard mixture. The RI values are those found in the database entry for the top-ranked analyte based on direct match factor (DMF) similarity. Figure 10.3b illustrates a reliable model for preliminary RI calibration without standardization, which was achieved by removing three entries (IDs #8, #18, and #21) where analyst supervision identified inconsistencies in the incorporated RI values [52].

When the high compositional complexity of processed foods complicates the clear identification of components based on retention patterns and MS spectra, specialized data processing tools can be beneficial. In cases where

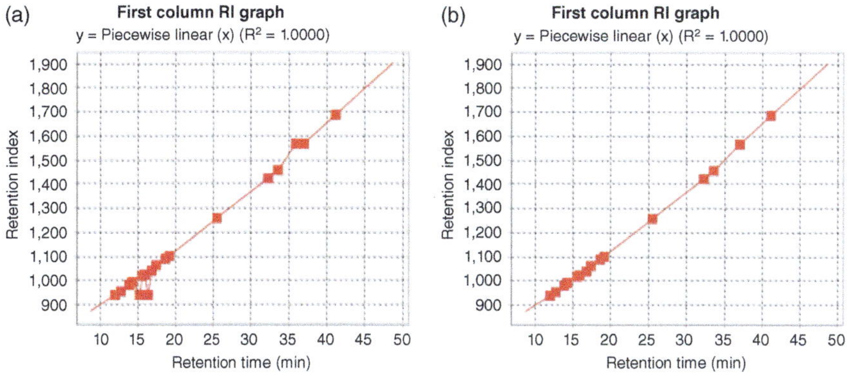

Figure 10.3 Linear retention index calibration based on model fitting to the observed relation between ${}^1t_R$ and the top-ranked analyte, resulted by DMF similarity. (a) 18 of Top 21 analytes (3 #1 hits lacked library RI); (b) 15 of Top 21 analytes consistent with piecewise linear RI calibration model. *Source:* Reichenbach et al. [52] / MDPI / CC BY 4.0.

homologous patterns overlap, selective extraction of diagnostic MS fragments/signatures from the third analytical dimension using scripting functions allows for the effective isolation of individual components and classes, along with 2D visualization [15, 35, 53, 54]. Figure 10.4a presents a contour plot image of volatiles sampled by HS-SPME on dry leaves of a black tea sample. In Figure 10.4b, pink-colored circles highlight 2D peaks where the scripting function: [(Relative(43) > 0.98)&(Relative(57) > 0.78)&(Relative(71) > 0.68)&(Relative (85) > 0.50)] was verified – normal and isoalkanes. In Figure 10.4c, green- and cyano-colored circles highlight 2D peaks where the scripting function: [AND ((Relative(41) > 0.70)&(Relative(44) > 0.70)&(Relative(57) > 0.50)),((Relative(41) >0.90)&(Relative(55) > 0.70)&(Relative(69) > 0.40)&(Relative(83) > 0.40))] was verified – saturated and unsaturated-aldehydes. CLIC™ Expression (GC-Image, LLC Lincoln, NE, USA) enables easy scripting and facilitates homologs visualization in pseudo-colorized images.

MS is a crucial component of MDA platforms for food analysis. Beyond its role in confirming analytes' identity, MS enhances specificity in the comparative analysis of extensive sample sets and supports reliable untargeted investigations [11]. Notably, MS platforms utilizing variable energy EI (e.g., Tandem Ionization™) [55–57], "soft" ionization through laser photoionization [58], or chemical ionization (CI) [59] offer valuable capabilities. By switching between two ionization energies [55] and employing a specially designed ion source, these platforms achieve efficient ionization with minimal sensitivity loss, even at low energies (e.g., 10–20 eV) [56, 60–64]. Applications of these techniques in food analysis include the study of volatiles and potent odorants in cocoa [56] and

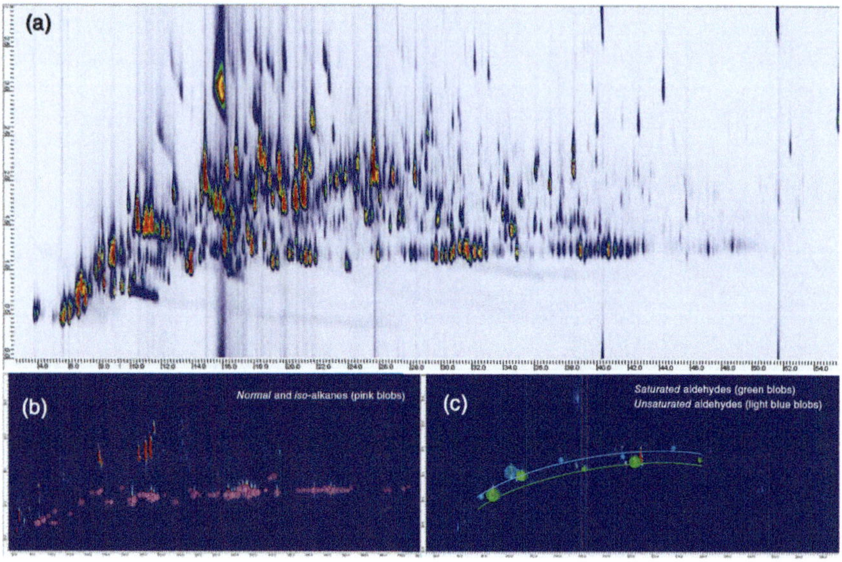

Figure 10.4 (a) Contour plot image of volatiles sampled by HS-SPME on dry leaves of a black tea sample; (b) pink-colored circles highlight 2D peaks where the scripting function: [(Relative(43) > 0.98)&(Relative(57) > 0.78)&(Relative(71) > 0.68)&(Relative(85) > 0.50)] is verified for normal and isoalkanes; (c) green- and cyano-colored circles highlight 2D peaks where the scripting function: [AND((Relative(41) > 0.70)&(Relative(44) > 0.70)&(Relative(57) > 0.50)),((Relative(41) > 0.90)&(Relative(55) > 0.70)&(Relative(69) > 0.40)&(Relative(83) > 0.40))] is verified for saturated and unsaturated-aldehydes. CLIC™ Expression (GC-Image, LLC Lincoln, NE, USA) enables easy scripting and facilitates homologs visualization in pseudo-colorized images. *Source:* Magagna et al. [53] / with permission of Elsevier.

extra-virgin olive oil [65], as well as primary metabolite silyl derivatives from hazelnuts [57] and FAMEs in biological samples [66].

TOFMS analyzers are commonly used MS detection platforms in combination with GC×GC. However, only a few of these TOF systems achieve high mass resolution, mass accuracy, and rapid scan speeds. Although examples of high-resolution TOFMS (HR TOFMS) application in food are limited, their potential for accurate identification and database construction for complex samples not fully represented in commercial databases is significant. Efforts by Tranchida and colleagues in this area include studies of the unsaponifiable fractions of milk fat from various mammals (cow, buffalo, ewe, and goat) [67, 68] and edible oils [69]. The advantages of HR TOFMS were particularly evident for the sterol class, where both group-type retention ordering and HR spectra facilitated reliable identification.

Conversely, more conventional single quadrupole instruments are suitable for MS detection in GC×GC applications, offering adequate frequency and sensitivity, though with some limitations in extending the mass range to capture a sufficient number of data points per peak.

10.1.3 Food-Omics Fields and Opportunities to Advance Existing Knowledge on Food Composition and Properties

Omics generally refers to the comprehensive characterization and quantification of biological molecules to understand the structure, function, and dynamics of organisms [70]. The first omics discipline, i.e., genomics, focused on studying entire genomes and was conceptually distinct from genetics, which examines individual gene variants. Today, various omics disciplines span genetics, biology, and chemistry. In the context of food, many omics fields aim to link chemical patterns and compositions with various properties, including nutritional density, sensory characteristics, biological activity, and the effects of external factors such as climate and processing practices on food. They also explore the impact of bacterial metabolism on food composition, particularly in fermented products.

A comprehensive and accurate food chemical code enables a deeper understanding of several phenomena, including (i) identifying key quality markers in raw materials and finished products; (ii) detecting bioactive compounds in food products, extracts, and biological fluids; (iii) finding biomarkers related to dietary interventions; and (iv) assessing human response indicators related to metabolic changes induced by different food groups and patterns [71].

Within the broader field of foodomics, defined as "*a discipline that examines food and nutrition through advanced omics technologies to enhance consumer health, well-being, and confidence*" [72], numerous specific omics disciplines have emerged. These specialized fields focus on particular interrelationships but employ similar investigation strategies, methodologies, and protocols.

Among these disciplines, food metabolomics focuses on elucidating the relationships between the chemical composition of food and external factors such as crop botanical traits and origin, harvesting location, climate effects, postharvest treatments, storage conditions, and shelf life. This emerging field highlights the pivotal role of GC×GC [73–76], which enhances the identification of chemical compounds by providing more comprehensive candidate lists with greater intrinsic information.

Closely related to food metabolomics is nutrimetabolomics, which merges nutritional science with metabolomics. This field explores "*the effects of specific diets, foods, nutrients, microorganisms, or bioactive compounds on the human metabolome*" [77], aiming to deepen our understanding of how dietary patterns impact human health and how the food metabolome (i.e., "*the portion of the human metabolome derived from the digestion and transformation of foods and their components*" [78]) interacts with human physiology. It examines both (i) nutrients that influence metabolic responses in humans and (ii) nonnutrients that may interact with metabolic pathways and receptors, affecting human metabolism.

When investigating food's hedonic properties, sensomics seeks to "*molecularize flavor entities of nature*," creating a unique odor and taste profile that aids in food recognition and enhances consumer enjoyment [79]. Sensomics and related fields like flavoromics [80] establish a foundation for the objective and rational assessment of food based on chemoreceptive events in the nose and mouth.

When all volatile metabolites including also degradation products and exogenous compounds not strictly derived from metabolic processes, such as environmental contaminants and substances produced by bacteria and molds, are targeted, food volatilomics is delineated. It focuses on the volatilome [81], or volatome [82], and by exploring its multiple chemical dimensions, more comprehensive insights into various phenomena compared to the metabolome alone are provided [83–85].

Although food safety and authenticity have not yet been fully integrated into an "omics" framework, they remain crucial aspects of food analysis. Food safety is a mandatory requirement for all products intended for human consumption. This review includes food safety due to GC×GC's potential in safety assessments and its ability to suggest measures to enhance food chain safety. Notable examples of food authentication and *identitation* are discussed in the referenced literature [86–88].

10.1.4 Approaches of Investigation and Terminology

In metabolomics, established investigation approaches include *profiling* and *fingerprinting* [89, 90]. Both strategies reveal compositional differences between samples, but they vary in scope. Profiling involves optimizing the analytical process to focus on known components of samples, following a reductionist approach. This method collects detailed data on analyte retention, mass spectra, and detector responses to enable qualitative and/or quantitative comparisons. Targeted profiling refers to this approach when it is directed toward a priori defined analytes (i.e., known components) [91]. Conversely, untargeted profiling (also referred to as nontargeted profiling) occurs when the analytical process generates distinctive features for all components independently by the chance to assign a correct identification, allowing a comprehensive evaluation of the sample's constituents [92, 93]. This approach could be also referred to as untargeted or nontargeted profiling.

An emerging strategy that integrates elements of both is *post-targeted analysis*, in which untargeted datasets are retrospectively mined to search for compounds of interest not initially included in the target list. This enables researchers to expand their scope of investigation without re-running samples, leveraging high-resolution data to bridge the gap between targeted and untargeted workflows.

Fingerprinting, as described by Fiehn [90] for metabolomics, differs conceptually from profiling. It is a high-throughput approach designed for effective cross-comparative analysis of samples, but it does not necessarily achieve the one-feature/one-component resolution typical of profiling. Instead, fingerprinting aims to capture a comprehensive set of features that ideally represent the chemical constituents of a sample. While profiling's information level – both qualitative and quantitative – is influenced by the analytical dimensions of the system, fingerprinting provides a broader overview.

Certain spectroscopic techniques, such as NMR and near-infrared spectroscopy (NIR), can deliver high levels of detail, though they may miss signals from trace and ultra-trace components. Recent advancements in MDA platforms have led to more stable and informative systems that enhance the concept of fingerprinting [12, 15].

2D peak patterns generated by techniques like GC×GC or LC×LC can be likened to human fingerprints, where *minutiae features* (e.g., ridge endings and bifurcations) are used for individual identification. In chromatographic fingerprinting, minutiae within 2D peak patterns are detected, aligned, and compared across multiple samples. Depending on the system's dimensionality, these 2D chromatographic features collect MD metadata, including retention times in both chromatographic dimensions (^1t_R, ^2t_R), detector responses, MS spectral signatures, and additional data related to olfactory quality or parallel detection. Chromatographic fingerprinting has the potential to unlock deeper insights within MDA datasets. Recent reviews have discussed its fundamental concepts and tools [12, 13, 15, 17].

The following section highlights selected research in food-omics and related fields that showcases how GC×GC enhances understanding of complex phenomena, provides more accurate and informative results, or offers new investigative perspectives on complex samples.

10.2 Applications of GC×GC in the Food-Omic Domains

10.2.1 Food-Metabolomics and Volatilomics

GC×GC-based food metabolomics can focus on volatile organic compounds (VOCs), with volatilomics [94]. Through precise profiling and fingerprinting of

VOCs, it is possible to gain deeper insights into food quality and spoilage, the impact of climate change on plant phenotypes, and the effects of pedoclimatic variations and geographical origin [34, 42, 43, 46, 76, 85, 95–101]. Similarly, the lipid fraction, analyzed through the profiling/fingerprinting of FAMEs, can provide valuable diagnostic information regarding food quality and origin [102–105]. For plant-based foods, primary metabolites – such as amino acids, sugars, low molecular weight acids, and amines – offer insights into the quality of primary materials [57, 75, 106, 107], harvesting practices [73, 75, 108], storage conditions [73, 75, 108], and the effects of processing and environmental stressors on crops [34, 42, 43, 46, 95–100].

Mack et al. [106] explored the effects of postharvest ripening on kiwifruit using untargeted metabolomics. Their study tracked metabolic changes across six stages of ripening and two nonmarketable stages characterized by significant water loss. Notable variations were observed in sugars, sugar-related substances, and organic acids throughout the ripening stages, with sugars increasing and organic acids decreasing. The study's multiplatform approach – combining GC×GC-QMS, GC-MS, and NMR – allowed for cross-validation of results. GC×GC-QMS proved to be the most sensitive platform, detecting the majority of primary metabolites in kiwifruit. Targeted GC-MS offered better profiling of the most concentrated analytes, such as main sugars and some organic acids, while NMR, despite being less sensitive, demonstrated superior reproducibility for certain analytes. Overall, the metabolome coverage was robust, with approximately 160 analytes reproducibly measured by GC×GC-QMS and GC-MS. Around 45% of these metabolites were identified using an in-house spectral database, and for unknown metabolites, chemical classes were hypothesized based on MS fragmentation patterns and their relative retention in the 2D space.

Using an untargeted multiplatform approach, the authors thoroughly investigated the kiwifruit metabolome's evolution through ripening and over-ripening stages. The combination of three complementary platforms allowed for a broader dynamic range of metabolite concentrations. Significant changes were observed in sugars and related compounds, which increased during ripening, and in organic acids, which decreased as over-ripening progressed. Other metabolites, like methanol and xylose, peaked during the mid-ripening stages, while galacturonic acid was found only in overripe fruits. This study underscores the utility of GC×GC-QMS as a powerful exploratory tool for untargeted food metabolomics.

More recently, Squara et al. combined different MDA platforms to collect information from volatile and nonvolatile metabolites to highlight quality traits and marker patterns in premium quality hazelnuts [76, 85]. The research examined the compositional complexity of the *Corylus avellana* L. metabolome by employing both untargeted and targeted metabolomic profiling methods to predict industrial quality. Using GC×GC and LC combined with high-resolution MS, the study characterized the nonvolatile (primary and specialized metabolites) and volatile components of hazelnuts. To improve classification performance, data fusion techniques, such as low-level data fusion and mid-level data fusion (MLDF), were utilized. Through principal component analysis and partial least squares discriminant analysis (PLS-DA), it is demonstrated that geographical origin and postharvest handling significantly influence the specialized metabolome, while storage conditions and duration predominantly affect the volatile metabolome. The findings indicated that MLDF strategies, especially supervised MLDF, offer superior predictive accuracy compared to single-fraction analyses. Notably, the study identified key metabolite patterns, such as aldehydes, alcohols, terpenes, and phenolic compounds, that were closely linked to quality attributes of hazelnuts. The integration of multiple analytical platforms and data fusion approaches holds significant potential for enhancing quality evaluation and optimizing storage and processing methods in the food industry.

In the context of extreme climate events such as drought and temperature fluctuations, GC×GC-TOFMS has proven to be a valuable addition to techniques targeting nonvolatile-specialized plant metabolites [109, 110]. Stilo et al. explored the primary metabolome of tea (*Camellia sinensis* L. Kuntze) to understand how plants adapt to varying climatic conditions, including drought, heavy rainfall, and temperature changes due to elevation [111]. By adapting an extraction and derivatization protocol used in biological fluids metabolomics, they achieved extensive coverage of tea's primary metabolites [112]. Out of 780 detected peak features across over 40 samples, 74 metabolites were tentatively identified based on available metadata and aligned with identification criteria shared within the metabolomics community [113, 114]. Seasonal (spring versus summer) and altitude-related variations were observed, with notable changes in amino acids, organic acids, polyalcohols, sugars, and some specialized metabolites such as flavan-3-ols and caffeine.

The study highlighted that during spring, the levels of alanine, aspartic acid, glycine, threonine, valine, phenylalanine, phosphoric acid, xylonic acid, and xylitol increased in tea plants. When comparing samples from two harvest years, fructose, glucose, maltose, and arabinose showed consistent up-regulation in spring teas. Altitude-related temperature variations led to increased levels of alanine, isoleucine, tyrosine, catechin, gallic acid, glycolic acid, malic acid, and ribonic acid in low-elevation teas.

The research, built on previous studies of unprocessed tea focusing on volatiles, specialized metabolites, and metals [109, 110, 115], provided deeper insights into the complex interactions between tea plants and climatic factors. The authors aimed to identify key primary metabolites and biomarkers to better understand

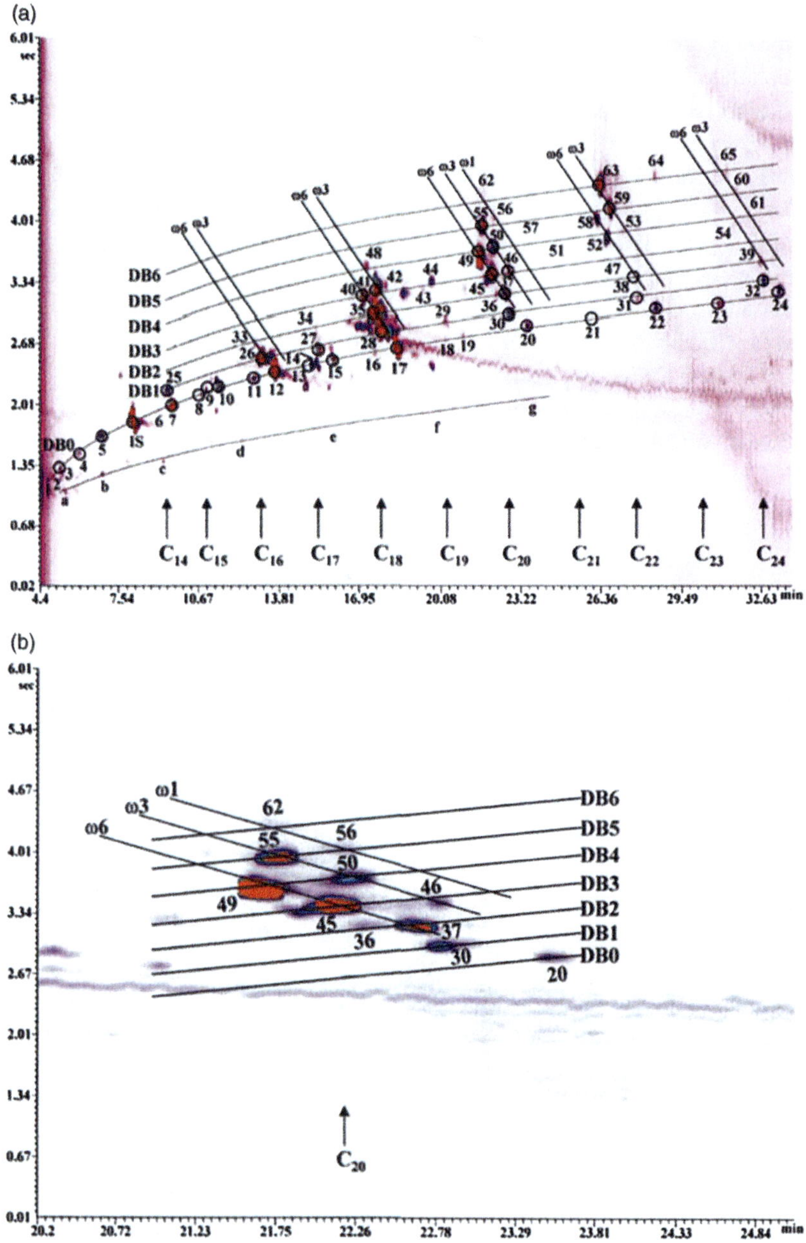

Figure 10.5 (a) 2D pattern of human plasma FAMEs within the C_4–C_{24} range. Analytes are aligned along seven diagonal bands, based on carbon number (CN) within C_{14} and C_{24} (zones indicated by lack arrows), and grouped while forming 2D subpatterns ordered based on double bonds (DB) number. (b) Enlarged area showing the C_{20} FAMEs group. *Source:* Tranchida et al. [118] / with permission of John Wiley & Sons.

the feedback loops between human activities and natural systems, particularly in addressing climate impacts on tea plants. The comprehensive data obtained through GC×GC-TOFMS, including known and unknown features, offer new avenues for food research. The extensive metadata – such as retention times in two dimensions, MS fragmentation patterns, and both absolute and relative responses – enables further exploration for specific discrimination and compound identification through (EI)MS spectra, ^{1}D retention indices, and retention pattern logic [11].

Accurate profiling and fingerprinting of fatty acids (FAs) represents another area where GC×GC-MS has provided a new dimension to research [116, 117]. Although the chemical diversity within lipid fractions is not exceedingly high, the presence of numerous FA isomers with closely similar retention characteristics necessitates high separation power and selectivity for thorough analysis. GC×GC offers both high separation capability and differential selectivity across two chromatographic dimensions, creating unique 2D patterns that assist in identifying and differentiating components lacking reference standards (such as odd-carbon-number or branched FAs). Tranchida and Mondello have focused their research on profiling and fingerprinting FAs from various sources, including plants, algae, and animals [102, 103].

Homologous FA series within the C_4–C_{24} range can be converted into FAMEs and analyzed using different column combinations. Figure 10.5a illustrates the 2D pattern of human plasma FAMEs analyzed with an apolar × polar column combination. In this pattern, analytes align along diagonal bands based on carbon number (CN) within the C_{14}–C_{24} range (indicated by black arrows) and are grouped into 2D subpatterns according to the number of double bonds (DB). Seven DB bands are marked and aligned across the 2D space in Figure 10.5a. Additional retention characteristics, such as the relative position of DBs, form diagonal retention bands. The ω-derivatives of $C_{20:5\omega3}$, $C_{20:4\omega3}$, and $C_{20:3\omega3}$ are shown in the magnified section of the chromatogram in Figure 10.5b. In Figure 10.5a,b, 36 FAMEs were identified using reference standards, while the remaining 29 components, for which no standards were available, were characterized based on the highly ordered nature of the 2D pattern [118].

10.2.2 Nutrimetabolomics

When integrating knowledge of food composition with nutritional science, the field of nutrimetabolomics becomes increasingly relevant, highlighting a promising path toward advanced nutritional care, targeted dietary treatments, and personalized nutrition [77]. This area of research faces significant challenges, particularly due to the high chemical complexity of biological fluids (such as urine, plasma, and saliva) and the difficulty in generating highly ordered patterns in the presence of many overlapping chemical dimensions. Furthermore, the vast concentration differences between various analytes and analyte classes can lead to system contamination, reduced method precision, and lower long-term reproducibility [119–121].

Consequently, much of the effort in GC×GC-MS-based nutrimetabolomics focuses on achieving optimal separation power in both chromatographic dimensions, ensuring sufficient column loadability and appropriate injection techniques to fully resolve detectable components across the dynamic concentration range, thereby enabling highly reliable absolute or relative quantitation. Recently, an interdisciplinary team of researchers from the FoodBAll Consortium, under the Joint Programming Initiative "A Healthy Diet for a Healthy Life" (JPI-HDHL) [77], reviewed nutrimetabolomics strategies, suggesting best practices and standardized protocols to achieve consistent, high-quality results.

In this complex research context, GC×GC has been effectively employed by Mack et al. [122] to identify robust biomarkers for coffee intake. The authors analyzed the volatile components in human urine collected over 24 hours from participants in the cross-sectional Karlsruhe Metabolomics and Nutrition (KarMeN) study, conducted at the Max Rubner-Institut in Karlsruhe, Germany, from 2011 to 2013. The samples underwent headspace solid-phase microextraction (HS-SPME) followed by GC×GC-QMS and a combined untargeted/targeted fingerprinting approach using pattern recognition algorithms. In total, 138 volatiles were reliably detected, with several showing strong positive correlations to coffee intake. Six key volatiles were identified as the most informative: 2-methyl-furan, guaiacol, 2-/3-methyl-butanoic acid, 2-vinylfuran, and 3,4-dimethyl-2,5-furandione, the latter emerging as the most promising robust biomarker.

A crucial aspect of this study was evaluating the most effective normalization method to account for variations in urine concentration, which provided valuable insights specific to the urine volatilome. The combined untargeted/targeted investigation workflow is depicted in Figure 10.6, showcasing the 2D raw data preprocessing steps, including batch correction using quality standards, normalization based on parameters such as creatinine content, osmolarity, urine volume, mass spectral total useful signal (MSTUS), and probabilistic quotient normalization (PQN), followed by data analysis to identify potential biomarkers through correlation and statistical significance assessments. Further details on the normalization methods and their impact on variation coefficients (CVs%) are discussed in Reference [122].

Another noteworthy study explored the metabolic transformation of dietary polyphenols by the colonic microbiota [123]. The researchers conducted a comparative analysis of *in vitro* microbial phenolic metabolite profiles from selected foods and beverages against those found in the urine of healthy individuals who

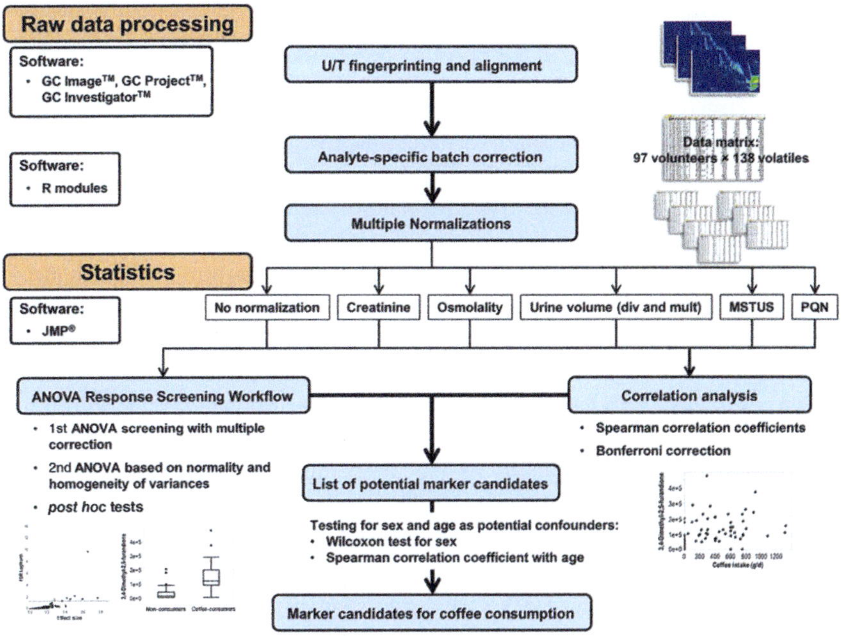

Figure 10.6 Combined untargeted and targeted fingerprinting investigation workflow, including 2D raw data pre-processing step (i.e., batch correction by quality standards and normalization based on different parameters: creatinine content, osmolarity, urine volume, MSTUS, and PQN) followed by data mining to reveal potential biomarkers through combined evaluation of correlation and statistical meaningfulness. *Source:* Mack et al. [122] / with permission of John Wiley & Sons.

participated in an 8-week dietary intervention involving the same food items. The study employed GC×GC-TOFMS in a targeted mode to identify known polyphenolic compounds. Green tea, which was the richest in polyphenols among the tested foods, exhibited significant derivatives of its primary metabolites (phenolic acids), characterized by modifications to the carbon backbone and extensive oxidation (primarily hydroxylation). Among these metabolites, 3,4-Dihydroxybenzoic acid; 3,4-Dihydroxyphenylacetic acid; 3,4-Dihydroxyphenylpropionic acid; 3-Hydroxybenzoic acid; 3-Hydroxyphenylacetic acid; 3-Hydroxyphenylpropionic acid; 4-Coumaric acid; 4-Methylcatechol; 4-Hydroxybenzoic acid; benzoic acid; gallic acid; ferulic acid; sinapic acid; enterodiol; and vanillic acid demonstrated a significant, positive correlation between urinary excretion and *in vitro* colon model results.

Additionally, the study found considerably higher levels of hippuric acid, 3-hydroxybenzoic acid, and ferulic acid in the urine of healthy subjects compared to those in the colon model, while the hepatic conversion model accounted for residual phenol-derived metabolites. This research provided valuable insights into the bioavailability of dietary polyphenols and clarified their connections to urinary metabolite patterns.

An untargeted investigation of urinary metabolite profiles provided deeper insights into the effects of solid versus liquid fructose formulations in animal models [64, 124]. Bressanello et al. [64] developed an analytical system using a one-dimensional column with a dual parallel secondary column setup, coupled to both MS and flame ionization detection (FID) (GC×2GC-MS/FID). This approach was designed to detect changes in the metabolic fingerprints of mice urine samples. The study involved mice on either a control diet or a fructose-enriched diet provided in liquid or solid form, with metabolite profiles tracked over a 12-week period for both individual subjects and subgroups.

The combined untargeted and targeted fingerprinting (UT fingerprinting) allowed for the coherent clustering of mice according to their dietary regimen [44, 63, 125]. Urinary metabolite profiles from mice given high doses of liquid fructose exhibited significant changes in concentrations of fructose, glucose, citric, pyruvic, malic, malonic, gluconic, cis-aconitic, succinic, and 2-ketoglutaric acids, along with glycine acyl derivatives (*N*-carboxyglycine, *N*-butyrylglycine, *N*-isovaleroylglycine, *N*-phenylacetylglycine) and hippuric acid. Post-targeting analysis revealed additional metabolites not initially identified, including *N*-acetyl glucosamine, *N*-acetyl glutamine, malonyl glycine, methyl malonyl glycine, and glutaric acid, with quantitative GC-MS confirming the significance of several potential biomarkers through precise quantification.

Visual feature fingerprinting, using comparative visualization of urine signatures at different time points, demonstrated individual metabolic variations and highlighted the potential for data processing to support personalized strategies. Figure 10.7a,b display pseudocolor images of mouse #40's urine after 6 (a) and 12 weeks (b) of liquid fructose supplementation, where red graphics indicate features identified by the untargeted/targeted workflow, and green graphics represent column bleeding or interferences, which were excluded from the analysis. Figure 10.7c presents a comparative visualization, color-coded to emphasize differences between the urine signatures at 12 weeks (reference pattern) and 6 weeks (analyzed pattern), with yellow areas indicating targeted analytes with notable response differences. Comparative visualization realized by UT fingerprinting based on image pattern recognition can be classified within the Artificial Intelligence tools as an augmented visualization approach [12, 44].

10.2.3 Sensomics and Artificial Intelligence Smelling

Sensomics is focused on deciphering the complex combination of aroma and taste-active molecules that interact with human chemosensory receptors and are

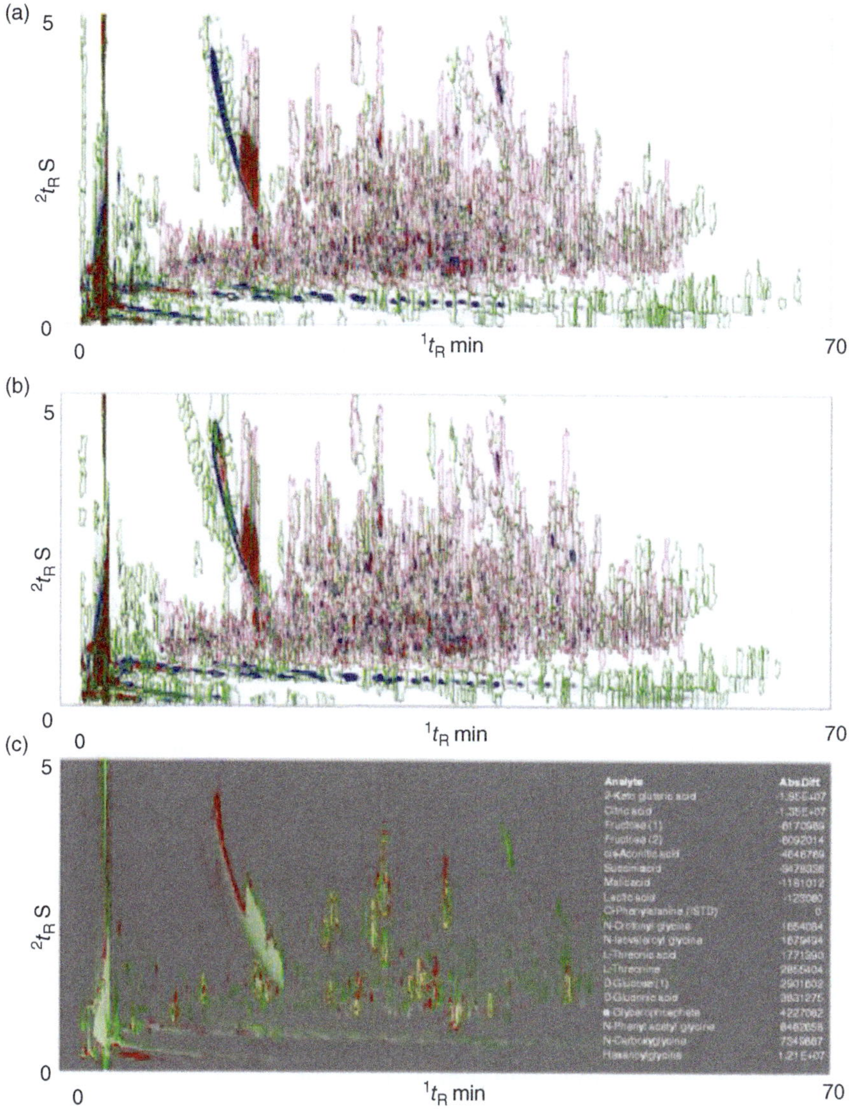

Figure 10.7 Visual features fingerprinting (MS signal full scan acquisition) was performed by comparing urine sample signatures at different time points during the diet intervention. Red graphics indicate untargeted or targeted features, while green graphics represent column bleeding or interferences that were excluded from the analysis. (a) Pseudocolor image of urine samples from mouse #40 after 6 weeks of diet intervention with liquid fructose supplementation, showing the analyzed pattern. (b) Pseudocolor image of urine samples from mouse #40 after 12 weeks of diet intervention with liquid fructose supplementation, depicting the reference pattern. (c) Comparative visualization with colorization to emphasize relative differences between patterns; yellow areas on the 2D space denote targeted analytes, with response differences listed. *Source:* Bressanello et al. [64] / with permission of Springer Nature.

processed by the brain to produce flavor perceptions [126]. Aroma alone accounts for approximately 80% of the hedonic experience of food [79]. This is primarily influenced by volatile compounds present in trace amounts (mg/kg or μg/kg) that typically have low solubility in water, low polarity, and molecular weights under 300 atomic mass units (amu). Odorants interact with odorant receptors in the olfactory epithelium, generating complex signal patterns, or the "Receptor Code," which the nervous system integrates to create olfactory perceptions [79, 127–129].

A crucial aspect of understanding the "aroma blueprint" or the chemical composition of food odors [79, 130] involves thoroughly characterizing all potential ligands and interfering/modulating compounds that form the Chemical Odor Code. This process follows a structured workflow that includes isolating, extracting, and concentrating volatile compounds; chromatographically separating them while using olfactometric screening to identify odor-active components; accurately quantifying them; and validating their aroma contributions through recombination and omission tests [79, 126, 131].

GC×GC platforms are well-suited for implementing the tools needed in sensomics. Significant progress in developing multidimensional chromatographic platforms has been made by Marriott and collaborators [132, 133]. These systems can switch between GC×GC and targeted multidimensional gas chromatography (MDGC) to achieve complete chromatographic resolution, particularly in complex regions where key odorants are present [134]. Additionally, Chin et al. [135] designed a platform that incorporates an orthogonal dimension of olfactometry to study aroma-active compounds in beverages like coffee and wine. This system combines gas chromatography-olfactometry (GC-O) with GC×GC and uses a combination of TOFMS, FID, and flame photometric detection in sulfur mode. This setup allows for the screening, separation, and identification of odor-active compounds effectively.

Conversely, off-line GC-O screenings present certain advantages. They can guide or refine fingerprinting toward regions in the 2D chromatogram that contain odor-active compounds, enhancing machine learning capabilities to discriminate samples based on characteristics likely linked to specific olfactory attributes. A case study by Gabetti et al. illustrates this approach, as they examined the distinctive aroma profiles of Piemonte peppermint essential oils using GC-O through aroma extract dilution analysis (AEDA), alongside GC×GC-TOFMS and variable ionization energy (Tandem Ionization™) [29]. By concentrating on chromatographic areas associated with strong creamy and sweet notes, detectable even at high dilution factors, they were able to enhance the performance of PLS-DA classification models. The authors suggested that a GC×GC-based AI-smelling tool could potentially make sensory-driven decisions without relying on a human sensory panel.

Beyond the multi-multidimensional GC×GC platforms, which play a crucial role in flavor research, conventional systems have now advanced enough to serve as essential tools in sensomics and flavoromics [136]. Nicolotti et al. [33] developed a rapid, sensomics-based expert system called SEBES, designed to predict key aroma compounds in foods without relying on all the individual steps of the traditional workflow [79, 126, 131]. Their method isolates and quantifies approximately 100 of the 226 known key food odorants (KFOs) identified by Dunkel et al. [79]. Using specialized data processing that includes odor thresholds (OTs) information, odor activity values were calculated by comparing concentrations to their respective OTs, thereby encoding the sample's combinatorial odor profile. The authors emphasized that GC×GC was vital for extending the method's dynamic range, enabling precise quantification of analytes present at vastly different concentrations in real samples. The system's efficiency and resolution also allowed for direct injection of crude extracts without additional sample preparation. The SEBES system was validated against traditional sensomics methods, demonstrating a high accuracy with a maximum quantification error of ±20%. The authors concluded that "*it is possible to characterize key food odorants with one single analytical platform and without using the human olfactory system, that is, by artificial intelligence smelling.*"

Numerous other studies have highlighted the key role of GC×GC in flavor analysis. Notably, several have focused on the direct analysis of volatile fractions using headspace (HS) approaches [40, 41, 46, 54, 63, 65, 96, 130, 137, 138]. An emerging area of application involves examining volatile compounds in extra-virgin olive (EVO) oil, where some odorants contribute to desirable attributes or aroma defects [63, 97–99]. Stilo et al. [65] investigated the complex volatile profile of EVO oil from Picual olives harvested at different ripening stages. They employed HS-SPME combined with GC×GC-TOFMS, utilizing both hard and soft ionization techniques (tandem ionization), identifying around 130 compounds out of over 450 detectable volatiles. For many of the target analytes, which play a significant role in the aroma profile of EVO oils [130, 139, 140], linearity in headspace sampling was achieved [141, 142], allowing for accurate quantification in a fully automated process. The use of tandem ionization TOFMS also enhanced the method's dynamic range and response linearity, thereby increasing its analytical capacity and informational value.

Recently, the concept of an AI-based smelling machine has been incorporated into a decision-making tool for the confectionery industry [85]. This approach, utilizing automated MHS-SPME with GC×GC-MS/FID, was used to precisely quantify key aroma compounds, as well as markers of rancidity and spoilage, in premium hazelnuts (*C. avellana* L.) over a 12-month period. Accurate quantification enabled the assessment of chemical quality traits for incoming materials, with results being consistent over time and across different laboratories. This approach facilitated objective decision-making and reliable predictions of sensory attributes (i.e., AI-driven sensomics), geographical origin, and storage duration of samples.

10.2.4 Food Safety

Mineral oil hydrocarbons (MOHs) are complex mixtures composed of various aliphatic hydrocarbons, including linear alkanes, branched isoalkanes, and cyclic compounds like cycloalkanes or naphthenes, collectively known as mineral oil saturated hydrocarbons (MOSHs). They also contain aromatic hydrocarbons (MOAHs), characterized by aromatic structures with one or more benzene rings and extensive alkyl substitutions. Some MOAH components are recognized for their mutagenic, toxic, and tumor-promoting properties.

The presence of MOSH and MOAH in raw food materials, semiprocessed, and finished products is often linked to contamination from external sources. Identifying these contamination sources is crucial for implementing safety measures throughout the food supply chain. However, accurately quantifying and characterizing MOH fractions presents significant challenges for analytical chemistry due to numerous interfering substances found in food. This necessitates the use of robust confirmatory methods to ensure reliability [143–145].

Currently, the most advanced methods for quantifying MOSH and MOAH in food involve LC either offline or online, combined with GC-FID [146]. However, the existence of hundreds to thousands of isomers results in an unresolved complex mixture (UCM) in GC, and the lack of a detection method with adequate specificity can compromise accurate quantitation even when using LC-GC [147, 148]. HR MS is valuable but not sufficient alone to distinctly separate exogenous MOSHs and MOAHs as mandated by EU Commission Decision 657/2002 [149] and recent updates [150].

In this context, the advantages of GC×GC became clear. Biedermann and Grob introduced chromatographic fingerprinting of MOHs using a medium polar × apolar column combination to differentiate compound classes based on retention patterns governed by their relative polarity and volatility. This approach also facilitates the identification of contamination sources through pattern recognition techniques [151]. Figure 10.8 illustrates the comparison between chromatographic profiles obtained by the standard LC-GC-FID method (left side) for MOSH and MOAH fractions and the corresponding GC×GC-TOFMS pseudocolor images of Asian rice suspected of MOH contamination.

The fingerprinting capabilities of GC×GC are demonstrated by the well-defined pattern of *n*-alkanes, ranging from *n*-C_{13} (an internal standard) to *n*-C_{25}, in uncontaminated rice samples. Branched paraffins, which show longer retention times in the second dimension, elute along the same line as *n*-alkanes, while oligomeric

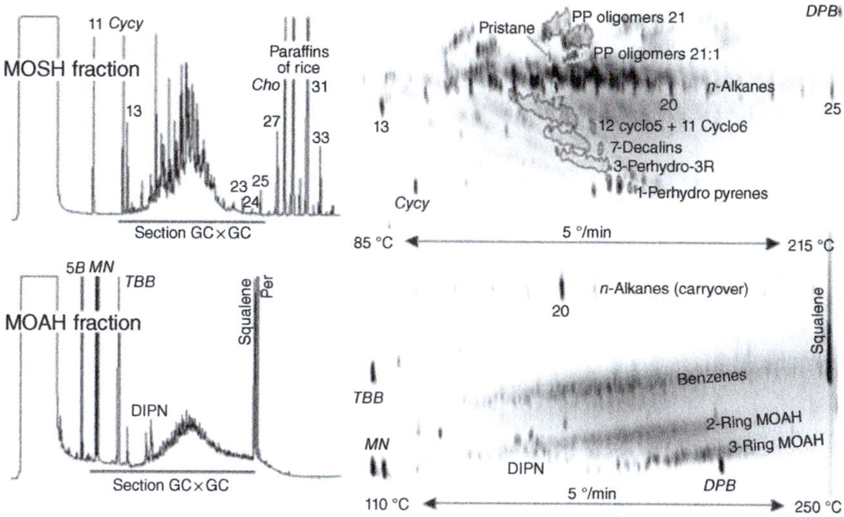

Figure 10.8 LC-GC-FID chromatograms (left side) of MOSH and MOAH fractions and the corresponding GC×GC-TOFMS contour plots (right side) from Asian rice samples. Compound assignments are (11) for *n*-undecane, (Cycy) for cyclohexyl cyclohexane, (Cho) for cholestane, (13) for *n*-tridecane, (20) for *n*-eicosane (DPB) for diphenoxy benzene, (5B) for *n*-pentyl benzene, (MN) for 1- and 2-methyl naphthalene, (TBB) for tri-tert-butyl benzene, (Per) for perylene. Acronyms are explicated in the text. *Source:* Biedermann and Grob [151] / with permission of Elsevier.

polyolefins (POSH), particularly C_{21} oligomers likely derived from polypropylene (PP) packaging, appear as a distinct cluster. Monounsaturated PP oligomers are identified between the *n*-alkanes and the POSH cluster. Additionally, slanted elution bands with shorter retention in the second dimension correspond to naphthenes and cyclic hydrocarbons, such as dodecyl cyclopentanes and undecyl cyclohexanes.

The lower section of Figure 10.8 displays both the 1D and 2D profiles of the MOAH fraction. While the LC-GC method results in an unresolved hump, GC×GC provides more structured elution patterns, clearly identifying various groups, including diisopropyl naphthalenes (DIPN), which may indicate contamination from recycled paperboard.

GC×GC-TOFMS plays a vital role in the precise quantification of MOSH and MOAH, offering opportunities for effective fingerprinting and contamination source identification. This technique could revolutionize the field by providing MD solutions that integrate online LC pre-fractionation with GC×GC and parallel detection using FID/MS [152]. Such an approach, which also supports untargeted investigations, could simultaneously detect nonintentionally added substances [145].

Looking ahead, GC×GC shows significant promise in the context of food safety for the monitoring of chlorinated paraffins (CPs), which are synthetic chlorinated *n*-alkanes used extensively in industry as lubricants, paint components, coating agents, and sealants. Depending on the length of their carbon chains, CPs are classified into short-chain chlorinated paraffins (SCCPs; C_{10-13}), medium-chain chlorinated paraffins (MCCPs; C_{14-17}), and long-chain chlorinated paraffins (LCCPs; C_{18-30}) [153]. Due to their potential toxicity, persistence, and bioaccumulation, SCCPs and MCCPs have attracted recent scrutiny from the European Food Safety Authority (EFSA) [154].

GC×GC coupled with low-resolution TOFMS has been effectively employed for the accurate quantification of SCCPs even in the absence of specific calibration standards for individual components. Researchers [154] created a calibration method linking the total response factor (RF) to chlorine (Cl) content based on peak volumes of quantifier ions for 24 congener groups. Results from both CP products and urban air samples showed significant variation among CP materials, with seasonal changes in urban air levels primarily driven by temperature fluctuations. The highest SCCP concentrations were found in summer, with lighter congeners (such as the C_{10} and Cl_6 groups) more prevalent in the gas phase, and heavier ones (like the C_{13} and Cl_7–Cl_8 groups) found mainly in the particle phase.

These advances, which combine accurate assessments with the identification of contamination sources through informative fingerprinting using GC×GC-TOFMS, could address the data gaps highlighted in the EFSA report [154]. The report noted limited data on SCCP and MCCP levels in certain fish species and suggested that dietary exposure may be underestimated due to the lack of comprehensive data on CPs in other foods. Systematic mapping of potential environmental contamination sources affecting the food chain could provide the robust data needed for a thorough risk assessment, enhancing food safety.

10.2.5 The Greenness Potential of GC×GC When Applied to Food Analysis

GC×GC offers notable greenness potential in food analysis when evaluated through the AGREE (Analytical Greenness) principles and metrics [155]. Its green profile is influenced by solvent and reagent use, energy consumption, and waste generation when analyzed in light of the information potential it provides. While GC×GC requires solvents for sample preparation, miniaturized techniques like HS-SPME or direct on-column sample injection can minimize, or even eliminate solvent volumes. The technique itself generates minimal waste, and adopting greener solvents or solvent-free preparation methods further enhances its sustainability. The use of nonhazardous or less toxic materials contributes positively to its overall environmental impact.

Energy consumption is a consideration for GC×GC due to the additional dimension of separation and the sophisticated instrumentation required. Although it might involve higher energy use compared to traditional GC methods, especially when thermal/cryogenic modulators are used, its efficiency in providing enhanced separation power can reduce the need for repeated analyses, lowering overall energy usage. The ability of GC×GC to analyze complex samples with fewer re-analyses offsets the energy demands associated with the technique.

GC×GC's capability to work with smaller sample volumes and its efficiency in providing comprehensive analytical data contribute to its greenness. When combined with advancements in technology, such as green solvents and effective waste management practices, GC×GC stands out as a valuable, sustainable analytical tool in the context of food quality and safety analysis.

10.3 Conclusions

Food-omics research demands a thorough and holistic approach to effectively capture the compositional complexity of samples and to establish reliable correlations with external stimuli and complex biological processes. Comprehensive two-dimensional chromatography offers a powerful tool to address these challenges, by handling the chemical complexity and varying concentration ranges of samples/fractions, thereby providing a solid foundation for hypothesis development. Enhanced by specialized data processing techniques, this approach further expands research capabilities and aligns with modern individualized and personalized research trends.

GC×GC-MS stands out as a pivotal analytical platform in food-omics research. Its extensive application has the potential to advance the field significantly, offering deeper insights into food quality, nutritional value, and health benefits. The demand for practical solutions in industry does not preclude the use of MDA platforms and omics approaches within industrial research contexts. A broader investigative scope enhances understanding of how processing, storage, fermentation, and biotransformation impact food quality and safety [44, 76, 85, 156, 157].

Initially, the full capabilities of GC×GC were not widely recognized, but its widespread use across various fields and the integration of interdisciplinary strategies have underscored its vital role. It has evolved from being an underutilized technique to becoming an indispensable tool in food analysis: "*from a technique that did not exist… to a technique that was missing*" [158].

List of Abbreviations

^{1}D	first dimension
1D GC	mono-dimensional gas chromatography
^{2}D	second dimension
CI-MS	chemical ionization
CN	carbon number
CPs	chlorinated paraffins
CV	variation coefficients
DB	double bonds
DHS	dynamic headspace
DMF	direct match factor
EI	electron ionization
EVOO	extra virgin olive oil
FAMEs	fatty acids methyl esters
FID	flame ionization detector
FPD	flame photometric detector
GC	gas chromatography
GC-O	GC-olfactometry
GC×GC	comprehensive two-dimensional gas chromatography
GC×2GC-MS/FID	comprehensive two-dimensional gas chromatography with parallel dual secondary column-dual detection by MS and FID
HR MS	high-resolution mass spectrometry
HS	headspace
HS-SPME	headspace solid-phase microextraction
HSSE	headspace sorptive extraction
IR	infrared
KFOs	key-food odorants
LC	liquid chromatography
LC×LC	two-dimensional comprehensive liquid chromatography
LCCPs	long-chain chlorinated paraffins
MCCPs	medium-chain chlorinated paraffins
MD	multidimensional
MDA	multidimensional analytical
MDGC	multidimensional gas chromatography
MLCS	modulated cryogenic system
MOHs	mineral oil hydrocarbons
MOAHs	mineral oil aromatic hydrocarbons
MOSHs	mineral oil saturated hydrocarbons

MS	mass spectrometry
MS/MS (or MS^n)	tandem MS
MSTUS	mass spectral total useful signal
NIAS	non-intentionally added substances
NMR	nuclear magnetic resonance
OAV	odor activity value
ORs	odorant receptors
PEG	polyethylene glycol
PDMS	polydimethylsiloxane
P_M	modulation period
POSH	oligomeric polyolefins
PP	polypropylene
PQN	probabilistic quotient normalization
QMS	quadrupole mass spectrometer
RF	response factor
RI	retention indices
SCCPs	short chain chlorinated paraffins
SCFAs	short-chain fatty acids
SFC	supercritical fluid chromatography
TLC	thin-layer chromatography
TOF	time of flight
UCM	unresolved complex mixture
UV	ultraviolet
UT	untargeted and targeted
VOCs	volatile organic compounds

References

1 Giddings, J.C. (1987). Concepts and comparisons in multidimensional separation. *J. High Resolut. Chromatogr.* 10 (5): 319–323.
2 Liu, Z. and Phillips, J.B. (1991). Comprehensive two-dimensional gas chromatography using an on-column thermal modulator interface. *J. Chromatogr. Sci.* 29 (6): 227–231.
3 Adahchour, M., Beens, J., Vreuls, R.J.J., and Brinkman, U.A.T. (2006). Recent developments in comprehensive two-dimensional gas chromatography (GC × GC). I. Introduction and instrumental set-up. *TrAC - Trends Anal. Chem.* 25 (5): 438–454.
4 Adahchour, M., Beens, J., Vreuls, R.J.J., and Brinkman, U.A.T. (2006). Recent developments in comprehensive two-dimensional gas chromatography

(GC × GC). II. Modulation and detection. *TrAC - Trends Anal. Chem.* 25 (6): 540–553.
5 Seeley, J.V. and Seeley, S.K. (2013). Multidimensional gas chromatography: fundamental advances and new applications. *Anal. Chem.* 85 (2): 557–578.
6 Prebihalo, S.E., Berrier, K.L., Freye, C.E. et al. (2018). Multidimensional gas chromatography: advances in instrumentation, chemometrics, and applications. *Anal. Chem.* 90 (1): 505–532.
7 Tranchida, P.Q., Purcaro, G., Dugo, P. et al. (2011). Modulators for comprehensive two-dimensional gas chromatography. *TrAC - Trends Anal. Chem.* 30 (9): 1437–1461.
8 Fiehn, O. (2002). Metabolomics - the link between genotypes and phenotypes. *Plant Mol. Biol.* 48 (1–2): 155–171.
9 Pierce, K.M., Kehimkar, B., Marney, L.C. et al. (2012). Review of chemometric analysis techniques for comprehensive two dimensional separations data. *J. Chromatogr. A* 1255: 3–11.
10 Pierce, K.M., Hoggard, J.C., Mohler, R.E., and Synovec, R.E. (2008). Recent advancements in comprehensive two-dimensional separations with chemometrics. *J. Chromatogr. A* 1184 (1–2): 341–352.
11 Stilo, F., Bicchi, C., Robbat, A. et al. (2021). Untargeted approaches in food-omics: the potential of comprehensive two-dimensional gas chromatography/mass spectrometry. *TrAC Trends Anal. Chem.* 135: 116162.
12 Caratti, A., Squara, S., Bicchi, C. et al. (2024). Boosting comprehensive two-dimensional chromatography with artificial intelligence: application to food-omics. *TrAC Trends Anal. Chem.* 174: 117669.
13 Trinklein, T.J., Cain, C.N., Ochoa, G.S. et al. (2023). Recent advances in GC×GC and chemometrics to address emerging challenges in nontargeted analysis. *Anal. Chem.* 95 (1): 264–286.
14 Pollo, B.J., Teixeira, C.A., Belinato, J.R. et al. (2021). Chemometrics, comprehensive two-dimensional gas chromatography and "omics" sciences: basic tools and recent applications. *TrAC - Trends Anal. Chem.* 134: 116111.
15 Stilo, F., Bicchi, C., Jimenez-Carvelo, A.M. et al. (2021). Chromatographic fingerprinting by comprehensive two-dimensional chromatography: fundamentals and tools. *TrAC - Trends Anal. Chem.* 134: 116133.
16 Bos, T.S., Knol, W.C., Molenaar, S.R.A. et al. (2020). Recent applications of chemometrics in one- and two-dimensional chromatography. *J. Sep. Sci.* 43 (9–10): 1678–1727.
17 Stefanuto, P.-H., Smolinska, A., and Focant, J.-F. (2021). Advanced chemometric and data handling tools for GC×GC-TOF-MS. *TrAC Trends Anal. Chem.* 139: 116251.

18 Peterson, R.T. (2008). Chemical biology and the limits of reductionism. *Nat. Chem. Biol.* 4 (11): 635–638.

19 Wilson, R.R. (1991). Analytical chemistry of biological systems. *Anal. Chem.* 63 (7): 379 A.

20 Adahchour, M., Beens, J., Vreuls, R.J.J., and Brinkman, U.A.T. (2006). Recent developments in comprehensive two-dimensional gas chromatography (GC × GC). III. Applications for petrochemicals and organohalogens. *TrAC - Trends Anal. Chem.* 25 (7): 726–741.

21 Tranchida, P.Q., Donato, P., Cacciola, F. et al. (2013). Potential of comprehensive chromatography in food analysis. *TrAC - Trends Anal. Chem.* 52: 186–205.

22 Marriott, P. and Shellie, R. (2002). Principles and applications of comprehensive two-dimensional gas chromatography. *TrAC - Trends Anal. Chem.* 21 (9–10): 573–583.

23 Adahchour, M., Beens, J., Vreuls, R.J.J., and Brinkman, U.A.T. (2006). Recent developments in comprehensive two-dimensional gas chromatography (GC × GC). IV. Further applications, conclusions and perspectives. *TrAC - Trends Anal. Chem.* 25 (8): 821–840.

24 Cordero, C., Kiefl, J., Schieberle, P. et al. (2015). Comprehensive two-dimensional gas chromatography and food sensory properties: potential and challenges. *Anal. Bioanal. Chem.* 407 (1): 169–191.

25 Cordero, C., Kiefl, J., Reichenbach, S.E., and Bicchi, C. (2019). Characterization of odorant patterns by comprehensive two-dimensional gas chromatography: a challenge in omic studies. *TrAC Trends Anal. Chem.* 113: 364–378.

26 Gruber, B., Weggler, B.A., Jaramillo, R. et al. (2018). Comprehensive two-dimensional gas chromatography in forensic science: a critical review of recent trends. *TrAC - Trends Anal. Chem.* 105: 292–301.

27 Giddings, J.C. (1995). Sample dimensionality: a predictor of order-disorder in component peak distribution in multidimensional separation. *J. Chromatogr. A* 703 (1–2): 3–15.

28 Tranchida, P.Q., Purcaro, G., Maimone, M., and Mondello, L. (2016). Impact of comprehensive two-dimensional gas chromatography with mass spectrometry on food analysis. *J. Sep. Sci.* 39 (1): 149–161.

29 Gabetti, E., Sgorbini, B., Stilo, F. et al. (2021). Chemical fingerprinting strategies based on comprehensive two-dimensional gas chromatography combined with gas chromatography-olfactometry to capture the unique signature of Piemonte peppermint essential oil (*Mentha ×piperita* var Italo-Mitcham). *J. Chromatogr. A* 1645: 462101.

30 Cordero, C., Schmarr, H.G., Reichenbach, S.E., and Bicchi, C. (2018). Current developments in analyzing food volatiles by multidimensional gas chromatographic techniques. *J. Agric. Food Chem.* 66 (10): 2226–2236.

31 Marriott, P.J. and Nolvachai, Y. (2021). Separation multidimensionality for improved sample characterization—is it worth the effort? *J. Agric. Food Chem.* 69 (39): 11535–11537.

32 d'Acampora Zellner, B., Casilli, A., Dugo, P. et al. (2007). Odour fingerprint acquisition by means of comprehensive two-dimensional gas chromatography-olfactometry and comprehensive two-dimensional gas chromatography/mass spectrometry. *J. Chromatogr. A* 1141 (2): 279–286.

33 Nicolotti, L., Mall, V., and Schieberle, P. (2019). Characterization of key aroma compounds in a commercial rum and an Australian red wine by means of a new sensomics-based expert system (SEBES) - an approach to use artificial intelligence in determining food odor codes. *J. Agric. Food Chem.* 67 (14): 4011–4022.

34 Chin, S.T., Eyres, G.T., and Marriott, P.J. (2015). Application of integrated comprehensive/multidimensional gas chromatography with mass spectrometry and olfactometry for aroma analysis in wine and coffee. *Food Chem.* 185: 355–361.

35 Jennerwein, M.K., Eschner, M., Gröger, T. et al. (2014). Complete group-type quantification of petroleum middle distillates based on comprehensive two-dimensional gas chromatography time-of-flight mass spectrometry (GC×GC-TOF MS) and visual basic scripting. *Energy & Fuels* 28 (9): 5670–5681.

36 Vendeuvre, C., Ruiz-Guerrero, R., Bertoncini, F. et al. (2007). Comprehensive two-dimensional gas chromatography for detailed characterisation of petroleum products. *Oil Gas Sci. Technol.* 62 (1): 43–55.

37 Yan, D.D., Wong, Y.F., Tedone, L. et al. (2017). Chemotyping of new hop (*Humulus lupulus* L.) genotypes using comprehensive two-dimensional gas chromatography with quadrupole accurate mass time-of-flight mass spectrometry. *J. Chromatogr. A* 1536: 110–121. https://doi.org/10.1016/j.chroma.2017.08.020.

38 Ryan, D., Shellie, R., Tranchida, P. et al. (2004). Analysis of roasted coffee bean volatiles by using comprehensive two-dimensional gas chromatography-time-of-flight mass spectrometry. *J. Chromatogr. A* 1054 (1–2): 57–65.

39 Mondello, L., Casilli, A., Tranchida, P.Q. et al. (2004). Comprehensive multidimensional GC for the characterization of roasted coffee beans. *J. Sep. Sci.* 27 (5–6): 442–450.

40 Magagna, F., Guglielmetti, A., Liberto, E. et al. (2017). Comprehensive chemical fingerprinting of high-quality cocoa at early stages of processing:

effectiveness of combined untargeted and targeted approaches for classification and discrimination. *J. Agric. Food Chem.* 65 (30): 6329–6341.

41 Magagna, F., Liberto, E., Reichenbach, S.E. et al. (2018). Advanced fingerprinting of high-quality cocoa: challenges in transferring methods from thermal to differential-flow modulated comprehensive two dimensional gas chromatography. *J. Chromatogr. A* 1536: 122–136.

42 Humston, E.M., Zhang, Y., Brabeck, G.F. et al. (2009). Development of a GC×GC-TOF MS method using SPME to determine volatile compounds in cacao beans. *J. Sep. Sci.* 32 (13): 2289–2295.

43 Humston, E.M., Knowles, J.D., McShea, A., and Synovec, R.E. (2010). Quantitative assessment of moisture damage for cacao bean quality using two-dimensional gas chromatography combined with time-of-flight mass spectrometry and chemometrics. *J. Chromatogr. A* 1217 (12): 1963–1970.

44 Caratti, A., Squara, S., Bicchi, C. et al. (2023). Augmented visualization by computer vision and chromatographic fingerprinting on comprehensive two-dimensional gas chromatographic patterns: unraveling diagnostic signatures in food volatilome. *J. Chromatogr. A* 1699: 464010.

45 Kiefl, J., Pollner, G., and Schieberle, P. (2013). Sensomics analysis of key hazelnut odorants (*Corylus avellana* L. 'Tonda Gentile') using comprehensive two-dimensional gas chromatography in combination with time-of-flight mass spectrometry (GC×GC-TOF-MS). *J. Agric. Food Chem.* 61 (22): 5226–5235.

46 Cordero, C., Liberto, E., Bicchi, C. et al. (2010). Profiling food volatiles by comprehensive two-dimensional ga schromatography coupled with mass spectrometry: advanced fingerprinting approaches for comparative analysis of the volatile fraction of roasted hazelnuts (*Corylus avellana* L.) from different ori. *J. Chromatogr. A* 1217 (37): 5848–5858.

47 Cordero, C., Bicchi, C., and Rubiolo, P. (2008). Group-type and fingerprint analysis of roasted food matrices (coffee and hazelnut samples) by comprehensive two-dimensional gas chromatography. *J. Agric. Food Chem.* 56 (17): 7655–7666.

48 National Institute of Standards and Technology (2005). *NIST/EPA/NIH Mass Spectral Library with Search Program: (Data Version: NIST 08, Software Version 2.0f)*. Gaithersburg, MD: National Institute of Standards and Technology (NIST).

49 McLafferty, F.W. (2020). *Wiley Registry of Mass Spectral Data*, 12e. Wiley.

50 Liberto, E., Cagliero, C., Sgorbini, B. et al. (2008). Enantiomer identification in the flavour and fragrance fields by "interactive" combination of linear retention indices from enantioselective gas chromatography and mass spectrometry. *J. Chromatogr. A* 1195 (1–2): 117–126.

51 d'Acampora Zellner, B., Bicchi, C., Dugo, P. et al. (2008). Linear retention indices in gas chromatographic analysis: a review. *Flavour Fragr. J.* 23 (5): 297–314.

52 Reichenbach, S.E., Tao, Q., Cordero, C., and Bicchi, C. (2019). A data-challenge case study of analyte detection and identification with comprehensive two-dimensional gas chromatography with mass spectrometry (GC×GC-MS). *Separations* 6 (3): 38.

53 Magagna, F., Cordero, C., Cagliero, C. et al. (2017). Black tea volatiles fingerprinting by comprehensive two-dimensional gas chromatography – mass spectrometry combined with high concentration capacity sample preparation techniques: toward a fully automated sensomic assessment. *Food Chem.* 225: 276–287.

54 Nicolotti, L., Cordero, C., Cagliero, C. et al. (2013). Quantitative fingerprinting by headspace-two-dimensional comprehensive gas chromatography-mass spectrometry of solid matrices: some challenging aspects of the exhaustive assessment of food volatiles. *Anal. Chim. Acta* 798: 115–125.

55 Markes International Ltd. (2016). *Select-eV: The Next Generation of Ion Source Technology*. Markes International Ltd.

56 Cordero, C., Guglielmetti, A., Bicchi, C. et al. (2019). Comprehensive two-dimensional gas chromatography coupled with time of flight mass spectrometry featuring tandem ionization: challenges and opportunities for accurate fingerprinting studies. *J. Chromatogr. A* 1597: 132–141.

57 Cialiè Rosso, M., Mazzucotelli, M., Bicchi, C. et al. (2020). Adding extra-dimensions to hazelnuts primary metabolome fingerprinting by comprehensive two-dimensional gas chromatography combined with time-of-flight mass spectrometry featuring tandem ionization: insights on the aroma potential. *J. Chromatogr. A* 1614 (460739): 1–11.

58 Welthagen, W., Mitschke, S., Muhlberger, F., and Zimmermann, R. (2007). One-dimensional and comprehensive two-dimensional gas chromatography coupled to soft photo ionization time-of-flight mass spectrometry: a two- and three-dimensional separation approach. *J. Chromatogr. A* 1150: 54–61.

59 Warren, C.R. (2013). Use of chemical ionization for GC-MS metabolite profiling. *Metabolomics* 9 (Suppl. 1): 110–120.

60 Alam, M.S., Stark, C., and Harrison, R.M. (2016). Using variable ionization energy time-of-flight mass spectrometry with comprehensive GC×GC to identify isomeric species. *Anal. Chem.* 88 (8): 4211–4220.

61 Dubois, L.M., Perrault, K.A., Stefanuto, P.H. et al. (2017). Thermal desorption comprehensive two-dimensional gas chromatography coupled to variable-energy electron ionization time-of-flight mass spectrometry for monitoring subtle changes in volatile organic compound profiles of human blood. *J. Chromatogr. A* 1501: 117–127.

62 Freye, C.E., Moore, N.R., and Synovec, R.E. (2018). Enhancing the chemical selectivity in discovery-based analysis with tandem ionization time-of-flight mass spectrometry detection for comprehensive two-dimensional gas chromatography. *J. Chromatogr. A* 1537: 99–108.

63 Magagna, F., Valverde-Som, L., Ruíz-Samblás, C. et al. (2016). Combined untargeted and targeted fingerprinting with comprehensive two-dimensional chromatography for volatiles and ripening indicators in olive oil. *Anal. Chim. Acta* 936: 245–258.

64 Bressanello, D., Liberto, E., Collino, M. et al. (2018). Combined untargeted and targeted fingerprinting by comprehensive two-dimensional gas chromatography: revealing fructose-induced changes in mice urinary metabolic signatures. *Anal. Bioanal. Chem.* 410 (11): 2723–2737.

65 Stilo, F., Liberto, E., Reichenbach, S.E. et al. (2021). Exploring the extra-virgin olive oil volatilome by adding extra dimensions to comprehensive two-dimensional gas chromatography and time-of-flight mass spectrometry featuring tandem ionization: validation of ripening markers in headspace linearity conditio. *J. AOAC Int.* 104 (2): 274–287.

66 Beccaria, M., Franchina, F.A., Nasir, M. et al. (2018). Investigation of mycobacteria fatty acid profile using different ionization energies in GC–MS. *Anal. Bioanal. Chem.* 410 (30): 7987–7996.

67 Tranchida, P.Q., Salivo, S., Bonaccorsi, I. et al. (2013). Analysis of the unsaponifiable fraction of lipids belonging to various milk-types by using comprehensive two-dimensional gas chromatography with dual mass spectrometry/flame ionization detection and with the support of high resolution time-of-flight mass. *J. Chromatogr. A* 1313: 194–201.

68 Tranchida, P.Q., Salivo, S., Franchina, F.A. et al. (2013). Qualitative and quantitative analysis of the unsaponifiable fraction of vegetable oils by using comprehensive 2D GC with dual MS/FID detection. *Anal. Bioanal. Chem.* 405 (13): 4655–4663.

69 Aloisi, I., Zoccali, M., Dugo, P. et al. (2020). Fingerprinting of the unsaponifiable fraction of vegetable oils by using cryogenically-modulated comprehensive two-dimensional gas chromatography-high resolution time-of-flight mass spectrometry. *Food Anal. Methods* 13 (7): 1523–1529.

70 Herrero, M., Simó, C., García-Cañas, V. et al. (2012). Foodomics: MS-based strategies in modern food science and nutrition. *Mass Spectrometry Reviews* 31 (1): 49–69. https://doi.org/10.1002/mas.20335.

71 Stilo, F., Bicchi, C., Reichenbach, S.E., and Cordero, C. (2021). Comprehensive two-dimensional gas chromatography as a boosting technology in food-omic investigations. *J. Sep. Sci.* 44 (8): 1592–1611.

72 Capozzi, F. and Bordoni, A. (2013). Foodomics: a new comprehensive approach to food and nutrition. *Genes Nutr.* 8 (1): 1–4.

73 Weinert, C.H., Egert, B., and Kulling, S.E. (2015). On the applicability of comprehensive two-dimensional gas chromatography combined with a fast-scanning quadrupole mass spectrometer for untargeted large-scale metabolomics. *J. Chromatogr. A* 1405: 156–167.

74 Jacobs, D.M., van den Berg, M.A., and Hall, R.D. (2021). Towards superior plant-based foods using metabolomics. *Curr. Opin. Biotechnol.* 70: 23–28.

75 Romo-Pérez, M.L., Weinert, C.H., Häußler, M. et al. (2020). Metabolite profiling of onion landraces and the cold storage effect. *Plant Physiol. Biochem.* 146: 428–437.

76 Squara, S., Caratti, A., Fina, A. et al. (2024). Artificial intelligence decision making tools in food metabolomics: data fusion unravels synergies within the hazelnut (*Corylus avellana* L.) metabolome and improves quality prediction. *Food Res. Int.* 194: 114873.

77 Ulaszewska, M.M., Weinert, C.H., Trimigno, A. et al. (2019). Nutrimetabolomics: an integrative action for metabolomic analyses in human nutritional studies. *Mol. Nutr. Food Res.* 63 (1): 1800384.

78 Scalbert, A., Brennan, L., Manach, C. et al. (2014). The food metabolome: a window over dietary exposure. *Am. J. Clin. Nutr.* 99 (6): 1286–1308.

79 Dunkel, A., Steinhaus, M., Kotthoff, M. et al. (2014). Nature's chemical signatures in human olfaction: a foodborne perspective for future biotechnology. *Angew. Chem. Int. Ed.* 53 (28): 7124–7143.

80 Charve, J., Chen, C., Hegeman, A.D., and Reineccius, G.A. (2011). Evaluation of instrumental methods for the untargeted analysis of chemical stimuli of orange juice flavour. *Flavour Fragr. J.* 26 (6): 429–440.

81 Broza, Y.Y., Mochalski, P., Ruzsanyi, V. et al. (2015). Hybrid volatolomics and disease detection. *Angew. Chem. Int. Ed. Engl.* 54 (38): 11036–11048.

82 Phillips, M., Cataneo, R.N., Chaturvedi, A. et al. (2013). Detection of an extended human volatome with comprehensive two-dimensional gas chromatography time-of-flight mass spectrometry. *PLoS One* 8 (9): e75274.

83 Lytou, A.E., Panagou, E.Z., and Nychas, G.-J.E. (2019). Volatilomics for food quality and authentication. *Curr. Opin. Food Sci.* 28: 88–95.

84 Noshad, M., Behbahani, B.A., and Karabagias, I.K. (2023). Volatilomic with chemometrics: a toward authentication approach for food authenticity control. *Eur. Food Res. Technol.* 249: 2215–2226.

85 Squara, S., Caratti, A., Fina, A. et al. (2023). Artificial Intelligence decision-making tools based on comprehensive two-dimensional gas chromatography data: the challenge of quantitative volatilomics in food quality assessment. *J. Chromatogr. A* 1700: 464041.

86 Purcaro, G., Barp, L., Beccaria, M., and Conte, L.S. (2015). Fingerprinting of vegetable oil minor components by multidimensional comprehensive gas chromatography with dual detection. *Anal. Bioanal. Chem.* 407 (1): 309–319.

87 Purcaro, G., Barp, L., Beccaria, M., and Conte, L.S. (2016). Characterisation of minor components in vegetable oil by comprehensive gas chromatography with dual detection. *Food Chem.* 212: 730–738.

88 Cuadros-Rodríguez, L., Ruiz-Samblás, C., Valverde-Som, L. et al. (2016). Chromatographic fingerprinting: an innovative approach for food "identitation" and food authentication – a tutorial. *Anal. Chim. Acta* 909: 9–23.

89 Horning, E.C. and Horning, M.G. (1971). Human metabolic profiles obtained by GC and GC/MS. *J. Chromatogr. Sci.* 9 (3): 129–140.

90 Fiehn, O. (2001). Combining genomics, metabolome analysis, and biochemical modelling to understand metabolic networks. *Comp. Funct. Genomics* 2 (3): 155–168.

91 Beale, D.J., Pinu, F.R., Kouremenos, K.A. et al. (2018). Review of recent developments in GC–MS approaches to metabolomics-based research. *Metabolomics* 14 (11): 152.

92 Wolfender, J.L., Marti, G., Thomas, A., and Bertrand, S. (2015). Current approaches and challenges for the metabolite profiling of complex natural extracts. *J. Chromatogr. A* 1382: 136–164.

93 Alonso, A., Marsal, S., and Julià, A. (2015). Analytical methods in untargeted metabolomics: state of the art in 2015. *Front. Bioeng. Biotechnol.* 3: 23.

94 Bicchi, C. and Maffei, M. (2012). The plant volatilome: methods of analysis. In: *High-Throughput Phenotyping in Plants: Methods in Molecular Biology (Methods and Protocols)*, 289–310. Totowa, NJ: Humana Press.

95 Wong, Y.F., Perlmutter, P., and Marriott, P.J. (2017). Untargeted metabolic profiling of Eucalyptus spp. leaf oils using comprehensive two-dimensional gas chromatography with high resolution mass spectrometry: expanding the metabolic coverage. *Metabolomics* 13 (5): 1–17.

96 Stilo, F., Liberto, E., Reichenbach, S.E. et al. (2019). Untargeted and targeted fingerprinting of extra virgin olive oil volatiles by comprehensive two-dimensional gas chromatography with mass spectrometry: challenges in long-term studies. *J. Agric. Food Chem.* 67 (18): 5289–5302.

97 Da Ros, A., Masuero, D., Riccadonna, S. et al. (2019). Complementary untargeted and targeted metabolomics for differentiation of extra virgin olive oils of different origin of purchase based on volatile and phenolic composition and sensory quality. *Molecules* 24 (16): 1–17.

98 Vaz-Freire, L.T., da Silva, M.D.R.G., and Freitas, A.M.C. (2009). Comprehensive two-dimensional gas chromatography for fingerprint pattern recognition in olive oils produced by two different techniques in Portuguese olive varieties Galega Vulgar, Cobrançosa e Carrasquenha. *Anal. Chim. Acta* 633 (2): 263–270.

99 Cajka, T., Riddellova, K., Klimankova, E. et al. (2010). Traceability of olive oil based on volatiles pattern and multivariate analysis. *Food Chem.* 121 (1): 282–289.

100 Stanimirova, I., Üstün, B., Cajka, T. et al. (2010). Tracing the geographical origin of honeys based on volatile compounds profiles assessment using pattern recognition techniques. *Food Chem.* 118 (1): 171–176.

101 Sudol, P.E., Galletta, M., Tranchida, P.Q. et al. (2022). Untargeted profiling and differentiation of geographical variants of wine samples using headspace solid-phase microextraction flow-modulated comprehensive two-dimensional gas chromatography with the support of tile-based Fisher ratio analysis. *J. Chromatogr. A* 1662: 462735.

102 Tranchida, P.Q., Giannino, A., Mondello, M. et al. (2008). Elucidation of fatty acid profiles in vegetable oils exploiting group-type patterning and enhanced sensitivity of comprehensive two-dimensional gas chromatography. *J. Sep. Sci.* 31 (10): 1797–1802.

103 Tranchida, P.Q., Donato, P., Dugo, G. et al. (2007). Comprehensive chromatographic methods for the analysis of lipids. *TrAC - Trends Anal. Chem.* 26 (3): 191–205.

104 Fina, A., Mascrez, S., Beccaria, M. et al. (2022). A high throughput method for fatty acid profiling using simultaneous microwave-assisted extraction and derivatization followed by reversed fill/flush flow modulation comprehensive multidimensional gas chromatography. *Adv. Sample Prep.* 4: 100039.

105 Ferrara, D., Cescon, M., Giacoppo, G. et al. (2024). Composition and nutritional values of fatty acids in marine organisms by one-step microwave-assisted extraction/derivatization and comprehensive two-dimensional gas chromatography - flame ionization detector. *J. Chromatogr. B* 1236: 124074.

106 Mack, C., Wefers, D., Schuster, P. et al. (2017). Untargeted multi-platform analysis of the metabolome and the non-starch polysaccharides of kiwifruit during postharvest ripening. *Postharvest Biol. Technol.* 125: 65–76.

107 Kanbar, A., Weinert, C.H., Kottutz, D. et al. (2024). Cold tolerance of woodland strawberry (*Fragaria vesca*) is linked to cold box factor 4 and the dehydrin Xero2. *J. Exp. Bot.* 75: 5857–5879.

108 Schmarr, H.-G. and Bernhardt, J. (2010). Profiling analysis of volatile compounds from fruits using comprehensive two-dimensional gas chromatography and image processing techniques. *J. Chromatogr. A* 1217 (4): 565–574.

109 Kfoury, N., Morimoto, J., Kern, A. et al. (2018). Striking changes in tea metabolites due to elevational effects. *Food Chem.* 264: 334–341.

110 Kfoury, N., Scott, E.R., Orians, C.M. et al. (2019). Plant-climate interaction effects: changes in the relative distribution and concentration of the volatile tea leaf metabolome in 2014–2016. *Front. Plant Sci.* 10 (November): 1–10.

111 Stilo, F., Tredici, G., Bicchi, C. et al. (2020). Climate and processing effects on tea (*Camellia sinensis* L. Kuntze) metabolome: accurate profiling and fingerprinting by comprehensive two-dimensional gas chromatography/time-of-flight mass spectrometry. *Molecules* 25 (10): 2447.

112 Chan, E.C.Y., Pasikanti, K.K., and Nicholson, J.K. (2011). Global urinary metabolic profiling procedures using gas chromatography-mass spectrometry. *Nat. Protoc.* 6 (10): 1483–1499.

113 Scalbert, A., Brennan, L., Fiehn, O. et al. (2009). Mass-spectrometry-based metabolomics: limitations and recommendations for future progress with particular focus on nutrition research. *Metabolomics* 5 (4): 435–458.

114 Alygizakis, N., Lestremau, F., Gago-Ferrero, P. et al. (2023). Towards a harmonized identification scoring system in LC-HRMS/MS based non-target screening (NTS) of emerging contaminants. *TrAC - Trends Anal. Chem.* 159: 116944.

115 Morimoto, J., Rosso, M.C., Kfoury, N. et al. (2019). Untargeted/targeted 2D gas chromatography/mass spectrometry detection of the total volatile tea metabolome. *Molecules* 24 (20): 1–14.

116 Delmonte, P., Fardin-Kia, A.R., and Rader, J.I. (2013). Separation of fatty acid methyl esters by GC-online hydrogenation × GC. *Anal. Chem.* 85 (3): 1517–1524.

117 Tranchida, P.Q., Franchina, F.A., Dugo, P., and Mondello, L. (2012). A flow-modulated comprehensive gas chromatography-mass spectrometry method for the analysis of fatty acid profiles in marine and biological samples. *J. Chromatogr. A* 1255: 171–176.

118 Tranchida, P.Q., Costa, R., Donato, P. et al. (2008). Acquisition of deeper knowledge onthe human plasma fatty acid profile exploiting comprehensive 2-D GC. *J. Sep. Sci.* 31 (19): 3347–3351.

119 Squara, S., Manig, F., Henle, T. et al. (2023). Extending the breadth of saliva metabolome fingerprinting by smart template strategies and effective pattern realignment on comprehensive two-dimensional gas chromatographic data. *Anal. Bioanal. Chem.* 415 (13): 2493–2509.

120 Cialiè Rosso, M., Stilo, F., Squara, S. et al. (2021). Exploring extra dimensions to capture saliva metabolite fingerprints from metabolically healthy and unhealthy obese patients by comprehensive two-dimensional gas chromatography featuring Tandem Ionization mass spectrometry. *Anal. Bioanal. Chem.* 413 (2): 403–418.

121 Almstetter, M.F., Oefner, P.J., and Dettmer, K. (2012). Comprehensive two-dimensional gas chromatography in metabolomics. *Anal. Bioanal. Chem.* 402 (6): 1993–2013.

122 Mack, C.I., Egert, B., Liberto, E. et al. (2019). Robust markers of coffee consumption identified among the volatile organic compounds in human urine. *Mol. Nutr. Food Res.* 63 (10): 1801060.

123 Vetrani, C., Rivellese, A.A., Annuzzi, G. et al. (2016). Metabolic transformations of dietary polyphenols: comparison between in vitro colonic and hepatic models and in vivo urinary metabolites. *J. Nutr. Biochem.* 33: 111–118.

124 Mastrocola, R., Cento, A., Chiazza, F.A.U.S.T.O. et al. (2017). Toxicological impact of high fructose intake on gut microbiota and liver/intestine integrity: any differences between solid and liquid formulations? *Diabetologia* 60: S245–S245.

125 Cialiè Rosso, M., Stilo, F., Bicchi, C. et al. (2021). Combined untargeted and targeted fingerprinting by comprehensive two-dimensional gas chromatography to track compositional changes on hazelnut primary metabolome during roasting. *Appl. Sci.* 11 (2): 525.

126 Cialiè Rosso, M., Stilo, F., Bicchi, C. et al. (2021). Combined untargeted and targeted fingerprinting by comprehensive two-dimensional gas chromatography to track compositional changes on hazelnut primary metabolome during roasting. *Applied Sciences* 11 (2): 525. https://doi.org/10.3390/app11020525.

127 Firestein, S. (2001). How the olfactory system makes sense of scent. *Nature* 413 (6852): 211–218.

128 Breer, H., Fleischer, J., and Strotmann, J. (2006). The sense of smell: multiple olfactory subsystems. *Cell. Mol. Life Sci.* 63 (13): 1465–1475.

129 Audouze, K., Tromelin, A., Le Bon, A.M. et al. (2014). Identification of odorant-receptor interactions by global mapping of the human odorome. *PLoS One* 9 (4): e93037.

130 Purcaro, G., Cordero, C., Liberto, E. et al. (2014). Toward a definition of blueprint of virgin olive oil by comprehensive two-dimensional gas chromatography. *J. Chromatogr. A* 1334: 101–111.

131 Grosch, W. (2001). Evaluation of the key odorants of foods by dilution experiments, aroma models and omission. *Chem. Senses* 26 (5): 533–545.

132 Kulsing, C., Nolvachai, Y., and Marriott, P.J. (2020). Concepts, selectivity options and experimental design approaches in multidimensional and comprehensive two-dimensional gas chromatography. *TrAC - Trends Anal. Chem.* 130: 115995.

133 Chin, S.T., Eyres, G.T., and Marriott, P.J. (2012). System design for integrated comprehensive and multidimensional gas chromatography with mass spectrometry and olfactometry. *Anal. Chem.* 84 (21): 9154–9162.

134 Maikhunthod, B., Morrison, P.D., Small, D.M., and Marriott, P.J. (2010). Development of a switchable multidimensional/comprehensive two-dimensional gas chromatographic analytical system. *J. Chromatogr. A* 1217 (9): 1522–1529.

135 Chin, S.T., Eyres, G.T., and Marriott, P.J. (2011). Identification of potent odourants in wine and brewed coffee using gas chromatography-olfactometry and comprehensive two-dimensional gas chromatography. *J. Chromatogr. A* 1218 (42): 7487–7498.

136 Gracka, A., Jelen, H.H., Majcher, M. et al. (2016). Flavoromics approach in monitoring changes in volatile compounds of virgin rapeseed oil caused by seed roasting. *J. Chromatogr. A* 1428: 292–304.

137 Dugo, G., Franchina, F.A., Scandinaro, M.R. et al. (2014). Elucidation of the volatile composition of marsala wines by using comprehensive two-dimensional gas chromatography. *Food Chem.* 142: 262–268.

138 Stilo, F., Liberto, E., Spigolon, N. et al. (2021). An effective chromatographic fingerprinting workflow based on comprehensive two-dimensional gas chromatography – mass spectrometry to establish volatiles patterns discriminative of spoiled hazelnuts (*Corylus avellana* L.). *Food Chem.* 340: 128135.

139 Morales, M.T., Luna, G., and Aparicio, R. (2005). Comparative study of virgin olive oil sensory defects. *Food Chem.* 91 (2): 293–301.

140 Oliver-Pozo, C., Trypidis, D., Aparicio, R. et al. (2019). Implementing dynamic headspace with SPME sampling of virgin olive oil volatiles: optimization, quality analytical study, and performance testing. *J. Agric. Food Chem.* 67 (7): 2086–2097.

141 Kolb, B. and Ettre, L.S. (2006). *Static Headspace – Gas Chromatography: Theory and Practice*, 2e. New York: Wiley.

142 Stilo, F., Cordero, C., Sgorbini, B. et al. (2019). Highly informative fingerprinting of extra-virgin olive oil volatiles: the role of high concentration-capacity sampling in combination with comprehensive two-dimensional gas chromatography. *Separations* 6 (3): 34.

143 Biedermann, M., Munoz, C., and Grob, K. (2020). Epoxidation for the analysis of the mineral oil aromatic hydrocarbons in food. An update. *J. Chromatogr. A* 1624: 461236.

144 Zoccali, M., Barp, L., Beccaria, M. et al. (2016). Improvement of mineral oil saturated and aromatic hydrocarbons determination in edible oil by liquid-liquid-gas chromatography with dual detection. *J. Sep. Sci.* 39 (3): 623–631.

145 Biedermann, M. and Grob, K. (2019). Advantages of comprehensive two-dimensional gas chromatography for comprehensive analysis of potential migrants from food contact materials. *Anal. Chim. Acta* 1057: 11–17.

146 Weber, S., Schrag, K., Mildau, G. et al. (2018). Analytical methods for the determination of mineral oil saturated hydrocarbons (MOSH) and mineral oil aromatic hydrocarbons (MOAH)—a short review. *Anal. Chem. Insights* 13: 1177390118777757.

147 Biedermann, M. and Grob, K. (2012). On-line coupled high performance liquid chromatography-gas chromatography for the analysis of contamination by mineral oil. Part 2: migration from paperboard into dry foods: interpretation of chromatograms. *J. Chromatogr. A* 1255: 76–99.

148 Biedermann, M., Munoz, C., and Grob, K. (2017). Update of on-line coupled liquid chromatography – gas chromatography for the analysis of mineral oil hydrocarbons in foods and cosmetics. *J. Chromatogr. A* 1521: 140–149.

149 Commission Decision 657/2002 (2002). Commission Decision 2002/657/EC implementing Council Directive 96/23/EC concerning the performance of analytical methods and the interpretation of results. *Off. J. Eur. Union* L221 (23 May 1996): 8–36.

150 European Commission (2021). Commission Implementing Regulation (EU) 2021/808 of 22 March 2021 on the performance of analytical methods for residues of pharmacologically active substances used in food-producing animals and on the interpretation of results as well as on the methods to. *Off. J. Eur. Union* 180 (401): 84–109.

151 Biedermann, M. and Grob, K. (2015). Comprehensive two-dimensional gas chromatography for characterizing mineral oils in foods and distinguishing them from synthetic hydrocarbons. *J. Chromatogr. A* 1375: 146–153.

152 Pantò, S., Collard, M., and Purcaro, G. (2020). Comprehensive gas chromatography coupled to simultaneous dual detection (TOF-MS/FID) as a confirmatory method for MOSH and MOAH determination in food. *Curr. Trends in Mass Spectrom.* 15–20.

153 Zou, Y., Niu, S., Dong, L. et al. (2018). Determination of short-chain chlorinated paraffins using comprehensive two-dimensional gas chromatography coupled with low resolution mass spectrometry. *J. Chromatogr. A* 1581–1582: 135–143.

154 Schrenk, D., Bignami, M., Bodin, L. et al. (2020). Risk assessment of chlorinated paraffins in feed and food. *EFSA J.* 18 (3): 1–220.

155 Tobiszewski, M. (2016). Metrics for green analytical chemistry. *Anal. Methods* 8: 2993–2999.

156 Bauwens, G., Cavaco Soares, A., Lacoste, F. et al. (2023). Investigation of the effect of refining on the presence of targeted mineral oil aromatic hydrocarbons in coconut oil. *Food Addit. Contam. Part A* 40 (3): 392–403.

157 Bauwens, G., Pantó, S., and Purcaro, G. (2021). Mineral oil saturated and aromatic hydrocarbons quantification: mono- and two-dimensional approaches. *J. Chromatogr. A* 1643: 462044.

158 Cordero, C. and Bicchi, C. (2020). Third GC×GC-MS school - Sic et simpliciter - mass spectrometry group Italian chemical society. *Division of Mass Spectrometry - Italian Chemical Society*. https://www.spettrometriadimassa.it/scuole_pratiche/3GCGCMSschool_web/.

11

Oil Refinery and Chemical Recycling of Plastics

Do Hyun Lee[1], *Varaha Jayarama Krishna Jonnalagedda*[1,#], *and Hilal Ezgi Toraman*[1,2,3]

[1] *Department of Energy and Mineral Engineering, Pennsylvania State University, University Park, PA, USA*
[2] *Department of Chemical Engineering, Pennsylvania State University, University Park, PA, USA*
[3] *Institute of Energy and the Environment, Pennsylvania State University, University Park, PA, USA*

11.1 Introduction

The global demand for energy is steadily rising due to the expansion of industrialization [1]. Hydrocarbon fuels, particularly fossil fuels, are widely utilized worldwide because of their higher heating value (HHV) in the liquid phase (e.g., gasoline: 47,300 kJ/kg, diesel: 44,800 kJ/kg) [2, 3]. Common hydrocarbon fuels include gasoline (C_4 to C_{12}), used in internal combustion engines; diesel fuels (C_8 to C_{21}), used in diesel engines, and jet fuels, which consist of kerosene-type (C_8 to C_{16}) or naphtha-type (C_5 to C_{15}) hydrocarbons for jet engines [3].

Fuel generation has recently expanded beyond conventional sources to include sustainable alternatives [4]. Growing concerns about climate change and efforts to mitigate its effects have brought increased focus on alternative energy sources [5]. Consequently, alternative and blended sources are becoming more favored for fuel production [6]. Chemical recycling techniques such as pyrolysis, hydrogenolysis, and liquefaction are gaining prominence for their significant potential in reducing plastic waste and converting it into valuable products, including fuels [7–9]. To optimize process variables and the desired feedstock composition, a detailed compositional analysis of fuels produced from waste is essential [10].

Regardless of the fuel source, analytical methods for investigating fuel composition are crucial [11]. Given that most fuels consist of numerous complex compounds, accurate and precise characterization is critical [11, 12]. This need

[#] Present address: Indian Institute of Technology Dharwad, Karnataka, India.

has become increasingly evident with the rising demand for fuels designed for high-performance engines and those that produce lower sulfur or nitrogen oxides emissions. The goal of the crude oil refinery sector is to produce fuels of the highest quality while minimizing process challenges, such as catalyst deactivation, poisoning, and operational issues caused by heteroatoms. These issues include corrosion, compatibility problems, and gum production in a motor engine [13, 14]. Additionally, refineries must comply with environmental regulations set by organizations such as the Environmental Protection Agency (EPA) in the United States and the European Environment Agency (EEA) in the European Union [13].

In chemical recycling research, the relationship between fuel composition and process factors such as feedstock, catalysts, temperatures, reactor flow rate, and heating rates remains largely unexplored [15]. For many years, gas chromatography (GC) has been the primary instrument for analyzing fuel composition, helping to meet strict quality standards [11]. Traditionally, the separation and characterization of chemical compounds were conducted using one-dimensional GC, which separates compounds based on their relative vapor pressures and affinity to the stationary phase [16]. This technique has been instrumental in the fundamental analysis of polymer pyrolysis [17–19] and advancing plastic recycling technology [20]. In GC, a variety of detectors are used to meet diverse analytical requirements [21]. Co-elution presents a significant challenge in the detection of complex analytes in GC [22, 23]. This issue can be potentially addressed by enhancing the column peak capacity, which can significantly improve the separation of analytes [23–25].

Since the 1990s, two-dimensional comprehensive GC (GC×GC) has been used to improve the detection limits, separation performance, and column peak capacity of GC [26, 27]. GC×GC employs two different columns connected in series via a modulator. The role of the modulator is to regulate the flow of the sample from the first column to the second column [10, 16, 27]. Compared to GC, GC×GC offers a more detailed product distribution of fuels due to its improved resolution achieved through modulation and the use of columns differing in terms of volatility and polarity [10, 28]. These advanced capabilities have made GC×GC valuable in various disciplines, such as pharmaceuticals [29], pollutant detection [30], forensic science [31], and petrochemicals [32]. It has become particularly important in oil refineries [33, 34] and has recently been used to evaluate sustainable fuels, addressing rising environmental standards [35–38].

This chapter introduces the application of GC×GC in fuel analysis, focusing particularly on the detailed characterization of complex mixtures. It emphasizes how this technique enhances our understanding of the chemical composition of fuel, which is essential for developing advanced upgrading strategies. This chapter presents case studies in two key areas: oil refineries and the chemical recycling of plastic waste. Finally, it highlights the critical role of GC×GC in analytical chemistry by addressing current challenges and exploring its potential future applications across various fields.

11.2 Comprehensive Two-Dimensional GC (GC×GC)

11.2.1 Overview

Liu and Phillips [26] reported the first GC×GC system, which is now gaining popularity as an effective method for achieving high-resolution separation of complex chemical mixtures. Figure 11.1 provides an overview of a generic GC×GC system, highlighting key components such as columns, modulators, and detectors, which are discussed in the following sections. For the technical discussion of the separation technique and MS, please see dedicated Chapters 3 and 4, respectively.

11.2.2 Components

Figure 11.2 shows the schematic of the GC×GC components. Selecting an appropriate column combination is essential for the successful operation of GC×GC [28, 39, 40]. In the first column, compounds are often separated based on differences in boiling point, while in the second column, separation is achieved based on differences in polarity. This arrangement is referred to as the normal phase configuration [41, 42]. Alternatively, the column phase configuration can be inverted, which is known as the reversed phase configuration. In specific cases involving the separation of complex mixtures, where polarity plays a more significant role than volatility in the initial separation phase, the reverse phase configuration can be beneficial [28, 43].

Another critical component of GC×GC, besides columns, is the modulator. It enables the rapid and efficient transfer of analytes from the first-dimension column to the second-dimension column for further separation [44]. Factors such as filling and flushing times, flow rates, and column lengths can influence the modulator's operation depending on the type of the modulator [40, 44]. There are two main types of modulators: thermal modulators and flow modulators [45]. Thermal modulators work by rapidly heating and cooling a section of the column to trap and release analytes [45]. They offer higher resolution and sensitivity, making them particularly well-suited for complex mixtures. In contrast, flow modulators are well-suited to handle thermally unstable (labile) samples as they do not involve any temperature cycling, thereby minimizing the risk of sample degradation. They operate by regulating the flow within the GC×GC

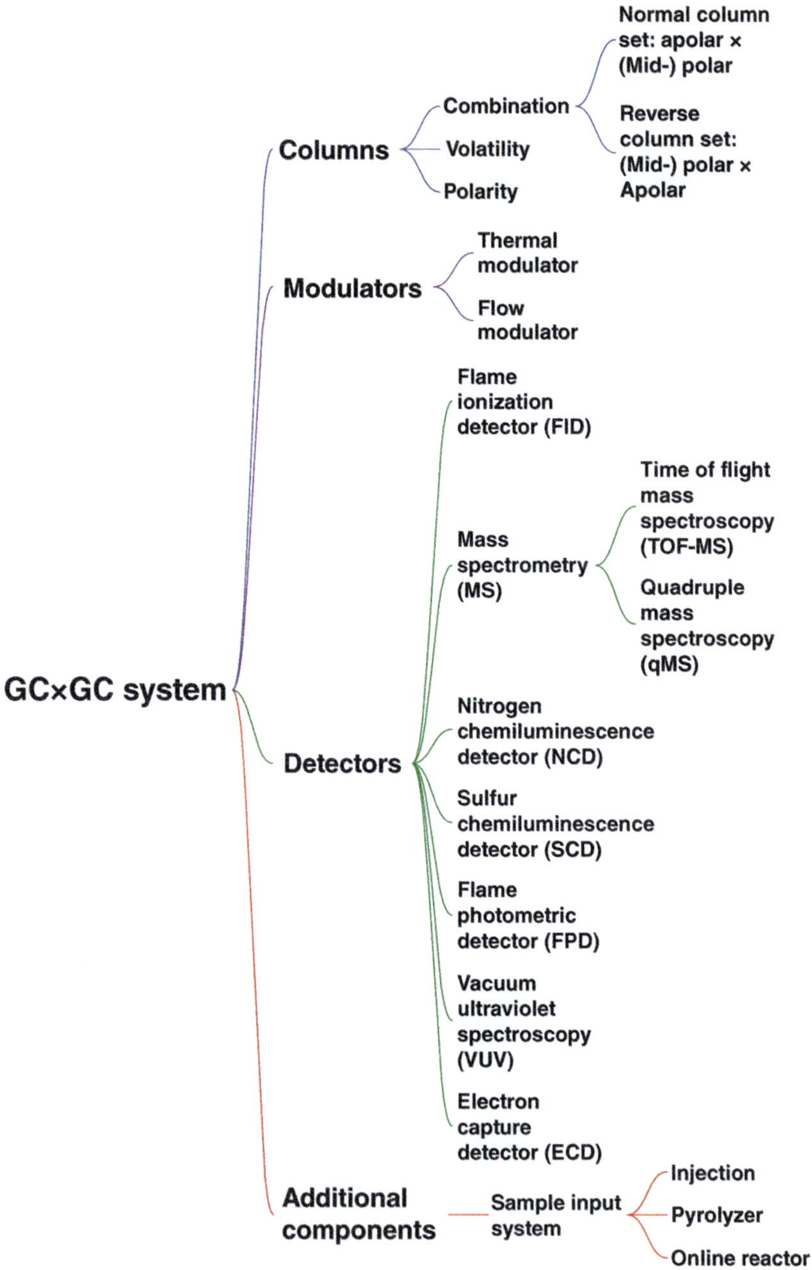

Figure 11.1 Overview of a GC×GC system, highlighting key components such as columns, modulators, and detectors.

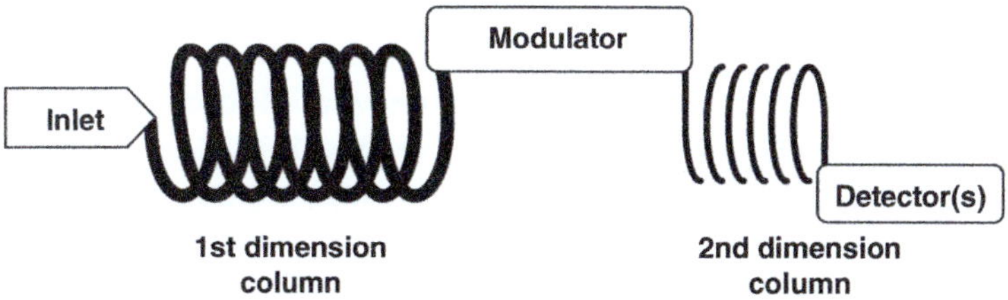

Figure 11.2 Scheme of GC×GC components.

system using a valve mechanism, allowing the transfer of analytes from the first-dimensional column to the second-dimensional column [45]. Additionally, unlike thermal modulators, flow modulators do not require cryogenic cooling with liquid nitrogen or carbon dioxide. The choice between flow and thermal modulators should be guided by the specific requirements of the sample being analyzed [45–47].

Detectors are essential components of GC×GC systems, alongside columns and modulators. When selecting detectors, the chemical composition of the target chemicals and the sample matrix must be considered [10]. Detectors in GC are specifically designed to respond to particular chemical properties of the analytes. Section 11.2.3 provides a detailed discussion of the types of detectors used in GC×GC setups based on their applications.

11.2.3 Types of Detectors

GC×GC can be connected to several detector types, each selected based on specific analytical purposes, as explained below:

(a) **Flame ionization detector (FID):** FID operates by combusting organic compounds in a hydrogen flame, producing ions. The generation of these ions is directly proportional to the concentration of carbon atoms in the organic sample. FID provides the highest response for hydrocarbon samples, with the response being proportional to the number of carbon atoms present. However, it gives a lower response for samples containing heteroatoms such as sulfur, nitrogen, and oxygen [12, 48].

(b) **Sulfur chemiluminescence detector (SCD):** This detector is utilized for analyzing sulfur compounds. It detects sulfur compounds by measuring the emitted light resulting from the reaction between sulfur monoxide and ozone in the sample. This reaction produces excited sulfur dioxide, which emits light as it returns to its ground state [48, 49].

(c) **Nitrogen chemiluminescence detector (NCD):** This detector operates by allowing nitrogen monoxide to react with ozone, forming nitrogen dioxide in an excited state. The light emitted as nitrogen dioxide returns to its ground state is measured, and this intensity is directly correlated with nitrogen concentration. NCD is a valuable tool for analyzing nitrogen-containing compounds in a variety of sample types. It is commonly used in industries such as petrochemicals, food and beverage flavor analysis, and environmental monitoring [48].

(d) **Electron capture detector (ECD):** This detector uses a compact chamber containing electrodes and a radioactive source (^{63}Ni) to ionize the carrier gas [48]. When a sample with electronegative functional groups, such as halogens, phosphorus, and nitro groups, enters the detector, analyte molecules capture electrons, reducing the measured current between the electrodes. By monitoring how analyte compounds interact with electrons, the ECD detects changes in the system's current with high sensitivity. This type of detector is frequently used to identify and analyze substances with high electron affinity [50].

(e) **Mass spectroscopy (MS):** MS is an analytical technique used to determine the molecular mass, chemical formula, and structure of a compound. A typical mass spectrometer consists of three main components: an ionization source, a mass analyzer, and a detector. The ionization source converts sample molecules into ions, commonly through bombardment with high-energy electrons (e.g., electron ionization). These ions are then introduced into the mass analyzer, where they are separated based on their mass-to-charge ratio (m/z). The separated ions reach the detector, which measures their abundance and intensity. The resulting data are displayed as a mass spectrum, in which peaks represent ions with specific m/z values. In GC-MS systems, the most commonly used mass analyzers are Quadrupole MS (QMS) and time-of-flight MS (TOFMS) [16]. These analyzers differ in mass range, resolution, and operating principles. TOFMS is ideal for analytes with broad spectral ranges and higher molecular weights [40], whereas QMS is more suited for targeted compound analysis [40]. Mass spectra obtained using these analyzers can be employed to identify hydrocarbons and heteroatom-containing compounds in samples from unknown or complex sources [51, 52].

(f) **Flame photometric detector (FPD):** FPD detects the intensity of the fluorescence emitted by heteroatom-containing molecules. Analytes ignite in a hydrogen-rich flame after passing through an inert gas-filled column. The detector records the specific wavelengths of light emitted during combustion. FPD is used to detect phosphorus, sulfur, and chlorine molecules, as these elements emit characteristic light upon detection [38, 44, 53].

(g) **Vacuum ultraviolet spectroscopy (VUV):** VUV spectroscopy operates within the UV band, detecting activity between 120 and 240 nm [54]. It provides information on time, absorbance, and wavelength [54].

Due to its broad absorption range, VUV spectroscopy effectively distinguishes individual components. It has been applied in analyses of permanent gases, health-related applications, and oil refinery processes [55].

11.3 Comprehensive Exploration of Oil Refining Processes

11.3.1 Technological Evolution and Impact of GC×GC in Oil Refineries

The efficient utilization of existing energy resources is essential to meeting global energy demands, which are primarily fulfilled by fossil fuels, particularly crude oil. Oil refineries employ several processes, such as hydroprocessing, catalysis, and distillation, to improve the quality of fuel products.

To the best of the author's knowledge, the first application of GC×GC was reported in 1993, when the technique was used to analyze commercial kerosene. This analysis detected over 6000 distinct peaks, demonstrating the technique's high peak capacity [56]. Chemical groups, including aliphatics, aromatics, and naphthalene derivatives, were identified by categorizing peak bands based on retention times [56].

GC×GC offers numerous benefits in oil refineries, as highlighted in various studies [14, 33, 57]. This advanced technique has proven valuable in evaluating the composition and quality of complex oil products. Besides, continuous advancements were made to the critical components related to GC×GC such as column design [58], modulators [59, 60], detectors [61], and data analysis techniques [62] to ensure that there is significant improvement in the resolution, identification, and accurate quantification of oil products from refinery processes. Figure 11.3 shows a comprehensive map of GC×GC analysis of oil refinery products, illustrating two key aspects: sample types and detector technologies. Sample types are broadly classified into conventional or alternative fuels, while detectors are categorized into MS, FID, SCD, NCD, FPD, and VUV.

11.3.2 Advancements in Oil Product Analysis using GC×GC

Distillates and commercial products are primary analytes in GC×GC analysis. Initially applied to commercial oils, the application of GC×GC has diversified significantly. This section highlights the key and emerging uses of this method.

Table 11.1 illustrates the application of GC×GC systems in oil refinery product analysis. For instance, FID is primarily used to quantify the chemical composition of commercial products, such as diesel [70], gasoline [112], jet fuels [188],

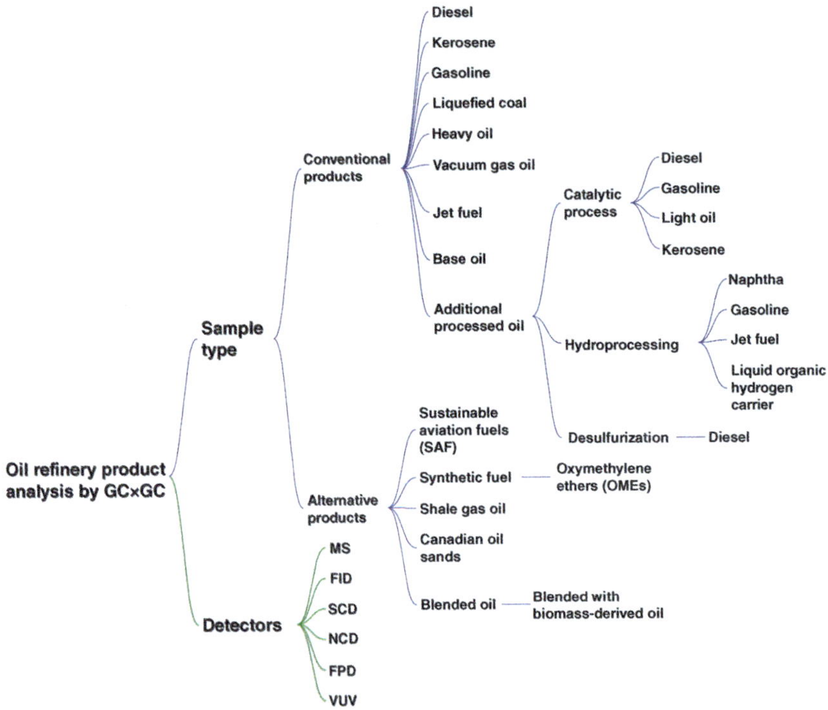

Figure 11.3 Comprehensive overview of oil refinery product analysis using GC×GC.

gas oil [144], and distillates [201]. Additionally, detectors such as MS, SCD, NCD, FPD, and VUV have enhanced resolution and enabled the identification of trace compounds crucial for assessing oil product quality [221, 222].

Detecting trace heteroatom compounds in oil is essential for addressing environmental concerns, as these chemicals contribute to air pollution and negatively impact refinery processes and engine efficiency. Early applications of GC×GC-NCD in diesel fuel analysis increased sensitivity and resolved co-elution issues, enabling the detection of nitrogen-containing compounds like indoles and carbazoles (20 ppm in total), which are neutral nitrogen compounds found in diesel [69]. Hydrodesulfurized middle distillates were also analyzed to detect basic and neutral nitrogen compounds that affect catalyst stability [84].

Djokic et al. [144] developed a method to identify and measure sulfur-containing compounds found in complex hydrocarbon matrices by utilizing the results obtained from GC×GC coupled with SCD and FID along with a refinery gas analyzer (RGA). The developed method was applied to the steam cracking of hydrocarbons and vacuum gas oil (VGO). It was observed that steam cracking of

Table 11.1 Application of GC×GC systems in oil refinery product analysis.

Analytes	Detectors	Modulators	Year	References
Commercial kerosene	FID	Thermal modulator	1993	[56]
Catalytic cracking heavy cycle oil and hydrotreated products	FID	Thermal modulator	1997	[34]
Heavy gas oils	FID	Thermal modulator	1998	[63]
Nonaromatic hydrocarbon solvents, kerosenes, crude oils, olefin-containing saturated oils	FID	Thermal modulator	2000	[64]
JP-TS jet fuel	FID	Thermal modulator	2000	[65]
Light mineral oil condensate	FID	Thermal modulator	2003	[66]
Fuel oil no.6 (heavy fuel oil)	FID	Flow modulator	2004	[67]
Jet fuels	FID	Flow modulator	2004	[68]
Diesel fuels	NCD	Thermal modulator	2004	[69]
Diesel fuels polycyclic aromatic hydrocarbons	FID	Thermal modulator	2005	[70]
Gasoline	FID	Flow modulator	2005	[71]
Gasoline	FID	Flow modulator	2005	[72]
Kerosene	FID	Thermal modulator	2005	[73]
Diesel fuels	MS with supersonic molecular beam	Flow modulator	2006	[74]
Diesel fuels	SCD	Thermal modulator	2006	[75]
Liquefied coal	TOFMS	Thermal modulator	2007	[76]
Petroleum source rocks (with pyrolyzer)	FID, SCD	Thermal modulator	2007	[77]

(Continued)

Table 11.1 (Continued)

Analytes	Detectors	Modulators	Year	References
Biodiesel, petroleum diesel, and biodiesel/petroleum diesel blends	FID	Flow modulator	2007	[78]
Lubricant base oils	FID	Thermal modulator	2007	[79]
E85 standard fuels (ethanol with gasoline)	FID	Flow modulator	2008	[80]
Diesel fuels	FID	Flow modulator	2008	[81]
Petroleum diesel blended with fatty acid from biomass oil and alcohols	TOFMS, FID	Thermal modulator	2008	[82]
Biodiesel and biodiesel blends	FID	Thermal modulator	2008	[83]
Diesel cuts (distillation, cokefaction, FCC, hydroconversion), liquefied coal	NCD	Thermal modulator	2009	[84]
Motor oils	TOFMS, FID	Thermal modulator	2009	[85]
Diesel cuts (light cycle oil, liquified coke)	FID	Thermal modulator	2010	[86]
Vacuum gas oils (distillation, coking, hydrotreatment, hydrocracking)	FID	Thermal modulator	2010	[87]
Vacuum gas oils	TOFMS, FID	Thermal modulator	2010	[88]
Vacuum gas oils (distillation, hydrotreatment, hydrocracking, mixture with gas oil and desaphalted oils), desaphalted oil	FID	Thermal modulator	2010	[89]
Military fog oil	TOFMS, FID	Thermal modulator	2010	[90]
Motor oil of a biodiesel fueled-car, biodiesel	TOFMS, FID	Thermal modulator	2010	[91]

Table 11.1 (Continued)

Analytes	Detectors	Modulators	Year	References
87-octane unleaded gasoline, mixed hydrocarbon with dialkyl phosphates	TOFMS, FID	Thermal modulator	2010	[92]
Diesel fuels	(EI)MS	Thermal modulator	2010	[93]
Fischer–Tropsch oil	TOFMS, FID	Not mentioned	2010	[94]
Heavy gas oil residues	TOFMS	Thermal modulator	2010	[95]
Ceara basin oil	TOFMS	Thermal modulator	2011	[96]
Extra heavy gas oil	TOFMS	Thermal modulator	2011	[97]
Diesel, mineral oil distillate	TOFMS	Thermal modulator	2011	[98]
Gasoline	FID	Thermal modulator	2011	[99]
Industrial distillated petroleum samples, standard with phosphates	TOFMS	Thermal modulator	2011	[100]
Vacuum gas oil	SCD	Thermal modulator	2011	[101]
Heavy fuel oil	TOFMS	Thermal modulator	2011	[102]
Diesel, blend with biodiesel	MS	Not mentioned	2011	[103]
Naphthenic acids refined from petroleum	TOFMS	Thermal modulator	2011	[104]
Straight run gas oil (SRGO), hydrodesulfurized SRGO ($CoMo/Al_2O_3$ catalyst)	QMS, FPD	Thermal modulator	2011	[53]
Gas oil cuts	TOFMS	Thermal modulator	2012	[105]
Light cycle oil (LCO), hydrotreated LCO (Pt-Pd on supports [Y Zeolite, HY Zeolite, FCC, amorphous γ-alumina])	MS, FID	Flow modulator	2012	[106]

(Continued)

Table 11.1 (Continued)

Analytes	Detectors	Modulators	Year	References
Vacuum gas oil	TOFMS, FID	Thermal modulator	2012	[107]
Special Antarctic blend diesel	FID	Flow modulator	2012	[62]
Kerosene oils	TOFMS, FID	Thermal modulator	2012	[108]
Gas oils	SCD	Thermal modulator	2012	[109]
Coal liquefaction middle distillate	FID	Not mentioned	2012	[58]
Kerosene, JP8-Jet fuel	FID	Flow modulator	2012	[110]
Vacuum gas oil with supercritical fluid chromatography	FID, SCD	Thermal modulator	2013	[111]
Gasoline, kerosene oils	QMS, FID	Flow modulator	2013	[112]
Gasoline, reformate, fluid catalytic cracked oils	QMS, FID	Flow modulator	2013	[113]
Middle distillate samples	TOFMS, NCD	Thermal modulator	2013	[114]
Diesel samples	NCD	Thermal modulator	2013	[115]
Heavy gas oil	TOFMS	Thermal modulator	2013	[116]
FCC gasoline (heteropolyacid/SiO_2)	MS, FPD	Thermal modulator	2013	[117]
Hydrodesulfurized product oil (CoMo on nano-γ alumina and micro-γ alumina)	SCD	Not mentioned	2013	[118]
Petroleum samples (trace organic sulfur compound included)	TOFMS	Thermal modulator	2014	[119]
Rocket propulsion fuel	TOFMS	Thermal modulator	2014	[120]
Kerosene (mild hydrocracking, distillate, hydrocracking, vegetable oil-blended)	TOFMS	Thermal modulator	2014	[121]

Table 11.1 (Continued)

Analytes	Detectors	Modulators	Year	References
JP-8 jet fuel, oxidative desulfurized JP-8 oils	TOFMS, SCD	Not mentioned	2014	[122]
Petroleum middle distillates (Jet fuel A1, Diesel blended with biodiesel, light heating oil)	TOFMS	Thermal modulator	2014	[123]
Hydrotreated diesel with oxidative desulfurization	FID, SCD	Thermal modulator	2015	[13]
Vacuum gas oil	FID	Thermal modulator, flow modulator	2015	[47]
Diesel fuel	FID, BID	Flow modulator	2015	[124]
Arabic heavy gas oil, sulfur-removed heavy gas oil	QqQMS	Flow modulator	2015	[60]
Base oil samples	FID	Thermal modulator	2015	[125]
Rocket propulsion (RP-1) fuel	TOFMS	Thermal modulator	2015	[126]
Commercial diesel fuel	QMS	Thermal modulator	2015	[127]
Ultralow sulfur diesel samples	TOFMS	Thermal modulator	2015	[128]
Light diesel	FID, (ENCI)TOFMS, ECD	Thermal modulator	2015	[129]
Ethanol blended with gasoline (E85)	FID	Flow modulator	2015	[59]
Heavy fuel oil	TOFMS	Thermal modulator	2016	[130]
Rocket propulsion (RP-1) fuel	TOFMS, FID	Diaphragm valve	2016	[131]
Diesel fuel	TOFMS	Diaphragm valve	2016	[132]
Diesel, catalytic pressureless depolymerization fuel (solid recovered fuel from demolition waste, municipal solid waste, lime, catalyst)	TOFMS, FID	Thermal modulator	2016	[133]

(Continued)

Table 11.1 (Continued)

Analytes	Detectors	Modulators	Year	References
JP8 Jet fuel	QMS	Flow modulator	2016	[134]
Reformate (petroleum-refinery reforming)	FID	Flow modulator	2016	[135]
Oxidized kerosene-based jet fuel	FID	Flow modulator	2016	[27]
Diesel fuel	TOFMS	Thermal modulator	2016	[136]
Gasoline, kerosene, adulterants	FID	Thermal modulator	2016	[137]
Diesel, diesel blended with biodiesel	FID	Thermal modulator	2016	[138]
High temperature-Fischer–Tropsch light oil, Online-hydrogenated sample in GC×GC (Pt or Pd acetyl-acetonate salt)	TOFMS	Thermal modulator	2016	[139]
Diesel, crude oil, phytane, crocetane	TOFMS	Thermal modulator	2016	[140]
Diesel, kerosene	FID	Thermal modulator	2016	[141]
Shale oil (Fisher assay pyrolysis, Mahogany zone)	MS	Thermal modulator	2017	[142]
Straight-run gas-oil	FID	Thermal modulator	2017	[143]
Steam cracked vacuum gas oil (on-line analysis)	FID, SCD, NCD	Thermal modulator	2017	[144]
Straight-run distillation	(FI)TOFMS	Thermal modulator	2017	[145]
Standard middle distillates (diesel [petrochemical, blended with biodiesel, heating oil, Jet fuel A1] modern high-performance diesel [mixed with hydrated vegetable oil])	TOFMS	Thermal modulator	2017	[146]
Thermal-stressed jet fuels	TOFMS, FPD	Thermal modulator	2017	[147]

Table 11.1 (Continued)

Analytes	Detectors	Modulators	Year	References
Petroleum distillate, Hop oil, GLC-85	MS, FID	Thermal modulator	2017	[148]
Kerosene	FID	Thermal modulator	2017	[149]
Hydrogenated heavy paraffinic fraction and, oxidized heavy paraffinic fraction	TOFMS	Thermal modulator	2017	[150]
Diesel fuel	TOFMS	Thermal modulator	2017	[151]
Pyrolyzed Estonian oil shale	TOFMS, FID	Thermal modulator	2017	[38]
Synthetic fuels (liquefied coal, hydrotreated vegetable oil, Jet A-1, Readyjet, ATJ-SKA, farnesane, ATJ-SPK)	MS	Thermal modulator	2017	[152]
Aviation hydrocarbon fuel samples	FID, MS	Thermal modulator	2017	[153]
Diesel fuels	FID, MS	Thermal modulator	2017	[154]
Aviation fuels (biomass-derived, syngas-derived, petroleum, oxygenated fuels)	TOFMS	Thermal modulator	2017	[155]
Heavy fuel oil	TOFMS	Thermal modulator	2017	[156]
Kerosene, diesel	FID	Flow modulator	2018	[157]
Steam cracking pyrolysis fuel oil, Arabian crude oil	TOFMS, FID	Thermal modulator	2018	[158]
Diesel fuel	TOFMS	Flow modulator	2018	[159]
Hydrotreated slurry oil (fluid catalytic cracking)	TOFMS	Thermal modulator	2018	[160]
Steam cracked oils – pyrolysis fuel oil, pyrolysis gasoline (naphtha, vacuum gas oil),	TOFMS, FID	Thermal modulator	2018	[32]

(Continued)

Table 11.1 (Continued)

Analytes	Detectors	Modulators	Year	References
Aviation fuels, diesel fuels, blend (petroleum, synthetic, bio-derived)	FID	Thermal modulator	2018	[161]
Catalytic upgraded oil (vacuum gas oil, kerosene, and model oxygenated compounds from biomass pyrolysis [acetic acid, guaiacol, and sorbitol], E-Cat Y-zeolite catalyst for cracking and upgrading)	TOFMS	Thermal modulator	2018	[162]
Diesel fuels (blended diesel with atmospheric heavy naphtha, intermediate light diesel, hydrodesulfurized heavy diesel oil, light cycle oil, light coker gas oil, heavy diesel, oxidative desulfurization with $Na_2WO_4 \cdot 2H_2O$ as catalyst)	MS, FPD	Thermal modulator	2019	[163]
Diesel fuels (unstable, stable)	NCD	Thermal modulator	2019	[164]
Base oil samples (solvent-free manufactured, hydroprocessed)	TOFMS	Thermal modulator	2019	[165]
Diesel fuel blended with biodiesel (B7)	TOFMS	Thermal modulator	2019	[166]
Bunker fuels (connected with TGA)	TOFMS	Thermal modulator	2019	[167]
Group II base oils	TOFMS	Thermal modulator	2019	[168]
Additional distilled vacuum gas oil	TOFMS, FID	Thermal modulator	2019	[169]
Petroleum-based diesel fuel	TOFMS, FID	Thermal modulator	2019	[170]
Jet fuels (petroleum-derived fuel, biofuel, synthetic fuel, blended)	TOFMS, FID	Thermal modulator	2019	[171]

Table 11.1 (Continued)

Analytes	Detectors	Modulators	Year	References
Surrogated fuel (RP-3 Kerosene)	TOFMS	Thermal modulator	2019	[172]
Diesel fuels	TOFMS	Thermal modulator (solid-state modulator)	2019	[173]
Aerospace kerosene products (Jet-A, JP-5, JP-8, RP-1, RP-2)	TOFMS	Thermal modulator	2020	[174]
Oxidative desulfurization fuels (bitumen-derived gas oil)	TOFMS, FID, SCD	Thermal modulator	2020	[175]
Gasoline, light gas oil	FID	Flow modulator	2020	[176]
Basic nitrogen compounds (straight run vacuum gas oil)	TOFMS	Thermal modulator	2020	[177]
Coal tars	MS	Thermal modulator	2020	[178]
Vacuum distillates, vacuum gas oil	FID	Thermal modulator	2020	[179]
Model bitumen (heavy crude oil)	TOFMS	Thermal modulator	2020	[180]
Diesel fuel with programmable temperature vaporizer	FID	Flow modulator	2020	[181]
Diesel fuel	TOFMS	Thermal modulator	2020	[182]
Kerosene, unleaded petrol	FID	Flow modulator	2020	[183]
Diesel	TOFMS	Flow modulator	2020	[184]
Oligomerized 1-butene (HZSM-5, hierZ-MFI)	TOFMS	Thermal modulator	2020	[185]
Diesel fuels	FID	Flow modulator	2020	[186]
Jet fuel (GC^3-TOFMS)	TOFMS	Thermal modulator (1st junction), flow modulator (2nd junction)	2020	[187]

(Continued)

Table 11.1 (Continued)

Analytes	Detectors	Modulators	Year	References
Jet fuel F-24 sample	FID	Thermal modulator	2020	[188]
Diesel fuel	MS, VUV	Thermal modulator	2020	[189]
Diesel, biodiesel blended with diesel	QMS	Flow modulator	2021	[190]
Diesel fuel	FID	Thermal modulator (solid-state modulator)	2021	[191]
Gas oils (straight run, light cycle oil, coker gas oil, ebullated-bed gas oil, hydrotreated gas oil)	MS, NCD	Thermal modulator	2021	[192]
Slurry oils (hydrotreated, vacuum distillation, distilled, solvent extracted)	TOFMS	Not mentioned	2021	[193]
Gas oils (straight-run gas oil, light cycle oils (LCO) from fluid catalytic cracking, coker gas oils, hydroconverted gas oils, hydrotreated gas oils)	FID, VUV	Flow modulator	2021	[61]
Diesel, diesel feeds (heavy kerosene, light cycle oil, light coker gas oil, post-treatment unit samples)	FID	Thermal modulator	2021	[194]
Jet A	FID	Thermal modulator	2021	[195]
JP8 Jet fuel with sulfur-containing compounds	TOFMS	Thermal modulator	2021	[196]
Sustainable aviation fuel candidates (Syntroleum FT-SPK, Dynamic Fuels HEFA-SPK, Sasol FT-SPK, UOP HEFA-SPK, Gevo ATJ, Lanzatech ETJ)	MS, FID	Flow modulator	2021	[197]

Table 11.1 (Continued)

Analytes	Detectors	Modulators	Year	References
Aviation fuels (RP-3 fuel, thermally stressed, hydrotreated)	MS, FID	Thermal modulator	2021	[198]
Hydroprocessed diesel fraction (canola oil blended with heavy vacuum gas oil, NiMo/Al_2O_3 and CoMo/Al_2O_3)	FID, SCD	Thermal modulator	2021	[199]
Diesel fuel (with non-native compounds)	TOFMS	Thermal modulator	2022	[200]
Distillation cuts (naphtha, kerosene, diesel, gas oil)	TOFMS, FID	Thermal modulator	2022	[201]
Pyrolyzed shale oil	TOFMS	Thermal modulator	2022	[202]
Pyrolysis oil (kerogens extracted from shale oil)	FID	Thermal modulator	2022	[203]
JP-10 (high-energy-density fuels)	MS, FID	Thermal modulator	2022	[204]
Gas oils (straight-run gas oil, light cycle oils from fluid catalytic cracking units, coker gas oils [CK GO], hydroconverted gas oils, hydrotreated gas oils)	VUV	Flow modulator	2022	[205]
Jet fuel (JP8 with sulfur-containing compounds)	TOFMS	Thermal modulator	2022	[206]
Jet fuel (JP8 with sulfur-containing compounds, J1800A, JP4 with sulfur-containing compounds)	TOFMS	Thermal modulator	2022	[207]
Middle and low temperature coal tar	TOFMS	Thermal modulator	2022	[208]
Catalytic pyrolysis oil (woody biomass, Miscanthus) with vacuum gas oil (Pt/TiO_2, or zeolite)	TOFMS	Thermal modulator	2022	[209]

(Continued)

Table 11.1 (Continued)

Analytes	Detectors	Modulators	Year	References
Hydrotreated oil (stabilized fast pyrolysis oil (Ni-based catalyst), stabilized deoxygenated pyrolysis oil (CoMo catalyst), hydrothermal liquefaction oil (pine wood) with vacuum gas oil, sulfided NiMo containing Y-zeolite)	FID, VUV, MS	Flow modulator (FID, VUV), thermal modulator (MS)	2022	[36]
Fluid catalytic cracked oil (water-soluble bio-oil from fast pyrolysis oil with vacuum gas oil, E-cat)	MS	Thermal modulator	2022	[210]
Aerospace fuels (RP-1, RP-2, JP-5, JP-7, JP-8, Jet-A, UL-RP-1, RP-TS-5)	TOFMS	Thermal modulator	2023	[211]
Fuels (heavy fuel oil, vacuum residue oil) with pyroprobe	TOFMS, FID, SCD	Thermal modulator	2023	[212]
Diesel sample	TOFMS	Thermal modulator	2023	[213]
Pyrolysates (OME-2)	QMS, FID	Not mentioned	2023	[214]
Water-saturated jet fuels (Jet A, F-24, and JP-5)	MS, FID	Flow modulator	2023	[215]
Aviation fuels (Jet A-1, sustainable aviation fuels [SPK-HEFA]), biodiesel	TOFMS, FID	Thermal modulator	2023	[37]
Jet fuels (Jet A1, Jet A2, Jet A3, F-24), synthetic jet fuel (ATJ), six varied cetane number specialty jet fuels (CN30, CN35, CN40, CN45, CN50, CN55)	TOFMS, FID	Thermal modulator	2023	[216]
Conventional and synthetic fuels (gasoline, Jet-A1, diesel, HEFA [hydroprocessed esters and fatty acids], Fisher-Tropsch crude oil, methanol to gasoline)	QMS, FID	Thermal modulator	2023	[57]

Table 11.1 (Continued)

Analytes	Detectors	Modulators	Year	References
JP-8 jet fuel spiked with 10 polar compounds	TOFMS	Thermal modulator	2023	[217]
Pyrolysis fuel oil (Hydrogenation over NiMoAl catalyst, Dehydrogenation over Pd/C, Isomerized hydrogenated oil over $AlCl_3$)	MS	Flow modulator	2023	[218]
JP-4 jet fuel spiked with 38 compounds at 400 ppm each	TOFMS	Thermal modulator (1st junction), flow modulator (2nd junction)	2023	[219]
Thermal-dissoluted coals	TOFMS	Thermal modulator	2024	[220]

VGO converted a large fraction of dibenzothiophenes, while benzothiophenes were found to be less reactive. The major sulfur product detected using GC×GC-SCD was hydrogen sulfide, while alkyl homologues of thiophene were also observed. The authors observed the occurrence of roof-tiles in GC×GC-SCD chromatograms, which helped in identification and separation of compounds such as highly branched thiophenes and benzothiophenes [144].

Nitrogen and sulfur compounds are known for their catalyst deactivation effects. GC×GC-MS has been employed to analyze catalyst performance in light-cycle oil hydrocracking processes. Soluble coke extracted from the deactivated catalysts was found to contain heteroatom compounds such as amines, quinolones, indoles, and thiazoles, all detected using GC×GC-MS [106]. These studies showed that the chemical composition of soluble coke is influenced by catalyst properties, such as acidity and porosity [106]. Recently, research has expanded to include pyrolysis product analysis of residual fuels, aiming to reduce the environmental impact of heavy residues and improve fuel quality [212]. Figure 11.4 shows the distribution of pyrolysis products from heavy fuel oils (HFOs) and vacuum residue oils (VROs) using GC×GC coupled with FID, SCD, and MS. An algorithm was developed for peak screening and postprocessing, classifying pyrolysis products into 26 molecular classes [212].

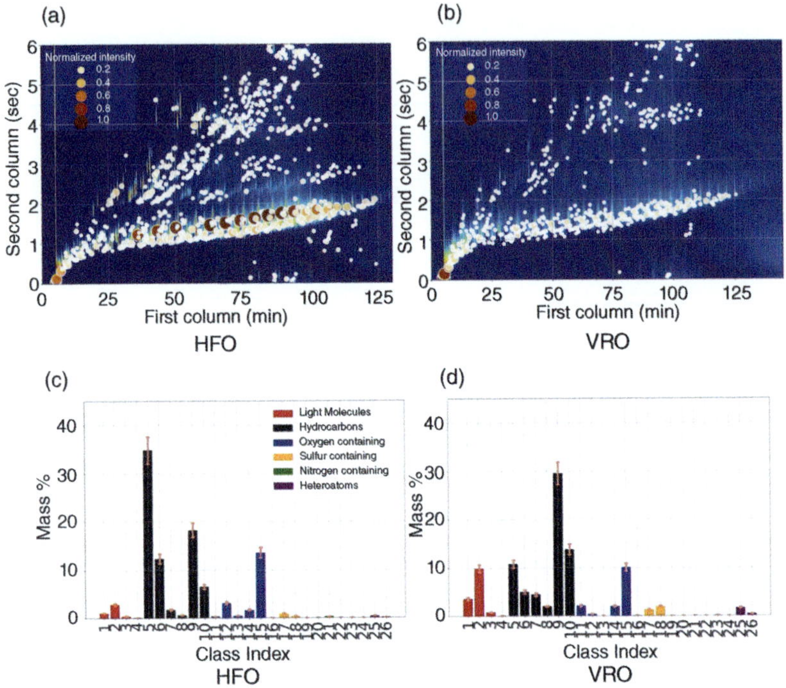

Figure 11.4 (a) Total ion chromatograms (TICs) recorded from TOFMS as a bubble plot and overlaying on FID chromatogram for HFO, (b) TICs – recorded from TOFMS as a bubble plot and overlaying on FID chromatogram for VRO, (c) products distribution of HFO, and (d) products distribution of VRO. *Source:* Reproduced in Colleoni et al. [212, 223] with permission from Elsevier.

As part of a medium-term strategy to reduce greenhouse gas emissions, sustainable aviation fuels (SAF) hold significant promise in the category of synthetic fuels. To ensure safety and engine functionality, a thorough investigation of their chemical properties is essential [197]. Figure 11.5a presents a comparison between the properties of SAF and the average properties of Jet A fuel. Figure 11.5b shows the distillation range of the SAF, determined using the boiling point distribution test technique by GC (ASTM D2887), highlighting the range of conventional jet fuel and the limitations of jet fuel properties [197]. The outcomes of the property prediction are illustrated in Figure 11.5c, where blue dots represent the results, and the green-shaded area depicts the ASTM D7566-based limits alongside the typical jet fuel range. An improved GC×GC-FID method was recently developed by Ikonen et al. [37] to enhance the precision of analyzing synthetic paraffinic kerosene derived from hydroprocessed esters and fatty acids (HEFA) and to improve its accuracy. This method has undergone

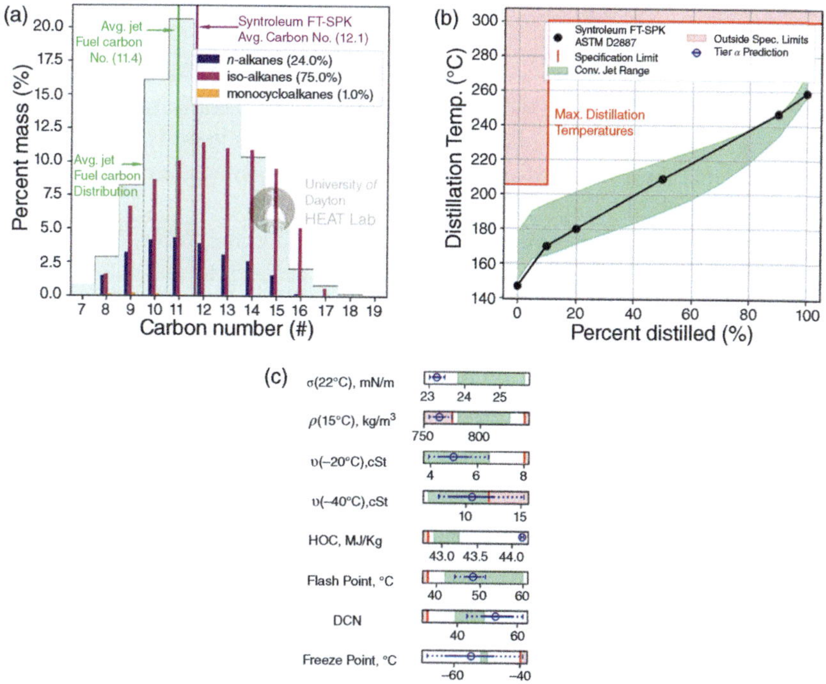

Figure 11.5 Panel plot for Syntroleum's Fischer–Tropsch Synthetic Paraffinic Kerosene (FT-SPK-Syntroleum, POSF 5018) (a) percent mass of identified chemicals as a function of carbon number and chemical group, compared with a typical jet fuel, (b) distillation properties, and (c) tier α property predictions. *Source:* Yang et al. [197] / with permission of Elsevier.

extensive testing to confirm its accuracy, precision, limitations, and selectivity [37]. Both traditional and synthetic jet fuels were used in the testing process, and the results verified that this procedure aligns with the customary standard (ASTM D2425) for compatibility with synthetic jet fuels [37].

Oxymethylene ethers (OMEs) are a promising class of synthetic compounds capable of replacing fossil fuels while reducing greenhouse gas and particulate matter emissions. OMEs have been tested for compatibility with diesel engines, as improvements are necessary for them to meet fuel standards for diesel engines [214]. The significance of carbene chemistry was explored through the analysis of the thermal decomposition of OME-2 at high temperatures. Using GC×GC in combination with QMS and FID, product identification and quantification were accomplished. The reactor temperature varied between 850 and 1150 K, with residence times ranging from 400 to 850 ms. The primary products identified included dimethoxymethane, formaldehyde, methyl formate, methane,

CO_2, CO, and H_2 [214]. Melder et al. [57] developed a unified analytical procedure for diverse fuels by comparing the emission patterns of synthetic fuels, including HEFA, Fischer–Tropsch crude oil (FT), and methane to gas (MtoG) fuels, with those of conventional fuels [57]. This method enables the classification of detected compounds into up to 20 chemical groups, as depicted in Figure 11.6, facilitating the evaluation of nonstandard analytes and predicting combustion properties for specific samples [57].

The extraction of hydrocarbons from shale has become a key focus in oil refinery research, driven by advancements in extraction technologies and increasing demand for alternative energy sources [224]. Ristic et al. [38] demonstrated that pyrolyzing oil shale, or crude oil extracted from shale rock, produces shale oil containing olefins (17.5 wt%), cycloalkanes (16.8 wt%) and oxygenated compounds such as benzenediols (13.4 wt%), aliphatic ketones (9.2 wt%), hydrobenzofurans (9.7 wt%), and phenols (4.1 wt%), as identified using GC×GC. By resolving the peak overlaps, pyrolysis shale oil was shown to contain detectable oxygen compounds [38]. Optimizing shale oil's potential as an energy source necessitates a detailed analysis of its chemical composition. Cui et al. [202] highlighted the advantages of GC×GC-MS for identifying cycloalkanes and heteroatoms, such as phenols, thiophenes, and pyridines in shale oil. When kerogens extracted from shale oil were pyrolyzed and analyzed using GC×GC, it was shown that the maximum hydrocarbon yield in the range of 22.2–24.4 wt% was obtained at 300 °C [203].

To address sustainability and meet ongoing energy demands, the compatibility and properties of hydrocarbon mixtures derived from biomass and oil refinery products have been extensively studied. Blending biofuel with diesel and aviation gasoline is an emerging trend. Vozka et al. [161, 171] quantified carbon, hydrogen, and density in blended fuels, while Xu et al. [210] identified and quantified trace oxygenates in these blends using GC×GC-MS. Jarvis et al. [162] investigated hydrocarbon product distribution when biomass model compounds were co-fed with fluid catalytic cracking (FCC) feed in the presence of an equilibrium zeolite catalyst (E-Cat). The study, conducted in a Davison Circulating Riser (DCR) reactor, used acetic acid, guaiacol, and sorbitan monooleate as biomass model compounds. GC×GC-TOFMS analysis showed that the model compounds were converted during the reaction, with nearly identical product distributions for each model compound. Furthermore, the presence of model compounds in the FCC did not significantly affect the DCR's performance [162]. Similarly, vacuum petrol oil was co-processed with bio-liquids from hydrothermal liquefaction or processed pyrolysis oil through hydrocracking [36]. The selectivity of middle distillates (27.5–69.8%) was assessed using GC×GC, showing conversion rates (38.8–84.9%) for polyaromatics, monoaromatics, and saturated hydrocarbons in the products, marginally influenced by processed bio-oil properties.

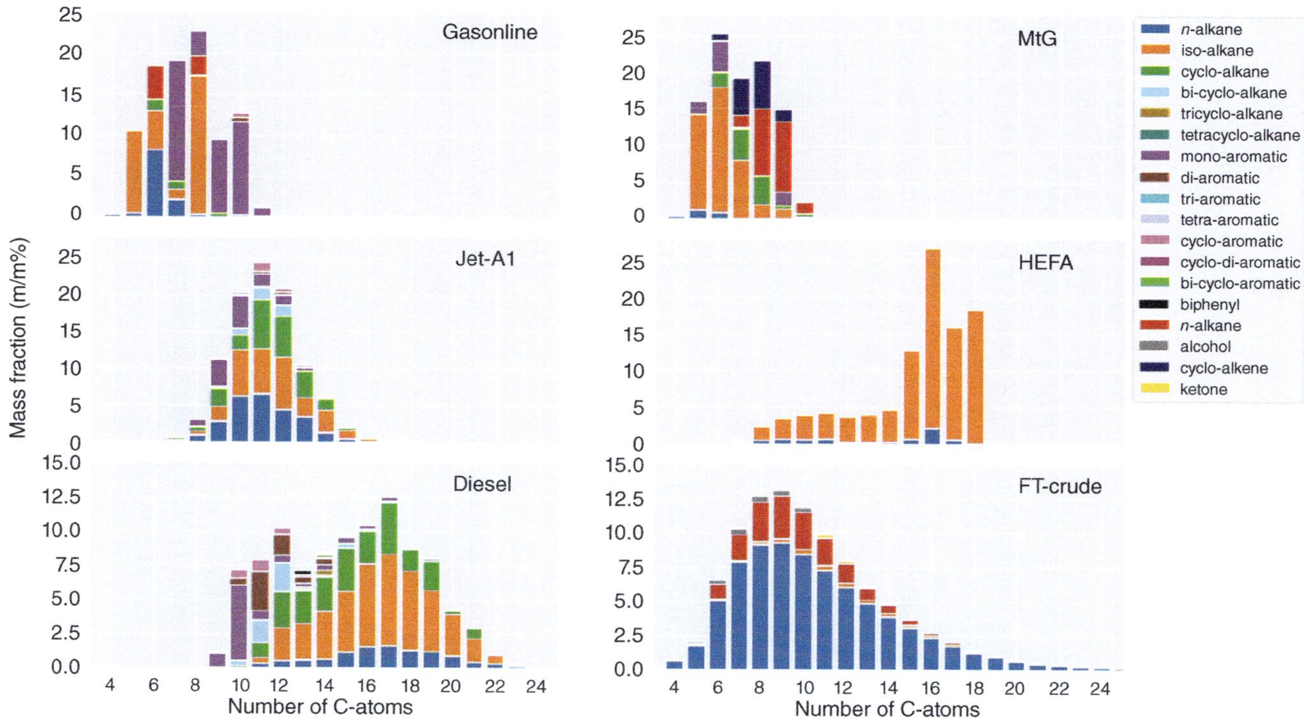

Figure 11.6 Comparison of the characteristics of conventional fuels (diesel, gasoline, jet-A1) and synthetic fuels (MtG, HEFA, FT). *Source:* Melder et al. [57] / with permission of American Chemical Society / CC BY 4.0.

Additionally, Gieleciak et al. [199] co-hydroprocessed heavy vacuum oil from Canadian oil sands with canola oil, producing a blend containing *m*-heptadecane and *n*-octadecane. This process improved the suitability of the oil as diesel fuel by enhancing the physical and combustion properties.

11.4 Advancements in Chemical Recycling of Plastics

11.4.1 Technological Innovation in the Recycling of Plastics

While oil refineries continue to play a dominant role in today's energy infrastructure, there is growing interest in developing more sustainable methods for fuel production. This shift is driven by advances in the circular plastic economy, where used polymers are efficiently and economically transformed into value-added chemicals, materials, and fuel molecules. Plastics are most commonly recycled via mechanical recycling, which involves steps such as sorting, cleaning, shredding, melting, and pelletizing. However, mechanical recycling is generally limited to clean, homogeneous plastic streams such as postindustrial waste. The presence of impurities can significantly reduce the quality of recycled products. Furthermore, with each recycling cycle, polymer quality deteriorates due to structural breakdown [225].

In contrast, chemical recycling can accommodate contaminated or mixed plastic waste streams such as household waste. This approach enables the conversion of diverse plastic mixtures into plastic monomers, value-added chemicals, and fuel molecules [21, 226, 227]. Common chemical recycling techniques include pyrolysis [21], hydrogenolysis [228], hydroformylation [229], olefin metathesis [230], and hydrothermal liquefaction [231]. Although GC is frequently employed in chemical recycling processes for product analysis, it has certain limitations. GC chromatograms can be very complex with overlapping peaks, especially when dealing with polymer mixtures [232]. This leads to incomplete and inaccurate identification and quantification of products. Besides, the detection of products present in minor amounts is quite challenging using GC. On the other hand, GC×GC has emerged as a potential technique in plastic recycling research due to its capacity to enhance separation and detection, and thus enable thorough characterization of chemically recycled polymer products, and ensure precise quality control [28, 232]. Section 11.4.2 discusses in detail the potential applications of GC×GC in plastic recycling processes.

11.4.2 Application of GC×GC in Plastic Recycling

The application of GC×GC to the chemical recycling of plastics has been explored recently, as shown in Table 11.2 and Figure 11.7. The substantial societal impact

Table 11.2 Evolution of GC×GC techniques in plastic recycling applications.

Analytes	Detectors	Modulators	Year	References
Pyrolyzed gasoline and polyolefin copolymer (with pyrolyzer)	Mini PDD	Thermal modulator	2006	[233]
Plastic waste pyrolysis oil	MS, SCD, NCD, FID	Thermal modulator	2014	[28]
Landfill and plastic pyrolysis oil	MS	Thermal modulator	2017	[234]
Plastic and cellulose	MS	Thermal modulator	2017	[235]
Pyrolysis-catalytic cracking with $CoCO_3$ catalyst	TOFMS	Thermal modulator	2018	[236]
Pyrolysis oil from waste HDPE ($CoCO_3$ catalyst)	TOFMS	Thermal modulator	2018	[237]
Automotive shredder residues-derived fuels	TOFMS, FID, SCD, NCD	Thermal modulator	2019	[238]
PVC with biomass	TOFMS	Thermal modulator	2020	[239]
HDPE pyrolysis oil from pilot-scale reactor	QMS, FID	Thermal modulator	2022	[240]
Hydrocracking of LDPE (NH_4-ZSM-5, NH_4-MOR, NH_4-FER, and H-USY)	MS, FID	Thermal modulator	2022	[241]
Pyrolyzer with LDPE, postconsumer mixed polyolefinic waste (steam treated FCC, HZSM-5 additives, E-cat)	TOFMS, FID	Thermal modulator	2022	[227]
Steam cracking of distilled pyrolysis oil from mixed plastic waste	QMS, SCD, NCD, FID	Thermal modulator	2022	[242]

(*Continued*)

Table 11.2 (Continued)

Analytes	Detectors	Modulators	Year	References
Pyrolysis oil from polyolefinic postconsumer plastic wastes	QMS, SCD, NCD, FID	Thermal modulator	2022	[243]
Virgin and recycled PET	QTOFMS	Split state modulator	2022	[244]
Used PS pyrolysis oil from a pilot-scale reactor	TOFMS, FID	Thermal modulator	2022	[245]
LDPE (SAPO-34, HZSM-5)	FID	Thermal modulator	2022	[246]
Pyrolysis oil from polyamide 6 (PA6) and beech wood	MS	Thermal modulator	2022	[247]
Mixed polyolefinic wastes (ZSM-5, SAPO-34, ZSM-22, SAPO-11, Al-MCM-41)	TOFMS, FID	Thermal modulator	2023	[248]
Pyrolysis oil from polyolefin wastes	QMS-FID	Thermal modulator	2023	[249]
Pyrolysis oil from mixed plastics	TOFMS	Thermal modulator	2023	[35]
High-boiling point samples (catalytic cracking gas oil, vacuum gas oil, catalytic cracking, and thermal cracking products [slack wax], pyrolysis product sludge sediments of municipal treatment facilities, and polyethylene)	MS, FID	Flow modulator	2023	[250]
Recycled PET	QTOFMS	Thermal modulator	2023	[251]
Virgin and recycled PET	QTOFMS	Thermal modulator	2023	[252]

Table 11.2 (Continued)

Analytes	Detectors	Modulators	Year	References
Hydroformylated pyrolysis oils from polyolefin waste	FID	Not specific mention	2023	[229]
Untreated and hydrotreated pyrolysis oil from mixed plastics	TOFMS, FID	Thermal modulator	2023	[228]
Pyrolysis oil (PET/PP/PE/Paper)	TOFMS, FID	Thermal modulator	2024	[253]
Pyrolyzer (PP)	TOFMS, FID	Flow modulator	2024	[254]
Pyrolyzer (PE)	TOFMS, FID	Flow modulator	2024	[255]
Pyrolyzer (PP, metal[Fe, Co, Ni, Cu, and Ce]-exchanged zeolites)	TOFMS, FID	Flow modulator	2024	[256]
Pyrolyzer (PP, ZSM-5, steam-treated phosphorus impregnated ZSM-5, mesoporous steam-treated phosphorus impregnated ZSM-5)	TOFMS, FID	Thermal modulator	2024	[257]
Pyrolyzer (Plastic additives [Irganox 1010, Irganox 1076, and BHT])	TOFMS, FID	Flow modulator	2024	[258]
Pyrolyzer (Plastic additives [Irgafos 168, zinc stearate])	TOFMS, FID	Flow modulator	2024	[259]
Pyrolysis oil (LDPE, LLDPE, HDPE, PP; virgin & postconsumer recycled)	FID	Not specific mention	2024	[260]
Pyrolyzer (PP, mesoporous HZSM-5)	TOFMS, FID	Thermal modulator	2024	[261]

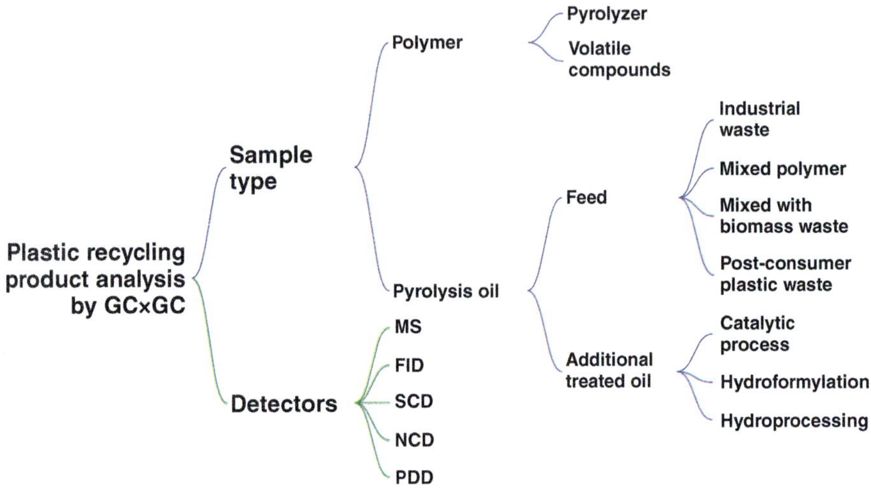

Figure 11.7 Comprehensive overview of plastic recycling product analysis using GC×GC.

of these studies should not be overlooked, although the number of studies that concentrate on the chemical recycling of polymers is lower than those focusing on oil refineries and biomass. The number of publications is increasing with time, which is indicative of the increasing interest and progress in this field.

GC×GC was utilized primarily to study the chemical composition of products obtained through the recycling of plastics [262]. GC×GC coupled with a miniaturized pulsed discharge detector (PDD) was utilized to study the pyrolysis of gasoline and polyethylene copolymer. Compared to GC, the chromatogram exhibited superior separation capabilities for two distinct categories: saturated hydrocarbons and aromatic compounds [233]. Toraman et al. [28] carried out a detailed compositional characterization of plastic waste pyrolysis oil using GC×GC coupled to four different detectors, viz., MS, FID, NCD, and SCD. Figure 11.8a–d shows GC×GC-NCD color plots, while Figure 11.8e–h shows GC×GC-SCD color plots for plastic waste pyrolysis oil. The authors studied the combination of different columns in both normal and reversed configurations. To detect nitrogen and sulfur-containing compounds in plastic waste pyrolysis oil, an apolar column (RTX-1 PONA or MXT-1) and a mid-polar column (BPX-50) were the best choice of column combination. Hydrocarbon groups like n-paraffins, isoparaffins, olefins and naphthenes, monoaromatics, naphthenoaromatics, diaromatics, naphthenodiaromatics, triaromatics, naphthenotriaromatics, and tetra-aromatics were identified and quantified in the plastic pyrolysis oil. Additionally, nitrogen-containing compounds such as nitriles, pyridines, quinolines, indole, and caprolactam, sulfur-containing compounds such as

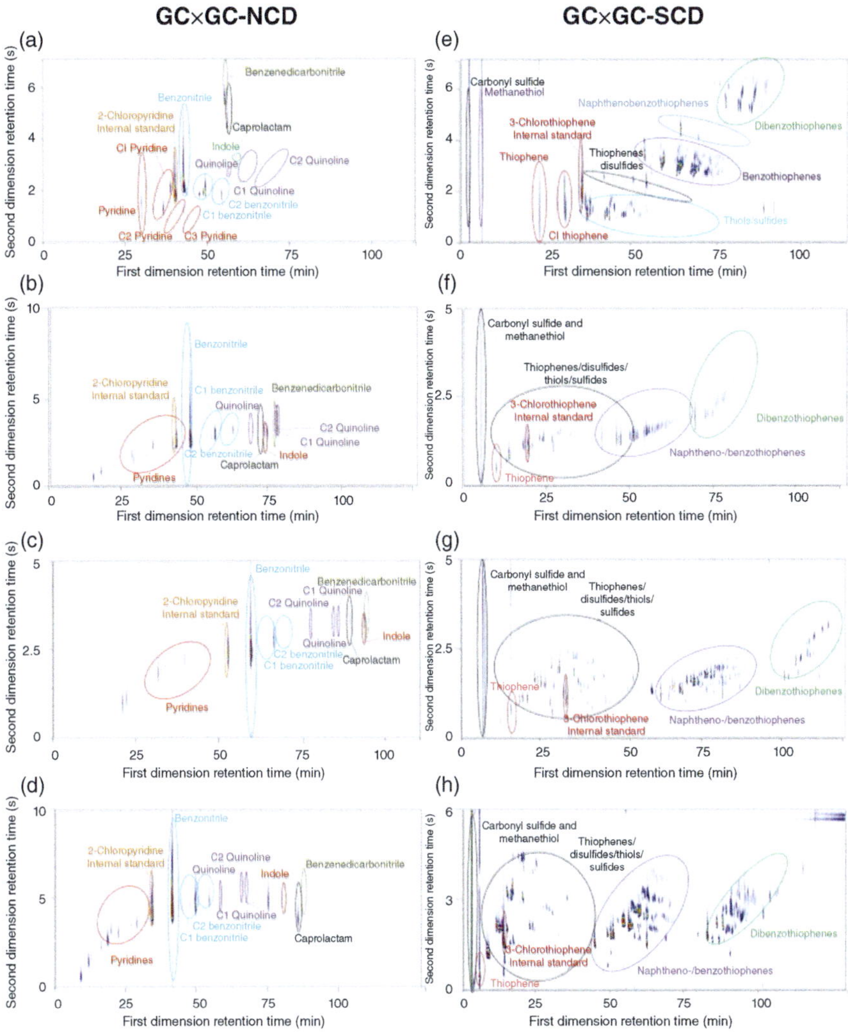

Figure 11.8 GC×GC color plots of the pyrolysis oil using RTX-1 PONA/BPX-50 with NCD (a), BPX-50/Rxi-5ms with NCD (b), Stabilwax/Rxi-5ms with NCD (c), IL-59/Rxi-5ms with NCD (d), RTX-1 PONA/BPX-50 with SCD (e), BPX-50/Rxi-5ms with SCD (f), Stabilwax/Rxi-5ms with SCD (g), and IL-59/Rxi-5ms with SCD (h) column and detector combination. *Source:* Toraman et al. [28] / with permission of Elsevier.

thiophenes/disulfides, thiols/sulfides, benzothiophenes, and dibenzothiophenes, and oxygen-containing compounds such as ketones, phenols, aldehydes, and ethers were also detected. Pure hydrocarbons with 91 wt% occupied the major fraction of pyrolysis oil with 66 wt% of aromatics followed by 13 wt% of paraffins and iso-paraffins, while olefins and naphthenes occupied 12 wt%. Furthermore, 6.5 wt% of nitrogen compounds, 0.6 wt% of sulfur compounds, and 2.5 wt% of oxygen compounds were present in the plastic pyrolysis oil.

Recently, Perez et al. [232] studied the product distribution of pyrolysis oil obtained from plastic feedstocks using two different analytical methods, viz., GC and GC×GC. Polymer samples such as PP, HDPE, PS, and their mixtures in 1:1 wt./wt. were subjected to pyrolysis at 500 °C using a micropyrolyzer coupled with GC and GC×GC. The study showed that GC×GC identified 5.3, 2.8, 4.8, 4.1, and 3 times more peaks than GC in the pyrolysis of HDPE, PP, PS, HDPE–PS, and PP–PS, respectively. Furthermore, this study also demonstrated that co-pyrolysis of HDPE–PS and PP–PS resulted in an increased experimental yield of toluene, with values of 51.3% for HDPE–PS and 29.7% for PP–PS compared to the additive yield, highlighting the interaction effect among polymer mixtures.

Tipler et al. [238] studied the composition of fuel produced from the combustion of automotive shredder residues (ASRs) using GC×GC coupled with FID and TOFMS. ASRs mainly composed of materials such as polyolefins, rubbers, foams, styrenic plastics, and other thermoplastics such as polyvinyl chloride (PVC) and polycarbonate (PC) that were utilized to generate liquid fuels. The GC×GC analysis showed that the ASR fuel contained 4.86% *n*-paraffins, 11.63% isoparaffins, 29.61% olefins, 24.95% aromatics, 12.95 naphthenes, and 16% oxygenates [238].

Dong et al. [251] utilized GC×GC to study the presence of volatile contaminants (VCs) present in polymers to evaluate their impact on human health. The results of a comparison of various methods for detecting VCs in recycled poly(ethylene terephthalate) (rPET) flakes showed that head-space solid phase micro-extraction, in conjunction with GC×GC-QTOFMS, was considered as a relatively more effective method compared to headspace (HS)-GC-MS and solid phase micro extraction (SPME)-GC-MS. This method was successfully applied to analyze rPET flakes from various recycling plants in China. A total of 212 VCs were identified, and the potential sources were food, plastic, and cosmetics. It was observed that 45 VCs are high-priority compounds with toxicity levels IV or V and may pose a potential risk to human health [251].

The effect of pyrolysis variables, viz., temperature, sample size, particle size, and carrier gas flow rate on the product distribution of polypropylene (PP) [254] and high-density polyethylene (HDPE) [255], was studied using a micropyrolyzer connected to GC×GC-FID/TOFMS for validating the pyrolysis conditions that can minimize transport effects and secondary gas-phase reactions. Around 600 peaks were identified in the pyrolysis of PP [254] belonging to chemical classes such as paraffins, olefins, naphthenes, diolefins, cycloolefins, cyclodiolefins, triene, cyclotrienes, and bicyclic and aromatic hydrocarbons. It was observed that the product distribution for pyrolysis of PP at 460, 530, and 600 °C was statistically different, with total yield of C_3 to C_{35} ranging from 39.88 ± 2.08 to 66.61 ± 3.26 wt%. Nevertheless, statistical differences in product distribution were not observed within the employed Box-Behnken design (BBD) design space for variables such as sample size (50–150 μg), particle size (53–125, 125–300, and >300 μm), and carrier gas flow rate (100–300 mL/min), suggesting minimization of transport effects and secondary gas phase reactions. Furthermore, statistically different product distribution was observed when pyrolysis of PP was conducted at 460, 530, and 600 °C under boundary conditions (100 mL/min, 150 μg) and extreme conditions (30 mL/min, 800 μg) [254]. In the pyrolysis of HDPE [255], GC×GC detected around 500 peaks with the carbon number range of C_3 to C_{28}. The identified peaks were classified into several chemical classes such as paraffins, olefins, diolefins, and cyclic and aromatic hydrocarbons among others [255]. Similar to PP pyrolysis [254], the pyrolysis of HDPE also resulted in statistical differences for temperature (480–600 °C), while no statistical differences were observed for particle size (53–125 to >300 μm), sample size (50–150 μg), and carrier gas flow rate (100–300 mL/min). Unlike PP pyrolysis [254], no statistical differences in product distribution were observed for HDPE under boundary conditions of 100 mL/min and 150 μg. However, the authors reported a statistically different product distribution when a very high sample size (1000 μg) and low carrier gas flow rate (50 mL/min) were employed. Through these studies, the authors [254, 255] demonstrated that GC×GC facilitated the comprehensive identification and quantification of chemical species that can help in the formulation of detailed and precise reaction pathways for the development of (micro)-kinetic models for plastic pyrolysis.

It is essential to understand the distribution of pyrolysis products from additives to assess the pyrolysis process of plastic waste. The pyrolysis product distributions of two commonly used antioxidants, Irgafos 168 and zinc stearate (ZnSt), were investigated using GC-MS and GC×GC-FID/TOFMS [259]. For Irgafos 168, GC×GC identified 15 chemical classes, with phenols and their derivatives showing the highest yield of 9.51 wt%. For ZnSt, 16 chemical classes were identified, with olefins having the highest yield at 9.73 wt%. In contrast, GC was able to identify only 4 and 11 chemical classes for Irgafos 168 and ZnSt, respectively. Additionally, a separate study compared the pyrolysis product compositions of Irganox 1010, Irganox 1076, and BHT using GC-MS and GC×GC-FID/TOFMS and reported detailed compositional analysis [258]. GC×GC-FID/TOFMS analysis of Irganox 1010 enabled the identification and quantification of 209 products, significantly exceeding the 45 products identified by GC-MS. The quantity of products identified by GC×GC for Irganox 1076 was fivefold greater. The GC-MS method for the pyrolysis of BHT demonstrated that a total of 7 decomposition products were identified. By employing the GC×GC method, a total of 169 compounds were identified with a total yield of 14.50 wt%.

The investigation of additional methods to improve the quality of pyrolysis oil such as catalysis, hydrotreatment, steam cracking, and distillation has expanded the scope of pyrolysis product analysis [263]. Bozkurt et al. [256] used GC×GC to study the catalytic pyrolysis of PP using metal-exchanged HZSM-5 catalysts (Fe, Co, Ni, Cu, Ce) in a micropyrolysis reactor connected to GC×GC-FID/TOFMS. Among the tested catalysts, Ni-exchanged catalysts exhibited a lower Brønsted/Lewis acid ratio and enhanced Lewis acid sites while maintaining a total acidity of approximately 1 mmol/g of catalyst. These Ni-exchanged catalysts enhanced the yields of benzene, toluene, and xylene (BTX) to a range of 28–38 wt%, in contrast to 22 wt% using the parent HZSM-5 [256]. Eschenbacher et al. [227] employed steam-treated FCC and HZSM-5 catalysts to enhance C_2–C_4 olefin selectivity during the pyrolysis of polymers such as LDPE and mixed polyolefin waste. The highest C_2–C_4 olefin yield (69 wt%) was obtained when the FCC catalyst was mixed with an HZSM-5 as an additive that had been steam-treated for 4 hours at 815 °C [227]. Lazzari et al. [228] reported that the hydrotreated plastic pyrolysis oil showed patterns similar to naphthenes and fewer peaks of diolefins and branched olefins compared to the original pyrolysis oil. Furthermore, the examination of GC-VUV corroborated the findings of GC×GC derived from plastic pyrolysis oil [228].

Ureel et al. [246] investigated an active learning framework called Gaussian N-Dimensional Active Learning Framework (GandALF) to evaluate its effectiveness in predicting the yield of light olefins (C_2–C_4) during the catalytic pyrolysis of LDPE using GC×GC coupled with FID [246]. Maximizing the light olefin (C_2–C_4) yield from LDPE *ex situ* catalytic pyrolysis was the objective function of the active learning approach. The study examined the effect of parameters such as catalyst type, temperature, flow rate, and space-time on light olefin yields. The results showed that HZSM-5 significantly outperformed SAPO-34, producing ethylene in the range of 7.2–40.0 wt% and propylene in the range of 16.8–41.6 wt% [246].

Steam cracking has been highlighted as a promising method to convert untreated plastic pyrolysis oil into base chemicals, such as light olefins and aromatics [263]. However, before feeding pyrolysis oil into a steam cracker, it is essential to quantify the level of metals and heteroatom impurities, including nitrogen, oxygen, phosphorus, chlorine, and sulfur. These impurities can degrade product quality and cause operational issues such as fouling, corrosion, catalyst poisoning, and coke formation [263]. Kusenberg et al. [242] produced naphtha by distilling plastic pyrolysis oil derived from two different waste sources: marine-littered plastics and municipal plastic waste. The chemical composition of the naphtha obtained from steam cracking of plastic pyrolysis oil was evaluated using GC×GC coupled with various detectors (FID, NCD, and SCD) and multiple analytic instruments (elemental analysis, combustion ion chromatography, liquid density, and inductively coupled plasma [ICP] spectroscopies) [242]. Distillates from municipal plastic waste generated more coke than fossil fuel-derived naphtha and marine litter-derived naphtha. This was attributed to the higher aromatic content in pyrolysis oil from municipal plastic waste [242]. Using GC×GC, nitrogen and sulfur impurities were detected, with nitrogen levels from both plastic waste samples exceeding industrial thresholds. Due to the high aromatic content of naphtha derived from municipal plastic waste, steam cracking of this feedstock produced a lower yield of C_1–C_4 hydrocarbons but a higher yield of gasoline (C_5–C_9) and fuel oil (C_{10+}) compared to the other two samples [242].

GC×GC has been recently employed to characterize products obtained from the hydroformylation of pyrolysis oil derived from polyolefin waste [229]. Li et al. [229] demonstrated that the olefins in polyolefin pyrolysis oil can be converted into mono- and di-alcohols through hydroformylation followed by hydrogenation. The authors utilized GC×GC to achieve a detailed product distribution, showing that the pyrolysis oil contained a broad range of chemical classes, including monoolefins (34–50 wt%), paraffins (20–39 wt%), monoaromatics (11–23 wt%), diolefins (6–13 wt%), and dicycloaromatics (<5 wt%), with carbon numbers ranging from C_5 to C_{40}. Furthermore, the authors demonstrated that the peaks in the GC×GC chromatogram corresponding to alcohols and aldehydes indicated the formation of aldehydes during the hydroformylation of pyrolysis oil [229].

11.5 Discussions and Conclusions

GC×GC has significantly enhanced the molecular level characterization of compounds across refinery feedstocks, refined products, and plastic pyrolysis oils. While most research has focused on analyzing hydrocarbon distributions, considerable efforts have also been made to identify and quantify trace-level compounds, particularly heteroatom-containing compounds. This technique plays a vital role in improving our understanding of pyrolysis reaction networks, optimizing catalysts, and developing robust (micro)-kinetic models.

Below are some key benefits and challenges associated with GC×GC:

- **Enhanced resolution capability:** GC×GC offers superior resolution compared to traditional GC. This enables improved separation of co-eluting compounds, thereby enhancing the accuracy of compound identification in complex feedstocks.
- **Comprehensive chemical characterization:** GC×GC enables the detailed identification of hundreds of chemical species in a single experimental run,

including low-abundance peaks associated with trace-level components. These species, such as additives, stabilizers, colorants, and adhesives, are commonly found in multilayer packaging materials. The superior resolving power of GC×GC is particularly valuable for detecting such trace components, which are often missed in traditional GC but may significantly impact product quality, catalyst performance, and downstream upgrading processes.

- **Kinetic model development:** The comprehensive identification and accurate quantification of chemical species provided by GC×GC are instrumental in developing and validating (micro)kinetic models for recycling and upgrading applications.
- **Complexity of data analysis:** GC×GC generates vast amounts of data that require time-consuming and intricate data analysis. Advanced computational tools and algorithms for pattern recognition and data processing are essential to streamline this process.
- **Advanced statistical analysis:** Interpreting the large datasets produced by GC×GC can be challenging. Multivariate statistical techniques, such as principal component analysis (PCA) combined with clustering algorithms, can identify trends and relationships via dimensionality reduction. Machine learning methods are better suited for subsequent classification or predictive tasks rather than direct chemical interpretation.
- **GC×GC optimization:** The sample type influences the choice of columns, column combinations, and method development in GC×GC. Key parameters, including but not limited to oven heating rate, modulation period, and final temperature, must be optimized. Proper column selection is critical for accurate identification and quantification, while minimizing issues like tailing and co-elution. Additionally, GC×GC is invaluable for evaluating and optimizing catalysts based on detailed product distribution and yield data.
- **Instrument cost and maintenance:** GC×GC systems are relatively expensive and more challenging to operate and maintain compared to traditional GC systems. Addressing these challenges requires proper training programs and the development of cost-effective solutions to make these instruments more accessible.

In summary, GC×GC is a powerful tool for fuel processing and chemical recycling of plastics. When coupled with appropriate detectors, it enables detailed product characterization and ultimately provides critical insights into reaction mechanisms. The growing demand for sustainable fuels and chemicals requires thorough compositional evaluation to assess both environmental impact and process performance metrics. Effective use of GC×GC supports comprehensive data analysis and informed decision-making toward long-term sustainability objectives while ensuring process adaptability, and reinforcing the foundations of a strong economy.

References

1 Hussain, J. and Zhou, K. (2022). Globalization, industrialization, and urbanization in belt and road initiative countries: implications for environmental sustainability and energy demand. *Environ. Sci. Pollut. Res.* 29 (53): 80549–80567.

2 Jie, X., Gonzalez-Cortes, S., Xiao, T. et al. (2019). The decarbonisation of petroleum and other fossil hydrocarbon fuels for the facile production and safe storage of hydrogen. *Energy Environ. Sci.* 12 (1): 238–249.

3 Demirel, Y. and Demirel, Y. (2012). Energy and energy types. In: *Energy: Production, Conversion, Storage, Conservation, and Coupling* (ed. Y. Demirel), 27–70. Springer London.

4 Zang, G., Sun, P., Elgowainy, A. et al. (2021). Life cycle analysis of electrofuels: Fischer–Tropsch fuel production from hydrogen and corn ethanol byproduct CO_2. *Environ. Sci. Technol.* 55 (6): 3888–3897.

5 Megía, P.J., Vizcaíno, A.J., Calles, J.A., and Carrero, A. (2021). Hydrogen production technologies: from fossil fuels toward renewable sources. A mini review. *Energy Fuel* 35 (20): 16403–16415.

6 Fan, J., Gephart, J., Marker, T. et al. (2016). Carbon footprint analysis of gasoline and diesel from forest residues and corn stover using integrated hydropyrolysis and hydroconversion. *ACS Sustain. Chem. Eng.* 4 (1): 284–290.

7 Bozkurt, O.D., Okonsky, S.T., Alexopoulos, K., and Toraman, H.E. (2022). Chapter Four - Catalytic conversion of SPW and products upgrading. In: *Advances in Chemical Engineering*, vol. 60 (ed. D. Moscatelli and M. Pelucchi), 117–168. Academic Press.

8 Perez, B.A., Jayarama Krishna, J.V., and Toraman, H.E. (2023). Insights into co-pyrolysis of polyethylene terephthalate and polyamide 6 mixture through experiments, kinetic modeling and machine learning. *Chem. Eng. J.* 468: 143637.

9 Ellis, L.D., Rorrer, N.A., Sullivan, K.P. et al. (2021). Chemical and biological catalysis for plastics recycling and upcycling. *Nat. Catal.* 4 (7): 539–556.

10 Perez, B.A., Jonnalagedda, V.J.K., and Toraman, H.E. (2022). Chapter Five - characterization of SPW pyrolysis oils: products spectra and opportunities. In: *Advances in Chemical Engineering*, vol. 60, 169–214. Academic Press.

11 Wilkes, H. (2010). Methods of hydrocarbon analysis, in Davide Moscatelli and Matteo Pelucchi *Handbook of Hydrocarbon and Lipid Microbiology* (eds. Timmis, K.N.), Springer Berlin Heidelberg Berlin, Heidelberg, pp. 49–65.

12 Pollo, B.J., Teixeira, C.A., Belinato, J.R. et al. (2021). Chemometrics, comprehensive two-dimensional gas chromatography and "omics" sciences: basic tools and recent applications. *TrAC Trends Anal. Chem.* 134: 116111.

13 Adam, F., Muller, H., Al-Hajji, A. et al. (2015). Oxidative desulfurization process monitoring using comprehensive two-dimensional gas chromatography and fourier transform ion cyclotron resonance mass spectrometry. *Energy Fuel* 29 (4): 2312–2318.

14 Machado, M.E. (2019). Comprehensive two-dimensional gas chromatography for the analysis of nitrogen-containing compounds in fossil fuels: a review. *Talanta* 198: 263–276.

15 Hinton, Z.R., Talley, M.R., Kots, P.A. et al. (2022). Innovations toward the valorization of plastics waste. *Annu. Rev. Mater. Res.* 52 (1): 249–280.

16 Adahchour, M. and Brinkman, U.A.T. (2014). Multidimensional and comprehensive two-dimensional gas chromatography. In: *Practical Gas Chromatography: A Comprehensive Reference* (ed. K. Dettmer-Wilde and W. Engewald), 461–502. Berlin, Heidelberg: Springer Berlin Heidelberg.

17 Mastalski, I., Sidhu, N., Zolghadr, A. et al. (2023). Intrinsic millisecond kinetics of polyethylene pyrolysis via pulse-heated analysis of solid reactions. *Chem. Mater.* 35 (9): 3628–3639.

18 Rejman, S., Vollmer, I., Werny, M.J. et al. (2023). Transport limitations in polyolefin cracking at the single catalyst particle level. *Chem. Sci.* 14 (37): 10068–10080.

19 Jayarama Krishna, J.V., Korobeinichev, O.P., and Vinu, R. (2019). Isothermal fast pyrolysis kinetics of synthetic polymers using analytical Pyroprobe. *J. Anal. Appl. Pyrolysis* 139: 48–58.

20 Xu, Z., Munyaneza, N.E., Zhang, Q. et al. (2023). Chemical upcycling of polyethylene, polypropylene, and mixtures to high-value surfactants. *Science* 381 (6658): 666–671.

21 Okonsky, S.T., Krishna, J.V.J., and Toraman, H.E. (2022). Catalytic co-pyrolysis of LDPE and PET with HZSM-5, H-beta, and HY: experiments and kinetic modelling. *React. Chem. Eng.* 7 (10): 2175–2191.

22 Kumar, S. and Dutta, S. (2021). Utility of comprehensive GC×GC-TOFMS in elucidation of aromatic hydrocarbon biomarkers. *Fuel* 283: 118890.

23 Fitz, B.D., Wilson, R.B., Parsons, B.A. et al. (2012). Fast, high peak capacity separations in comprehensive two-dimensional gas chromatography with time-of-flight mass spectrometry. *J. Chromatogr. A* 1266: 116–123.

24 Blumberg, L.M., David, F., Klee, M.S., and Sandra, P. (2008). Comparison of one-dimensional and comprehensive two-dimensional separations by gas chromatography. *J. Chromatogr. A* 1188 (1): 2–16.

25 Giddings, J.C. (1967). Maximum number of components resolvable by gel filtration and other elution chromatographic methods. *Anal. Chem.* 39 (8): 1027–1028.

26 Liu, Z. and Phillips, J.B. (1991). Comprehensive two-dimensional gas chromatography using an on-column thermal modulator interface. *J. Chromatogr. Sci.* 29 (6): 227–231.

27 Kulsing, C., Nolvachai, Y., Rawson, P. et al. (2016). Continuum in MDGC technology: from classical multidimensional to comprehensive two-dimensional gas chromatography. *Anal. Chem.* 88 (7): 3529–3538.

28 Toraman, H., Dijkmans, T., Djokic, M. et al. (2014). Detailed compositional characterization of plastic waste pyrolysis oil by comprehensive two-dimensional gas-chromatography coupled to multiple detectors. *J. Chromatogr. A* 1359: 237–246.

29 Vuckovic, D., de Lannoy, I., Gien, B. et al. (2011). in vivo solid-phase microextraction for single rodent pharmacokinetics studies of carbamazepine and carbamazepine-10,11-epoxide in mice. *J. Chromatogr. A* 1218 (21): 3367–3375.

30 Hoh, E., Lehotay, S.J., Mastovska, K. et al. (2009). Capabilities of direct sample introduction–comprehensive two-dimensional gas chromatography–time-of-flight mass spectrometry to analyze organic chemicals of interest in fish oils. *Environ. Sci. Technol.* 43 (9): 3240–3247.

31 Dubois, L.M., Aczon, S., Focant, J.-F., and Perrault, K.A. (2020). Translation of a one-dimensional to a comprehensive two-dimensional gas chromatography method with dual-channel detection for volatile organic compound measurement in forensic applications. *Anal. Chem.* 92 (14): 10091–10098.

32 Ristic, N.D., Djokic, M.R., Delbeke, E. et al. (2018). Compositional characterization of pyrolysis fuel oil from naphtha and vacuum gas oil. *Energy Fuel* 32 (2): 1276–1286.

33 Beens, J. and Brinkman, U. (2000). The role of gas chromatography in compositional analyses in the petroleum industry. *TrAC Trends Anal. Chem.* 19 (4): 260–275.

34 Blomberg, J., Schoenmakers, P., Beens, J., and Tijssen, R. (1997). Comprehensive two-dimensional gas chromatography (GC × GC) and its applicability to the characterization of complex (petrochemical) mixtures. *J. High Resolut. Chromatogr.* 20 (10): 539–544.

35 Beccaria, M., Piparo, M., Zou, Y. et al. (2023). Analysis of mixed plastic pyrolysis oil by comprehensive two-dimensional gas chromatography coupled with low- and high-resolution time-of-flight mass spectrometry with the support of soft ionization. *Talanta* 252: 123799.

36 Bouzouita, D., Lelevic, A., Lorentz, C. et al. (2022). Co-processing bio-liquids with vacuum gasoil through hydrocracking. *Appl Catal B* 304: 120911.

37 Ikonen, K.E., Li, W., Ruonakangas, A. et al. (2023). Different hydrocarbon types in jet fuel as determined by GC × GC/FID - an alternative method to D2425. *Energy Fuel* 37 (9): 6629–6641.

38 Ristic, N.D., Djokic, M.R., Konist, A. et al. (2017). Quantitative compositional analysis of Estonian shale oil using comprehensive two dimensional gas chromatography. *Fuel Process. Technol.* 167: 241–249.

39 Ramos, L. and Brinkman, U.A.T. (2009). Chapter 1 multidimensionality in gas chromatography: general concepts. In: *Comprehensive Analytical Chemistry*, vol. 55 (ed. L. Ramos), 3–14. Elsevier.

40 Semard, G., Adahchour, M., and Focant, J.-F. (2009). Chapter 2 basic instrumentation for GC × GC. In: *Comprehensive Analytical Chemistry*, vol. 55 (ed. L. Ramos), 15–48. Elsevier.

41 Mommers, J. and van der Wal, S. (2021). Column selection and optimization for comprehensive two-dimensional gas chromatography: a review. *Crit. Rev. Anal. Chem.* 51 (2): 183–202.

42 Zanella, D., Focant, J.-F., and Franchina, F.A. (2021). 30th anniversary of comprehensive two-dimensional gas chromatography: latest advances. *Anal. Sci. Adv.* 2 (3–4): 213–224.

43 Dao Thi, H., Djokic, M.R., and Van Geem, K.M. (2021). Detailed group-type characterization of plastic-waste pyrolysis oils: by comprehensive two-dimensional gas chromatography including linear, branched, and di-olefins. *Separations* 8 (7): 103.

44 Milani, N.B.L., van Gilst, E., Pirok, B.W.J., and Schoenmakers, P.J. (2023). Comprehensive two-dimensional gas chromatography—a discussion on recent innovations. *J. Sep. Sci.* 46 (21): 2300304.

45 Bahaghighat, H.D., Freye, C.E., and Synovec, R.E. (2019). Recent advances in modulator technology for comprehensive two dimensional gas chromatography. *TrAC Trends Anal. Chem.* 113: 379–391.

46 Semard, G., Gouin, C., Bourdet, J. et al. (2011). Comparative study of differential flow and cryogenic modulators comprehensive two-dimensional gas chromatography systems for the detailed analysis of light cycle oil. *J. Chromatogr. A* 1218 (21): 3146–3152.

47 Duhamel, C., Cardinael, P., Peulon-Agasse, V. et al. (2015). Comparison of cryogenic and differential flow (forward and reverse fill/flush) modulators and applications to the analysis of heavy petroleum cuts by high-temperature comprehensive gas chromatography. *J. Chromatogr. A* 1387: 95–103.

48 von Mühlen, C., Khummueng, W., Alcaraz Zini, C. et al. (2006). Detector technologies for comprehensive two-dimensional gas chromatography. *J. Sep. Sci.* 29 (12): 1909–1921.

49 Pu, S., Pan, Y., Zhang, L., and Lv, Y. (2023). Recent advances in chemiluminescence and cataluminescence for the detection of volatile sulfur compounds. *Appl. Spectrosc. Rev.* 58 (6): 401–427.

50 Shellie, R.A. and Heng, W.S. (2023). Gas chromatography. In: *Encyclopedia of Forensic Sciences*, 3e (ed. M.M. Houck), 1–9. Oxford: Elsevier.

51 Trinklein, T.J., Cain, C.N., Ochoa, G.S. et al. (2023). Recent advances in GC × GC and chemometrics to address emerging challenges in nontargeted analysis. *Anal. Chem.* 95 (1): 264–286.

52 Ieda, T. and Hashimoto, S. (2023). GC × GC and computational strategies for detecting and analyzing environmental contaminants. *TrAC Trends Anal. Chem.* 165: 117118.

53 Toussaint, G., Lorentz, C., Vrinat, M., and Geantet, C. (2011). Comprehensive 2D chromatography with mass spectrometry: a powerful tool for following the hydrotreatment of a straight run gas oil. *Anal. Methods* 3 (12): 2743–2748.

54 Kaplitz, A.S. and Schug, K.A. (2023). Gas chromatography—vacuum ultraviolet spectroscopy in petroleum and fuel analysis. *Anal. Sci. Adv.* 4 (5–6): 220–231.

55 Dunkle, M.N., Pijcke, P., Winniford, B., and Bellos, G. (2019). Quantification of the composition of liquid hydrocarbon streams: Comparing the GC-VUV to DHA and GC × GC. *J. Chromatogr. A* 1587: 239–246.

56 Venkatramani, C.J. and Phillips, J.B. (1993). Comprehensive two-dimensional gas chromatography applied to the analysis of complex mixtures. *J. Microcolumn Sep.* 5 (6): 511–516.

57 Melder, J., Zinsmeister, J., Grein, T. et al. (2023). Comprehensive two-dimensional gas chromatography: a universal method for composition-based prediction of emission characteristics of complex fuels. *Energy Fuel* 37 (6): 4580–4595.

58 Omais, B., Courtiade, M., Charon, N. et al. (2012). Reversal of elution order in a single second dimension by changing the first column nature in comprehensive two-dimensional gas chromatography. *J. Chromatogr. A* 1255 (SI): 196–201.

59 Seeley, J.V. and Seeley, S.K. (2015). Comprehensive two-dimensional gas chromatography with pattern modulation. *J. Chromatogr. A* 1421: 114–122.

60 Franchina, F.A., Machado, M.E., Tranchida, P.Q. et al. (2015). Determination of aromatic sulphur compounds in heavy gas oil by using (low-)flow modulated comprehensive two-dimensional gas chromatography-triple quadrupole mass spectrometry. *J. Chromatogr. A* 1387: 86–94.

61 Lelevic, A., Geantet, C., Moreaud, M. et al. (2021). Quantitative analysis of hydrocarbons in gas oils by two-dimensional comprehensive gas chromatography with vacuum ultraviolet detection. *Energy Fuel* 35 (17): 13766–13775.

62 Harvey, P.M.A. and Shellie, R.A. (2012). Data reduction in comprehensive two-dimensional gas chromatography for rapid and repeatable automated data analysis. *Anal. Chem.* 84 (15): 6501–6507.

63 Beens, J., Boelens, H., Tijssen, R., and Blomberg, J. (1998). Quantitative aspects of comprehensive two-dimensional gas chromatography (GC × GC). *J. High Resolut. Chromatogr.* 21 (1): 47–54.

64 Beens, J., Blomberg, J., and Schoenmakers, P. (2000). Proper tuning of comprehensive two-dimensional gas chromatography (GC × GC) to optimize the separation of complex oil fractions. *J. High Resolut. Chromatogr.* 23 (3): 182–188.

65 Fraga, C., Prazen, B., and Synovec, R. (2000). Comprehensive two-dimensional gas chromatography and chemometrics for the high-speed quantitative analysis of aromatic isomers in a jet fuel using the standard addition method and an objective retention time alignment algorithm. *Anal. Chem.* 72 (17): 4154–4162.

66 Burger, B., Snyman, T., Burger, W., and van Rooyen, W. (2003). Thermal modulator array for analyte modulation and comprehensive two-dimensional gas chromatography. *J. Sep. Sci.* 26 (1–2): 123–128.

67 Cai, H. and Stearns, S. (2004). Partial modulation method via pulsed flow modulator for comprehensive two-dimensional gas chromatography. *Anal. Chem.* 76 (20): 6064–6076.

68 Johnson, K., Prazen, B., Young, D., and Synovec, R. (2004). Quantification of naphthalenes in jet fuel with GC × GC/Tri-PLS and windowed rank minimization retention time alignment. *J. Sep. Sci.* 27 (5–6): 410–416.

69 Wang, F., Robbins, W., and Greaney, M. (2004). Speciation of nitrogen-containing compounds in diesel fuel by comprehensive two-dimensional gas chromatography. *J. Sep. Sci.* 27 (5–6): 468–472.

70 Arey, J., Nelson, R., Xu, L., and Reddy, C. (2005). Using comprehensive two-dimensional gas chromatography retention indices to estimate environmental partitioning properties for a complete set of diesel fuel hydrocarbons. *Anal. Chem.* 77 (22): 7172–7182.

71 Diehl, J.W. and Di Sanzo, F.P. (2005). Determination of aromatic hydrocarbons in gasolines by flow modulated comprehensive two-dimensional gas chromatography. *J. Chromatogr. A* 1080 (2): 157–165.

72 Micyus, N.J., McCurry, J.D., and Seeley, J.V. (2005). Analysis of aromatic compounds in gasoline with flow-switching comprehensive two-dimensional gas chromatography. *J. Chromatogr. A* 1086 (1): 115–121.

73 Vendeuvre, C., Bertoncini, F., Thiébaut, D. et al. (2005). Evaluation of a retention model in comprehensive two-dimensional gas chromatography. *J. Sep. Sci.* 28 (11): 1129–1136.

74 Kochman, M., Gordin, A., Alon, T., and Amirav, A. (2006). Flow modulation comprehensive two-dimensional gas chromatography-mass spectrometry with a supersonic molecular beam. *J. Chromatogr. A* 1129 (1): 95–104.

75 Ruiz-Guerrero, R., Vendeuvre, C., Thiebaut, D. et al. (2006). Comparison of comprehensive two-dimensional gas chromatography coupled with sulfur-chemiluminescence detector to standard methods for speciation of sulfur-containing compounds in middle distillates. *J. Chromatogr. Sci.* 44 (9): 566–573.

76 Hamilton, J.F., Lewis, A.C., Millan, M. et al. (2007). Comprehensive two-dimensional gas chromatography coupled to time-of-flight mass spectrometry of coal liquids produced during a coal liquefaction process. *Energy Fuel* 21 (1): 286–294.

77 Wang, F.C.-Y. and Walters, C.C. (2007). Pyrolysis comprehensive two-dimensional gas chromatography study of petroleum source rock. *Anal. Chem.* 79 (15): 5642–5650.

78 Seeley, J.V., Seeley, S.K., Libby, E.K., and McCurry, J.D. (2007). Analysis of biodiesel/petroleum diesel blends with comprehensive two-dimensional gas chromatography. *J. Chromatogr. Sci.* 45 (10): 650–656.

79 Wang, F.C.-Y. and Zhang, L. (2007). Chemical composition of group II lubricant oil studied by high-resolution gas chromatography and comprehensive two-dimensional gas chromatography. *Energy Fuel* 21 (6): 3477–3483.

80 Seeley, J.V., Libby, E.M., Seeley, S.K., and McCurry, J.D. (2008). Comprehensive two-dimensional gas chromatography analysis of high-ethanol containing motor fuels. *J. Sep. Sci.* 31 (19): 3337–3346.

81 Seeley, J.V., Seeley, S.K., Libby, E.K. et al. (2008). Comprehensive two-dimensional gas chromatography using a high-temperature phosphonium ionic liquid column. *Anal. Bioanal. Chem.* 390 (1): 323–332.

82 Adam, F., Bertoncini, F., Coupard, V. et al. (2008). Using comprehensive two-dimensional gas chromatography for the analysis of oxygenates in middle distillates: I. Determination of the nature of biodiesels blend in diesel fuel. *J. Chromatogr. A* 1186 (1–2): 236–244.

83 Tiyapongpattana, W., Wilairat, P., and Marriott, P.J. (2008). Characterization of biodiesel and biodiesel blends using comprehensive two-dimensional gas chromatography. *J. Sep. Sci.* 31 (14): 2640–2649.

84 Adam, F., Bertoncini, F., Dartiguelongue, C. et al. (2009). Comprehensive two-dimensional gas chromatography for basic and neutral nitrogen speciation in middle distillates. *Fuel* 88 (5): 938–946.

85 Mao, D., Van De Weghe, H., Lookman, R. et al. (2009). Resolving the unresolved complex mixture in motor oils using high-performance liquid chromatography followed by comprehensive two-dimensional gas chromatography. *Fuel* 88 (2): 312–318.

86 Adam, F., Thiebaut, D., Bertoncini, F. et al. (2010). Supercritical fluid chromatography hyphenated with twin comprehensive two-dimensional gas chromatography for ultimate analysis of middle distillates. *J. Chromatogr. A* 1217 (8): 1386–1394.

87 Dutriez, T., Courtiade, M., Thiebaut, D. et al. (2010). Advances in quantitative analysis of heavy petroleum fractions by liquid chromatography-high-temperature comprehensive two-dimensional gas chromatography: breakthrough for conversion processes. *Energy Fuel* 24 (8): 4430–4438.

88 Dutriez, T., Courtiade, M., Thiebaut, D. et al. (2010). Extended characterization of a vacuum gas oil by offline LC-high-temperature comprehensive two-dimensional gas chromatography. *J. Sep. Sci.* 33 (12): 1787–1796.

89 Dutriez, T., Courtiade, M., Thiebaut, D. et al. (2010). Improved hydrocarbons analysis of heavy petroleum fractions by high temperature comprehensive two-dimensional gas chromatography. *Fuel* 89 (9): 2338–2345.

90 Kohl, A., Cochran, J., and Cropek, D.M. (2010). Characterization of military fog oil by comprehensive two-dimensional gas chromatography. *J. Chromatogr. A* 1217 (4): 550–557.

91 Peacock, E.E., Arey, J.S., DeMello, J.A. et al. (2010). Molecular and isotopic analysis of motor oil from a biodiesel-driven vehicle. *Energy Fuel* 24 (2): 1037–1042.

92 Rosse, A.D. and Harynuk, J.J. (2010). Methodology for the derivatization and quantification of dialkyl phosphate esters in petroleum samples. *Anal. Methods* 2 (8): 1176–1179.

93 Saraji-Bozorgzad, M.R., Eschner, M., Groeger, T.M. et al. (2010). Highly resolved online organic-chemical speciation of evolved gases from thermal analysis devices by cryogenically modulated fast gas chromatography coupled to single photon ionization mass spectrometry. *Anal. Chem.* 82 (23): 9644–9653.

94 van der Westhuizen, R., Crous, R., de Villiers, A., and Sandra, P. (2010). Comprehensive two-dimensional gas chromatography for the analysis of Fischer-Tropsch oil products. *J. Chromatogr. A* 1217 (52): 8334–8339.

95 von Muehlen, C., de Oliveira, E.C., Zini, C.A. et al. (2010). Characterization of nitrogen-containing compounds in heavy gas oil petroleum fractions using comprehensive two-dimensional gas chromatography coupled to time-of-flight mass spectrometry. *Energy Fuel* 24 (6): 3572–3580.

96 Aguiar, A., Aguiar, H.G.M., Azevedo, D.A., and Aquino Neto, F.R. (2011). Identification of methylhopane and methylmoretane series in Ceara basin oils, Brazil, using comprehensive two-dimensional gas chromatography coupled to time-of-flight mass spectrometry. *Energy Fuel* 25 (3): 1060–1065.

97 Ávila, B.M.F., Pereira, R., Gomes, A.O., and Azevedo, D.A. (2011). Chemical characterization of aromatic compounds in extra heavy gas oil by comprehensive two-dimensional gas chromatography coupled to time-of-flight mass spectrometry. *J. Chromatogr. A* 1218 (21): 3208–3216.

98 Eschner, M.S., Groeger, T.M., Horvath, T. et al. (2011). Quasi-simultaneous acquisition of hard electron ionization and soft single-photon ionization mass spectra during GC/MS analysis by rapid switching between both ionization methods: analytical concept, setup, and application on diesel fuel. *Anal. Chem.* 83 (10): 3865–3872.

99 Fonseca de Godoy, L.A., Pedroso, M.P., Ferreira, E.C. et al. (2011). Prediction of the physicochemical properties of gasoline by comprehensive two-dimensional gas chromatography and multivariate data processing. *J. Chromatogr. A* 1218 (12): 1663–1667.

100 Harynuk, J.J., Rossé, A.D., and McGarvey, G.B. (2011). Study of alkyl phosphates in industrial petroleum mixtures by comprehensive two-dimensional gas chromatography time-of-flight mass spectrometry. *Anal. Bioanal. Chem.* 401 (8): 2415–2422.

101 Mahe, L., Dutriez, T., Courtiade, M. et al. (2011). Global approach for the selection of high temperature comprehensive two-dimensional gas chromatography experimental conditions and quantitative analysis in regards to sulfur-containing compounds in heavy petroleum cuts. *J. Chromatogr. A* 1218 (3): 534–544.

102 Parastar, H., Radovic, J.R., Jalali-Heravi, M. et al. (2011). Resolution and quantification of complex mixtures of polycyclic aromatic hydrocarbons in heavy fuel oil sample by means of GC × GC-TOFMS combined to multivariate curve resolution. *Anal. Chem.* 83 (24): 9289–9297.

103 Pierce, K.M. and Schale, S.P. (2011). Predicting percent composition of blends of biodiesel and conventional diesel using gas chromatography-mass spectrometry, comprehensive two-dimensional gas chromatography-mass spectrometry, and partial least squares analysis. *Talanta* 83 (4, SI): 1254–1259.

104 Rowland, S.J., West, C.E., Scarlett, A.G., and Jones, D. (2011). Identification of individual acids in a commercial sample of naphthenic acids from petroleum by two-dimensional comprehensive gas chromatography/mass spectrometry. *Rapid Commun. Mass Spectrom.* 25 (12): 1741–1751.

105 Avila, B.M.F., Vaz, B.G., Pereira, R. et al. (2012). Comprehensive chemical composition of gas oil cuts using two-dimensional gas chromatography with time-of-flight mass spectrometry and electrospray ionization coupled to Fourier transform ion cyclotron resonance mass spectrometry. *Energy Fuel* 26 (8): 5069–5079.

106 Castano, P., Gutierrez, A., Hita, I. et al. (2012). Deactivating species deposited on Pt-Pd catalysts in the hydrocracking of light-cycle oil. *Energy Fuel* 26 (3): 1509–1519.

107 Dutriez, T., Courtiade, M., Ponthus, J. et al. (2012). Complementarity of Fourier transform ion cyclotron resonance mass spectrometry and high temperature comprehensive two-dimensional gas chromatography for the characterization of resin fractions from vacuum gas oils. *Fuel* 96 (1): 108–119.

108 Lissitsyna, K., Huertas, S., Morales, R. et al. (2012). Determination of trace levels of fatty acid methyl esters in aviation fuel by GC × GC-FID and comparison with the reference GC-MS method. *Chromatographia* 75 (21–22): 1319–1325.

109 Muller, H., Adam, F.M., Panda, S.K. et al. (2012). Evaluation of quantitative sulfur speciation in gas oils by Fourier transform ion cyclotron resonance mass spectrometry: validation by comprehensive two-dimensional gas chromatography. *J. Am. Soc. Mass Spectrom.* 23 (5): 806–815.

110 Seeley, J.V., Bates, C.T., McCurry, J.D., and Seeley, S.K. (2012). Stationary phase selection and comprehensive two-dimensional gas chromatographic analysis of trace biodiesel in petroleum-based fuel. *Selected Papers from the 35th International Symposium on Capillary Chromatography, the 26th International Symposium on MicroScale Bioseparations and the 8th GC × GC Symposium*, San Diego, CA, USA (1–5 May 2011), 1226, pp. 103–109.

111 Dutriez, T., Thiebaut, D., Courtiade, M. et al. (2013). Application to SFC-GC × GC to heavy petroleum fractions analysis. *Fuel* 104: 583–592.

112 Krupcik, J., Gorovenko, R., Spanik, I. et al. (2013). Flow-modulated comprehensive two-dimensional gas chromatography with simultaneous flame ionization and quadrupole mass spectrometric detection. *J. Chromatogr. A* 1280: 104–111.

113 Krupcik, J., Gorovenko, R., Spanik, I. et al. (2013). On the use of ionic liquid capillary columns for analysis of aromatic hydrocarbons in low-boiling petrochemical products by one-dimensional and comprehensive two-dimensional gas chromatography. *J. Chromatogr. A* 1301: 225–236.

114 Lissitsyna, K., Huertas, S., Quintero, L.C., and Polo, L.M. (2013). Novel simple method for quantitation of nitrogen compounds in middle distillates using solid phase extraction and comprehensive two-dimensional gas chromatography. *Fuel* 104: 752–757.

115 Lissitsyna, K., Huertas, S., Quintero, L.C., and Polo, L.M. (2013). Quantitation method of *N,N′*-disalicylidene-1,2-propanediamine by comprehensive two-dimensional gas chromatography coupled to a nitrogen chemiluminescence detector. *J. Sep. Sci.* 36 (11): 1768–1763.

116 Machado, M.E., de Menezes, E.W., Bregles, L.P. et al. (2013). Palladium(II) chemically bonded to silica surface applied to the separation and identification of polycyclic aromatic sulfur heterocycles in heavy oil. *J. Sep. Sci.* 36 (9–10): 1636–1643.

117 Obame, H., Toussaint, G., Laurenti, D. et al. (2013). Comprehensive GC × GC characterization of the catalytic alkylation of thiophenic compounds in a FCC gasoline. *Top. Catal.* 56 (18): 1731–1739.

118 Rashidi, F., Sasaki, T., Rashidi, A.M. et al. (2013). Ultradeep hydrodesulfurization of diesel fuels using highly efficient nanoalumina-supported catalysts: impact of support, phosphorus, and/or boron on the structure and catalytic activity. *J. Catal.* 299: 321–335.

119 Avila, B.M.F., Pereira, V.B., Gomes, A.O., and Azevedo, D.A. (2014). Speciation of organic sulfur compounds using comprehensive two-dimensional gas chromatography coupled to time-of-flight mass spectrometry: a powerful tool for petroleum refining. *Fuel* 126: 188–193.

120 Kehimkar, B., Hoggard, J.C., Marney, L.C. et al. (2014). Correlation of rocket propulsion fuel properties with chemical composition using comprehensive two-dimensional gas chromatography with time-of-flight mass spectrometry followed by partial least squares regression analysis. *J. Chromatogr. A* 1327: 132–140.

121 Lissitsyna, K., Huertas, S., Quintero, L.C., and Polo, L.M. (2014). PIONA analysis of kerosene by comprehensive two-dimensional gas chromatography coupled to time of flight mass spectrometry. *Fuel* 116: 716–722.

122 Timko, M.T., Schmois, E., Patwardhan, P. et al. (2014). Response of different types of sulfur compounds to oxidative desulfurization of jet fuel. *Energy Fuel* 28 (5): 2977–2983.

123 Jennerwein, M.K., Eschner, M., Groeger, T. et al. (2014). Complete group-type quantification of petroleum middle distillates based on comprehensive two-dimensional gas chromatography time-of-flight mass spectrometry (GC × GC-TOFMS) and visual basic scripting. *Energy Fuel* 28 (9): 5670–5681.

124 Franchina, F.A., Maimone, M., Sciarrone, D. et al. (2015). Evaluation of a novel helium ionization detector within the context of (low-)flow modulation comprehensive two-dimensional gas chromatography. *J. Chromatogr. A* 1402: 102–109.

125 Hourani, N., Muller, H., Adam, F.M. et al. (2015). Structural level characterization of base oils using advanced analytical techniques. *Energy Fuel* 29 (5): 2962–2970.

126 Kehimkar, B., Parsons, B.A., Hoggard, J.C. et al. (2015). Modeling RP-1 fuel advanced distillation data using comprehensive two-dimensional gas chromatography coupled with time-of-flight mass spectrometry and partial least squares analysis. *Anal. Bioanal. Chem.* 407 (1): 321–330.

127 Maciel, G.P.S., Machado, M.E., da Cunha, M.E. et al. (2015). Quantification of nitrogen compounds in diesel fuel samples by comprehensive two-dimensional gas chromatography coupled with quadrupole mass spectrometry. *J. Sep. Sci.* 38 (23): 4071–4077.

128 Parsons, B.A., Marney, L.C., Siegler, W.C. et al. (2015). Tile-based Fisher ratio analysis of comprehensive two-dimensional gas chromatography time-of-flight mass spectrometry (GC × GC-TOFMS) data using a null distribution approach. *Anal. Chem.* 87 (7): 3812–3819.

129 Samanipour, S., Dimitriou-Christidis, P., Gros, J. et al. (2015). Analyte quantification with comprehensive two-dimensional gas chromatography: assessment of methods for baseline correction, peak delineation, and matrix effect elimination for real samples. *J. Chromatogr. A* 1375: 123–139.

130 Eftekhari, A. and Parastar, H. (2016). Multivariate analytical figures of merit as a metric for evaluation of quantitative measurements using comprehensive two-dimensional gas chromatography-mass spectrometry. *J. Chromatogr. A* 1466: 155–165.

131 Freye, C.E., Fitz, B.D., Billingsley, M.C., and Synovec, R.E. (2016). Partial least squares analysis of rocket propulsion fuel data using diaphragm valve-based comprehensive two-dimensional gas chromatography coupled with flame ionization detection. *Talanta* 153: 203–210.

132 Freye, C.E. and Synovec, R.E. (2016). High temperature diaphragm valve-based comprehensive two-dimensional gas chromatography with time-of-flight mass spectrometry. *Talanta* 161: 675–680.

133 Gonzalez-Quiroga, A., Djokic, M.R., Van Geem, K.M., and Marin, G.B. (2016). Conversion of solid waste to diesel via catalytic pressureless depolymerization: pilot scale production and detailed compositional characterization. *Energy Fuel* 30 (10): 8292–8303.

134 Keshet, U., Fialkov, A.B., Alon, T., and Amirav, A. (2016). A new pulsed flow modulation GC × GC-MS with cold EI system and its application for jet fuel analysis. *Chromatographia* 79 (11–12): 741–754.

135 Krupcik, J., Gorovenko, R., Spanik, I. et al. (2016). Comparison of the performance of forward fill/flush and reverse fill/flush flow modulation in comprehensive two-dimensional gas chromatography. *J. Chromatogr. A* 1466: 113–128.

136 Mostafa, A. and Gorecki, T. (2016). Development and design of a single-stage cryogenic modulator for comprehensive two-dimensional gas chromatography. *Anal. Chem.* 88 (10): 5414–5423.

137 Parastar, H., Mostafapour, S., and Azimi, G. (2016). Quality assessment of gasoline using comprehensive two-dimensional gas chromatography combined with unfolded partial least squares: a reliable approach for the detection of gasoline adulteration. *J. Sep. Sci.* 39 (2): 367–374.

138 Parsons, B.A., Pinkerton, D.K., Wright, B.W., and Synovec, R.E. (2016). Chemical characterization of the acid alteration of diesel fuel: non-targeted analysis by two-dimensional gas chromatography coupled with time-of-flight mass spectrometry with tile-based Fisher ratio and combinatorial threshold determination. *J. Chromatogr. A* 1440: 179–190.

139 Potgieter, H., Bekker, R., Govender, A., and Rohwer, E. (2016). Two-dimensional gas chromatography-online hydrogenation for improved characterization of petrochemical samples. *J. Chromatogr. A* 1445: 118–125.

140 Spaak, G., Nelson, R.K., Reddy, C.M. et al. (2016). Advances on the separation of crocetane and phytane using GC-MS and GC × GC-TOFMS. *Org. Geochem.* 98: 176–182.

141 Zhang, C., Park, R.A., and Anderson, J.L. (2016). Crosslinked structurally-tuned polymeric ionic liquids as stationary phases for the analysis of hydrocarbons in kerosene and diesel fuels by comprehensive two-dimensional gas chromatography. *J. Chromatogr. A* 1440: 160–171.

142 Cho, Y., Birdwell, J.E., Hur, M. et al. (2017). Extension of the analytical window for characterizing aromatic compounds in oils using a comprehensive suite of high-resolution mass spectrometry techniques and double bond equivalence versus carbon number plot. *Energy Fuel* 31 (8): 7874–7883.

143 Couprie, C., Duval, L., Moreaud, M. et al. (2017). BARCHAN: blob alignment for robust chromatographic analysis. *J. Chromatogr. A* 1484: 65–72.

144 Djokic, M.R., Ristic, N.D., Olahova, N. et al. (2017). Quantitative on-line analysis of sulfur compounds in complex hydrocarbon matrices. *J. Chromatogr. A* 1509: 102–113.

145 Genuit, W. and Chaabani, H. (2017). Comprehensive two-dimensional gas chromatography-field ionization time-of-flight mass spectrometry (GC × GC-FI-TOFMS) for detailed hydrocarbon middle distillate analysis. *Int. J. Mass Spectrom.* 413 (SI): 27–32.

146 Jennerwein, M.K., Sutherland, A.C., Eschner, M. et al. (2017). Quantitative analysis of modern fuels derived from middle distillates - the impact of diverse compositions on standard methods evaluated by an offline hyphenation of HPLC-refractive index detection with GC × GC-TOFMS. *Fuel* 187: 16–25.

147 Kulsing, C., Rawson, P., Webster, R.L. et al. (2017). Group-type analysis of hydrocarbons and sulfur compounds in thermally stressed Merox jet fuel samples. *Energy Fuel* 31 (9): 8978–8984.

148 Mucedola, V., Vieira, L.C.S., Pierone, D. et al. (2017). Thermal desorption modulation for comprehensive two-dimensional gas chromatography using a simple and inexpensive segmented-loop fluidic interface. *Talanta* 164: 470–476.

149 Nan, H., Zhang, C., O'Brien, R.A. et al. (2017). Lipidic ionic liquid stationary phases for the separation of aliphatic hydrocarbons by comprehensive two-dimensional gas chromatography. *J. Chromatogr. A* 1481: 127–136.

150 Potgieter, H., Bekker, R., Beigley, J., and Rohwer, E. (2017). Analysis of oxidised heavy paraffininc products by high temperature comprehensive two-dimensional gas chromatography. *J. Chromatogr. A* 1509: 123–131.

151 Reaser, B.C., Wright, B.W., and Synovec, R.E. (2017). Using receiver operating characteristic curves to optimize discovery-based software with comprehensive two-dimensional gas chromatography with time-of-flight mass spectrometry. *Anal. Chem.* 89 (6): 3606–3612.

152 Scheuermann, S.S., Forster, S., and Eibl, S. (2017). In-depth interpretation of mid-infrared spectra of various synthetic fuels for the chemometric prediction of aviation fuel blend properties. *Energy Fuel* 31 (3): 2934–2943.

153 Shi, X., Li, H., Song, Z. et al. (2017). Quantitative composition-property relationship of aviation hydrocarbon fuel based on comprehensive two-dimensional gas chromatography with mass spectrometry and flame ionization detector. *Fuel* 200: 395–406.

154 Wang, F.C.-Y. (2017). Comprehensive three-dimensional gas chromatography mass spectrometry separation of diesel. *J. Chromatogr. A* 1489: 126–133.

155 Webster, R.L., Rawson, P.M., Kulsing, C. et al. (2017). Investigation of the thermal oxidation of conventional and alternate aviation fuels with comprehensive two-dimensional gas chromatography accurate mass quadrupole time-of-flight mass spectrometry. *Energy Fuel* 31 (5): 4886–4894.

156 Zarghani, M. and Parastar, H. (2017). Joint approximate diagonalization of eigenmatrices as a high-throughput approach for analysis of hyphenated and comprehensive two-dimensional gas chromatographic data. *J. Chromatogr. A* 1524: 188–201.

157 Chakravarthy, R., Acharya, C., Savalia, A. et al. (2018). Property prediction of diesel fuel based on the composition analysis data by two-dimensional gas chromatography. *Energy Fuel* 32 (3): 3760–3774.

158 Djokic, M.R., Muller, H., Ristic, N.D. et al. (2018). Combined characterization using HT-GC × GC-FID and FT-ICR MS: a pyrolysis fuel oil case study. *Fuel Process. Technol.* 182: 15–25.

159 Freye, C.E., Moores, N.R., and Synovec, R.E. (2018). Enhancing the chemical selectivity in discovery-based analysis with tandem ionization time-of-flight mass spectrometry detection for comprehensive two-dimensional gas chromatography. *J. Chromatogr. A* 1537: 99–108.

160 Lin, C., Sang, J., Chen, S. et al. (2018). Thermal treatment of FCC slurry oil under hydrogen: correlation of hydrogen transfer ability with carbonization performance of the fractions. *Fuel* 233: 805–815.

161 Vozka, P., Mo, H., Simacek, P., and Kilaz, G. (2018). Middle distillates hydrogen content via GC × GC-FID. *Talanta* 186: 140–146.

162 Jarvis, M.W., Olstad, J., Parent, Y. et al. (2018). Catalytic upgrading of biomass pyrolysis oxygenates with vacuum gas oil using a davison circulating riser reactor. *Energy Fuel* 32 (2): 1733–1743.

163 Bourane, A., Koseoglu, O., Al-Hajji, A. et al. (2019). Assessment of a three step process using tungsten catalyzed hydrogen peroxide-based oxidative desulfurization for commercial diesel fuels. *React. Kinet. Mech. Catal.* 126 (1): 365–382.

164 Deese, R.D., Morris, R.E., Metz, A.E. et al. (2019). Characterization of organic nitrogen compounds and their impact on the stability of marginally stable diesel fuels. *Energy Fuel* 33 (7): 6659–6669.

165 Giri, A., Coutriade, M., Racaud, A. et al. (2019). Compositional elucidation of heavy petroleum base oil by GC × GC-EI/PI/CI/FI-TOFMS. *J. Mass Spectrom.* 54 (2): 148–157.

166 Jennerwein, M., Eschner, M., Wilharm, T. et al. (2019). Evaluation of reversed phase versus normal phase column combination for the quantitative analysis of common commercial available middle distillates using GC × GC-TOFMS and visual basic script. *Fuel* 235: 336–338.

167 Kaefer, U., Groeger, T.M., Rohbogner, C. et al. (2019). Detailed chemical characterization of bunker fuels by high-resolution time-of-flight mass spectrometry hyphenated to GC × GC and thermal analysis. *Energy Fuel* 33 (11): 10745–10755.

168 Manheim, J., Wehde, K., Zhang, W.T.J. et al. (2019). Identification and quantitation of linear alkanes in lubricant base oils by using GC×GC/EI TOF mass spectrometry. *J. Am. Soc. Mass Spectrom.* 30 (12): 2670–2677.

169 Qian, K. and Wang, F.C. (2019). Compositional analysis of heavy petroleum distillates by comprehensive two-dimensional gas chromatography, field ionization and high-resolution mass spectrometry. *J. Am. Soc. Mass Spectrom.* 30 (12): 2785–2794.

170 Vozka, P. and Kilaz, G. (2019). How to obtain a detailed chemical composition for middle distillates via GC × GC-FID without the need of GC × GC-TOF/MS. *Fuel* 247: 368–377.

171 Vozka, P., Modereger, B.A., Park, A.C. et al. (2019). Jet fuel density via GC × GC-FID. *Fuel* 235: 1052–1060.

172 Wu, Z., Mao, Y., Raza, M. et al. (2019). Surrogate fuels for RP-3 kerosene formulated by emulating molecular structures, functional groups, physical and chemical properties. *Combust. Flame* 208: 388–401.

173 Zoccali, M., Giocastro, B., Tranchida, P.Q., and Mondello, L. (2019). Use of a recently developed thermal modulator within the context of comprehensive two-dimensional gas chromatography combined with time-of-flight mass spectrometry: gas flow optimization aspects. *J. Sep. Sci.* 42 (3): 691–697.

174 Berrier, K.L., Freye, C.E., Billingsley, M.C., and Synovec, R.E. (2020). Predictive modeling of aerospace fuel properties using comprehensive two-dimensional gas chromatography with time-of-flight mass spectrometry and partial least squares analysis. *Energy Fuel* 34 (4): 4084–4094.

175 Gieleciak, R., Litz, K.E., Rankin, J.P. et al. (2020). Analysis of hydrocarbon compositional changes during oxidative desulfurization of bitumen-derived gas oil. *Energy Fuel* 34 (10): 12088–12102.

176 Guan, X., Luong, J., Yu, Z., and Jiang, H. (2020). Quasi-stop-flow modulation strategy for comprehensive two-dimensional gas chromatography. *Anal. Chem.* 92 (9): 6251–6256.

177 Lai, T., Mao, Y., Wang, W. et al. (2020). Characterization of basic nitrogen compounds isolated with $FeCl_3$ in vacuum gas oil and its hydrotreated product. *Fuel* 262: 116523.

178 Liu, J., Ahmad, F., Zhang, Q. et al. (2020). Interactive tools to assist convenient group-type identification and comparison of low-temperature coal tar using GC × GC-MS. *Fuel* 278: 118314.

179 Muller, H., Alawani, N.A., and Adam, F.M. (2020). Innate sulfur compounds as an internal standard for determining vacuum gas oil compositions by APPI FT-ICR MS. *Energy Fuel* 34 (7): 8260–8273.

180 Neumann, A., Kaefer, U., Groeger, T. et al. (2020). Investigation of aging processes in bitumen at the molecular level with high-resolution Fourier-transform ion cyclotron mass spectrometry and two-dimensional gas chromatography mass spectrometry. *Energy Fuel* 34 (9): 10641–10654.

181 Nyiri, Z., Voros-Palya, D., Novak, M., and Eke, Z. (2020). Coupling of large volume injection with flow modulated two-dimensional gas chromatography. *Talanta* 216: 120984.

182 Ochoa, G.S., Prebihalo, S.E., Reaser, B.C. et al. (2020). Statistical inference of mass channel purity from Fisher ratio analysis using comprehensive two-dimensional gas chromatography with time of flight mass spectrometry data. *J. Chromatogr. A* 1627: 461401.

183 Pandohee, J., Hughes, J.G., Pearson, J.R., and Jones, O.A.H. (2020). Chemical fingerprinting of petrochemicals for arson investigations using two-dimensional gas chromatography - flame ionisation detection and multivariate analysis. *Sci. Justice* 60 (4): 381–387.

184 Schoneich, S., Trinklein, T.J., Warren, C.G., and Synovec, R.E. (2020). A systematic investigation of comprehensive two-dimensional gas chromatography time-of-flight mass spectrometry with dynamic pressure gradient modulation for high peak capacity separations. *Anal. Chim. Acta* 1134: 115–124.

185 Silva, A.F., Neves, P., Rocha, S.M. et al. (2020). Optimization of continuous-flow heterogeneous catalytic oligomerization of 1-butene by design of experiments and response surface methodology. *Fuel* 259: 116256.

186 Sudol, P.E., Gough, D.V., Prebihalo, S.E., and Synovec, R.E. (2020). Impact of data bin size on the classification of diesel fuels using comprehensive two-dimensional gas chromatography with principal component analysis. *Talanta* 206: 120239.

187 Trinklein, T.J., Schoneich, S., Sudol, P.E. et al. (2020). Total-transfer comprehensive three-dimensional gas chromatography with time-of-flight mass spectrometry. *J. Chromatogr. A* 1634: 461654.

188 Vozka, P. and Kilaz, G. (2020). Determination of jet fuel system icing inhibitor by GC × GC-FID. *Talanta* 218: 121146.

189 Wang, F.C.-Y. (2020). Comprehensive two-dimensional gas chromatography hyphenated with a vacuum ultraviolet spectrometer to analyze diesel—a three-dimensional separation (GC × GC × VUV) approach. *Energy Fuel* 34 (7): 8012–8017.

190 Castiblanco, E.B., Quero, R.F., de Oliveira, A.M. et al. (2021). Improving selective channel occlusion of complex hydrocarbons and fatty acid methyl esters in urea crystals by using an expendable 3D-printed microfluidic device for sample preparation in untargeted petroleomics. *Anal. Chim. Acta* 1160: 338425.

191 Giocastro, B., Zoccali, M., Tranchida, P.Q., and Mondello, L. (2021). Evaluation of different internal diameter coated modulation columns within the context of solid-state modulation. *J. Sep. Sci.* 44 (9): 1931–1938.

192 Guillemant, J., Lacoue-Negre, M., Berlioz-Barbier, A. et al. (2021). Towards a new pseudo-quantitative approach to evaluate the ionization response of nitrogen compounds in complex matrices. *Sci. Rep.* 11 (1): 6417.

193 Jiao, S., Guo, A., Wang, F. et al. (2021). Sequential pretreatments of an FCC slurry oil sample for preparation of feedstocks for high-value solid carbon materials. *Fuel* 285: 119169.

194 Machakanur, S., Savalia, A., and Bhakthavatsalam, V. (2021). Multivariate statistics for summarizing diesel feeds for flammability attributes using comprehensive two-dimensional gas chromatography. *J. Sep. Sci.* 44 (15): 2941–2949.

195 Modereger, B.A., Nowling, S.E., Zhang, W.T.J. et al. (2021). Influence of storage conditions, sample handling, sample preparation, and repeated analysis on the measured flash point and hydrocarbon composition of jet A. *Fuel* 305: 121399.

196 Sudol, P.E., Ochoa, G.S., and Synovec, R.E. (2021). Investigation of the limit of discovery using tile-based Fisher ratio analysis with comprehensive two-dimensional gas chromatography time-of-flight mass spectrometry. *J. Chromatogr. A* 1644: 462092.

197 Yang, Z., Kosir, S., Stachler, R. et al. (2021). A GC × GC tier α combustor operability prescreening method for sustainable aviation fuel candidates. *Fuel* 292: 120345.

198 Yuan, S., Li, H., Liu, Z. et al. (2021). Measurement of non-hindered and hindered phenolic species in aviation fuels via tandem-SPE with comprehensive GC × GC-MS/FID. *Fuel* 287: 119561.

199 Gieleciak, R., Farooqi, H., and Chen, J. (2021). Detailed characterization of diesel fractions from co-hydroprocessing vegetable oil and petroleum heavy vacuum gas oil blends. *Energy Fuel* 35 (21): 17721–17738.

200 Cain, C.N., Trinklein, T.J., Ochoa, G.S., and Synovec, R.E. (2022). Tile-based pairwise analysis of GC × GC-TOFMS data to facilitate analyte discovery and mass spectrum purification. *Anal. Chem.* 94 (14): 5658–5666.

201 Coutinho, D.M., Franca, D., Vanini, G. et al. (2022). Understanding the molecular composition of petroleum and its distillation cuts. *Fuel* 311: 122594.

202 Cui, D., Yin, H., Liu, Y. et al. (2022). Effect of final pyrolysis temperature on the composition and structure of shale oil: synergistic use of multiple analysis and testing methods. *Energy* 252: 124062.

203 Hou, L., He, K., Zhai, J. et al. (2022). Compositional kinetics for hydrocarbon evolution in the pyrolysis of the Chang 7 organic matter: implications for in-situ conversion of oil shale. *J. Anal. Appl. Pyrolysis* 162: 105434.

204 Jia, T., Liu, Q., Zou, J.-J. et al. (2022). The dynamics and mechanism of JP-10 thermal oxidative deposition. *Fuel* 321: 124093.

205 Lelevic, A., Geantet, C., Moreaud, M. et al. (2022). Quantification of hydrocarbons in gas oils by GC×GC-VUV: comparison with other techniques. *Energy Fuel* 36 (18): 10860–10869.

206 Ochoa, G.S., Sudol, P.E., Trinklein, T.J., and Synovec, R.E. (2022). Class comparison enabled mass spectrum purification for comprehensive two-dimensional gas chromatography with time-of-flight mass spectrometry. *Talanta* 236: 122844.

207 Sudol, P.E., Ochoa, G.S., Cain, C.N., and Synovec, R.E. (2022). Tile-based variance rank initiated-unsupervised sample indexing for comprehensive two-dimensional gas chromatography-time-of-flight mass spectrometry. *Anal. Chim. Acta* 1209: 339847.

208 Zhu, Y., Zheng, H., Tian, F. et al. (2022). Characterization of nitrogen-containing compounds in coal tar and its subfractions by comprehensive two-dimensional GC × GC-TOF and ESI FT-ICR mass spectrometry hbased on new separation method. *Fuel Process. Technol.* 227: 107125.

209 Seiser, R., Olstad, J.L., Magrini, K.A. et al. (2022). Co-processing catalytic fast pyrolysis oil in an FCC reactor. *Biomass Bioenergy* 163: 106484.

210 Xu, Y., Zhao, Y., Guo, X., and Fang, Y. (2022). Characterization of trace oxygenates from co-processing of bio-oils with petroleum-derived feed with GC × GC-MS. *Energy Fuel* 36 (19): 12076–12084.

211 Cain, C.N., Ochoa, G.S., and Synovec, R.E. (2023). Enhancing partial least squares modeling of comprehensive two-dimensional gas chromatography time-of-flight mass spectrometry data by tile-based variance ranking. *J. Chromatogr. A* 1694: https://doi.org/10.1016/j.chroma.2023.463920.

212 Colleoni, E., Samaras, V.G., Guida, P. et al. (2023). Unraveling the complexity of pyrolysates from residual fuels by Py-GC × GC-FID/SCD/TOF-MS with an innovative data processing method. *J. Anal. Appl. Pyrolysis* 175: 106204.

213 Corbally, M.A., Hinz, N.S., and Freye, C.E. (2023). Comprehensive two-dimensional gas chromatography under low-pressure conditions. *J. Chromatogr. A* 1705: 464203.

214 De Ras, K., Kusenberg, M., Thybaut, J.W., and Van Geem, K.M. (2023). Unraveling the carbene chemistry of oxymethylene ethers: experimental investigation and kinetic modeling of the high-temperature pyrolysis of OME-2. *Proc. Combust. Inst.* 39 (1): 125–133.

215 Fu, J. (2023). Dissolved water in jet fuels: a low-temperature quality and water solubility study. *Fuel* 331: 125950.

216 Mehta, J.M., Lynch, P.T., Mayhew, E.K., and Brezinsky, K. (2023). Evaluation of chemical functional group composition of jet fuels using two-dimensional gas chromatography. *Energy Fuel* 37 (3): 2294–2306.

217 Ochoa, G.S. and Synovec, R.E. (2023). Investigating analyte breakthrough under non-linear isotherm conditions during solid phase extraction facilitated by non-targeted analysis with comprehensive two-dimensional gas chromatography time-of-flight mass spectrometry. *Talanta* 259: 124525.

218 Samoilov, V.O., Sultanova, M.U., Borisov, R.S. et al. (2023). The production of a liquid organic hydrogen carrier (LOHC) and jet fuel fractions by hydroprocessing of pyrolysis fuel oil (PFO). *Int. J. Hydrog. Energy* 49: 1386–1400.

219 Schoneich, S., Cain, C.N., Sudol, P.E., and Synovec, R.E. (2023). Enabling cuboid-based fisher ratio analysis using total-transfer comprehensive three-dimensional gas chromatography with time-of-flight mass spectrometry. *J. Chromatogr. A* 1708: 464341.

220 Xu, Y.-Y., Fan, X., Ma, F.-Y. et al. (2024). Molecular and geochemical characteristics of lignite in different plant sources evaluated using unsupervised analyses. *Fuel* 357 (B): 129850.

221 Han, Y., Zhang, Y., Xu, C., and Hsu, C.S. (2018). Molecular characterization of sulfur-containing compounds in petroleum. *Fuel* 221: 144–158.

222 Heyne, J., Bell, D., Feldhausen, J. et al. (2022). Towards fuel composition and properties from Two-dimensional gas chromatography with flame ionization and vacuum ultraviolet spectroscopy. *Fuel* 312: 122709.

223 Colleoni, E., Samaras, V.G., Guida, P. et al. (2024). Corrigendum to “unraveling the complexity of pyrolysates from residual fuels by Py-GC × GC-FID/SCD/TOF-MS with an innovative data processing method”

[*J. Anal. Appl. Pyrolysis* 175 (2023) 106204]. *J. Anal. Appl. Pyrolysis* 178: 106404.

224 Wang, Q., Chen, X., Jha, A.N., and Rogers, H. (2014). Natural gas from shale formation – the evolution, evidences and challenges of shale gas revolution in United States. *Renew. Sust. Energ. Rev.* 30: 1–28.

225 Dogu, O., Pelucchi, M., Van de Vijver, R. et al. (2021). The chemistry of chemical recycling of solid plastic waste via pyrolysis and gasification: state-of-the-art, challenges, and future directions. *Prog. Energy Combust. Sci.* 84: 100901.

226 Nicholson, S.R., Rorrer, J.E., Singh, A. et al. (2022). The critical role of process analysis in chemical recycling and upcycling of waste plastics. *Annu. Rev. Chem. Biomol. Eng.* 13 (1): 301–324.

227 Eschenbacher, A., Varghese, R.J., Abbas-Abadi, M.S., and Van Geem, K.M. (2022). Maximizing light olefins and aromatics as high value base chemicals via single step catalytic conversion of plastic waste. *Chem. Eng. J.* 428: 132087.

228 Lazzari, E., Piparo, M., Mase, C. et al. (2023). Chemical elucidation of recycled plastic pyrolysis oils by means of GC × GC-PI-TOF-MS and GC-VUV. *J. Anal. Appl. Pyrolysis* 176: 106224.

229 Li, H., Wu, J., Jiang, Z. et al. (2023). Hydroformylation of pyrolysis oils to aldehydes and alcohols from polyolefin waste. *Science* 381 (6658): 660–666.

230 Kim, D., Hinton, Z.R., Bai, P. et al. (2022). Metathesis, molecular redistribution of alkanes, and the chemical upgrading of low-density polyethylene. *Appl Catal B* 318: 121873.

231 Seshasayee, M.S. and Savage, P.E. (2020). Oil from plastic via hydrothermal liquefaction: production and characterization. *Appl. Energy* 278: 115673.

232 Perez, B.A., Krishna, J.J., and Toraman, H.E. (2024). Characterization of polyolefins-based pyrolysis oils: a comparison between one-dimensional gas chromatography and two-dimensional gas chromatography. *J. Chromatogr. A* 1739: 465510.

233 Winniford, B.L., Sun, K., Griffith, J.F., and Luong, J.C. (2006). Universal and discriminative detection using a miniaturized pulsed discharge detector in comprehensive two-dimensional GC. *J. Sep. Sci.* 29 (17): 2664–2670.

234 Ware, R., Rowland, S., Rodgers, R., and Marshall, A. (2017). Advanced chemical characterization of pyrolysis oils from landfill waste, recycled plastics, and forestry residue. *Energy Fuel* 31 (8): 8210–8216.

235 Sophonrat, N., Sandström, L., Johansson, A., and Yang, W. (2017). Co-pyrolysis of mixed plastics and cellulose: an interaction study by Py-GC × GC/MS. *Energy Fuel* 31 (10): 11078–11090.

236 Singh, M. (2018). Waste and virgin high-density poly(ethylene) into renewable hydrocarbons fuel by pyrolysis-catalytic cracking with a $CoCO_3$ catalyst. *J. Anal. Appl. Pyrolysis* 134: 150–161.

237 Singh, M., Kumar, S., and Sarker, M. (2018). Waste HD-PE plastic, deformation into liquid hydrocarbon fuel using pyrolysis-catalytic cracking with a $CuCO_3$ catalyst. *Sustain. Energy Fuels* 2 (5): 1057–1068.

238 Tipler, S., Mergulhao, C.S., Vanhove, G. et al. (2019). Ignition study of an oxygenated and high-alkene light petroleum fraction produced from automotive shredder residues. *Energy Fuel* 33 (6): 5664–5672.

239 Kumagai, S., Matsukami, A., Kabashima, F. et al. (2020). Combining pyrolysis-two-dimensional gas chromatography-time-of-flight mass spectrometry with hierarchical cluster analysis for rapid identification of pyrolytic interactions: case study of co-pyrolysis of PVC and biomass components. *Process Saf. Environ. Prot.* 143: 91–100.

240 Abbas-Abadi, M., Zayoud, A., Kusenberg, M. et al. (2022). Thermochemical recycling of end-of-life and virgin HDPE: a pilot-scale study. *J. Anal. Appl. Pyrolysis* 166: 105614.

241 Costa, C.S., Dao Thi, H., Van Geem, K.M. et al. (2022). Assessment of acidity and the zeolite porous structure on hydrocracking of HDPE. *Sustain. Energy Fuels* 6 (15): 3611–3625.

242 Kusenberg, M., Faussone, G.C., Thi, H.D. et al. (2022). Maximizing olefin production via steam cracking of distilled pyrolysis oils from difficult-to-recycle municipal plastic waste and marine litter. *Sci. Total Environ.* 838: 156092.

243 Kusenberg, M., Zayoud, A., Roosen, M. et al. (2022). A comprehensive experimental investigation of plastic waste pyrolysis oil quality and its dependence on the plastic waste composition. *Fuel Process. Technol.* 227: 107090.

244 Li, H., Wu, X., Wu, S. et al. (2022). Machine learning directed discrimination of virgin and recycled poly(ethylene terephthalate) based on non-targeted analysis of volatile organic compounds. *J. Hazard. Mater.* 436: 129116.

245 Zayoud, A., Dao Thi, H., Kusenberg, M. et al. (2022). Pyrolysis of end-of-life polystyrene in a pilot-scale reactor: maximizing styrene production. *Waste Manag.* 139: 85–95.

246 Ureel, Y., Dobbelaere, M.R., Akin, O. et al. (2022). Active learning-based exploration of the catalytic pyrolysis of plastic waste. *Fuel* 328: 125340.

247 de Rezende Locatel, W., Mohabeer, C., Laurenti, D. et al. (2022). Co-pyrolysis of beech wood and polyamide-6: impact of plastic concentration and wood/plastic synergistic effects. *J. Anal. Appl. Pyrolysis* 168: 105779.

248 Akin, O., Varghese, R.J., Eschenbacher, A. et al. (2023). Chemical recycling of plastic waste to monomers: effect of catalyst contact time, acidity and pore size on olefin recovery in ex-situ catalytic pyrolysis of polyolefin waste. *J. Anal. Appl. Pyrolysis* 172: 106036.

249 Abbas-Abadi, M., Kusenberg, M., Zayoud, A. et al. (2023). Thermal pyrolysis of waste versus virgin polyolefin feedstocks: the role of pressure, temperature and waste composition. *Waste Manag.* 165: 108–118.

250 Dolgushev, P.A. and Shashkov, M.V. (2023). High-temperature two-dimensional gas chromatography with flow modulator for analyzing complex objects of catalytic processing. *Chromatographia* 86 (4): 267–283.

251 Dong, B., Wu, X., Wu, S. et al. (2023). Occurrence of volatile contaminants in recycled poly(ethylene terephthalate) by HS-SPME-GC × GC-QTOF-MS combined with chemometrics for authenticity assessment of geographical recycling regions. *J. Hazard. Mater.* 445: 130407.

252 Hao, T., Xu, X., Lin, Q. et al. (2022). Rapid discrimination of recycled and virgin poly(ethylene terephthalate) based on non-targeted screening of semi-volatile organic compounds using a novel method of DSI/GC × GC-Q-TOF-MS coupled with various chemometrics. *Food Packag. Shelf Life* 34: 100978.

253 Konarova, M., Batalha, N., Fraga, G. et al. (2024). Integrating PET chemical recycling with pyrolysis of mixed plastic waste via pressureless alkaline depolymerization in a hydrocarbon solvent. *Waste Manag.* 174: 24–30.

254 Perez, B.A. and Toraman, H.E. (2024). Investigating primary decomposition of polypropylene through detailed compositional analysis using two-dimensional gas chromatography and principal component analysis. *J. Anal. Appl. Pyrolysis* 177: 106376.

255 Jayarama Krishna, J.V., Perez, B.A., and Toraman, H.E. (2024). Parametric study of polyethylene primary decomposition using a micropyrolyzer coupled with two-dimensional gas chromatography. *ACS Sustain. Chem. Eng.* 12 (19): 7508–7518.

256 Bozkurt, O.D. and Toraman, H.E. (2024). Conversion of polypropylene into light hydrocarbons and aromatics by metal exchanged zeolite catalysts. *Langmuir* 40 (18): 9636–9650.

257 Akin, O., He, Q., Yazdani, P. et al. (2024). Tailored HZSM-5 catalyst modification via phosphorus impregnation and mesopore introduction for selective catalytic conversion of polypropylene into light olefins. *J. Anal. Appl. Pyrolysis* 181: 106592.

258 Khan, R., Perez, B.A., and Toraman, H.E. (2024). Comparative analysis of additive decomposition using one-dimensional and two-dimensional gas chromatography: part I - Irganox 1010, Irganox 1076, and BHT. *J. Chromatogr. A* 1732: 465243.

259 Khan, R., Perez, B.A., and Toraman, H.E. (2024). Comparative analysis of additive decomposition using one-dimensional and two–dimensional gas chromatography: part II-Irgafos 168 and zinc stearate. *J. Chromatogr. A* 1732: 465244.

260 Wu, J., Jiang, Z., Cecon, V.S. et al. (2024). The effects of polyolefin structure and source on pyrolysis-derived plastic oil composition. *Green Chem.* 26 (24): 11908–11923.

261 He, Q., Akin, O., Ureel, Y. et al. (2024). Enhancing catalytic pyrolysis of polypropylene using mesopore-modified HZSM-5 catalysts: insights and strategies for improved performance. *Front. Chem. Eng.* 6: 1439400.

262 Zanella, D., Romagnoli, M., Malcangi, S. et al. (2023). The contribution of high-resolution GC separations in plastic recycling research. *Anal. Bioanal. Chem.* 415 (13): 2343–2355.

263 Kusenberg, M., Eschenbacher, A., Djokic, M.R. et al. (2022). Opportunities and challenges for the application of post-consumer plastic waste pyrolysis oils as steam cracker feedstocks: to decontaminate or not to decontaminate? *Waste Manag.* 138: 83–115.

Index

h

i

n

u

Printed and bound by CPI Group (UK) Ltd, Croydon, CR0 4YY
24/08/2026
14947830-0004